CHATHAM HOUSE SPECIAL PAPER

EUROPE'S FUTURE IN SPACE

CHATHAM HOUSE SPECIAL PAPER

EUROPE'S FUTURE IN SPACE

A Joint Policy Report

Forschungsinstitut der Deutschen Gesellschaft für Auswärtige Politik (Bonn)
Institut Français des Relations Internationales (Paris)
Istituto Affari Internazionali (Rome)
Nederlands Instituut voor Internationale Betrekkingen 'Clingendael' (The Hague)
Royal Institute of International Affairs (London)

The Royal Institute of International Affairs

Routledge & Kegan Paul
London, New York and Andover

The Royal Institute of International Affairs is an independent body which promotes the rigorous study of international questions and does not express opinions of its own. The opinions expressed in this publication are the responsibility of the authors.

First published 1988
by Routledge & Kegan Paul Ltd
11 New Fetter Lane, London EC4P 4EE
29 West 35th Street, New York, NY 10001, USA, and
North Way, Andover, Hants SP10 5BE

Reproduced from copy supplied by
Stephen Austin and Sons Ltd and
printed in Great Britain by
Redwood Burn Limited,
Trowbridge, Wiltshire.

Library of Congress Cataloging-in-Publication Data

Europe's future in space : a joint policy report [Forschungsinstitut der Deutschen Gesellschaft für Auswärtige Politik et al.].
p. cm. – (Chatham House special paper)

ISBN 0-7102-1444-8

1. Astronautics and state – Europe.
I. Deutschen Gesellschaft für Auswärtige Politik, Forschungsinstitut.
II. Royal Institute of International Affairs.
III. Series.
TL789.8.E9E886 1988
387.8'094 – dc19 *87-27009*
CIP

CONTENTS

CONTRIBUTORS

This report was prepared by the following members of the participating institutes:

Forschungsinstitut der Deutschen Gesellschaft für Auswärtige Politik (DGAP)
Institut Français des Relations Internationales (IFRI)
Istituto Affari Internazionali (IAI)
Nederlands Instituut voor Internationale Betrekkingen 'Clingendael'
Royal Institute of International Affairs (RIIA)

Admiral Sir James Eberle, Director, RIIA
Professor Karl Kaiser, Director, DGAP
Eva Kulesza, IFRI
Dr Pierre Lellouche, IFRI
Dr Cesare Merlini, President, IAI
Dr Thierry de Montbrial, Director, IFRI
Professor Henk Neuman, Director, Clingendael
Dr Stefano Silvestri, IAI
Dr Helen Wallace, RIIA
Dr Stephan Frhr von Welck, DGAP

Several experts assisted our deliberations by submitting written papers to our meetings:

Dr Jacques Battistella, Matra
Professor J. A. M. Bleeker, Utrecht University and SRON
Dr Roberto Cajati, IAI
Dr Wolfgang Finke, previously in the Federal Ministry for Research and Technology, Bonn

Contributors

Carol Geldart, RIIA, rapporteur
Keith Hayward, North Staffordshire Polytechnic, rapporteur
Dr Hans M. Kappler, MBB-Erno, Bremen
Dr Mahindra Naraine, Lancaster University
Dr Geoffrey Pardoe, General Technology Systems Ltd

Many other experts helped us by commenting on drafts at meetings in Colleferro and London:

Ing. Stefano Abba', Space Division, Aeritalia
Peter W. Bauer, Planning Staff, Foreign Office, Bonn
Ing. Giorgio Bianchi, Selenia-Spazio
Dr Walter Brado, Head, Office of the Director-General, European Space Agency
Professor Carlo Buongiorno, Ministry for Scientific Research and Technology, Rome
Wing Cdr Christopher Carrington, Ministry of Defence, London
Jean-Pierre Contzen, Director-General, Joint Research Centre, Commission of the European Communities
Ing. Giuseppe Grande, SNIA BPD
Dr David Green, Butterworth Scientific
Ken Inglis, British National Space Centre
Pat Knight, British Aerospace
Jack Leeming, British National Space Centre
Dr Peter Linssen, Dutch National Aerospace Institute
Reinhard Loosch, Federal Ministry for Research and Technology, Bonn
Professor Luigi Napolitano, University of Naples and Telespazio
Dipl.-Ing. Wolfgang Pittelkow, Dornier System GmbH, Friedrichshafen
Captain Naquet Radiguet, Ministry of Defence, Paris
Y. Rebillard, Centre National d'Etudes Spatiales
Dr Trevor Roberts, British National Space Centre
Dr Alain Simon, Centre National d'Etudes Spatiales
Rear-Admiral John Slater, Ministry of Defence, London

PREFACE

The five institutes of international affairs of France, Germany, Italy, the Netherlands and the United Kingdom, building upon their tradition of close cooperation and previous joint studies (see, in particular, *The European Community: Progress or Decline?*, 1983), have here come together to analyse the challenges which space presents to Western Europe, to review Europe's achievements and potential, and to develop proposals for action.

The institutes were able to enlist the support of the Commission of the European Communities, the Fritz Thyssen Stiftung (Cologne), the Algemene Loterij Nederland (The Hague) and the Sijthof Estate (The Hague). Our institutes would like to express sincere thanks for their help, without which this collective work would not have been accomplished.

A working group, composed of authors augmented by advisers, met in Paris, Bonn, Collefero, London, Friedrichshafen and Barcelona to draft this joint report. We owe a special debt of gratitude to Dr Helen Wallace, who cheerfully accepted the extra burden of creating a modicum of coherence out of rough texts originating from a variety of disciplines and linguistic backgrounds. Warm thanks are also due to the many people who have typed, translated and edited successive versions of the manuscript.

The report falls into three parts. Part I reviews the relevance of space to Europe, then examines the space programmes of the world's major space powers and concludes by analysing West European achievements at both the national and the multilateral level. Part II offers a detailed analysis of all the possible uses of space, such as communications and security, and of the available means, such as launchers, satellites and space stations. Part III draws up a balance-sheet of Europe's strengths and weaknesses in space, defines European priorities for future action and concludes with our recommendations for a European space policy.

The reader who is short of time may wish to concentrate on the Summary, on the analysis of the relevance of space in Part I and on the three sections of Part III, in particular the recommendations. A glossary has been provided for the non-specialist.

Monetary units are expressed, wherever possible, in European Accounting Units (EAU), which is the accounting unit used by ESA and is approximately equivalent to the European Currency Unit (ECU).

We would like to make it clear that the opinions expressed in this report are merely those of the individuals responsible as authors and not those of the institutes with which they are associated.

Bonn, The Hague, London,
Paris, Rome, September 1987

J.E.
K.K.
H.N.
C.M.
Th. de M.

SUMMARY

The extent to which nations or groups of nations are able to explore and exploit space will increasingly determine their standing and influence in the world of tomorrow. The five institutes of international relations of France, Germany, Italy, the Netherlands and the United Kingdom, after reviewing Europe's record in space, reach the conclusion that Western Europe can take pride in a number of impressive achievements. But it does not achieve its full potential. A quantitative and qualitative jump is necessary towards a truly collective space policy, if Western Europe is to have anything more than a walk-on part in the world.

The global economy and the nature of international society are changing rapidly, precisely for the reason that satellites transform the flow and availability of information. Among the various applications of space, *telecommunications* was the first, after weather forecasting, to demonstrate the commercial value of space. New techniques and technologies are now being fast developed, and will be put to several uses, including search-and-rescue and direct broadcasting. Space technologies foster the flexibility and innovation required to create new jobs and promote economic modernization.

Remote sensing permits both the specialized mapping of land masses and the monitoring of the atmosphere and of the physical conditions on the sea's surface, thus providing an inventory of a wide range of resources. Satellites will in the future help to satisfy the economic and social needs of developing countries. *Space science* is continually adding to mankind's knowledge of the origin and evolution of our solar system. Microgravity conditions permit experiments and processes which will subsequently have applications on Earth. The *industrialization of space* will provide important economic opportunities.

The uses of space for *security* purposes are self-evident. Satellite-based technologies are already essential instruments for reliable military communications, for the transparency of the strategic balance and for the verification of arms control agreements. These technologies may well prove to be a key to international stability.

Nations or regions wanting to strengthen their contribution to the *cultural life* of the world will find a new opportunity to do so through the use of space.

From a brief survey of the space programmes of the main *non-European countries*, it appears that the two leading space powers, the Soviet Union and the United States, spare no effort or expense to forge ahead in this area of human endeavour. Japan, and also less affluent powers like the People's Republic of China and India, have dynamic space policies; they are perfectly open about their ultimate goal of becoming independent.

An analysis of the space programmes of France, the Federal Republic of Germany, Italy, the Netherlands and the United Kingdom shows that *European space activities* have so far been carried on within a double framework, part national – serving particular political, economic or security purposes – part collaborative, first in ELDO and ESRO and then, since the mid-1970s, under the aegis of the European Space Agency, ESA, a unique example of a multinational space organization. ESA's contribution to the development of a collective European space capability has been fundamental.

In Western Europe, scientific research has been a primary field of space exploration and one in which European states and organizations have been particularly successful. Western Europe has also managed, in a relatively short time and with limited means, to achieve notable results in space technology and space industry, of which the development of a family of indigenous European launchers is an essential part.

On balance, however, it seems clear that European space efforts fall considerably short of a truly effective joint policy. No single West European nation can muster sufficient resources to become what only Europe collectively can be: an autonomous space power.

Europe is not held back by scientific obstacles or intrinsic inferiority. Its financial, technological, industrial and human *potential* is on the same scale as that of the United States or the Soviet Union. The problems are those of approach, purpose and the best use of resources. There is a persistent fragmentation of effort and a relative shortage – compared with the dominant powers – of resources devoted to space activities. The objectives lack sharpness of definiton, and the time perspective is too limited. The public as an end-user is not sufficiently aware of the extent of services provided by current space activities, especially communications, meteorology and other information flows. Few people realize that in many European countries the savings resulting from satellite use for telephone links often exceed the total cost of national civilian space programmes. There is no real appreciation among the broad public of the value of Europe's presence in space. This sometimes shades into a wider public disenchantment with high-technology ventures, both military and civilian.

The nub of the problem is, however, that it would be extremely short-sighted for Western Europe to look at the matter only from the accountant's point of view. The *balance-sheet* must include the foreign policy and security dimensions for a Europe moving into the twenty-first century, the cultural impact and links with developing countries. Space offers an area in which two of the most important aspirations of Western Europe come together: unification and the promotion of high technology. Access to crucial technologies via other powers cannot be taken for granted. European security interests are not coterminous with those of the United States, and an independent source of reconnaissance data is a necessary complement to US-derived information. The course of action to be taken is outlined in the final part of this report.

The ESA member states have committed themselves to achieve *autonomy in space*. This should mean that Europe develops a capability to reach, operate in and return from space, and to do so not on sufferance of friend or foe, but according to its own perception of what is to the common good. This requires indigenous launchers for manned and unmanned transport into space, an independent space station that could be manned and that would be a highly visible symbol of Europe's capabilities and common will, and an adequate system of relay and other satellites and ground infrastructure. Implementation of ESA's Programme of Action is, therefore, the indispensable minimum.

But *more is needed*. Europe should, in order to have its own eyes and ears in space, develop and operate a joint reconnaissance satellite system in low earth orbit for safeguarding its political and diplomatic interests, as well as for arms control verification, the monitoring of military movements and crisis control. This system should be based on existing remote-sensing techniques (optical and microwave) and could be complemented by a jointly operated electronic intelligence-gathering package.

In order to protect its satellites against potential Asat threats, Europe should actively prepare for and participate in negotiations on a code of conduct for the peaceful uses of space, based on a regime of confidence-building measures and rules for space use. At the same time, Europe should study means of protection for its satellites in the context of the right of self-defence.

Europe should maintain a *strong scientific programme* to improve its position in the exploration of the universe. The ESA programme 'Horizon 2000' points in the right direction and deserves full support by the European governments.

Greater *economic efficiency* should be sought by urgent efforts to improve the competitiveness of the space sector and to liberalize markets. Europe must deregulate, unify its internal market, europeanize public procurement, consolidate viable transnational consortia of space companies, supply a larger share of the home market for the ground segment, and improve its capacity to bid successfully for international contracts.

Over time, ESA should consolidate its role as the central policy-making body for Europe in the civilian field, as well as representing Western Europe vis-à-vis the rest of the world.

The development of European security collaboration in space should not wait for, but will eventually require, an appropriate institutional framework. A strengthened WEU may become the focus of such collaboration.

The implementation of the recommendations outlined in this report would require approximately the doubling in volume terms of the current level of annual European expenditure up to the year 2000. This would provide the resources for our recommended ESA plan, a new European security programme and continued national expenditure. This figure of on average about 4.5 billion EAU per year over 13 years, while considerable in absolute terms, would represent only about one-quarter to one-fifth of current US spending – a reasonably modest price to pay for European autonomy.

These and other practical measures will achieve their goals only if they reflect the political will of West European governments to look – and to have their publics look – upon space policy as crucial for the future definition of Europe's political role vis-à-vis the other major space powers and, indeed, the world.

Part I
WESTERN EUROPE AND SPACE POLICY

1

THE RELEVANCE OF SPACE TO EUROPE

Space: a European horizon

Western Europe has to confront some hard decisions about whether it wishes to operate as a major actor in space, a field of human endeavour which will play a growing role in determining influence, prosperity, technological achievement and security in the global environment of the twenty-first century. If Europe, where the roots of modern technological civilization lie, does not realize its potential in this field in contributing significantly to the conquest of mankind's last empty space, it abdicates as a major actor in world politics. Influence on the future international space regime will be wielded only by those who have real space assets. The challenge of space can be met only with a common European effort. No single West European nation can muster sufficient resources to be an autonomous space power. Indeed, space is a major area in which Europe can consolidate a common identity and develop its unity.

The global economy and the nature of international society are changing rapidly, not least because new means of communication and information have developed, in which satellite links play a crucial part. In the field of security, satellite-based technologies open the way to increasingly sophisticated military communications, reconnaissance, intelligence and verification of arms control agreements. The exploration of space thus creates new economic opportunities, promotes industrial innovation, makes new services possible and provides new instruments for foreign, cultural and overseas development policies.

Europe has come to a crossroads in its space policy. A quantitative and qualitative jump in European space efforts would be required to respond to the challenges outlined here and to harness the benefits which lie ahead. Western Europe has achieved impressive successes, including the Ariane launcher, the 1986 Giotto mission and many other satellites, but its all-round performance woefully lags behind the space efforts of the United States and the Soviet Union, not to mention the emerging challenge of competition from China, India and Japan. Only 40 of the total of 3,500 satellites in orbit at the end of 1986 were European. Political, economic and organizational constraints, including persistent national blinkers, prevent a better European performance. The continuation of European policy at current levels is likely to make Western Europe marginal in the space politics of the next century. The Europeans initially missed many opportunities in space. Thanks to the European Space Agency (ESA), much ground has been recovered. But Europe still lacks the all-round capabilities to maximize its potential.

Part of the explanation for Europe's limited efforts thus far has to do with a narrowness of approach in most European countries. Only the French consistently and with foresight have pursued an active and wide-ranging policy. So far both national policies and European collaboration, notably through ESA, have produced a disjointed bundle of space activities for scientific, technological, and economic reasons with, to date, very few military applications. A more wide-ranging strategy to exploit space would require more sharply defined objectives and a more long-term perspective. ESA members have declared a commitment to achieve European autonomy in space, but this aim requires not only a greater investment of resources but also a broader conception to link the scientific, technological and economic dimensions to the future role of Western Europe in the world, in the fields of security, cultural impact and links with developing countries. Public awareness of the economic and technological dimensions is limited, knowledge about the political and security implications confined to a few experts. Few people realize the relevance of space to Europe's political future and security. Without wider public appreciation of what a European space policy might achieve, there is little prospect of a more ambitious and more rounded approach emerging.

Scientific and cultural endeavours

The driving forces of the American and Soviet space programmes are military and symbolic, fuelled by enthusiasm and imagination. In Europe, by contrast, scientific research was from the outset a primary motivation of space exploration and one in which European organizations and states have been particularly successful. Modern space technology has revolutionized the scientific exploration of space at an astonishing pace. Satellites, re-usable laboratories, telescopes stationed in space and exploration satellites probing deep into space have opened new vistas for scientific research. Scientific experiments conducted in the fields of biology, physics or medicine offer new insights, and some of these may well have significant practical applications. To maintain a European foothold in these areas would require further technological development. But we should note that the enthusiasm of European scientists has not been matched in the wider policy or public debate.

The exploration of space has produced dramatically new means of communication. Not only can satellites transmit extraordinary amounts of data at speed and decreasing cost, but they have established a network of communications that transcends national frontiers and continents and indeed the entire globe. Satellites provide direct access to other societies. Hence the use of space is likely to play an increasingly important role in the fields of influencing public opinion, advertising products, and exercising cultural influence. Europe has traditionally excelled in the cultural field. The use of space offers a new opportunity to strengthen its contribution to the cultural life of the globe. But without a collective approach Europe cannot realize its potential.

Economic and social benefits

The use of space already provides new fields of economic activity and growing markets. The space industry has become an important sector of technology, parts of which are already commercially profitable. There is considerable commercial activity in the construction, sale and operation of satellites, and space transportation has become an important service in which Europe competes successfully through Arianespace. Communications satellite systems are furthest down the road to profitability; observations from space of what happens on Earth are the next in line to have commercial application. In the longer term the production in space of new pharmaceutical and biotechnological products as well as of new materials may prove economically viable.

In all of these fields Europeans have scored considerable successes, but in increasingly intense competition with other leading technological nations. Only through further efforts can Europe avoid technological dependency and be assured of a continuing, even increasing share of world markets, actual and potential. As in other fields of technology, access to know-how and the ability to exploit it efficiently will be necessary to prevent Europe from being held to ransom by others.

But equally important, space technologies shade into many other key areas of high technology with both civilian and military relevance: advanced avionics, information technology, robotics, new materials and many more. Thus this range of space-related industries has a still wider economic and technological significance, not least in the management of large and complex systems. We should not look at space technologies in isolation; taken together, they foster the flexibility and innovation required to create new jobs and to promote economic modernization. Space therefore helps the European economy in competing with Japan and the United States. Space projects induce scientific and industrial cooperation and thus underpin European integration.

Services

Few people realize how the use of space has revolutionized the service sector. Modern satellite-based communications systems have become an indispensable part of the modern information society in which masses of data, television images, telex, telephone, etc., are transmitted with growing rapidity and in increasing amounts. Europeans are major producers of these systems, and modern European society and culture are unthinkable without them. Not only has the access to new data, information and cultural events been dramatically improved but costs have been decreased significantly through modernization. Few people realize that in many

European countries the savings resulting from satellite use for telephone links often exceed the total cost of national civilian space programmes. The use of space-based information transmission is likely to increase in the future, creating new links across frontiers and a multitude of new economic activities. Moreover, there is the possibility in the longer term that orbital planes which enter space for portions of their trajectory may drastically shorten travel time between continents.

Observation satellites have significantly improved services which are of great use for individuals, economic activities and transport, as well as defence. Weather forecasting is perhaps the best known application, and other applications include crop forecasting or, in the field of environmental policy, techniques for monitoring emissions and for early warning in case of threatening damage to the environment. The images of the stricken nuclear reactor at Chernobyl vividly illustrated just how far space technology had advanced.

In all of these areas European countries and organizations have developed significant skills and activities. The rate of progress has been breathtaking and is likely to produce a number of new fields of economic activity, with services to people and whole economic sectors, for improving standards of living, environmental conditions and the cultural richness of life.

Europe's role and influence in world politics

The political role of states or groups of states in the global society of the twenty-first century will be significantly influenced by their ability and willingness to explore and use space, to develop the necessary technologies and to create the required industrial infrastructure. If West Europeans aspire to a serious influence in international politics, they have to take account of this and to recognize just how much importance other countries, especially but not only the superpowers, attach to enhancing their own capabilities in space. The conduct of cultural, economic and political affairs will be linked with the use of space, as will the conduct of security policy and the pursuit of arms control and disarmament.

Countries or regional groups like Europe which participate in the exploration and use of space will increase their power in the pursuit of their interests. Europe today remains among the technologically advanced regions of the globe. It shares that status with the United States, Japan and the Soviet Union. There is no reason why in the coming century only the United States and the Soviet Union – and possibly Japan, China or India – should exploit the full range of space activities.

This is not to say that Europe should simply imitate the two superpowers. In the economic domain, competition will remain a characteristic of the global society, even if states are tied together in alliances or systems of economic cooperation as is the case between Western Europe, the United States and Japan. Access to crucial technologies via other countries cannot be taken for granted, access which is being increasingly restricted in the case of the United States and which is effectively closed to Europe with regard to the Soviet Union.

The Americans and the Soviets have no preordained right to monopolize manned space exploration. Europeans have so far provided an occasional guest on their spacecraft, but have also shown themselves capable of taking the first steps towards manned space flight. A rounded space programme implies some element of manned exploration, and Europeans can further pursue this goal, even though the scope and the methods may well be different from those of the United States and the Soviet Union.

European attention, which tends to focus on the European Community (EC) in matters of European unification, has barely noticed that space has become an area of highly successful European cooperation and integration. Thanks to the work of ESA, European countries have successfully pooled their resources in joint ventures with which all Europeans, in particular European youth, can identify. Barriers have been removed, industries and laboratories have merged their efforts, genuinely joint facilities operate successfully and new projects open the possibility for additional and enhanced cooperation. Space offers an area in which two of the most important aspirations of Western Europe come together: unification and the promotion of high technology.

ESA has so far been Europe's primary instrument for developing its role in space and is likely so to remain. But other organizations with different interests and memberships have a role to play. Eutelsat and Eumetsat provide some collective services for telecommunications and meteorology. The EC has relevant competences in the areas of technology *and* industrial and trade policies. For commercial exploitation special organizations are required, as already achieved for Ariane. An extension of European space collaboration in the field of security would require the use of some framework other than ESA, such as WEU. But developments beyond the current level of cooperation would depend on a new European consensus and a willingness to override narrow national interests, both commercial and military.

Europe's goal should be to achieve autonomy in space. This would require indigenous launchers for manned and unmanned transport into space, an independent space station that could be manned, and an adequate system of relay satellites and ground infrastructure. Such autonomy is not to be confused with total self-sufficiency: an autonomous European space policy would still be open to international cooperation not only with the United States, but with other countries as well, notably the Soviet Union, Japan, and developing countries. But as the history of American-European space cooperation has shown, the United States will recognize Europe as an interesting partner, and grant it terms that correspond to European interests, only if Europe has a significant space capability of its own.

We should also recognize the importance of setting the international rules for the use of space. Given the growing relevance of space to the activities of states, notably in the areas of power politics and security, Europeans have a keen interest in contributing to an international space regime that enhances stability. European influence will depend both on the extent of European capabilities in space and on the degree to which policy is collectively shaped.

Space and European security

The controversies surrounding the Strategic Defense Initiative (SDI) of the United States have somewhat blurred the real issues that space poses for European security. Fears of this further 'militarization of space', for which not Europe, but the two superpowers are responsible, may lead Europeans to a self-denying regimen. It is one thing to resist calls for space weaponry, but quite another to opt out of the use of space to safeguard security. Europeans should be made aware of the relevance of space policy both for arms control and for supporting existing conventional military capabilities.

Major dimensions of modern arms control would be impossible without the technological possibilities that the use of space offers. Recent and likely future agreements on arms control and disarmament require verification, in which remote-sensing satellites will necessarily play a major part. These will need strengthening, as further attempts are made to constrain and reduce nuclear and conventional arsenals. So far only the two superpowers possess a satellite capacity adequate to verify agreements on arms control and disarmament. All information available to Europeans is of American origin, and not all information is passed on. The Soviet Union does not share any data at all. Europe is technologically capable of building up a satellite capacity of its own. European security interests are not coterminous with those of the US, and an independent source of reconnaissance data could be useful to complement US-derived information.

Similar considerations apply to confidence-building measures. If these are to work effectively and to be expanded as a means of decreasing tension and enhancing stability, the observation of military developments in Eastern Europe through satellites will be very important. A European capacity in this area of data collection might therefore serve to improve transparency and calculability in East-West relations.

The use of space has taken on great importance for the system of nuclear deterrence between the superpowers, a system which, of course, also protects Western Europe. But space is increasingly used for other defence purposes. Optical and electronic reconnaissance satellites help to provide constant information about wide-ranging military developments and deployments, including early warning and crisis monitoring. Moreover, satellites for weather prediction, navigation and military communication are of growing importance for the defence of Europe. Weather satellites are the only ones for which Europe has an effective system of its own. France and Britain have some independent capabilities and, as for the rest, they and their fellow Europeans have to rely to varying degrees on Nato or directly on the US.

In a number of European countries the non-civilian use of space is viewed with apprehension. An enhanced European programme could cause controversy. The propaganda battle surrounding SDI has made a rational debate more difficult, but such a debate is nevertheless necessary. If Europeans are to become more responsible for their own security, they cannot avoid considering the use of space. Even sceptics are likely to recognize its relevance in the field of arms control and disarmament. But the questions need to be assessed more broadly.

Space and developing countries

The use of space is increasingly relevant to the needs of developing countries. Satellites offer new possibilities to assess harvests, to identify raw material deposits, to predict weather, to survey environmental developments or to predict natural catastrophies. Perhaps even more important, communications satellites provide a relatively fast and cost-effective means of pursuing national integration and regional cooperation, modernizing agriculture, and expanding education and health care. Europeans have an opportunity to help developing countries to acquire the means to exploit these technologies.

Towards a new European perspective on space

So far in Europe space exploration has been seen and justified to democratic publics and parliaments primarily from the point of view of its use for scientific research, technological advance and economic gains. To be sure, even these advantages of space use are not fully appreciated. But a justification of further space exploration in disjointed scientific, technological and economic terms is unlikely to generate support for the larger public expenditure which European autonomy in space implies or for the range of applications which are likely to be relevant in the future.

The time has come for a qualitative jump in European thinking on space. The publics in Europe need to become more aware of the political dimensions of the use of space. When decisions of great magnitude on space activities are at stake, it is short-sighted to look only at the costs and benefits in financial, technological and economic terms. The balance-sheet must include the political and security dimensions of a Europe moving into the twenty-first century.

This report is therefore designed as a contribution to greater understanding of both the imperatives for European action and the policy issues involved. It deliberately ranges across a wide spectrum of uses of space and pinpoints those that are likely to become increasingly relevant to Europe's role in the world over the next decades and into the twenty-first century. It seeks to identify both the options and the constraints. The thrust of the argument points to the need for a far more vigorous European space policy in the future. But none of this will be achieved unless Europeans have the imagination and ambition to harness effectively their considerable potential for making the most of space.

2

PARTNERS AND COMPETITORS

Current and potential space policies for Western Europe need to be seen against the backcloth of the space programmes being developed elsewhere in the world. The American and Soviet programmes are of course the longest established and the most ambitious across the range of civilian and military applications. But equally importantly many other countries are in the process of developing their own capabilities for a mixture of civilian and military reasons. Several countries include in their aims the pursuit of autonomy in launch capabilities and in both space and ground segments, with the ability to compete for international business. This section of the report briefly surveys the most relevant features of some of these programmes; it pays particular attention to the American and Soviet programmes and gives an overview of the fast developing programmes of Japan, India and China. Further statistical details are given in the tables and figures at the end of the report.

(a) UNITED STATES

A brief survey can hardly do justice to the long history of the American space programme, let alone to the difficult choices currently facing American policy-makers. Successive American governments have sought to drive forward ambitious projects from scientific exploration through economic exploitation to innovative and often controversial defence and security applications and with major projects in manned missions. From a West European perspective, the evaluation of American space capabilities is especially important because of the extent to which they exceed in scale and in market share the efforts of other Western countries, and because of the transatlantic defence and security relationship, in which space-based technologies have acquired increasing importance. There are many lessons, both positive and negative, for Europeans to derive from the history, operation and aspirations of American policy.

Objectives

The objectives of the American space programme have been complex. To be sure, the development of military rocket and missile technology underpinned the early experiments, and these were driven by a determination to keep a military edge vis-à-vis the Soviet Union. But in the 1950s under President Eisenhower there was a strong, and indeed organizationally separate, civilian dimension with scientific and technological aims. It was the shock of the Soviet success with Sputnik in October 1957, ahead of the first successful launch by the US of a satellite in January 1958, which led to a more integrated policy and more determined competition with the USSR. But even so, in 1958, the new National Aeronautics and Space Administration (Nasa) formally had only a civilian responsibility, while the Department of Defense (DoD) retained charge of the military programme with intense competition among the three services. Crucially, though, Nasa won control of manned missions.

In 1961 the civil programme enjoyed a major boost when President Kennedy endorsed the race through the Apollo programme for the first landing on the moon (achieved in 1969) in the belief that an American 'victory' would be a key symbol of both technological prowess and international stature. These high ambitions captured the public imagination and had a major impact on US prestige throughout the world, though there were always vociferous critics. But it is also important to underscore the early priority attributed to satellite applications for communications, reconnaissance and scientific experiments. Comsat was created in 1962 to promote the commercial development within the US of communications in satellites, and the 'open skies' policy was embraced. The Americans took the initiative which led to the creation in 1965 of Intelsat. They had a dominating role in its original subscription arrangements and were to prove the major industrial beneficiaries. Under President Johnson increases of funding were agreed to reinforce the scientific, technological and industrial dimensions, and a vigorous space industry developed fast.

In the 1970s, however, the ambitious goals and projected funding attracted recurrent Congressional criticism and sharp debate among experts. The ensuing constraints led to efforts by President Nixon to trim Nasa's budget and to focus attention on exploitable applications of space technologies (both launchers and satellites). Nonetheless, the determined, if less visible, efforts of the DoD achieved increased levels of funding for military applications of space not only to support the American defence forces and to provide intelligence, but also to maintain a R&D capability for new applications. The civil programme lost momentum after the Apollo programme and was to suffer from over-reliance on the Shuttle programme as its centrepiece. By the 1980s and Reagan's 1982 space review, the balance of government attention had tilted towards the military, paving the way for the announcement of the SDI in March 1983. Meanwhile Nasa, under some severe constraints, sought to retain an ambitious manned flight programme and new ventures such as the 'international space station' (announced in July 1984) and deep space probes.

Programmes and projects

Launchers

The US has employed ten major types of launcher in over fifty versions over some thirty years of space activity. Of these only four types are still available for use. This proliferation was partly a result of inter-service rivalry in the early years and led to much duplication. Nasa initially had to rely on the military and their contractors, and the available expendable launch vehicles (ELVs) were nearly all based on military missiles. The wide use of the Atlas and Titan launchers, both of which were ICBMs, grew out of the close cooperation between Nasa and the DoD, in particular the Air Force, which from 1962 had been given the primary responsibility for the armed services space efforts. The US also tried over many years and at great cost to apply nuclear and electrical propulsion systems to space travel, but budgetary constraints prevented their development. Proposals abounded for alternatives to ELVs, both space planes and other re-usable vehicles. The large number of launches (over 800 in the first 25 years) increased reliability from about 30 per cent to some 95 per cent. By the end of 1986 there had been some 869 launches, of which 113 failed.

The Saturn series of heavy-lift launchers was used for the Apollo lunar programme. Saturn 1 had a capacity of 9 tonnes to low earth orbit and Saturn 5, the most powerful launcher until very recently, almost 100 tonnes to low earth orbit (cf. Ariane 5 – 15 tonnes, Shuttle – 30 tonnes, and Soviet Proton – 27 tonnes, all to low earth orbit, but now allegedly capped by the new Soviet Energiya). The Saturn vehicles were officially scrapped at the end of the Apollo and Skylab missions.

At the end of the Apollo era, the high launch costs rekindled interest in re-usable vehicles, one of the discussed missions for which was Nasa's proposed space station. Nasa proposed a fully re-usable vehicle for the 1970s, but Congress vehemently opposed the costs. Instead, in 1972, President Nixon approved the Shuttle programme, a simpler and cheaper but only partly re-usable vehicle, for all users including the military. The Shuttle was to establish a new space system that would allow easier and lower-cost access to space and permit a wide range of space activities. The intention was to move away from ELVs to total dependence on the Shuttle not only by Nasa but also by the DoD, which was to use 40 per cent of Shuttle flights. Delays with the programme and high cost overruns led the DoD to retain the option of the Titan launcher; meanwhile Nasa was able to fall back on the Delta and Atlas-Centaur, though neither was optimized for commercial payloads. (See Table 6.)

The complexity of Shuttle operations and the vulnerability arising from dependence on a single vehicle with technical deficiencies was dramatically confirmed by the Challenger disaster of January 1986. This has led to modification of old ELVs and the development of new ones. The DoD has contracted for a heavier Delta vehicle to launch its Navstar satellites; it will also be available to the commercial market. Most recently the DoD has requested funding for a new heavy-lift vehicle (some 60 tonnes to low earth orbit), perhaps based on Saturn 5. The Air Force is requesting proposals for an Advanced Launch System which will permit production and launch on a routine basis. The target is to reduce launch costs by a factor of ten to lift payloads of up to 100 tonnes into low earth orbit. In the meantime, there is a backlog of American satellites awaiting launch and of satellites from elsewhere which had booked space on Shuttle flights.

Unmanned missions

Since the early 1960s the United States has developed a huge array of satellites – about 1,000 by the end of 1986 – for both civilian and military purposes, with large private-sector involvement in civil applications and an impressive share of world markets. Experiments and procurement have been handled by the DoD and Nasa both separately and in combination, though the DoD has been responsible for over 70 per cent of payloads. Some satellites perform more than one function, and so no classification is entirely satisfactory. Table 7 summarizes the range and approximate numbers of missions.

Manned missions

Since 1961 the United States has carried out 54 manned flights in five series of programmes. The major manned programme was the Apollo programme. There were eleven flights in this programme from 1968 to 1972 (six of which landed men on the Moon using the Saturn 5 launcher) at a cost of $25 billion.

Although the initial lunar landings captured the imagination of people all over the world, interest within the US waned after the first few successes, to be revived only temporarily by the ill-fated Apollo 13. Indeed Harris and Gallup polls during the 1960s showed American opinion to be equivocal about spending large sums of money on manned missions. Shortly after President Kennedy's announcement in 1963 of the $40 billion project to put Americans on the moon, 58 per cent of respondents opposed the project, with only 33 per cent in favour. In only one poll, taken in 1965, did those in favour – 45 per cent – outnumber the 42 per cent who were opposed. By 1967 support for a lunar mission had dropped to 34 per cent. Changing political and social priorities, against the backcloth of the Vietnam war, had undermined support for Nasa's manned space programme. Growing Congressional opposition to the lunar programme resulted in the cancellation of the last three lunar missions. If funding is an indicator of commitment, support was at its most vigorous in the immediate period after Sputnik, peaked in 1964 and precipitously decreased after that date.

Nasa then commenced the Skylab programme as an attempt to determine the long-term effects of manned presence in space and to conduct experiments in space processing, Earth resources photography and astronomy observations. The workshop was a converted Saturn 4 rocket casing, and three flights of three-man crews were sent in 1973. More workshops were planned, but general disenchantment with space activities precluded them. The cost of the programme was $2.5 billion. In July 1975 an Apollo spacecraft docked with a Soviet Soyuz craft, at a mission cost of $225 million. No further manned flights were made until the Shuttle flights in April 1981. Twenty-three shuttle flights followed before the Challenger accident in January 1986 halted the programme. Details of the Shuttle programme and the proposed new space station are in Chapter 6 of this report.

Budget and resources

It is difficult to be precise on the total allocations for space activities because the military expenditure is not fully recorded. Moreover, spending patterns have been erratic and objectives have changed. It is generally held that over the past thirty years overall expenditure has been in the order of $160 billion, of which Nasa has spent $100 billion. Nasa funding increased rapidly from $0.5 billion a year in 1959 to a peak of $5.5 billion in 1964. There was then a dramatic fall in funding – as interest in space waned – until 1974, when expenditure levelled out. In 1985 the Nasa budget was much the same as in 1962. In 1986 allocations were some $4.6 billion. Although spending on military programmes had been generally at a much lower level than in the civilian programme, it steadily increased and by 1981 exceeded civilian expenditure. By 1986 it was well over double that for the civilian programme, and the emphasis on SDI will, it is assumed, result in significant increases in the level of funding for military programmes. Private-sector expenditure has been predominantly on communication satellites, but has amounted to less than 1 per cent of total funding.

Nasa's funding has been a recurrent source of controversy, notably in Congressional committees. Its budgets have been very susceptible to cuts, a feature that has jeopardized many international cooperative ventures. In contrast, the high profile and powerful lobbies associated with the military, as well as the primary importance of national security and anxieties about Soviet efforts, have ensured high and increased funding for military programmes.

At the 1964 peak of Nasa funding, some 400,000 people were employed in civilian space activities, of which 10 per cent were employed by Nasa; by 1974 this number had declined to 100,000. There has been a slight increase recently because of space station activities, but this has been offset both by a tailing off of domestic demand for satellites and by delays in launches. Eurospace, for example, estimates that some 136,000 people are involved in primary civilian space activities, with an additional 116,000 said to be involved in military programmes. About one million people may be associated with secondary and related space activities.

Organization

Space policy has been managed on a pluralist basis in the United States. Nasa, founded in 1958, has been the most visible agency with responsibilities for R&D, big projects and launch facilities both directly and through its associated research centres. But the DoD, and to varying degrees the three armed services, have maintained their own programmes. Nasa largely absorbed the Army and Navy space capabilities, but the DoD and the Air Force continued to operate separately with various liaison mechanisms – especially the National Aeronautics and Space Council – to coordinate with Nasa. Successive attempts have been made throughout the history of American space activities to forge a coherent and overall space policy with high-level presidential reviews and keen attention from Congress. Funding has fluctuated partly in response to this political scrutiny, partly because it has depended on the apparent 'success' of particular programmes, and partly as a result of the other expenditure commitments and priorities of successive administrations. None of these debates has yet produced a single framework of policy or an entirely coherent organization of capabilities, even though the boundaries between civil and military applications have become increasingly blurred, especially with the Shuttle programme.

Nasa's work has sometimes fallen within the presidential domain, commanding a high profile, but at other periods suffered a fall from grace. Problems have arisen from typical inter-agency conflicts. Some space functions have been removed from Nasa: for example, the transfer of Earth observation and weather to the National Oceanic and Atmospheric Administration (NOAA). In November 1984 President Reagan authorized the creation of a Unified Space Command with a consolidated operations centre at Colorado Springs. This absorbed the Air Force's and Navy's independent space commands and took charge of orbital surveillance, the anti-satellite (Asat) programme and GPS/Navstar, as well as the space-related aspects of SDI.

Over the years there have also been weaknesses of management and tight technological constraints. Given the combination of ambitious symbolic goals for the American space programme and fluctuating funding, it is perhaps not surprising that there should have been temptations to cut some of the technological corners, or that the DoD should prudently have preferred to retain some independent capabilities. Experts in the field were already well aware of the problems, but it took the tragic Challenger accident to reveal the consequential stresses and strains, not least in Nasa's relationships with its contractors. In the ensuing months, anguished debates and evaluative reports edged the Americans towards a reappraisal of both policy and management.

On the other hand, at the commercial end of the scale, the American programme benefited from an entrepreneurial and innovative business culture, not only among producers of space systems but crucially in the buoyancy of demand for space-derived services. Here Nasa and the private sector, helped by increasing market liberalization and the Technology Utilization Programme, have tried to forge a partnership to develop commercial applications. However, Nasa has not always had a good sense of user needs, and deregulation came too late for AT&T (for example) to capture an early foothold. It should also be noted that the large community of space specialists in American universities is now suffering from the cut in Nasa's civilian research programmes.

Foreign policy dimension

In space issues, as in so many areas of international cooperation and management, the United States has played a leading role among Western countries. One element has been collaborative, through the opportunities it has offered to other countries to take part in American-led missions. The great adventures of scientific exploration were from early on identified as opportunities for international participation, even to the extent of one US/USSR joint mission.

The succession of manned missions also opened up collaboration with the Spacelab project and, recently, the American-inspired 'international space station'. Not surprisingly, such issues as the balance of partnership, mission definition, mission control and provision of space and ground segments have all raised awkward choices for the partners of the United States concerning the trade-off between access to the project and dependency. The US has not always been an easy partner.

Another source of controversy has been international regulation, in which, for example, the United States both took a lead in establishing Intelsat and has subsequently played a predominant role. This reflects the level of American investment or subscription, but also the sharp competitive edge of American companies in tendering for Intelsat contracts. Current US attitudes to Intelsat and Landsat may reduce their value to European countries and LDCs, particularly if US companies are allowed to compete with Intelsat. If, however, tariffs were reduced as a result, then consumers could benefit. What is clear is the powerful role of the Americans in chasing international markets (actual and potential) in an important area of high technology. Significantly, the American government, especially under the Reagan administration, has drawn back from using the United Nations (UN) as a forum for regulating the use of space.

Security policy dimension

American defence and security policy has become increasingly dependent on space-based systems. For a long time the major concentration of DoD space activities has been on the development of communications and intelligence-gathering facilities through space technologies. The US relies on space for accurate strategic intelligence and the verification of arms control agreements. Space has also become an essential element in the command and control of globally deployed US forces. Seventy per cent of all peacetime long-haul communications are handled by satellite. The GPS/Navstar system will provide a revolutionary degree of positional accuracy for SSBNs, aircraft, ships and ground forces, even down to individual soldiers. This will enhance US (and Nato) all-weather capability for refuelling, target acquisition and ordnance delivery. Early-warning satellites give the US extra time for crisis management. Meteorological satellites provide the only real-time weather updates, again of considerable value for tactical and strategic operations.

While much of this activity has been dedicated to American military requirements, international collaboration has also been developed both within Nato and through privileged relationships with particular partners. Thus the Nato system of communications satellites has rested so far on American capabilities. As for intelligence-gathering, recurrent and close collaboration has persisted between the United States and the United Kingdom, while more intermittent and *ad hoc* cooperation has been practised with other Alliance partners. In these cases the extent of access to US data has been conditional and dependent on American assessments of what was relevant and desirable.

The announcement of the SDI by President Reagan in March 1983 added a new dimension to the US military space programme. SDI envisaged a mixture of space- and earth-based anti-ballistic-missile systems. This project, highly controversial both politically and technologically, appeared to provide a major injection of money into almost all aspects of space technology. Some of the R&D performed under the SDI umbrella will enhance US space capabilities, as well as contributing to a general improvement in US technology. America's allies feared that if they were not able to participate at this new 'frontier of science', there would be detrimental effects on their own technological standing. Aware of these fears, and of the need to marshal the best of the allied research effort to meet the challenges of SDI, the US Department of Defense in March 1985 invited the Nato allies, as well as Australia, Israel, Japan and South Korea, to participate directly in the SDI research programme. Memoranda of Understanding (MOU) on SDI participation have since been signed on a bilateral basis with the United Kingdom, the Federal Republic of Germany, Israel, Italy and Japan. SDI research contracts have also been awarded in France, Belgium and the Netherlands. The US presented the SDI as a major opportunity for cooperation within Nato. To date, however, the reality has failed to live up to the promise.

(b) SOVIET UNION

The Soviet commitment to space started in the mid-1950s, with developments in ballistic missiles. The first Soviet ICBM, the SS-6, was fired in August 1957, while the first satellite, Sputnik 1 – the first man-made object ever put into orbit – was launched by a slightly modified version of the same missile on 4 October 1957. Since that time, the Soviet Union has conducted steadily growing space programmes to serve scientific, economic and, above all, national security purposes. With the world's highest rate of successful launches (91 missions for 114 payloads in 1986) and dramatically improving technological know-how in almost every sector of space activity, and with the launch of Energiya and the success of Mir, the USSR has even overtaken the United States as the world's leading space power. On the other hand, the USSR has to contend with the facts of geography and orbital geometry, which dictate that their launches require more power per payload than any Western launch. The programme is also handicapped by problems with component reliability and miniaturization.

Objectives

Soviet involvement in space has been shaped by a complex series of motivations during the past three decades or so. First and foremost, Soviet space programmes serve the dual political goal of enhancing the international prestige and power of the Soviet Union and stimulating the national pride of the Soviet population. Space has also been systematically used as a propaganda theme. Soviet exploitation of space for peaceful purposes is contrasted with American militarization of space, especially in the aftermath of SDI.

In the mid-1960s, however, the idea took hold of exploiting the applications of the various services of space systems, by building up a comprehensive, multi-purpose satellite network. Today, even though civilian application satellites are still less numerous than the military ones, space-based systems certainly represent an important contribution to the Soviet national economy, especially in the fields of agriculture, forestry, water resources and meteorology. Moreover, the establishment of a permanent presence in space has become a major driving force of Soviet space programmes. Consequently, the setting up of modular space stations, together with the goal of extending the length of time humans can stay in orbit, has been given the highest priority. Already the Soviets have spent nearly 100,000 hours in space.

Although the Soviet Union presents its space activities as peaceful, it is obvious that military uses of space are an essential element within its military doctrine and its strategic posture. Not only has the Soviet Union built up an impressive network of space systems to support military functions; it has also been involved in two space weapons programmes, namely, the fractional orbital bombardment system (FOBS) in the late 1960s, and the anti-satellite co-orbital interceptor, which can hit targets in low earth orbit (LEO). The second of these programmes started tests in 1968 and has been judged operational since the early 1980s.

Programmes and projects

Launchers

In terms of launch capabilities, the Soviet Union is today in a very favourable position in comparison with other space powers. Its space industry produces routinely about one hundred launchers a year. In 1986 the USSR was able to carry out 91 launches, an average of one launch every four days.

So far the Soviet Union has used eight types of operational launch vehicle. All are adaptations of long-range ballistic missiles, with the exception of the Proton vehicle. Interestingly, since the early 1980s, Proton has been the only Soviet space rocket put forward for commercial use by potential foreign customers. No Western country has yet taken up this offer, essentially for political reasons. However, an Indian payload is due to be put into orbit by Proton in autumn 1987. Commercial companies in the West have considered following suit, and are giving increasing attention to both the vehicle's operational reliability (reportedly only 7 failures out of 97 missions) and the attractive financial terms offered by the Soviets. Until recently Proton was the Soviet rocket with the heaviest lift capacity for the launching of orbital infrastructure elements, lunar and interplanetary probes and geostationary satellites.

Another Soviet launcher – the 'A' or Semyorka – is derived from the SS-6 intercontinental missile, and is still the mainstay of the Soviet programme. In operation since 1978, it has been used more times than any other orbital transportation vehicle in the world, essentially for manned space missions. In addition, a new Soviet booster of medium-lift capacity has been flight-tested since 1985. Called the SL-X-16 by US specialists, it has a reported payload-to-LEO capability of 15 tonnes.

Most recently, on 15 May 1987, the Soviets succeeded in the first flight-test of Energiya, a new heavy-lift launcher, which has a reported payload-to-LEO capability of well over 100 tonnes – more than the US Saturn 5 rocket, previously the world's most powerful launcher. Significantly, it can launch more than three times the payload of the US Shuttle. It is expected to carry into orbit modules of the Soviet space station and heavy military satellites, and to contribute eventually to manned missions to the Moon or Mars. The development of Energiya has confirmed the Soviet lead in space.

The Soviets are also developing two re-usable spacecraft: the first perhaps comparable to the US Shuttle, and the second a much smaller 'spaceplane'. The Soviet heavy shuttle is reportedly undergoing tests at the Tyuratam space centre and could be flown piggyback on the Energiya booster as soon as 1988. The 'spaceplane' has made four successful orbital flights in a sub-scale version and could become operational by 1990.

Unmanned space missions

The total number of satellites launched annually by the Soviet Union is approximately three times greater than that of all other countries combined: 114 payloads were put into orbit in 1986. But this spectacular number is essentially explained by Soviet satellites still being less sophisticated than Western ones and staying in orbit for a shorter period of time. The most important Soviet satellite programme, the Kosmos series, which started in 1962, has become a blanket label covering all Soviet military missions. So far more than 1,800 Kosmos satellites have been put into orbit.

Currently the Soviet Union launches an average of 30 military photo-reconnaissance satellites a year, including area surveillance satellites, which have been used several times for crisis monitoring. It operates a six-satellite constellation of electronic intelligence (Elint) satellites, as well as a large network of satellites for military command, control and communications. The latter may belong to either the Kosmos or Molniya 1 (highly elliptical orbit) categories. Soviet early-warning satellites use 12-hour orbits to facilitate the largest possible ground coverage with a relatively small number of satellites. For ocean reconnaissance missions, two basic satellites are used by the Soviet military: active radar-equipped, nuclear-powered, ocean reconnaissance satellites (Rorsat); and passive, sensor-equipped Elint ocean reconnaissance satellites (Eorsat). The Soviet Union employs two constellations of navigation satellites, comparable to the US Transit system. Recently it has begun operating a new generation of Navsats, called Glonass, akin to the US Navstar system.

Partners and competitors

In the field of civilian applications, the Soviets have a wide satellite telecommunications network (the Molniya, Reduga, Ekran and Gorizont systems), and earth and ocean resources detection capabilities, as well as meteorological satellite systems (Meteor).

Scientific space missions represent the smallest proportion of the Soviet space effort. Among the most recent scientific space programmes, the interplanetary Vega 1 and Vega 2 probes constitute undoubtedly one of the greatest achievements. The Soviet Union is preparing to launch a dozen or more important scientific space missions, including at least ten Earth-orbit science payloads and two missions to Mars. Many of the Soviet space science programmes are carried on in collaboration with West European as well as Soviet-bloc nations.

Manned space missions

Since the early 1970s, missions involving space stations have represented the bulk of Soviet manned space programmes. Seven Salyut-series space stations had been launched between 1971 and 1982, and the new Mir station has operated since February 1986.

In the early period, there were apparently two different space station programmes, one each for military (Salyut 3 and 5) and civilian (Salyut 4 and 6) purposes. It seems, however, that such a distinction has been abandoned since the launching of Salyut 7. The Soviets probably believe that it is more efficient and cost-effective to develop new stations that combine civilian and military capabilities.

On 19 February 1986 the Soviets sent into orbit a new space station called Mir. The station, similar in size to the Salyuts, has six docking ports to which modular station segments can be attached. This confirms Soviet willingness to build up huge space complexes by joining successively to the main element several multi-purpose (laboratory, scientific, residential) modules. It is probable that within the next few years the Soviet Union will be able to deploy space configurations that will be permanently occupied by the crews.

Two long-duration missions have already been sent to the Mir station: the Kizim-Soloviev 125-day flight in 1986 and the Romanenko-Laveikin mission, initiated on 6 February 1987. The latter has set up a new manned space flight duration record of well over 200 days in orbit and has successfully conducted several extra-vehicular activities, as well as Earth and space observation and laboratory activities.

All manned missions launched since the beginning of 1987 have used the modernized Soyuz T11 space transportation vehicle, the computational capability of which has been greatly increased in comparison with the previous versions of the Soyuz-series spacecraft. Resupplying and refuelling of the orbital stations while in space is assured by the Progress-series vehicles, first launched in January 1978.

Budget and resources

Only rough estimates are available of the budget and resources which underpin Soviet programmes and of the breakdown between civilian and military applications. Comparisons with the US are also imperfect because of the large private-sector involvement in American satellite operations. One study by the US Congress hypothesized an annual budget of $23 billion in 1985, perhaps as much as 1.5 per cent of Soviet GNP. The same study suggested that some 600,000 people, mainly engineers and technicians, were involved in the programme. Key features of the Soviet economy and society partly account for the lesser emphasis on commercial applications.

Organization

The Soviet military plays a fundamental role within the organizational framework of Soviet space programmes. Although details of the Soviet organizations for civil space development and production are scanty, both military and space programmes are managed by the Main Production Administration for Strategic Missiles and Space Systems on behalf of the Soviet Ministry for General Machine-Building. The Strategic Rocket Forces are responsible for all Soviet space launchings and exercise operational control over the three Soviet launching-sites of Tyuratam (Baikonur), Plesetsk and Kapustin Yar. The two branches of the Air Defence Forces, the VPRO (Anti-Missile Defence Forces) and VPKO (Anti-Space Defence Forces) are in charge of (space-based) anti-missile defence and anti-satellite warfare programmes. Finally, Soviet space personnel training facilities are administered by the Air Force, and the majority of Soviet cosmonauts are military pilots.

On the civilian side, the Space Research Institute of the Academy of Sciences plays an essential role in the conception and coordination of Soviet civilian applications as well as scientific space activities. In 1985 a new institution, Glavkosmos, was established, apparently in order to simplify the administrative structures of civilian space programmes and to enhance their efficiency. Glavkosmos is also in charge of negotiating cooperative space agreements with Western countries. Intercosmos has a similar role as far as cooperation with socialist countries is concerned.

The Soviets disclose little information on the industrial infrastructure of their space activities. It is believed that the Ministry of Defence has a close relationship with, and probably also actual control over, the industries which manufacture launchers and other Soviet space hardware.

Foreign policy dimension

Soviet attitudes towards international cooperation have three major characteristics. First, political considerations predominate in determining the limits of cooperation. Thus cooperation with Western countries develops during periods of detente and decreases when there are growing East-West tensions. Second, cooperation with Eastern Europe and the Third World almost certainly has political purposes. But third, in contrast, the Soviets have been keen to cooperate in such areas as the scientific exploration of space and human space flight.

Organizational arrangements for space cooperation between the USSR and other socialist countries were created within the Intercosmos programme, set up in 1967, and the international organization Intersputnik, established in 1971. Intercosmos covers a broad range of space-related problems in physics, meteorology, communications, biology, medicine and environmental uses. The primary purpose of Intersputnik is to provide a system of space communications for all allies of the Soviet Union. Space cooperation has a role as an integrating force for unity in the Soviet bloc.

The USSR also cooperates extensively with India, in the building and launching of Indian satellites, in manned space flight, and by helping the Indians to develop their own facilities and skills for space monitoring and probably launching facilities. Among West European countries, France has been the most important partner of the Soviet Union in space cooperation. There have been continuous exchanges between the two countries which culminated in 1982 in the participation of a French astronaut in the Soyuz T-6/Salyut 5 mission. A new Franco-Soviet agreement on space cooperation in manned space flight was concluded in October 1986. Soviet bilateral cooperation with other non-communist countries has been minimal for a long time. In 1986, however, a Soviet-British space agreement was reached. Joint Soviet-ESA space endeavours, such as the exchange of data collected by the Soviet Vega 1 and 2 missions and the European Giotto probe, have not yet been formalized.

Soviet-US cooperation is the most dependent on the general evolution of international relations. It was at its apogee during the period of detente, with the joint Apollo-Soyuz flight of July 1985, but more recently it has been seriously affected by the Soviet invasion of Afghanistan and above all by the consequences of SDI. In May 1987 a new Soviet-US space agreement on civilian research was signed, giving American scientists access to the Soviet unmanned probe of Mars and its Moon photos in 1988.

The Soviet Union has been among the major actors in the UN Committee on Peaceful Uses of Outer Space (Copuos). On some issues, such as the Moon Treaty and direct broadcasting, sharp differences with Western countries have arisen. Finally, the Soviets carry on multilateral cooperation in space-related matters within other international organizations, such as Inmarsat, ITU, WMO, Cospar and the International Astronautical Federation.

Security policy dimension

Despite official disavowals, the Soviet space programme is dominated by military concerns. Throughout the 1970s and 1980s, over 75 per cent of Soviet space missions have been defence-related, and many known to have two distinct configurations, and cosmonauts have carried out reconnaissance, C3I and Elint tasks. Like the US, the USSR relies on satellites for political and military reconnaissance, and makes use of space for navigation, Elint and general C3I tasks. Equally, the USSR has deployed a constellation of early-warning satellites. It has also put considerable effort into ocean surveillance in both radar and electronic forms. A Soviet Asat capability has been demonstrated, but to date its success rate has been poor and incapable of reaching the bulk of Western space assets. Finally, while the Soviet Union has not revealed a programme comparable to SDI – indeed, the evidence is that it is apprehensive about matching such systems either technologically or financially – it is developing laser and particle-beam technology and has a well- established ABM programme.

The significance of space for Soviet security, however, must be placed in the context of Soviet strategic thinking and military doctrine. In peacetime, the Soviet Union undoubtedly values space-derived intelligence as highly as the USA. Global crises and military activity are invariably reflected in the pattern of its satellite launches. It has a special interest in tracking Western naval operations and uses satellites for European and global communications. Even the KGB has its own dedicated systems. However, the Soviet Union's primary military interests lie within the Eurasian land mass. In the event of war, even the Soviet Navy will fight mainly from home waters. The bulk of Soviet SSBNs will launch from defined locations and will have less need of satellite navigation systems. Furthermore, communications for tactical and strategic forces will rely more on conventional means of transmission. In short, the USSR would suffer less from any loss of space-based systems than the US and Nato. The Soviet Union's routinized launch capability would provide a useful surge and replacement facility, but, in terms of wartime and military operations, space is auxiliary to mainstream Soviet activity. This is not to say, however, that the USSR would not derive a military advantage from maintaining a superior space capability or, more significantly, by degrading the West's increasingly vital satellite systems.

It is difficult to assess the impact of SDI on Soviet space programmes. On the one hand it is certain that the Soviets are trying to find an appropriate technological as well as a political response to this new challenge. However, there is no firm evidence that they have a space-based strategic defence programme similar to that of the United States.

What the Soviets fear above all is the financial and economic consequences of a new arms race in space, given their scarce resources and Gorbachev's proposed economic reforms. The Soviet Union may continue to oppose US military space policy by diplomatic means, including its proposal to establish a new international organization devoted exclusively to peaceful uses of space. If this proves unsuccessful, it could relatively easily orient its own space activities towards an even more intense military programme, given the already blurred line between military and civil applications.

(c) JAPAN

Since the mid-1950s, Japan has developed a space programme through which it has established itself as a major force in the application of space technology for scientific and economic purposes. Although the programme is still modest by American standards, it is comparable in scope and breadth to the West European space programme. It has the potential to become one of the major space programmes in the world for the commercial exploitation of space. Already Japan regards itself as No. 3 among the leading space powers – behind the two world powers but ahead of Western Europe.

Objectives

The objectives of the Japanese space programme are primarily economic and are part of the country's overall aim of establishing and strengthening its competitive position on the world market for advanced technology. So far, security considerations have had little overt impact on Japan's space policy.

Initially Japan's space programme relied heavily on cooperation with the US and access to American technology. This prevented transfers of space technology to third countries or launches of third-country payloads. However, Japan's long-term objective is to achieve technological as well as political autonomy in space. This reflects the underlying conviction that as a major industrial country Japan must establish an independent presence in space if it wants to remain a competitive player on the world markets of tomorrow and a political power during the next decade.

Programmes and projects

Since 1970 Japan has developed, launched and operated 15 scientific and experimental satellites and, since 1975, 20 application satellites, which are used for telecommunication, broadcasting, television and meteorology. The launch of the first domestic remote-sensing satellite is planned during 1987.

Japan has developed its own space transportation systems: small solid propellant launchers (M-series) to carry scientific and test satellites of up to 750 kg into low earth orbit; and medium-sized liquid and solid propellant launch vehicles (N-series) to carry application satellites of up to 350 kg into geostationary orbit. Most of these launchers have been produced under licence from US companies. A large expendable launch vehicle is under development (H-series) which uses advanced liquid oxygen/hydrogen propellant technology to carry payloads of up to 550 kg into geostationary orbit. The first experimental launch of the H-1 launcher took place in August 1986. The Japanese space programme envisages the launch of an enlarged H-2 version for the transport of payloads of up to 2,000 kg into geostationary orbit (comparable to Ariane 3) by 1990. The Japanese intend to offer commercial launch services on the world market.

Japanese plans include active involvement in the 'international space station' project, and negotiations are under way with Nasa. In the long run, they aim to build and operate an autonomous space station to be used for commercial purposes and served by a re-usable manned space vehicle. The underlying goal is to establish an autonomous capability for using the specific conditions of space for commercial purposes.

As for the ground segments necessary for using payloads in outer space, Japanese suppliers have been especially active in developing highly advanced technology. As a result they have become a major player in the ground segment market, including within Western Europe.

Japan now has a broad and technologically advanced infrastructure for its space programme. It has two major space centres with launch facilities, one for scientific, the other for space application activities, and a number of additional development centres and ground facilities, including a major space centre at Tsukuba Science City.

Budget and resources

Although Japan's space programme is comparable to that of Western Europe in scope and range, the annual budget is much less. This suggests that the programme is more efficient and that the amount of private money spent on space activities is considerably higher than in Europe. Annual public expenditure for 1986 totals about 117 billion yen, or 640 million European Accounting Units (EAU), divided into some 620 million EAU for the space application programme, 10 million EAU for the space science programme, and another 10 million EAU for all other agencies involved in space activities. All in all, there is a high degree of efficiency and cost-effectiveness in the Japanese programme. In addition, Japanese industry spent roughly 250 million EAU on space applications in 1986.

Organization

The fifteen-year space plan, which runs up to the year 2000, and the space budget are established by the Space Activities Commission, an advisory committee to the Prime Minister under the chairmanship of the Science and Technology Agency. The programme is mainly implemented by the National Space Development Agency (Nasda), which has prime responsibility for the practical applications of space developments, and by the Institute of Space and Astronautical Sciences (ISAS), which promotes activities in the field of space science and has prime responsibility for research, development and operation of scientific satellites. Nasda is supervised by the Science and Technology Agency; ISAS is under the supervision of the Ministry for Education. The Ministry for Post and Telecommunications is responsible for the use of communications, and the Ministry of Transport for the use of weather satellites by the Japan Meteorological Agency.

Industry is heavily engaged in the space programme. As early as 1961 it established a Space Activities Promotion Council, which seeks to coordinate the space activities of the major companies and to cooperate with the government through the Ministry of International Trade and Industry. Some 60 companies are members of the Council. More recently the Keidanren, the umbrella association of Japanese industry, and the six largest trade companies have formed the Space Station Promotion Council, which coordinates the interests of more than 230 companies in Japanese participation in the development, operation and utilization of a permanent space station. Since the financing of the space programme has to be justified on economic grounds, the active interest of industry is essential. Most companies are looking beyond the short-term profits to their potential profits from the push in industrial innovation, which they expect to result in the long run.

Foreign policy dimension

The Japanese have relied heavily on cooperation with the United States. The conditions attached to this dependence have impeded

Japanese exploitation of the world market. Consequently national policy aims for greater technological independence from the US, mainly in satellite and launcher technologies. In other respects there is a low profile stance vis-à-vis other countries which avoids giving the impression that Japan is making a strong move into future space markets. A similarly low profile policy is pursued in the UN Copuos. However, in the more technically oriented specialized organizations, such as Intelsat, Inmarsat, ITU and WMO, Japan is more proactive.

Security policy dimension

The early years of Japan's space programme were marked by a self-imposed reluctance to fund technologies that might be perceived as having military applications. This self-restraint with regard to dual-use space technology (mainly launcher technology) was abandoned late in the 1960s when Japan became heavily involved in the development of modern launcher technology under licence to American companies.

Today, Japan's space programme is civilian, though with military options, which could be developed relatively quickly. The government has recently decided to allow Japanese companies and research establishments to participate in the SDI research programme. A formal agreement with the United States was signed in September 1986.

(d) CHINA

The People's Republic of China began developing rockets in the 1950s. Today China belongs to the small club of 'space powers' which have successfully launched their own satellites, using indigenously developed launchers. China is now offering space transportation services on the world market. Swedish and US companies have already signed agreements for the launch of their satellites with Chinese rockets.

Objectives

The objectives of the Chinese space programme are primarily political and military, and geared to the establishment of a strategic deterrence capability, especially vis-à-vis the Soviet Union. However, since the mid-1970s economic policy considerations have played an increasing role. The Chinese leadership is convinced that the promotion of space research and technology will help to increase the country's scientific and technological potential and lead to major improvements in the telecommunications infrastructure, weather prediction and the exploitation of natural resources. All three types of satellite can be used for military as well as civilian purposes.

Programmes and projects

Like those of the US and the Soviet Union, Chinese launchers have been developed from military rockets – thus indicating that China has a significant ICBM capability. The launchers currently at China's disposal – Long March 2 and 3 – are able to transport payloads of up to 2,000 kg into LEO or up to 1,300 kg into GEO. The upper stage of Long March 3 uses a high-efficiency cryogenic fuel. The plans are to increase Long March 3's geostationary booster capacity to 2,500 kg in the coming years and to double launch frequency – at present 3 to 4 launches per year. In September 1981 China launched its first multiple payload: three satellites were launched using one rocket. The launches take place at two launch sites: one for low orbital satellites with Long March 2, and one for geostationary satellites with Long March 3.

Altogether 19 satellites were launched between 1970 and 1986. Payloads from some 8 of these satellites have been recovered, suggesting that these had been on reconnaissance missions. In April 1984 the first television communications satellite (STW-1) was launched into GEO. However, the capacity of this satellite and its successor is not sufficient to satisfy China's needs in the field of telecommunications. Hence the launch of a considerably larger communications satellite (STW-3) is planned for 1987. A further increase in capacity is planned for the 1990s.

The launch of an indigenously developed weather satellite for polar orbit is expected in 1988, and the development of weather satellites for GEO with high resolution is planned for the 1990s.

A manned space programme was initiated during 1979/80, but discontinued because of financial constraints. It has recently been revived with the aim of achieving manned missions in the 1990s.

Budget and resources

There are no official data available on the budget. Western experts have estimated it to be in the order of 3 billion EAU. In 1986 a workforce of roughly 100,000 people was thought to be engaged in the Chinese space programme, a quarter of whom were engineers and researchers.

Organization

The fundamental goals of policy and the budget are determined by the State Council, which consists of the chairmen of 8 State Commissions as well as 36 Ministers under the leadership of the Prime Minister. The Ministry of Astronautics is responsible for the formulation and implementation of the space programme. Three further institutions are directly subordinate: the Chinese Academy of Space Technology, the Shanghai Astronautics Bureau, and the China Great Wall Industry Corporation.

Foreign policy dimension

In its early phase the Chinese space programme was modelled closely on that of the Soviet Union. Since the Cultural Revolution, the Chinese have sought technical and economic cooperation with the West in the exploration and use of space. They also cooperate with some socialist countries, but not the Soviet Union.

China is actively involved in space matters in the UN, in particular in the Copuos and the Conference on Disarmament. Here the Chinese government has claimed a role as the representative of the developing countries. China is a member of the WMO, ITU, Intelsat and Inmarsat.

Security policy dimension

A military orientation permeates Chinese policy in space, and a key purpose has been to support Chinese ballistic and submarine-launched missiles. This accounts for the emphasis on the development of military communications and reconnaissance satellites.

At the same time, however, China actively opposes the militarization of space in the United Nations, most specifically the development and testing of 'space weapons' and the stationing of such systems in space. China's criticism focuses in this context on SDI. It is apparent that behind this political position is the conviction that for technological and economic reasons China itself will not be capable of stationing weapons in space in the foreseeable future.

(e) INDIA

India has been engaged in the exploration and use of space since 1961. The principal characteristics of India's space policy are long-term planning, straightforward implementation and concentration on a few clearly identified objectives. This policy is based on a broad national consensus.

Objectives

The objectives of the Indian space programme are primarily of a socio-economic nature: namely, to make significant contributions to the solution of problems which are characteristic of day-to-day life, such as frequent natural disasters, mass illiteracy or lack of raw materials.

In addition, the programme has political motivations: it is designed to strengthen India's position as the dominant political power in South Asia, as a leading power within the group of developing countries, and as a political rival of China and an antagonist of Pakistan. Therefore, a principal objective is self-reliance, that is, the ability to explore and use outer space with indigenous launch vehicles and satellites without having to rely on technical assistance from other countries.

Programmes and projects

So far the results of India's space programme have been impressive, given the economic and social context in which the programme has been built. The country has developed a small communication satellite (650 kg) which was launched in 1981 by Ariane. The current emphasis is on a system of multi-purpose satellites – Insat – which can be used for telecommunications, television, broadcasting and meteorology. After an aborted launch in 1982, the first of these satellites (Insat 1B), developed and built according to Indian specifications in the US, was launched by the Shuttle and has been in operation since autumn 1983. A successor satellite (Insat 1C) has been ordered in the US and will be launched in 1988. The second generation of these multi-purpose satellites (Insat 2) is to be developed and built by India, and launched in 1990 with a non-Indian booster, and from 1993 with Indian launchers.

In addition, several small and experimental satellites (450 kg) for remote sensing have been developed in India and launched with

Soviet boosters in 1979 and 1981. A larger 850 kg version, called IRS, to be operated for surveys of natural resources and ocean monitoring, is currently in the final stage of development and will be launched in 1987 with a Soviet booster. Two further IRS satellites are due to be launched in 1989 and 1990.

India has a small inertially guided, solid propellant, launch vehicle (SLV-3) to carry payloads of up to 50 kg into low earth orbit. Since 1980 three rockets of this type have been successfully launched. The first launch of an enlarged version (ASLV – Augmented Satellite Launch Vehicle), for payloads of up to 150 kg in 400–500 km orbits, failed in March 1987. A second generation for 1,000 kg payloads into polar orbits (PSLV – Polar Satellite Launch Vehicle) is planned for the early 1990s.

Budget and resources

The Indian space budget for 1985/6 was 2.1 billion rupees, equivalent to about 250 million EAU, which amounted to slightly more than 0.4 per cent of the national budget. The Indian Space Research Organization employs some 11,000 people, about 4,000 of whom are engineers and technicians.

Organization

The Indian space programme is characterized by strong centralization. The Indian Space Research Organization (ISRO), established in 1969, is an integral part of the Department of Space (DOS), an agency directly responsible to the Prime Minister. The head of the DOS is also chairman of ISRO. Coordination with other space-related departments and agencies is carried out at cabinet level through the Indian Space Commission, which is also chaired by the head of the Department of Space. ISRO has four main space centres for development, production, launch and operation of rockets and satellites. The involvement of Indian industry in the programme is increasing, although more than 90 per cent of the development and construction is still conducted in government research centres.

Foreign policy dimension

Space policy plays an integral part in India's foreign policy. It enhances the country's leading role within the non-aligned movement as well as within the group of developing countries. India cooperates with all the major space countries (except Japan), especially with the Soviet Union, and takes an active part in the work of the UN Outer Space Committee, Intelsat, Inmarsat and WMO.

Security policy dimension

So far space policy has had little direct relevance to India's defence and security policy. However, as the programme matures and becomes more sophisticated, India will acquire the option of using space for defence purposes too. In this respect, the main official interest is likely to be in using space assets for reconnaissance and military communications, although, combined with Indian nuclear capability, they could provide the basis for a ballistic missile system. Either way the space programme is an important adjunct to Indian aspirations to be the leading political power in South Asia.

3

THE RECORD: NATIONAL PROGRAMMES

In contrast to the policies of the five countries just reviewed, European space activities have so far been conducted within a double framework. On the one hand, most European countries have been involved in collaborative space endeavours, first in ELDO and ESRO, then, since the mid-1970s, under the aegis of ESA; indeed, they have also conducted several space projects with space powers outside Europe. On the other hand, several European states have simultaneously built up national space programmes serving particular political, economic or security purposes. This structural ambivalence acts as an obstacle to the development of a comprehensive European space policy, though it reflects the political, economic and cultural diversity of Western Europe today.

The following presentation of the national space policies of five European countries – France, the Federal Republic of Germany, Italy, the Netherlands and the United Kingdom – reflects the present state of affairs. Information in the tables and figures at the end of the report fills in many of the details. The panorama of Europe's current space effort reveals a complex intertwining of persisting traditions oriented towards national interests, together with a growing consciousness of the need to strengthen European space policy on a collective basis.

(a) FRANCE

France was the first West European country – and, indeed, the third country in the world, after the USSR and the United States – to initiate a national space programme, from the early 1960s. At the outset goals of national independence in relation to both security interests and prestige certainly played an essential role. Initial efforts, aimed at building up a French launching capability, finally succeeded when, on 26 November 1965, the Diamant A launcher put into orbit the first French satellite, Asterix. This symbolized the French ambition of acquiring an attribute that only the greatest powers could afford: presence in and autonomous access to space. Indeed, involvement in space was another way for de Gaulle's France to affirm its independence vis-à-vis the other world powers, and its ability to succeed in a field apparently reserved for the US-Soviet duopoly.

The development of an autonomous civilian space capability with indigenous launchers was also closely linked to French efforts in the field of ballistic missile technology. Indeed, it benefited from the decision to build up an independent nuclear deterrent. Significantly, the Diamant-A programme, initiated and supervised by the civilian space agency, the Centre National d'Etudes Spatiales (CNES), was conducted by the Délégation Générale pour l'Armement, part of the Ministry of Defence, and used the hardware developed under the validation programme for ballistic missiles.

These goals of the 1960s, which rested on national ambition, have not entirely disapppeared. They are still reflected in the commitment to manned space flight, in the decision to develop the Hermes space vehicle, and most of all in the military programme. However, the French have realized that international collaboration has become necessary to the fulfilment of its space activities for scientific, technological and financial reasons. This realization is evident in the increased commitment to a large number of collaborative programmes.

Above all, commitment to the joint European space effort is an essential aspect of French space policy. This started with membership of ELDO/ESRO, but significantly France played a large part in the 'package deal' constructed in 1973 which led to the creation of ESA in 1975. The French convinced their partners of the need to develop a European launch capability on the basis of their own experience. The French government also offered launch facilities at Kourou in French Guiana as common ESA facilities for European missions. Largely because of its role in developing the Ariane launcher, France is the leading contributor to ESA. In 1986 its eventual contribution to ESA was 2.31 billion francs (337.03 million EAU), or 29.6 per cent of ESA's budget.

Successive French governments have actively promoted the development of advanced space technologies. They have acknowledged the economic spin-offs offered by space activities, and the growing importance of commercial space applications such as communications, direct broadcasting or Earth observation. Within the framework of French science and technology policy, this increased effort in sponsoring space industries is viewed as having had a positive impact on both employment and the international competitiveness of French industries, particularly in the sectors of electronics and computer engineering. Successes in space technology are perceived by the French public as a proof of the country's ability to hold its own in a highly competitive area of activity with great potential.

Consequently, there is wide public support for the national space programme. The public impact of space is undeniably greater than in other European countries, and this reflects its association with national pride. According to an opinion poll carried out in 1985, 85 per cent of French citizens interviewed were in favour of space activities. Public attitudes towards national achievements in space are similar to their approval of the 'force de frappe'. Space thus emerges as a natural extension of the independent deterrent. In addition, national successes in space are used more and more frequently by almost all the political parties as a theme to generate political support.

Programmes and projects

The major part of the French civilian effort in space is today developed on a collaborative basis, either bilaterally or multilaterally, mostly within ESA, but also with other space powers, such as the USSR and the United States.

France initiated the decision by ESA in 1973 to develop a European launcher on the basis of the L3S project of the CNES, the eventual Ariane, a fundamentally different conception from the Europa project, for which each country built one stage. The French are project leaders of Ariane and currently contribute a 59.25 per cent share of the capital of Arianespace from various companies and institutions.

In 1978 the CNES started preliminary studies on the new Ariane 5 launcher and Hermes manned spaceplane configuration. A technical proposal for Ariane 5 was presented to ESA member states in 1984 and endorsed by the January 1985 ESA ministerial conference. The French then pursued their studies of Hermes and invited other European governments to take part. Twelve countries in ESA have agreed to participate, with the French financial contribution initially at 39 per cent of the total.

As for application satellites, France has participated in the development and production of a series of ESA programmes, including ECS, Marecs, ERS-1 and Meteosat. However, two important programmes are conducted on an almost exclusively national basis.

In 1979 the Directorate-General for Telecommunications initiated the Télécom 1 programme, for which two satellites – Télécom 1A and Télécom 1B – were launched in 1984 and 1985 respectively, and are now operational. A third – Télécom 1C – is to be launched in November 1987. The programme has four major functions: digital telecommunications; conventional telephone and TV traffic between metropolitan France and the French overseas territories; video communication links; and communication links for government and military use.

The Spot programme is an ambitious project for Earth observation. The first satellite, Spot 1, was put into orbit on 21 February 1986, and with three further satellites is scheduled to operate for ten to twelve years. A new commercial company, Spot Image (actually a branch of the CNES), was set up to market the resulting images. Belgium (4 per cent) and Sweden (6 per cent) participate in the programme.

France has developed two satellite programmes with the Federal Republic of Germany. First, under the Symphonie programme, two telecommunications satellites were launched in 1974 and 1975. Second, in 1980 France and the Federal Republic signed an agreement to develop a pre-operational direct broadcasting system for each of the two countries, TDF-1 for France and TV-Sat for Germany. The TDF-1 and 2 programme has encountered difficulties linked to the political debate in France on the future of TV techniques and structures. Criticized by some liberal ministers as 'technically obsolete', the programme nonetheless was approved by the Prime Minister in February 1987. TDF-1 is to be launched in January 1988 and TDF-2 perhaps in mid-1989.

Other collaborative ventures include the French-US Argos and the Franco-British Eurostar. The French also participate with Canada, the US and the USSR in Cospas-Sarsat, a new localization and data collection programme for a worldwide search and rescue system.

Scientific programmes remain important, with extensive involvement in ESA missions (such as Exosat, Hipparcus and Giotto) and bilateral cooperation with the US and USSR. An important new French-US project, the Topex-Poseidon oceanographic satellites, is to be launched in the early 1990s. Franco-Soviet cooperation dates back to the mid-1960s, the first agreement having been in 1966. General de Gaulle was the first Western head of state to visit the Soviet Tyuratam (Baikonur) space centre. More recently, the Soviet Vega missions have used French drifting balloons for the study of Venus, and France will take part in the Soviet Phobos and Vesta missions.

France has also carried out two manned space missions, the first one, carrying Jean-Loup Chrétien, with the Soviet Soyuz-T-6/Salyut 5 in 1982, and the second one, carrying Patrick Baudry, with the American Shuttle in June 1985. Under an agreement signed in 1986, a French astronaut will take part in a Soviet Mir space station mission.

Budget and resources

France dedicates a relatively large share of GNP to space activities, more than 0.11 per cent per year for the last few years. The budget of the CNES largely represents the aggregate civilian effort. In 1986 the CNES devoted a total of about 5.848 billion francs (860 million EAU) to its various space activites. Of this sum, 37.99 per cent (2.221 billion francs) covered multilateral collaborative programmes; 21.06 per cent (1.231 billion francs) was spent on national programmes; 12.69 per cent (0.74 billion francs) went to bilateral programmes; 13.91 per cent (0.81 billion francs) was spent on technical support; and 14.35 per cent (0.84 billion francs) on administration. Some 11,500 people are employed by space industries, administration and research institutions. This is over a third of the total European work-force engaged in space activities.

Organization

Space policy-making is in practice confined to a relatively small group of high-level specialists. Political and administrative elites have been neither very interested nor significantly involved. The orientation of space policy is not a matter for nationwide debate, in contrast to how it is in America.

The CNES was established in 1962 as a primarily civilian research agency under the authority of the Ministry for Industry and Research. In 1986 it had a staff of some 2,000 people, of whom 70 per cent were engineers and technical specialists. It is responsible for managing and coordinating civilian space activities and for identifying trends in space technology.

The board of directors of the CNES operates within the framework of long-term space policy set by the President of the Republic and the government. Restricted sessions of the Council of Ministers are held annually or biennially to make major decisions on space policy. The space budget is supervised by the Ministry of Finance, with annual budgetary allocations voted on by the Parliament.

The CNES has a dual role, both policy-formulating and operational, an object of some criticism. Indeed, its legal statutes, adopted in 1961, no longer correspond to the realities of the space business, given the growing importance of commercial and military applications of space in comparison with purely scientific space activities. There has notably been a debate between the CNES and the Ministry of Defence over the division of labour between them and the allocation of responsibilities for developing military programmes. In 1986, a decision was finally taken: the Ministry of Defence's Délégation Générale pour l'Armement (DGA) is now responsible for the execution of all the military programmes; the CNES's contribution is limited to general technical support. Consequently, the current organization of the French space effort has become, in some ways, comparable to the Nasa-DoD model.

French achievements in space could not have come about without the vitality of the major industrial firms. In the launcher sector, experience in the field of ballistic missiles has been crucial, but equally relevant was the infrastructure already developed in the sectors of aeronautics and electronics. This is why the production of space systems is generally done by large companies active in other sectors of industry. For the same reason, the space industry is very concentrated. Two-thirds of French space products are manufactured by four large companies: Aerospatiale (general architecture of Ariane, satellite production), Matra (satellites), Alcatel-Thomson-Espace (satellite electronic equipment) and SEP (propulsion systems). The rest is produced by some 50 companies, mostly small or medium-size firms which specialize in components for space systems.

In 1985 the French space industry employed about 9,000 people: 50 per cent in the field of launchers; 45 per cent in satellite production; and 5 per cent in ground-based stations production. The total annual turnover of the space industry is over 4 billion francs. The major companies collaborate, in both satellite and launcher sectors, with other leading European space firms through consortia arrangements.

Military programmes

The military have been involved in French space programmes from the outset, but specific military programmes are much more recent. The evolution of both strategic concepts and modern warfare has given a growing importance to space as a new dimension of the military balance between the world powers, especially since SDI. The first French reaction to SDI was to reject the US project of an anti-ballistic-missile space shield as both unrealistic and destabilizing, especially for Western Europe. At the same time SDI served to trigger a realization in France of the military importance of space, reflected in President Mitterrand's speech in The Hague of February 1984, in which he claimed that a 'European space community would be the response best adapted to the military realities of tomorrow'. In spring 1985 France proposed a European civilian R&D programme – the Eureka project – some fields of which were similar to those covered by SDI. This attitude reflected conflicting desires to oppose SDI as a strategic concept, yet to maintain technological competitiveness vis-à-vis the Americans. The French government did not prevent French companies from participating in SDI industrial contracts. The new government elected in March 1986 was at first more favourable towards SDI, but has in practice maintained the same broad policy stance.

France remains opposed to the extension of the arms race to outer space by placing weapons in orbit, but recognizes the need for somc military space programmes to underpin its defence capabilities. French space arms control policy has been marked by two proposals. In 1978, at the first special disarmament session of the UN General Assembly, President Giscard d'Estaing made a proposal for the establishment of an international satellite monitoring agency (ISMA). In 1984, another proposal, this time for the limitation of anti-satellite systems (essentially by the prohibition of any high-orbit operational Asat and by the introduction of a series of multilateral guarantees protecting satellites other than those of the United States or of the Soviet Union), was formulated at the UN Disarmament Conference in Geneva.

The military effort has two main functions: first, to increase the credibility of the French nuclear deterrent, mainly by reconnaissance and early-warning capabilities; and, second, to support France's military presence and intervention capability in the regions of the world considered essential for the country's security (in particular Africa, the Indian Ocean and the South Pacific). The most important space applications for the military are thus communications and intelligence.

The Syracuse 1 network for military telecommunications is based on the civilian Télécom 1 satellite telecommunications system. It covers approximately one-third of the Earth's surface, and has been in operation since 1984. Ground- and sea-based stations guarantee the inter-connection of the space segment with other military communication networks. Syracuse 1 will be replaced by Syracuse 2 in the early 1990s. The present Syracuse 1 system has cost about 1.36 billion francs.

The French have sought to reduce the present dependence on military intelligence provided by US satellite systems; hence the research effort since the late 1970s to build up their own satellite reconnaissance system. This first project, Samro (Satellite Militaire de Reconnaissance Optique), was a slightly modified military version of Spot. In 1982 negotiations to promote a Franco-German satellite for military reconnaissance were initiated on the basis of the Samro concept, but proved abortive.

In February 1986 the French government authorized the development of Helios, a modified project for an optical reconnaissance system, with three or four satellites using the multi-mission space-based platform developed for Spot. The Ministry of Defence will be responsible, with inputs from the CNES, and hopes to operate Helios from 1993. Initial cost estimates were 12 billion francs (1.7 billion EAU), now revised down to 8 billion francs (1.1 billion EAU). The French authorities are open to collaboration on Helios with other West European countries. The Italian government decided in February 1987 to join the project and share 15 per cent of the costs, and the Spanish may follow with a 5 per cent share.

The decision on Helios dramatically increased the French military space budget between 1986 (697 million francs or 102.5 million EAU) and 1987 (1,707 million francs or 251 million EAU). To coordinate current and future military space activities, the Ministry of Defence set up in March 1985 a Space Study Group (Groupe d'Etudes Spatiales).

(b) GERMANY

The Federal Republic of Germany has engaged in the research and exploitation of space since the early 1960s. From the outset German policy rested on a resolve to create a joint European approach and to foster transatlantic cooperation. Military goals were not pursued. Until the late 1970s German space activities were almost entirely funded from the government's research and technology budget, with particular support for basic research and the development of areas promising solid economic returns. In the 1960s and 1970s, German space activities leaned rather strongly towards the US, especially during the critical period before ESA emerged. German-American space collaboration reflected technical factors and personal links, but it was also politically motivated. It reached its culmination with the federal government's decision to take the lead in developing the European Spacelab for the then new US Space Transport System (Shuttle) and to carry more than half of the costs.

Since the early 1980s the points of emphasis in Germany's space commitment have gradually, but clearly, shifted. Interest in the further development of an indigenous European space transport capacity, including manned systems, has increased. Cooperation with the United States is still endorsed, but rests on satisfying aspirations for a much stronger European role. This is clearly reflected in the negotiations over European participation in the space station project proposed by President Reagan. Discussion about German participation in European space-based systems for verification and reconnaissance has been reopened. The Ministry of Posts, the Weather Service and the media have now all begun to devote resources to space systems for their own use. Medium-range planning for the future envisages a substantial increase in state funding of space research, development, and operations. An internal reorganization of German activities is being discussed.

For a long time German involvement in space, and all that that implies, was more often passively accepted by successive governments than actively promoted. It is only recently that the developments in certain fields of high technology, including space technology, have received increased attention as an element of foreign policy. The decisions to strengthen the scientific programmes of ESA and to support Ariane 5, Columbus and Hermes all confirm this trend. The planned increases in the space budget calling for a simultaneous increase in the proportion of the budget allotted to European projects underline this new pro-European accent.

The parliamentary parties in the Bundestag have, with the exception of the Greens, spoken supportively on space matters but not enthusiastically. Space issues have not been controversial apart from the recent debate over SDI. The media in Germany were often more reserved and tended to take a negative view of projects apparently undertaken for reasons of prestige. This attitude has recently combined with a growing scepticism towards exotic technologies. As a result, the media have frequently portrayed a distorted image of German space activities. In clear contrast, German public opinion has repeatedly demonstrated a broad and positive interest in the opening of the space frontier and Europe's association with this, a trend that is likely to continue.

Programmes and projects

German space activities span the entire spectrum except military applications, and include the full capacity to organize and conduct manned missions, as the D-1 mission in autumn 1985 demonstrated. The first priority remains the scientific investigation of space, utilizing sounding rockets, satellites, space probes, and also the Spacelab system. German scientists have shown particular interest in the physics of the solar system and the clarification of solar-terrestrial relationships. The two Helios solar probes, which during their active life from 1974 to 1985 covered an entire solar cycle, provided important information in this field. German participation in the research satellites of ESA (and previously ESRO) made no small contribution to their success, as was demonstrated again, recently, by the camera and the particle experiment of Giotto. The Rosat project will be a major step forward in the systematic investigation of X-ray sources in space. It is already at an advanced stage with some American and British involvement. German institutes and occasionally German firms have made numerous contributions to other, especially American, space research projects.

As for Earth observation, strong German participation in the European Meteosat programme was followed by a substantial role in the ERS programme of ESA, including the industrial prime contractorship for ERS-1 and the provision of various optical and microwave sensor systems. Earth observation data are processed and evaluated on a large scale by the German DFVLR. In addition, a small but growing number of private enterprises are active in this field, and there is close cooperation with developing countries.

Germany is among the European pioneers in the use of satellites for telecommunications. The two Symphonie satellites, built by German and French industry, served for almost a decade in a range of experiments in telecommunications and transmissions. Subsequently the federal government took the initiative of building the first direct TV broadcasting satellite (DBS), and in 1980 France joined the system. The German TV-Sat is due to be brought to its GEO when an Ariane launch is available, and will be followed by TDF-1, its French twin, within a few months. The installation of an operational DBS system and a second TV satellite are planned. In addition, the German Postal Service has made considerable progress in developing its own satellite system – DFS Kopernikus – for telecommunications and TV distribution; it is due to operate in 1988. German industry is substantially involved in the manufacture of the ECS satellites for the Eutelsat system and the Marots/Marecs satellites for Inmarsat. It has also supplied subsystems for the fifth and sixth generation of Intelsat satellites. In cooperation with the DFVLR, German industry has made a most impressive contribution to the development of a satellite-based worldwide search and rescue system. A large-scale participation is planned in European systems for radio transmission between satellites and with space stations and platforms in low orbits.

Research on weightlessness has been of particular interest, especially since the decision to support Spacelab, and many studies and experiments have been carried out, including parabolic flights, sounding rockets and the German-built Shuttle Pallet Satellite. The majority of the European experiments conducted during the first Spacelab flight in 1983, with the German-born ESA astronaut Ulf Merbold, were prepared in German institutes. The Spacelab flight D-1 in autumn 1985, with ESA astronaut Wubbo Ockels and the German scientists Ernst Messerschmid and Reinhard Furrer, was entirely organized and controlled from the DFVLR control centre in Oberpfaffenhofen. Another mission, D-2, is now in preparation, but launcher problems will probably delay it until 1991. The Intospace company has been founded with the express purpose of stimulating industrial interest in such research work and providing advice.

For ten years (1973–82) Spacelab was the most important and most expensive project on space technology in the Federal Republic. Its development supplied an extraordinary amount of information and experience. The laboratory proved its worth under operating conditions. However, its use is now severely crippled by the Challenger accident. Led by German industry and with a German contribution of 38 per cent of the costs, ESA is now developing the unmanned space platform Eureca, designed primarily for experiments on weightlessness. Germany is taking the lead in the preparation of the Columbus programme.

Germany has been involved throughout the development of European launcher systems. It provided the third stage for the Europa launcher. The German share of 20 per cent in the Ariane programme was the second largest and will be slightly larger in the development of the new Ariane 5. After protracted debate the Federal Government eventually decided to participate in the Hermes preparatory programme with a share of 30 per cent. Recently a German proposal was put to ESA for a new generation of space transporters, the two-stage Sänger II. German interest in European space transport systems is increasingly evident.

Budget and resources

Germany has excellent foundations for an active space role in the field of science and a sound industrial capability. Government funding has been continuous and indeed higher than the European average, albeit lower than that of France and far lower than that of the US. In 1986 the Ministry of Research and Technology spent a total of 898.5 million marks (436 million EAU) on space research and technology; of this, 559.2 million marks (271.5 million EAU), over 62 per cent, went to ESA. Of the remaining sum, 247.7 million marks (120 million EAU) were earmarked for national research and development purposes, including the German contribution to bilateral and trilateral projects. A sum of 91.6 million marks (44 million EAU) was allocated to the relevant expenditures of the DFVLR. A further increase in the total expenditures is planned in the medium term.

In addition, the 1986 budget of the German Ministry of Posts included 777 million marks (377 million EAU) for space-based activities, and the German Weather Service appropriated 38 million marks (18.5 million EAU) for its participation in Meteosat. No data are available on the space expenditures of the German radio and television networks, publishing houses and other organizations, nor on expenditure by German companies. But we can assume that the respective amounts are relatively low as compared with total expenditures.

Organization

Since 1962 governmental responsibility for space research and the development of space technology has been in the hands of the Ministry of Research and Technology, though the Ministry's staff working in this field is quite small. It is supported by a larger working-group within the DFVLR. The DFVLR also runs the German Space Operations Centre, test facilities for rocket engines, a training centre for astronauts and special institutes for Earth observation, telecommunications and weightlessness. Space research in the strict sense is carried out in several institutes of the Max Planck Society and several university institutes. Furthermore, one of the three largest European space test centres, with extensive simulation and experimental facilities, is situated in the Federal Republic near Munich in the 'Industrieanlagen-Betriebsgesellschaft' (IABG). ESOC, ESA's Satellite Operations Centre, is located in Darmstadt. The headquarters of Eumetsat will soon be established in the immediate vicinity of ESOC. Finally, the German postal service has its own facilities for communication with and operation of satellites.

Taken as a whole the organization of space policy has thus far functioned without mishap, though critics maintain that it has often been far from optimal. Its status within the Ministry of Research and Technology as one among some dozen sectors causes increasing problems, since this neither does justice to the importance of the field nor does it guarantee adequate impact vis-à-vis Nasa, ESA, and CNES. However, there has at least been the advantage that German space activities have not suffered from excessive bureaucratization. Meanwhile, increased political interest at home and new developments in neighbouring countries, in particular the UK and Italy, have led the federal government to study options for reorganizing space activities. A preference is emerging for a more independent management agency.

The primary producers of space systems in the FRG are the aerospace industry (Dornier and MBB-Erno), the electrical and electronics industry (eg., AEG, ANT, SEL, Siemens), companies active in the fields of optics, precision mechanics, scientific instruments and related branches (eg., Kaiser-Threde, Teldix, Zeiss) and some specialist firms of mechanical engineering (eg., MAN). Some 6,000 highly skilled employees are directly involved in space activities ranging from the systems management of large projects to the delivery of single small elements. Greater industrial collaboration within Europe may diminish the problem of continuing dependence on US sources for key components.

Foreign and security policy dimensions

The European and transatlantic dimensions have consistently pervaded German space activities and made imperative the inclusion of foreign policy interests in German space policy. By the same token space activities influence German foreign policy. The strong Western orientation of German space policy for a long time prevented cooperation with the German Democratic Republic, other East European countries and the Soviet Union, and even now links are few and largely confined to space science. Within the West European framework, Franco-German cooperation has taken on a special quality, made concrete in the Symphonie satellite programme and the TV-Sat/TDF-1 programme. Together France and Germany have steadily provided more than half of the funding for European space cooperation within ESA. With regard to the two large development programmes of the 1970s and early 1980s – Spacelab and Ariane – their joint contribution was even higher: more than 70 per cent in the case of Spacelab and more than 80 per cent for Ariane. A similar though somewhat more equal distribution of financial burdens will apply for Hermes. Yet the GNPs of the two countries have always amounted to less than 45 per cent of the total GNP of all ESA member states. Indeed with the increase in membership of this organization, the figure has now dropped to 40 per cent.

So far the Federal Republic's activities in space have been limited to the civilian sector. In the early 1980s Germany, after accepting the French invitation to look at a common military reconnaissance system, then decided not to pursue it. Discussions about a variant of such a system are only just beginning. German industry is only minimally involved in Nato's military communications satellite. Under the American SDI programme, some small research contracts have been given to German industry under a bilateral memorandum of understanding, but these seem to offer little scope for expansion.

(c) ITALY

Italian interests in space have been largely determined by two factors. First, space is perceived as a competitive field of rapid technological development in sectors such as telecommunications, propulsion and remote sensing, and thus crucial in industrial and strategic terms. Second, Italy is keen to maintain its position as a full member of the family of technologically advanced nations. But the full array of space activities, some of which have uncertain or at best long-term returns, has appeared beyond the reach of the financial, industrial and scientific resources available at the national level. The guiding philosophy has thus been to rely on international collaboration: European first, then bilateral or multilateral with the US, and more broadly multilateral. This is evident in the structure of the National Space Plan (PSN), formulated since 1979 under the direction of the Ministry of Science and Technology. The PSN marked a shift from the planning of individual projects to a wider approach, sector by sector.

The outcome of this pragmatic approach is not unsatisfactory, especially in relation to the level of expenditure. First, a cooperative framework, stimulated by both private and public sectors that are particularly dynamic in this field, has meant that telecommunications have emerged as the *de facto* priority. Second, Italian industry and science have performed quite satisfactorily in a number of joint programmes within ESA and other international cooperative bodies: propulsion systems and space modules are good examples. Third, all this has been accompanied by diplomatic achievements, notably the success of the 1985 ESA ministerial conference in Rome.

Programmes and projects

National activities in the field of telecommunications started with the launch of the Sirio satellite programme on 25 August 1977. Today the most important telecommunications programme is represented by Italsat, the first domestic pre-operational satellite, with 11,000 new telephone channels and direct broadcast and data relay. It will use high-frequency bands for specialized services such as video-conferencing, high-density telephony and rapid facsimile transmission between computers in full network conditions. Italsat is scheduled for launch in 1988 and the programme will cost 595 billion lira. Outside the PSN budget, and under the auspices of the Ministry of Posts, two more satellites – the Sarit D for direct broadcast only and the Sarit M with telecommunication channels – are under consideration. Italy was fourth in the international market for ground stations from 1965 to 1982, with a 5 per cent market share (the United States had 39 per cent, Japan 37 per cent and France 13 per cent). New ground stations will be set up in the Italsat programme.

In remote sensing, the main Italian effort is in receiving stations and data processing. Techniques for processing and analysing remote-sensing data, methodologies aimed at producing maps of parameters for general application interest and new architectures of remote-sensing data pre-processing are being developed through the PSN. The PSN also distributes to Italian users the data supplied by Landsat, HCMM, Seasat, Nimbus and Spot satellites. In sensor technology Italy is cooperating with Germany in a project to develop a Synthetic Aperture Radar, SAR-X, at a cost of around 40 billion lira. This new instrument was scheduled to fly on a Shuttle mission in the 1990s.

The national weather service run by the Italian Air Force routinely uses satellite information, especially from Meteosat. A national network has been created to improve the utilization of satellite data and is centred at the Primary Data User Station (PDUS) in Rome. Five secondary data user stations are in operation.

Italian activities in space geodesy and geodynamics are carried out in cooperation with the United States and other countries. A ground station for satellite laser ranging has been installed at Matera in the Mezzogiorno, which (through a satellite network) will measure the motion of the Earth's surface. Italy will collaborate with the US in building Lageos 2, a new passive satellite, equipped with retroflectors.

The main PSN project in the field of propulsion systems is represented by the Iris programme (Italian Research Interim Stage). This is to be used in cooperation with the Shuttle to place into geo-transfer orbit payloads of the order of 900 kg. Iris consists of two main modules: the ISS (Iris Spinning Stage) and the ASE (Airborne Support Equipment). The former is an expendable propulsion module equipped with a solid rocket motor, whereas the latter is a re-usable module which is needed to operate the propulsion module. But Iris will not have the capacity independently to launch even small satellites. Two studies are under way at the moment: one to investigate an entirely Italian launcher; and a second to explore a joint venture with Volvo, perhaps using Ariane hardware.

As for scientific research, the PSN envisages two satellites: the Sax (X-ray Astronomy Satellite) and the TSS (Tethered Satellite System). The Sax is due to explore the universe in the radiation band between 2 and 200 kev, with a Shuttle launch scheduled originally for 1989. It will cost 172 billion lira by 1988. The Italian share in the TSS programme is expected to cost 82 billion lira by 1988. The TSS should be launched in 1987. In 1983 Italy performed five experiments on the first Spacelab flight. Another programme which deserves mention is the Italian/Nasa San Marco project for scientific research in the upper atmosphere.

Technological research is currently being carried out in the following fields: electronic technologies, space telecommunications, robotics, chemical and electronic propulsion, thermal control, and in-orbit altitude and control. In addition, Italian industry has recently entered the field of software research.

Launching sites at the San Marco range and the Malindi station in Kenya are managed by the Centro Ricerche Aerospaziale (CRA) of the University of Rome and by the Italian Air Force. A series of launches have been performed in cooperation with the US and with other countries during the past twenty years. The base has also been used for the San Marco scientific programme, and current plans envisage its continued use.

Military activities include telecommunications satellites, for which the Italian Air Force has been developing the AM-136 programme since 1980. This has a capacity of 12,000 telephone channels and will be used for a wide range of services. Its coverage will be centred on Italy and the Mediterranean area and will be oriented towards communications with mobile (aircraft and ships) and fixed stations (ministries, airports, harbours, etc.). Two satellites, the Sicral-1A and the Sicral-1B, are expected to be active and will be used not only for military and police services, but also for civilian emergencies. AM-136 will be put in GEO in 1991, at a cost of 1,006 billion lira. Italy has also decided to participate (15 per cent) in the French surveillance satellite Helios.

Italy is heavily involved in the major ESA projects. In the Olympus programme, it has prime responsibility for payload, structure, integration and test of propulsion equipment, with a total share of over 30 per cent. In Columbus, Italy holds prime responsibility for the pressurized module and a 25 per cent share of the total cost. In Ariane 5, Italy holds prime responsibility for the large solid boosters and will develop the turbo-pump for the HM-60 cryogenic motor, a participation level of 15 per cent. Italy participates significantly in Eureca (17.3 per cent), ERS-1 (10.61 per cent), the ECS programme (14 per cent), and the Ariane 3 (17.55 per cent) and Ariane 4 (7.75 per cent) launchers. As for Italian participation in ESA scientific research programmes, from a total of 240 experiments proposed by the European scientific community, 15 were Italian, 7 of which have been selected.

Budget and resources

The budget for space activities has increased steadily over the past few years. The budget for national activities in the period 1986–91 is expected to be approximately 1,870 billion lira. Over the same period Italy will spend 1,248 million EAU (about 1,720 billion lira) on ESA activities. Therefore, the overall aggregate annual budgetary outlay should reach the level of 600 billion lira a year. This amount is still considered quite low by some, while others assert that the capabilities of the space industry are already overstretched. They point out that to expand the space industry would necessitate the training and specialization of more personnel, which would take some time.

The Italian government established the PSN in 1979 in order to stimulate the space industry and to increase its level of participation in ESA initiatives. The Plan also now supports some activities (such as TSS and Iris) in sectors which are not covered by ESA programmes. From 1980 to 1986 the PSN distributed its funds by giving over 35 per cent to telecommunications and about 18 per cent each to propulsion programmes and scientific research. Approximately 86 per cent of all the contracts allocated from 1980 to 1986 by the PSN went to industry, while 9.6 per cent were allocated to the National Research Centre (CNR) and to various universities. Projected PSN spending for the period 1987–91 includes a smaller share for telecommunications (30 per cent) and propulsion (7 per cent), but increases funds for scientific research to 30 per cent.

Italy is the third largest contributor to ESA programmes. The economic return has been low in the past but is now increasing, and a recent assessment estimates the return coefficient to be around 0.80. The Italian aim is to reach 0.95. On the other hand, collaboration within the ESA framework is considered largely satisfactory, in so far as ESA programmes emphasize sectors which are consistent with Italian interests and priorities.

Organization

Italy is an active member of ESA, but at the same time is developing national and bilateral programmes. Relations with ESA are handled directly by the Ministry for Scientific and Technological Research. The CNR manages space activity at the national level, for which it has established an *ad hoc* structure, the Servizio Attività Spaziale (SAS).

The CNR organizes, coordinates and itself finances scientific research, and as a public body it reports directly to the Prime Minister's Office. The day-to-day direction and supervision of the CNR were made the responsibility of the Ministry for Scientific and Technological Research under a decree of 15 September 1979. The Ministry also has the task of coordinating national and ESA activities.

Italian space policy is formulated by the Ministry of Scientific and Technological Research, whose proposals are in turn reviewed by the Interministerial Committee for Economic Planning (CIPE), which assesses the programmes and the budget. Parliament must take the necessary steps to translate all this into law. The Ministry for Scientific and Technological Research has no financial resources of its own and must go through other ministries in order to obtain funding. ESA contributions, for example, are appropriated by the Ministry of Foreign Affairs. The PTT and the Ministry of Defence also deal with space activities.

Within the space industry Aeritalia, Selenia-Spazio and SNIA-BPD are the three most important companies. They are the prime contractors for several ESA and national projects. In particular Aeritalia, which belongs to the IRI group, is the prime contractor for the following programmes: TSS with Nasa, Lageos, Iris, AFPM (autonomous fluid physics module) and Hipparcos. Selenia-Spazio, a company of the Selenia-Elsag Group in IRI-STET, is the prime contractor for the development, production and integration at the systems level of the Italsat, Sarit and AM-136 satellites. It is also responsible at the systems level for the Olympus satellite. In the private sector SNIA-BPD is the largest company. It belongs to the Fiat group and is the prime contractor for solid propulsion used on launchers of the Ariane family, and is also responsible at the systems level for the expendable stage of Iris. In the field of liquid propulsion SNIA-BPD is responsible for the propulsion system of Olympus. Other state-owned, private and multinational companies are also involved in producing components and developing projects.

The number of employees involved in the space industry has steadily increased in the past few years. Between the years 1980 and 1984 the number of employees increased from 1,000 to 3,000, and it is expected to reach 5,000 by the end of 1987.

(d) THE NETHERLANDS

For a country like the Netherlands, there are, in the main, two reasons for devoting time, energy and, especially, money to space research and space technology: first, to maintain a level of scientific excellence in a field where a tradition of national proficiency exists, and, second, to contribute to long-term economic survival through participation in a basic innovative enterprise which fits Dutch industrial capabilities.

Since the days of Christian Huygens, Dutch astronomy has always flourished. Many of its practitioners have earned the respect of their colleagues in other countries. For centuries, Dutch astronomers have found friends and sites abroad for their programmes, as was illustrated again in June 1985 with the inauguration of the international observatory at La Palma. Leading Dutch astronomers like Van de Hulst and De Jager played a substantial role in the preparation of European space research programmes.

Under the combined stimuli of the International Geophysical Year (1957–8) and the launching of Sputnik, the Royal Academy of Sciences in 1959 came to the conclusion that coordinated national activities in the fields of geophysical and space research were necessary. An organization was established for the selection of national space experiments and for the acquisition of launching facilities for these experiments. It was soon realized that Dutch participation in the emerging forms of international cooperation were of the utmost importance: first, to give national industry a chance to be involved in this kind of advanced technology; and, second, to enable the users in the country to make use of the facilities in space.

The Netherlands was in 1962 one of the founder members of ESRO and ELDO. The Dutch started to reap the benefits of this international cooperation when, in the mid-1960s, Estec was established on Dutch soil, in the Noordwijk area. Estec is still the hub of ESA's research and technology.

Major projects

The Netherlands has, so far, participated in the setting up of nine scientific satellites: four in the American Nasa programme and five in the ESRO/ESA programmes. In 1966 Dutch astronomers and representatives of the industry submitted a plan for a Dutch astronomy satellite (ANS), in order both to further science and to strengthen the technological and competitive position of Dutch industry. At about the same time, the PTT and industry proposed the building of a national ground station for satellite communication (Burum 1). The construction of both ANS and Burum 1 made it necessary to increase government outlays for space research and technology and, consequently, to develop a coherent space policy for the Netherlands.

ANS was built within a period of five years by a consortium of Fokker-VVFW and Philips. Its successful launching on a Nasa

Scout rocket in August 1974 prompted the NIVR to re-examine the question of whether another predominantly national satellite could and should be built and launched. On the basis of proposals made by Dutch industry and in cooperation with Nasa (and later with the UK Science and Engineering Research Council, SERC), the NIVR developed plans for a satellite intended to observe the infra-red radiation of celestial bodies, the Infra-Red Astronomical Satellite (Iras). In 1976–7, the Netherlands government approved the Iras plans and authorized the NIVR to sign a memorandum of understanding with Nasa and SERC. Iras has become an outstanding success and has contributed considerably to the international reputation of the Dutch space industry and science. Within a timespan of ten months, the satellite has produced a wealth of astronomical data.

Iras was launched on 26 January 1983. It was built by US and Dutch industry; the UK provided the ground station and its operation. The Netherlands built the spacecraft and was responsible for systems integration and launch preparations. The US provided the infra-red telescope system, the launch vehicle and the ground data-processing.

Organization

Dutch space policy is developed by the Interdepartmental Committee on Space Research and Technology (ICR). The need for an executive agency which could implement this policy was fulfilled by extending the charter of the Netherlands Institute for Aircraft Development, which in 1969 was renamed the Netherlands Agency for Aerospace Programmes (NIVR). The industry-oriented NIVR is a foundation, similar to the science-oriented Space Research Organization of the Netherlands (SRON). Both organizations are represented in the ICR in an advisory capacity.

In the industrial sector the Fokker Aerospace Division has developed as the main Dutch contractor, operating either on its own or in cooperation with firms such as Dornier, ERNO, British Aerospace and Matra. It specializes in solar panels, structural, thermal and control systems, and robotics. Holland Signaal Apparaten has provided subsystems for international and ESA projects in the fields of data-processing and telecommunications. AT&T and Philips Telecommunications develop and construct small telecommunications ground stations. The National Aerospace

Laboratories (NRL), the Technical Research Organization and the Technical Physics Service, as well as a number of smaller organizations, are also involved.

In the early 1970s the Advisory Council for Science Policy suggested that the government limit its predominantly national experiments to ANS and otherwise seek international cooperation wherever possible. It also warned against fixing the financial contributions to European and other programmes at too low a level, since this would result in insignificant contracts for the industry. Without a tangible national effort the Netherlands would not long remain a worthwhile partner.

In 1973 an international firm of management consultants, at the behest of the Minister of Economic Affairs, confirmed that activity at the prevailing levels would result in real but modest gains. Industrial policy would have to emphasize electronics and aircraft construction. The firm suggested that the Dutch contribution to European programmes should be fixed as a proportion of national income and recommended approximately 5 per cent. This recommendation was not followed; the Dutch contribution was fixed at less than half that percentage.

In 1978 General Technology Systems of the United Kingdom were asked to investigate the implications of space technology for the Netherlands. The conclusion of GTS was that it would be necessary for the Dutch government to increase its space budget from around 80 million guilders a year in 1977 to 130 million a year in 1985 (1977 price level) in order to lift and maintain Dutch industry at a level comparable with European countries such as Germany, France, the United Kingdom and Italy. In other words, expenditure on space should be at the level of these countries when allowance was made for the difference in population and GNP per capita. Again, the recommendation was not followed; the Dutch contribution to European space programmes has fluctuated between 3 and 4 per cent, instead of being a percentage based on national income, which would have resulted in a 5–6 per cent share.

In May 1982, the Dutch government published a White Paper on space policies in the 1980s. The gist of it was that, while the government saw no immediate possibilities to reduce, let alone to terminate, its financial support for space research and technology, it looked to the producer industry and 'users' (i.e. ministries such as those of Transport, the Environment and Agriculture, and the business community) gradually to take over the role of major investors. Such a policy would be aimed primarily at practical applications of space for the various potential users in and outside the Netherlands. At the same time, the position of products and services of Dutch industry would be improved in both the European and the world market. The necessary technological developments would be fostered. The government's thesis, based on 1979 data, was that the Netherlands was not lagging unreasonably behind other countries, as far as outlays for R&D were concerned. Others argued, however, that the Dutch scientific and industrial infrastructure for space is more comparable in character to the French and the German capabilities than to those of the smaller countries, the space expenditures of which are similar to Dutch spending in terms of percentage of GNP.

The White Paper stated that the main emphasis in the following years would be placed on applications in the ESA context, such as communications and remote-sensing satellites. The Dutch contribution to Olympus would amount to 135 million guilders (12 per cent). The Netherlands would also contribute 4.5 million guilders towards the costs of ERS-2. National industry would benefit from these contributions. The White Paper also announced that the Netherlands would continue to participate in ESA's scientific programme, in particular research into microgravity, metallurgy and fluid dynamics. Experiments had been conducted in those areas in the German Spacelab mission in November 1985 by the first Dutch ESA astronaut, Dr Wubbo Ockels.

The Netherlands has little reason to be dissatisfied with its ESA membership. In terms of geographical distribution or the percentage of return on investment the country has, for the years 1984–7, a return coefficient of 1.16 in contracts placed. The total return coefficient for the period 1972–87 was 0.96. In 1986 the Netherlands contributed 35.9 million EAU spent on 3.5 per cent of ESA's income, as against 10.2 million EAU spent on national activities. ESA spent 114 million EAU or 9.9 per cent of its total disbursement in guilders.

Further discussion on Dutch space policy was stimulated by the preparation for and the results of the ministerial conference of ESA, held under Dutch chairmanship in Rome in January 1985. Since the conference decided to expand the activities of ESA, with a planned increase in expenditure of 60 per cent, the Netherlands had to look critically at what this implied. In the mandatory programmes it will continue to pay according to its GNP, but a decision has to be taken on the level of investment in the other programmes.

Policy dilemmas

The central dilemma for Dutch space policy today is that the expectation, expressed in the 1982 White Paper, that the 'users' would gradually take over the role of major investors has proved to be an illusion. There is a general appreciation of the fact that the country's economic future may stand or fall with technological innovation. But since a nation with limited resources is unable to do battle on all fronts simultaneously, a choice has to be made. Only if one takes a very imaginative long-term approach is it possible to argue that innovation derived from basic research, not just the improvement of existing technologies, offers the richest prospects of economic survival. But which basic innovation, in which field? The Netherlands has already, more or less, opted out of large computers, nuclear energy and technology for users. Space research, together perhaps with biotechnology, is one of the few remaining openings.

The Dutch government asked the NIVR and SRON for advice on these issues, and they responded in June 1985. The Ministries of Economic Affairs and of Education and Science organized in April 1986 a formal hearing with industry and with the scientific and operational 'user' communities. In the summer of 1986, the government decided to increase its space policy budget to 187 million guilders by 1990 (1990 price levels). Not all of this money will go to ESA; some will be used to finance 'collateral' national activities. Although this represents an increase of 20 per cent on current spending, it is below the 220 million guilders advocated as a minimum by industry, and well below the 235 million guilders (1984 price level) recommended by the NIVR. In all probability, therefore, the Dutch share in ESA's overall programme will continue to drop.

A plausible assumption is that the country will keep up its financial commitments under the mandatory programme. This will amount to approximately 59 million guilders in 1990. Contributions to new ESA programmes could, by the same year, total some 86 million guilders. This would leave approximately 42 million guilders for national activities (SRON, NRL, remote-sensing/telecommunications and Sax).

A point of some interest is that the Dutch Ministry of Defence has, for the first time, committed itself to contribute (for the time being in a modest way) to the national space budget. The reason it gave was that it considers itself a significant future user of high technology (telecommunications, Earth observation and/or position-finding).

(e) UNITED KINGDOM

The British government established in November 1985 a new British National Space Centre (BNSC), whose task was to identify a coherent national strategy for the exploitation of space technology. This marked a major reorientation in the UK and appeared to signal a higher priority for space policy than had been the case for many years. Britain was into the space business early on, but difficulties of a technical and financial kind, combined with a lack of robustness about the policy area in general, led to a long period of hesitation and fragmentation of effort. During the 1970s, however, the British government gradually became more involved in international and European collaborative efforts.

In the meantime two other factors served to maintain a British foothold in space activities. First, there was a recurrent commitment to developing defence-related applications of space, some in close association with the United States. Second, several British companies sought to establish themselves as prime contractors not only for British defence and science projects but for various international and European projects, particularly in the communications satellite field. But outside the highly specialized policy community the exploitation of space inspired little enthusiasm. Space policy did not fire the public imagination; it was not seen as glamorous or as a necessary and cost-effective attribute of a serious actor on the international stage. In other words, it lacked the motive forces which helped to drive the American and French programmes. The new British National Space Centre was designed to secure a basic change of approach.

The British had been among the early pioneers in the potential military use of space. The development of the Blue Streak was an ambitious and technically demanding but expensive programme to deliver nuclear weapons. After its cancellation as a national programme and the subsequent experiment with ELDO, the British opted out of launchers and came to rely first on American and later also on European launch capabilities. Instead, efforts by government and industry concentrated on satellites, their payloads, ground support facilities and some components for launchers. The rationale of this approach rested on the assumption that the UK did not need overall self-reliance across the range of space activities, but should rather concentrate on selective efforts in which British capabilities were strong and relatively cost-effective.

The early European orientation of British space policy deserves mention. The British government took the initiative which led to the creation of ELDO and later ESRO, even though ELDO's track record proved a disappointment and led to early British withdrawal. In the 1970s the British government chose (and the decision was not easily reached) to promote the creation of ESA and to participate actively, albeit relatively modestly. But that decision profoundly influenced the subsequent history of the UK's civilian space policy, during which a small national programme came to be dwarfed by the large concentration of effort on British participation in ESA programmes. From the early 1970s both government and the main British companies producing space systems and components have been locked into European collaboration through ESA and the industrial consortia. In 1980 the British government agreed to participate in the Ariane programme, although on only a modest scale.

On the civilian side both the main companies and government have continued to be involved in the production of scientific and communications satellites, the provision of ground stations and certain other equipment. The role of the scientific community in the UK has been very important in generating support for the development of space technology, albeit not always with commercial applications as a corollary. But the Ariel series of scientific satellites was a major development and generated demand for work from British contractors. This process has been mirrored in British participation in important scientific missions by ESA, notably the Giotto mission of 1986. The main thrust of British civilian efforts therefore came to depend on the rather diffuse momentum generated by the producer companies, the emerging prime users (eventually especially telecommunications) and the scientific community. Meanwhile a military programme was also developed, which concentrated on communications satellites.

Programmes and projects (updated in October 1987)
The British have now contemplated re-entering the launcher segment. British Aerospace (BAe) and Rolls-Royce are currently carrying out proof-of-concept studies of Hotol with government support. Hotol's design and innovative air-breathing engine are intended to provide an efficient and economic re-usable launch capability. Collaboration partners would be necessary for the next phase, most obviously from Europe. Other British companies have retained an involvement in subsystems for launchers. In the meantime some British satellites, including some of the Skynet series, are being adapted for launch by Ariane instead of by Shuttle. If Hotol is developed the UK will have a more rounded space capability. But increasingly it is recognized that this capability has to be effectively harnessed and must depend on sustained international collaboration. In this context the argument for promoting European autonomy is beginning to be articulated.

The British public and private sectors have over the years been involved in a range of civilian satellite applications and in the design of their overall systems. The Ariel programme of six successive satellites provided the British scientific community–especially through Ariel 5 and Ariel 6 – with an indigenous capability in X-ray astronomy and other investigations of cosmic radiation. British scientists and the industry have been associated with many international scientific ventures. They were heavily engaged in the Giotto mission to intercept Halley's Comet, with BAe as the prime contractor.

Communications satellites have been the primary field of interest. Here British industry had an early success in 1974 with Miranda and then established itself with one or other of the main companies in Europe as the leading contractor for successive generations of ESA communications satellites, including OTS, the ECS series (some for use by Eutelsat), the Marecs series and recently Olympus, a major project with potentially wide-ranging applications. Experience and technology derived from these satellites is also being harnessed for contributions to the Inmarsat series and to Eurostar, which has been developed jointly by BAe and Matra for Satcom International. Delays in plans within the UK to establish direct broadcasting have put back implementation of some new satellite programmes, and there is fierce competition with overseas suppliers for contracts. The first contract for the British Satellite Broadcasting consortium has recently been awarded to Hughes from the US, not to the British tender, on the basis of the financial terms and the delivery date promised by Hughes. As the use of satellite capacity becomes more cost-effective and economic, efforts are under way to develop new markets for satellite communication services in the business community, the mobile sector and in such fields as education.

A military programme was also developed, especially through the Skynet series of satellites. The UK was one of the pioneers of the concept of geostationary communications satellites, which were designed to replace the previous British global network for long-range telecommunications, of which the Royal Navy was a principal user. Skynet was thus designed to serve particular national purposes. But this programme drew on the persistent closeness of Anglo-American relations in the fields of military technological collaboration and shared defence intelligence, and was designed to complement both US and Nato systems. All this took place against the background of a defence budget faced with more demands for expenditure than it could satisfy.

The main indigenous capability remains the national Skynet series of military communications satellites. This series is now highly developed and the design was recently (January 1987) awarded a Nato contract, a major breakthrough for the companies involved, notably BAe and GEC-Marconi. For some space-based intelligence gathering, the British have so far relied on access to American systems. An independent British capability in electronic intelligence was the object of the rumoured Zircon project, now abandoned, although its mission was to complement, not duplicate, existing systems. So far, however, military programmes have not included any indigenous military reconnaissance capabilities or any European collaboration in this field. There is some British involvement in SDI under a bilateral memorandum of understanding.

As for remote sensing, the British have been involved in ESA programmes, such as Meteosat and ERS, with production capabilities concentrated on instrumentation and ground equipment. The National Remote Sensing Centre at Farnborough processes data and is developing a user-awareness strategy with industry; and techniques of radar sensing are under active investigation. Recent discussions over Columbus and the international space station have led to particular British interest in the polar platform and its primary utilization for remote sensing, with potential user needs very much in mind. Interestingly, the wide range of opportunities offered by a space station has excited a more active British interest in several previously underdeveloped fields.

Britain has developed ground segment production and been active in international markets for many years. This sector is attracting increased attention, partly in recognition of the importance of developing a competitive edge.

Budget

Expenditure on space is allocated through the budgets of the sponsoring departments. The Ministry of Defence declares annual spending of about £100 million on the national programme. The Zircon project would have represented a major increase. On the civil side, in 1985/6 the Department of Trade and Industry spent about £70 million, the Science and Engineering Research Council about £20 million and the Meteorological Office some £5 million – a total of £100 million. Of this some 80 per cent, a very high proportion, was committed to ESA programmes (15 per cent of the ESA budget), leaving a modest residual national programme. The total rose to £109 million in 1986/7, of which £79 million was dedicated to ESA. This last amount was concentrated on support for innovative applications in communications and Earth observation, longer-term research, mostly through the Royal Aircraft Establishment, Farnborough, and the Rutherford Appleton Laboratory (these two drawing on some 240 specialized staff), and the other science programmes, including the Ariel 5 X-ray astronomy satellite. Proposals recently under discussion included a significant increase in the civil space budget. Within this there was to be an enhanced commitment to ESA, but also a larger collateral national programme.

Organization

The BNSC is a small coordinating agency, with 40 central personnel staffed on secondment from government departments and industry, and with science and technology arms. Its first Director was Roy Gibson, former Director-General of ESA. Its creation has resulted from, first, accumulated pressures for a more targeted national policy; second, anticipation of a major round of international negotiations; third, the need to harness indigenous technological progress; and, fourth, recognition of a growing international market for space applications. It replaces the previous and decentralized policy framework in which the Department of Trade and Industry coordinated civil space policy through its space branch. The BNSC board includes members from the main organizations (public and private) with a direct interest in space and is intended also to improve liaison with the Ministry of Defence. The BNSC's remit covers the core ESA and national programmes as well as bilateral collaboration with other programmes, such as Nasa (in addition to Nasa/ESA), the Canadian Radarsat, China, India and the USSR. The Overseas Cooperation Unit of BNSC manages and looks to develop these activities.

The BNSC's first strategy document was recently before ministers. It argued the importance of making further investments in space technology and applications and of strengthening national capabilities to complement Britain's already intense involvement in European and international collaboration. Although it did not cover military sectors directly, the overlap of technology between the civil and military sides is increasingly recognized. The document called for a substantially increased expenditure in order to underpin a larger, more ambitious and more diverse range of space applications.

The proposed space plan did not win agreement from ministers. It proved difficult to align the military and civilian sides of the argument or to convince the Treasury of the merits of a substantial increase in public funding. The debate was much complicated by a new review of all public R&D which had not yet provided the criteria against which space could be judged. In July 1987 Mrs Thatcher announced that funding would remain constant, a decision modified in August by the announcement of interim funding to keep Britain in the preliminary studies of Hermes and Columbus. Despite surprisingly strong public pressure for the BNSC's plan, it was not accepted for the ESA ministerial meeting in November 1987.

Industrial infrastructure

Within the space industry the main companies – British Aerospace and GEC-Marconi – have established leading positions as prime contractors for many national, European and international satellites and payloads, with particular strengths in communications. Here Britain has the greatest depth and quality of industrial capabilities outside the USA. The same companies, along with others (including Ferranti, Logica, Racal-Decca and Thorn-EMI) providing sub-systems and equipment (including in the ground segment), have been actively involved in successful consortia for international contracts, again with particular strengths in communications, meteorology and space science, but also with emerging capabilities in remote sensing.

Perforce the space industry is very international in orientation in terms of contracts sought and won and of consortia links with companies in other countries. In civil applications industrial achievements have depended on establishing a quality edge and international markets rather than on the sustained direct patronage of the British government, though of course the government is a prime purchaser. A major change has occurred with the steps to deregulate telecommunications in the UK, a development which is both changing the operating context domestically and altering attitudes to the international arena.

Precise figures cannot easily be given on the economic weight of the British space industry. The companies which are the main producers of space systems have an annual turnover of about £220 million. They employ directly some 3,500 personnel. Roughly 100 British companies have a significant interest in space technology on the producer side, but their staff are often not working solely on space-related work. On the research side, there are some 500 staff in the university sector as well as upwards of 200 in government research establishments. If the operation of services dependent on space technology is included, then the economic relevance of the sector increases considerably. British Telecom, for example, already has the equivalent of around 1,000 full-time jobs in its satellite business and gross investments (in both satellites and the ground segment) of some £300 million.

4

WEST EUROPEAN COLLABORATION

First steps in collaboration

The first step towards European collaboration in space was taken in April 1960. Leading scientists from ten West European countries met in London and discussed the new possibilities. They agreed that the development of national space technologies would quickly overburden the economic capabilities of individual European countries and that it would, therefore, be preferable for Europe to pool its resources in order to set up an effective space research programme. Following their suggestion, at an intergovernmental conference held in November 1960 twelve nations signed the Meyrin Agreement, which set up the Commission Préparatoire Européenne pour les Recherches Spatiales (COPERS). At about the same time, the United Kingdom suggested the development of a European launcher.

These two initiatives were pursued almost simultaneously. In February 1964 the Convention of the European Space Vehicle Launcher Development Organization (ELDO) came into force, followed, in March of the same year, by the Convention of the European Space Research Organization (ESRO). Parties to the ESRO convention were Belgium, Denmark, the Federal Republic of Germany, France, Italy, the Netherlands, Spain, Sweden, Switzerland and the United Kingdom.

The initial years were not without their setbacks. On the whole, ESRO turned out to be more successful than ELDO. Part of the explanation was that the ESRO secretariat managed to achieve a certain autonomy in the implementation of its plan of action and in its organization, while ELDO was much more subject to the wishes and preferences of national governments. ESRO set up two main technical centres: the European Space Research and Technology Centre (Estec) in the Netherlands, and the European Space Data Centre (Esdac) in the Federal Republic.

The technological status of Europe was fully revealed in the design and development of its first scientific satellites: ESRO-I (polar ionosphere and auroral phenomena), ESRO-II (solar and galactic cosmic rays and particles) and Heos 1 (interplanetary magnetic field and solar particles). All three spacecraft were launched by American rockets in 1968. ELDO began the design, development and construction of a launcher using a British rocket as its first stage, a French rocket as its second stage and a third stage developed by Germany. But the programme was plagued by problems.

The development of the solar probe, Helios, was the first big project with transatlantic cooperation. These successes confirmed the scientists' enthusiasm and paved the way for the major decision to extend the missions of ESRO to application programmes (communications and meteorology in particular). Governments became increasingly interested in the social and economic benefits which they felt would accrue from the development of applications satellites. At the same time, the first real stirring of industrial interest became visible.

The dependence on American launch vehicles proved, however, to be a serious constraint. For example, the first European (i.e. Franco-German) communications satellite, Symphonie, was restricted by American launching conditions to experimental operation.

By the end of 1972, the European countries were confronted with a number of portentous decisions. The United States had accepted Europe as an emerging space power and had offered it participation in its post-Apollo programme, i.e., the development and construction of a Spacelab that would be carried by an American space shuttle in the early 1980s. The Federal Republic wanted to accept this American offer. France, however, mindful of the problems within ELDO, wanted priority for the development of a heavier launcher, Europa III. It found American commitments to provide launchers for European applications satellites insufficient and unsatisfactory. France wanted Europe ultimately to seek autonomy in the launcher market.

Eventually a compromise was worked out. The ELDO programme of launchers was discontinued. On the basis of a design by the French CNES, a European launcher, Ariane, was to be developed. Spacelab would be developed as a contribution to the US post-Apollo programme. It was also agreed to set up a single European Space Agency (ESA), entrusted with carrying out not only all tasks previously undertaken by ESRO and ELDO but also with new activities. The divergent priorities which the three member states attached at the outset to different programmes remained visible in the division of labour. The Federal Republic accorded first place to (and led the development of) Spacelab; France led the development of Ariane; and the United Kingdom opted for the European approach to maritime space communications, Marecs, and later L-Sat, subsequently renamed Olympus.

The creation of ESA

On 30 May 1975 the ten original ESRO countries, together with the Republic of Ireland, signed the treaty to establish the European Space Agency. Under the terms of its Convention, ratified in October 1980, ESA's task is 'to provide and to promote, for exclusively peaceful purposes, cooperation among European states in space research and technology and their space applications, with a view to their being used for scientific purposes and for operational space applications systems'. Austria and Norway have been full ESA members since 1 January 1987. Finland became an associated member on the same day. Canada concluded a cooperation agreement with ESA in 1979. ESA's headquarters is in Paris; it has various research and operational centres and a total staff of some 1,400.

It is also part of ESA's mission to elaborate a long-term European space policy and to recommend space objectives to member states. It coordinates the European space programme with national space programmes and seeks gradually, and as completely as possible, to integrate the latter into the former. Finally, it has to work out and put into operation the industrial policy appropriate to its own programme and to recommend to member states a coherent industrial policy. From this it follows that ESA's role in the development of European technology is far from negligible.

The policy of ESA is determined by its Council, which consists of representatives at official level from the member states and meets five times a year. The Council decides on scientific, technical, administrative and financial questions. The Ministerial Council meets infrequently. The first and main committee is the Science Programme Committee. Others include the Administrative and Finance Committee and the Industrial Policy Committee, while Programme Boards supervise individual programmes. There are two types of programme: mandatory and optional. Participation in the scientific programmes and contributions to the general budget (technological research, investments, data-processing, etc.), at a level proportionate to GNP shares, are mandatory for all members. The applications programmes, however, as well as the launcher and space infrastructure programmes, are optional. This coexistence of mandatory and optional programmes may well be one of the reasons why ESA works so smoothly. Each member state has one vote in the Council, except when the discussion is about an optional programme in which it has chosen not to participate. The level of financial resources to be put at the ESA's disposal for its mandatory activities in the next five years is decided upon by unanimous vote of the Council. Other decisions are taken with a simple or a two-thirds majority.

As a rule, the distribution of ESA contracts orients itself to the level of financial shares provided by the member states, for a long time reflected in usage of the term 'juste retour'. This has been defined not for individual contracts, but on a cumulative basis. Occasionally the question of 'industrial return' has given rise to problems. It is, of course, a difficult task to provide an equitable share of contracts to each member state while at the same time, for reasons of efficiency, awarding contracts through competitive tendering. In the larger countries national space development efforts often tend to give industry a certain competitive advantage, so that, from time to time, special measures become necessary to increase the industrial return for some of the smaller countries. (See Table 5.) ESA works on the principle of 'no profit, no loss'; as projects have approached commercialization, they have been handed over to industrial entities such as Arianespace.

ESA programmes

The first ESA applications satellites, Meteosat 1 (weather forecasting) and OTS (experimental telecommunications satellite) were launched in 1977 and 1978 respectively. Four new scientific spacecraft were placed into orbit: Cos B (gamma-ray astronomy) in 1975, ISEE-2 (a collaborative Nasa/ESA mission designed to study the magnetosphere and sun/earth relations) in 1977; Geos 2 (far magnetosphere) in 1978; and IUE (International Ultra-violet Explorer, a joint ESA/Nasa and UK project) also in 1978. A fifth scientific satellite, Exosat (X-ray observation), was added in 1983. (European programmes for satellites and vehicles are summarized in Table 2.)

On 24 December 1979, only six years after the European ministers confirmed that it was essential for Europe to develop its own launch capability, Ariane successfully completed its first test flight, the first of a series of four. Companies from ten Western European countries have cooperated in this, the most extensive ESA programme to date. The lion's share of the cost has been borne by France (63.87 per cent); Germany has contributed 20.72 per cent. In January 1980 the ESA countries decided to transfer the technical and commercial responsibilities for the production, marketing and launching of the eleventh and all successive Ariane rockets to Arianespace, an international private-law company established in France. French shareholdings are around 60 per cent, German 20 per cent, with the rest spread among the remaining ESA participants. Shares are held by space companies and banks. The launch services of Ariane are now offered worldwide by Arianespace. In addition, the way has been opened for Europe to achieve commercial space flight, mainly for communications.

Within Europe the demands of the space science community continue to push the state-of-the-art, both in the technology of spacecraft design and services, and in the instrumentation for the collection and retrieval of scientific data. The space science programme is making real breakthroughs. On 2 July 1985 the Giotto spacecraft, developed from Geos, was launched from Kourou in French Guiana. With remarkable technical precision, both in time and space, Giotto came very close to Halley's Comet on 14 March 1986. Other pioneering missions are currently under development for scheduled launches before the end of the decade, including Hipparcos (measurements of astrometric parameters of some 100,000 stars) and Ulysses, a joint ESA/Nasa solar polar mission. ISO (Infra-red Space Observatory) is scheduled for launching in 1992.

The first of ESA's maritime communications satellite programmes, Marecs, was approved in 1973. It is designed to reduce the isolation of ships at sea by linking them through ground stations directly to the international telephone and telex networks. Marecs A was launched in December 1981 and Marecs B2 in November 1984. These satellites are leased to Inmarsat.

ESA has also developed satellites that cover European requirements in the telephony, business services and broadcasting fields. The European Communications Satellite programme (ECS) followed the success of ESA's pre-operational satellite, OTS, which was launched in 1978. ECS-1, ECS-2, ECS-4 and ECS-5 were scheduled for launch by Ariane from 1983 onwards. The system is now operated by Eutelsat. Another communications programme, Olympus (the former L-Sat), is due for launch in 1987. This large multi-purpose satellite will extend European know-how in a field which has wide international repercussions, with global networks for teleconferences and the use of satellite links in the retrieval and transfer of information.

Some spacecraft are platforms from which to observe the earth and its immediate environment. ESA entered the Earth observation field in November 1977 with the launch of its first applications satellite, Meteosat 1, which remained in service till the end of 1985. Meteosat 2 was launched in June 1981. ESA operates a Meteorological Programme for Eumetsat (see below).

ESA's first remote-sensing satellite programme got under way in 1981, when plans were drawn up to launch a pre-operational observation satellite, ERS-1, in 1990. The satellite will carry a synthetic aperture radar sensor and is oriented towards ocean monitoring. Its expected lifetime is two to three years and it is intended to be the forerunner of a series that will become operational in the 1990s. Europe has its own network, Earthnet, for the acquisition, preprocessing and archiving of remote-sensing satellite data and for its distribution to the European user community, as part of a future European remote-sensing space programme. At present Earthnet is handling two US remote-sensing missions (Landsat, and Nimbus 7) and one Japanese mission (MOS-1). The Earthnet offices are at the European Space Research Institute (Esrin) in Frascati, Italy.

Spacelab is Europe's contribution to Nasa's Shuttle programme. It is a manned space laboratory, in which for the first time scientists and engineers as well as astronauts are able to work in low earth orbit. Between 28 November and 8 December 1983 a very successful ten-day mission took place, the culmination of ten years' work on the Spacelab programme. It proved that Spacelab is an excellent facility and an ideal tool for space research. Spacelab 1 marked the first flight into space of an ESA astronaut, Ulf Merbold from Germany.

Spacelab 3 (Mission Challenger 51-b) carried during its flight between 29 April and 5 May 1985 a predominantly microgravity science payload. The payload had an international flavour, with Europe represented by the CNES-sponsored Mercuric Iodide Crystal Growth experiment. ESA's Instrument Pointing System (high-precision stellar, solar and earth pointing) flew successfully from 29 July to 6 August 1985 with Spacelab 2. The first German-managed Spacelab mission, D1, took place between 30 October and 8 November 1985. During this flight, controlled by Nasa and the Oberpfaffenhofen Mission Control Centre, numerous experiments from France, the Federal Republic, Italy, the Netherlands, Spain and Switzerland were conducted.

After the loss of Challenger, Nasa issued a manifesto in October 1986 indicating that Shuttle operations were planned to resume on 18 February 1988. The first launch dedicated to Spacelab will not be until the first quarter of 1993. Since the size of the Shuttle fleet will be smaller and the launch turnaround slower, there will be a significant reduction in the amount of use Spacelab makes of it. Phase 3 of the ESA microgravity programme will be affected.

An analysis by Nasda, the Japanese Space Agency, shows that in the period 1978–82 the ratio of the total ESA budget to the staff complement managing it was higher than in any of the other organizations evaluated (Nasa, CNES, DFVLR, ISAS and Nasda itself). This notwithstanding the fact that ESA is an international organization, with all the attendant constraints.

ESA's future plans

Half-way through this decade, as key programmes reached completion, it was clear that a review of the future scope of ESA's activities would be timely. The smaller member states were keen to define the space activities for which ESA would remain the primary organization. Eventually, in January 1985, a ministerial conference met in Rome. On the assumption of a continuing Shuttle programme, it was decided to embark on the Columbus programme, as an optional programme in the field of in-orbit infrastructure and as a significant part of the 'international space station' programme proposed by the United States. The fully independent Columbus programme is also intended to be an important element of an autonomous European capability in automatic and manned orbital operations. Columbus consists of a manned module, permanently attached to the space station and carrying facilities for materials-science and life-science payloads. It has a man-tended free flyer (MTFF) and polar orbiting platforms to be serviced from the space station or by manned vehicles. The programme will also include the basis for an operational data-relay system.

The ESA programme agreed for the period 1985–95 also contained the following elements:

– the development of a new generation of powerful launchers, Ariane 5, equipped with the large cryogenic engine HM-60, to be completed by 1995;
– the continuation and extension of programmes in the fields of Earth observation (ERS-1 and follow-on missions for oceanographic and meteorological applications, a land applications project, participation in the development of the second generation of meteorological satellites, and studies aimed at solid earth, atmospheric and climatology missions), space telecommunications (advanced payload systems), and microgravity (enhanced utilization programme of Spacelab and Eureca leading to the utilization of Columbus and the international space station for microgravity R&D);
– the progressive increase of the level of funding of the mandatory scientific programme to reach 162 million EAU by 1989 (an annual increase of 5 per cent); and
– the progressive increase of ESA's overall level of funding to reach about 1,650 million EAU per year by 1990. The detailed costing of this plan is currently under review. (See Table 4.)

The Rome conference took note with interest of the French decision to pursue the proposed manned spaceplane, Hermes, and of their invitation to European partners to join the detailed studies. The conference invited France and associated partners to keep ESA informed of progress with these studies, with a view to including the Hermes programme, as soon as feasible, within the optional programmes. Over the following months, France advanced its studies on Hermes and presented a detailed dossier on the programme to

ESA's member states. At its 74th meeting in June 1986, the ESA Council approved the Europeanization of Hermes by adopting an enabling resolution for a preparatory programme.

ESA long-term plans include a range of programmes for the remainder of this century. In space science four missions are planned, the so-called 'cornerstones'. They will require much more precise control and positioning of satellites, with new inter-satellite communications for a cluster of satellites; new rendezvous techniques and drilling devices in order to bring back material samples from comets or asteroids; and new concepts of multiple telescopes with degrees of sensitivity not previously available to meet the demands of the spectroscopy missions. In the applications programmes, improving long-life reliability is a key factor. The telecommunications programme will be extended. After Olympus will come the development of data-relay satellites as part of the in-orbit infrastructure. Permanent or semi-permanent orbital systems are also envisaged for earth observation in the future.

The next ESA ministerial conference was scheduled after some delays for November 1987. The deliberations were to revolve around several important questions: the priorities of European space policy; the degree of European autonomy; and the pace at which to attain it. The conference had to determine the correct balance between mission-oriented programmes (such as earth observation, science, telecommunications) and infrastructure-oriented programmes (Ariane, Columbus, Hermes). And finally ministers had to decide whether they could accommodate their ESA ambitions within the budget agreed in Rome.

The collaborative context

ESA provides the foundations of European collaboration in space, and accounts for a huge proportion of the total public expenditure in Western Europe (see Figure 2). Indirectly it plays a vital catalytic role in fostering R&D programmes which may then be turned into commercial enterprises and operating services. Directly, however, ESA accounts for only part of the pattern of European collaboration in space and represents perhaps 20 per cent of the European space industry's business.

ESA's self-imposed limitation to 'peaceful uses' of outer space has meant that it has been excluded from discussions on any European defence and security programmes, such as the national programmes of France, Italy and the UK; from bilateral collaboration with the US, which is most fully developed in the British case; from specific collaborative projects, such as Helios (French-led with some Italian and Spanish involvement); and from the Nato Integrated Communications System (NICS). This last provides the European members of the integrated military structure with access to a system of permanent and secure communications via satellite links to Nato's American and European forces on land and sea. It is interoperable with both the American DSCS III and British Skynet 4 systems. The first generation was built in the US by Ford Aerospace. As for the next generation of Nato satellites, due to come into service in the 1990s, a first order has been placed with British Aerospace/Marconi, based on the Skynet design.

ESA's other deliberate limitation has been to concentrate on the pre-commercial phase of project development. Thus, once applications programmes have reached the point of take-off as self-sustaining services, new and functional operating entities have emerged. Eutelsat, the European Telecommunications Satellite Organization, was set up in provisional form by seventeen European PTTs in 1977. It was designed to operate, on a regional basis, two space segments: one for fixed service communications, which it retains; and the other for a mobile maritime service, subsequently transferred to Inmarsat. The Eutelsat organization became definitive in September 1985 and now has 26 members. Its activities are further discussed in Chapter 5a.

Eutelsat and ESA have agreed complementary roles. In effect ESA has been the R&D arm and thus far responsible for the procurement, launch and in-orbit support of satellites (first OTS and then the ECS series). The satellites have been leased to Eutelsat as the operator of the service which they provide for an enhanced European network – namely telephone, telegram, telex and broadcasting – accessible to members and other European countries. Thus, for example, Eutelsat provides the EBU with the capacity for Eurovision transmissions. Eutelsat is in the process of procuring its second generation of satellites, Eutelsat 2, and promoting studies for Eutelsat 3. It may also operate international segments of national satellites, such as the French Télécom 1. The ground segment is operated by the PTT signatories.

Eutelsat is required to strike a balance between serving the European users and maintaining a sound commercial basis. Designated signatories have investment shares and thus contribute to the capital endowment on a basis that reflects shifting patterns of usage. The tariffs charged to users pay for the operating costs of Eutelsat and provide a return on the capital investment. As deregulation and liberalization measures extend, so the role of Eutelsat will have to adapt, a point discussed in Chapter 7.

Eumetsat, the European Organization for the Exploitation of Meteorological Satellites, is a similar venture, though more limited in scope. It was spawned by ESA in 1983, when the Meteosat series of meteorological satellites began to produce an operational system. Eumetsat offers extensive coverage of Europe, Africa and the Atlantic on a regular basis, the pictures from which are carried on television across Europe. The signatories are the designated meteorological agencies of most West European countries. A further generation of weather satellites is currently planned by Eumetsat. This was the first earth observation programme sponsored by ESA to develop as an operational service.

The development of Eutelsat and Eumetsat rested on European decisions to provide European services with European procurement. But West European governments, public agencies and industries are also part of a global pattern of collaboration and competition. Designated signatories from West European countries, the PTTs, belong to Intelsat and Inmarsat. Intelsat was of course the first such international organization, created at a time when Europe had slender space capabilities of its own. Western Europe now holds a share of around 34 per cent in Intelsat, as well as operating the Eutelsat regional system. European companies regularly tender for Intelsat procurement contracts, though generally in association with American industrial partners. Although the European share of procurement has been low in relation to that of the US, the business generated has been significant.

The Inmarsat case is different. Inmarsat, the International Maritime Satellite Organization, created more recently in 1979, provides mobile communications systems for the shipping and offshore industries. It is attached to the IMO, based in London and has 48 members. It is now to provide the secretariat for the Cospas-Sarsat search and rescue satellite system. In this case the European option adopted as appropriate was to join in a broad international system. Inmarsat started out by leasing satellite capacity from elsewhere, including Marecs from ESA. It is now procuring its own satellites. Significantly the UK has been the most proactive European member, and British companies have played a leading role in supplying Inmarsat satellites.

The other major European organization with a growing involvement in space activities is the European Community. The EC's interest in space is fourfold. First, it provides a focus for developing some applications of space technology – remote sensing of crop forecasts is one example. Second, among various R&D programmes sponsored by the EC, several include a space element, directly or indirectly. The Race programme (Research and Development in Advanced Communications Technologies for Europe) is the clearest case, in that future developments in telecommunications are intimately bound up with satellite technologies and operating costs. In addition, some Esprit projects cover adjacent technologies in electronics, software, artificial intelligence and so on, as does Brite (Basic Research in Industrial Technologies for Europe) in areas such as materials.

Third, the EC has a central role in helping to define and monitor the market conditions within which the space industry operates. It plays a key role in international negotiations about the terms of trade, which are highly relevant for example to the debate about competitive pricing of launchers. The EC can and does legislate to promote common standards, to remove obstacles to trade, to monitor state aids to industry and to liberalize public procurement. The new emphasis on the need to create by 1992 a single European home market, endorsed in the Single European Act (SEA), has made this a key priority and highly pertinent to the fortunes of the European space industry and its customers.

Lastly, the EC has now acquired with the SEA a limited but explicit security dimension to European Political Cooperation which opens the way to a more active involvement. This has also to be seen in the context of the revival of Western European Union (WEU) as a focus for a European defence dialogue involving seven ESA members (Belgium, France, Germany, Italy, Luxembourg, the Netherlands and the UK). Several discussions have already taken place within WEU about its potential role in contributing to any space dimension to Europe's collective security.

West European collaboration

There is a dense and dynamic pattern of European collaboration relevant to Europe's future in space. As technologies mature, new markets develop and additional policy tasks are defined, so the firm foundations of ESA are being complemented by other collaborative frameworks.

Part II
USES AND MEANS

5

USES

This chapter summarizes the technical possibilities available, together with current and projected developments in space technology. It shows the range of functions which can already be performed, indicates those which can already be envisaged as viable and highlights the rapid pace of technological change.

(a) COMMUNICATIONS

The use of space technology for worldwide telecommunication links was the first demonstration of the potential commercial value of space. There is already a competitive global industry in the civilian sector, as well as a capability for important defence purposes. These latter are dealt with below in section 5e, though, of course, civilian and military applications benefit from many of the same technologies, operate within similar technical and financial limits, and may sometimes share facilities. Communication satellites provide secure, flexible and economic links to fixed and mobile users in all domains on Earth. This combination of attributes has user-convenience and is a stimulus to economic development. But it is not enough to rely on industry to develop its own business. In such a fast- moving and competitive environment other policies, including the regulatory aspects, must be established to the benefit of both producers and users. And a sensible division of labour must be found with other forms of telecommunications.

As an indication of scale, Intelsat has spent over $4.8 billion just on procuring satellites (let alone launcher costs). Of this some $3.8 billion has been spent on industry in the USA, and some $3.2 billion of that has been spent in California alone! Since the USA is the largest, but still a minority (25.7 per cent) shareholder among over 100 countries, there is clearly some scope for improved benefit to Western Europe, which in aggregate holds a 34 per cent shareholding.

Spacecraft as communication relays

The spacecraft is designed to carry a payload with transponders which receive and re-transmit, through appropriate antenna arrays, a wide variety of signals for many different purposes. Separate communication systems in the spacecraft monitor its overall performance and convey operational signals.

Such links are of course provided for scientific satellites and space probes, as well as for inter-satellite data-relay links. The demands on technology and performance are extremely high for these systems.

Fixed telecommunication services

Commercial telecommunication services were first carried by satellite in 1965, when the Early Bird satellite was successfully launched into GEO, with a capacity of only 240 telephone circuits or one TV channel. Developments in satellite technology and power generation have led to the current range of satellites with capacities of up to 30,000 circuits plus 3 TV channels. New technologies could increase capacity by a factor of four or more before the year 2000.

Intelsat, which owned Early Bird, was created on 20 August 1964. Today, the space segment of Intelsat comprises 16 geostationary satellites, which it owns and operates; the ground segment is composed of 650 stations owned by the designated telecommunications entity in each member country. Intelsat currently has 112 signatory and shareholding member countries, with some other countries associated as users only. This global network carries telephone, teletype, data, facsimile and television communications in more than 160 countries via 250 antennas. Other ground services support domestic or specialized operations, some on lease.

In addition, four regional systems with dedicated satellites of limited coverage area have been established. These are Eutelsat, for Western Europe; Intersputnik, serving the CMEA countries; the Indonesian Palapa B satellites, which also service the ASEAN countries; and Arabsat for the members of the Arab League.

Eutelsat is the West European organization of PTTs which builds and launches communications satellites, rather than relying on working through the Intelsat framework. Eutelsat provides capacity for telephoning, the European Broadcasting Union and *ad hoc* services, and also leases transponders to its signatories. Unlike Intelsat, which has its own R&D facilities, Eutelsat relies for its R&D on ESA. Eutelsat operates ECS on lease from ESA and is now developing a second-generation system.

Major policy issues arise about the most appropriate means of determining producer and user requirements in the commercialization of all these services, especially with the moves afoot to deregulate the telecommunications sector and competing technologies. Satellite telecommunication services are operated by national PTT or government-licensed common carriers which have conflicting interests as members of Intelsat, national services providers and participants in regional systems. Already several US companies have begun to develop separate systems for business traffic in competition with Intelsat.

Recent advances in optical fibre technology currently make cable systems the cheapest form of wide-band transmission for most high capacity fixed routes, although they require a large initial investment. Satellites also involve large investments because of their technical complexity and the high cost of launching them into orbit. However, for telecommunication networks that cover a number of low capacity routes, multi-point operations, inhospitable terrain, or a requirement for rapid installation including temporary service, satellites are likely to be a better option.

Recently, specialized business services have been introduced which employ Very Small Aperture Terminal systems (VSAT or microterminals) with antenna sizes typically in the range of 1.2 to 1.8 metres. Applications include corporate data, stock control, credit-card verification, electronic mail, etc. The VSAT stations currently cost under $10,000 and are designed to be installed on each of the sites of a single corporate user and work to a larger central hub station. The growth of these corporate networks in North America is spectacular, with approximately 30,000 VSAT terminals in service and orders for 100,000 terminals expected to be placed in the next twelve months. A similar growth pattern is anticipated in Europe by 1990.

Current communication satellites have an operational life of between seven and ten years. So far they have been replaced by models incorporating newer technology and usually with a higher capacity. The driving force for new developments centres on efforts to increase capacity and to reduce costs per circuit. ESA programmes are currently exploring improvements in component reliability; higher capacities using new frequency bands; highly directional antennas providing spot beams; higher EIRP; on-board systems for circuit-switching; direct inter-satellite links between satellites in different orbit positions; and design life in excess of ten years (already a requirement for the most advanced satellites, though shorter lifetimes are acceptable in some cases).

Maritime telecommunication services

An international satellite service for communication to ships at sea has been available commercially since the first call made over the Marisat system in July 1976. Inmarsat came into being on 16 July 1979. By January 1987, the organization included 48 countries through their designated PTT signatories, with the US having the largest shareholding of 29 per cent, and the West Europeans collectively 48 per cent of the total.

Inmarsat provides the satellite capacity for a range of communications services between 'earth' stations on ships, provided by the ship owner, and 'fixed earth' or coast stations, provided by national telecommunications administrations. As of mid-1986 there were sixteen operational shore stations, with fourteen more due to be completed by 1990, and 4,479 operational ship stations. The system has no provision for direct ship-to-ship communications, though this could be established via a shore station with a 'double hop' through the satellite.

Services provided include telephone, telex and other data services. Voice and telex channels are available on a dial-up basis from ships, and other data services are being developed. Subscribers on terrestrial networks may have to go via a maritime operator for voice channels to ships, but direct dialling is widely available for telex. Group-call facilities are available in the shore-to-ship direction, permitting a common telex message to be sent to a particular fleet, to a national group, or to all ships in a defined geographical area, this last service being particularly useful for meteorological broadcasts and navigational hazard warnings. Following recent successful trials with the liner *Queen Elizabeth II* by British Telecom using Intelsat capacity, Inmarsat has approved (initially as a one-year pilot) a television service to be specially equipped for ships from Comsat's shore stations in the USA.

The Inmarsat service is provided by three operational geostationary spacecraft and in-orbit spares, servicing the Atlantic, Pacific and Indian Ocean regions and providing coverage between the latitude of 70° North and 70° South, with the exception of a small part of the eastern Pacific. There are plans to complete this coverage, initially by changing some of the operational satellite longitudes. Ways to extend the service to the polar regions are under study, although its economic viability will have to be evaluated carefully.

The Inmarsat space segment has until now been provided by leasing spacecraft from Comsat (Marisat), from ESA (Marecs), and by leasing a maritime facility carried on some of the Intelsat V spacecraft. The second-generation systems will be owned and operated by Inmarsat, and the industrial procurement of spacecraft and ground control facilities is under way. The new satellites are expected to be in service from the first half of 1989. Inmarsat's capacity is estimated to be sufficient for the Atlantic region into the mid-1990s and for the other regions up to the end of the century. Ways of increasing the capacity in the Atlantic region are being studied.

Land mobile

Radio communication to moving vehicles on land – the Land Mobile Service – is a business growth area. The annual growth rate in Europe of 12 per cent in the late 1970s led to demand for services outstripping capacity. The competitors for the available channels include: military requirements; public services (police, fire, ambulance, utilities); civil defence; private networks (taxi, delivery, haulage, public transport); mobile subscribers in the Public Switched Network; radio-paging services; and Citizens' Band radio. These users are currently catered for by terrestrial radio services using VHF and UHF frequencies. In recent years many countries have released new channels for these services, but even so there remains a severe shortage of channels. The new 'cellular' radio systems are designed to permit a much greater degree of frequency re-use, a development which will remove this constraint to growth. There is no worldwide frequency allocation, although Inmarsat signatories are seeking an allocation at the ITU World Administrative Radio Conference (WARC).

No commercial satellite systems yet serve land mobiles, for which disadvantages arise because the current technology does not match the benefits of cellular radio systems, and the operating costs are higher. Cellular radio is likely to remain the preferred technology, at least for the public correspondence services. However, satellite systems could provide a wide-area overlay network in order to cover remote areas with no local terrestrial systems; serve long-haul vehicles that otherwise would need to be equipped for different local networks; offer a vehicle and cargo location system; and provide wide-area paging and message systems. This last is the most likely to emerge as the first land mobile service. ESA's Prosat programme includes experimental trials designed to solve several of the technical problems.

Aeronautical services

Proposals for aeronautical systems have concentrated on ground-to-air and air-to-ground links – for air traffic control, operational traffic and public correspondence. No significant requirement is evident for air-to-air communications circuits via satellite. A system designed for air-to-ground links could, of course, accommodate the occasional air-to-air circuit by 'double hop' via a ground station; or, if there is sufficient demand, by incorporating switching in the satellite. No such system is so far operational.

There is a resurgent interest in aeronautical satellite systems. Trials with Boeing 747 aircraft, equipped with a modified Inmarsat terminal, have demonstrated both voice and data communications. BTI, British Airways and Racal Decca also plan trials with a view to a commercial service. In October 1985 the Inmarsat Convention was amended to allow the provision of aeronautical services. The Inmarsat Council is committed to providing a full range of commercial aeronautical services from 1988/9, and has authorized investments in the development and adaptation of equipment. The International Civil Aeronautical Organization (ICAO) is studying the possible use of satellite systems for air traffic control.

Search and rescue

The elements of a search and rescue operation are emergency alerts; position data; identifying units that can provide assistance; and the coordination of the rescue attempt. Satellite techniques can play a part in all of these functions. For example, the 'alerting' function can be performed using automatic call processing through the Inmarsat system. The operations from the appropriate rescue coordination centre and subsequent communications could also be handled through Inmarsat. As a back-up, radio beacons to indicate the emergency position can operate through the polar orbiting Cospas-Sarsat system, or possibly through the Inmarsat system. Inmarsat is currently carrying out trials for non-polar regions.

Gradually from 1991 to 1997 new maritime techniques will be introduced in the Future Global Maritime Distress and Safety System. The essence of this system, in which Inmarsat will have a significant role to play, will be automated techniques for communication.

Direct broadcasting satellites (DBS)

The 1977 WARC reached agreement on the parameters for satellite systems to broadcast television programmes from the geostationary orbit for regions other than the Americas. Each country was allocated five channels for its national use in the band 12–12.5 Ghz, with a satellite-radiated power level high enough to enable a high-quality signal to be received on individual home receivers with antennas of some 60 cm in diameter. The satellites are spaced in orbit at 6° apart. The coverage of each country – 'footprint' – is closely defined to reduce interference in other countries in the same area. To minimize interference, use is made of spatial separation (orbit and ground), frequency allocation and orthogonal circular polarization.

The allocation of channels and orbit positions for the Americas was made in 1983. The different criteria for levels of interference and the involvement of fewer countries have permitted a larger number of channel allocations per country. All countries have a minimum of eight channels and the USA as many as 32. However, not all countries, particularly the less developed, will take up their allocation. At present, only those countries outside the Americas with a need to provide full national coverage (e.g. Australia, India, Scandinavia) or with governmental backing (e.g. France, West Germany) are initiating a DBS service. There are commercial doubts about the viability of a TV service that will restrict viewing subscribers to a nominal choice of only five channels, though in Central

Europe combined coverage from French and German satellites will give ten channels. In the USA and Canada the greater number of channels will enable a larger subscriber growth to be achieved and thereby alleviate the commercial risks.

Within Europe, the French and Germans have been especially quick to develop DBS under the WARC arrangements, and the British are now close behind. The interesting option for Europe is to establish pan-European channels. This involves the technical problems of arranging suitable frequency allocations and orbit positions within the existing plan format. But it also raises social, commercial and political issues, if multilingual programmes are envisaged and if costs have to be recovered for the provision of the service. Financing from advertising revenue is a further complication, since the advertising regulations and copyright rules vary significantly among European countries. Olympus (the large communication satellite being developed by ESA) will be used for widespread experiments in broadcasting.

Television distribution

The distribution of programme material for television to broadcast sites and to cable distributors is classified internationally as a fixed service. It is thus transmitted in the frequency bands allocated for operation by common carriers. Antennas with diameters of 1–2 metres are required to receive these signals. Satellite transmission is the most cost-effective medium for the distribution of programmes to multiple sites, and for outside broadcast events where a temporary link is required. Hence they are regularly used for European and indeed international events, whether for news coverage or for spectacles such as the Eurovision Song Contest. The satellite capacity available has prompted the emergence of a growing industry to provide television programmes. In Europe some 18 channels are in use with schedule times ranging from 1 to 24 hours a day. The Société Européenne des Satellites is planning to introduce a 16-channel Europe-wide service.

The cost of a receiver for satellite broadcasts has now come down to £1,000-£1,500, and is still falling. It is therefore within the purchasing power of a high proportion of the population of the developed world. An increasing number of individuals are equipping their houses with receivers. Their legal entitlement to do so is a regulatory issue in many countries, and in some it raises considerable problems. The operator of TV services has to generate demand and collect revenue either by subscription or by paid advertisements. The regulatory licensing and advertising issues make for lively political debate. Commercial considerations also probably require that TV distribution will ultimately have to be encrypted to prevent unauthorized access. Beyond these issues lies the crucial question of what kind of service the customer will actually want to use.

The prospects

Western Europe has demonstrated an excellent technical capability across the full spectrum of telecommunications satellites and has further potential to be exploited. Europe has built the foundations for a full involvement in all types of communications system and ground station. Good though the industrial skills in Europe are, however, the United States has had a much larger and more rapidly growing home market across the range of applications and has a volume demand within a protected environment. Some of the large US aerospace and electronics companies, with the considerable backing of large defence contracts, have not only learned good and reliable engineering, but have acquired practice in supplying spacecraft at competitive prices to world users. Aggressive sales techniques and strong government support from the outset have helped to establish operational systems within the US, wide entrepreneurial success in software as well as hardware supply, and a buoyant service industry. Interestingly the Canadian industry has also succeeded in winning major contracts for spacecraft in open international competition against both European and US tenders. And we must expect increasingly vigorous competition from Japan.

The European industry is increasingly active, and has been well supported by the technological base of ESA as well as by national expenditures. There is, however, as yet no overall European policy to improve the cohesiveness of complementary European companies, along the lines of Esprit in information technology. The reluctance of the European PTTs to develop operational systems on a continental scale is still a major constraint inhibiting investment in infrastructures (both private and public) and thereby limits market growth and international competitiveness. Attitudes are changing, but slowly, given the huge procurement power of the PTTs both directly and through bodies such as Intelsat and Eutelsat.

Deregulation may help considerably, and here Britain is some way ahead of the other Europeans. Equally some satellite producers have yet to develop a sensitivity to user needs and commercial viability.

It is not just a question of new technology to be gained and profit to be made. It is also a question of employment. A communication satellite is a new enabling tool and it has permitted a new global industry to emerge. The supply industries in Europe for satellites and launchers employ a few tens of thousands of people who are specialists in space. These industries in turn supply the communications industry, which serves business, entertainment, defence and other fields, which employ many hundreds of thousands of people in European countries, who in turn provide services for a combined population of over 250 million.

(b) REMOTE SENSING

Remote sensing is the second major application of space technology. Satellite-based Earth observation covers all activities relating to the acquisition from orbital altitudes of physical data concerning the terrestrial environment, i.e. land, seas and the atmosphere. Thus defined, Earth observation involves three main types of system. First, earth resources satellites allow the thematic mapping of exposed land masses and the return of useful data concerning topography, geology, hydrology, vegetation, habitats, pollution or, more generally, an inventory of a wide range of resources and the monitoring of changes. Second, weather satellites regularly monitor the atmosphere from either polar or geostationary orbit. Third, oceanographic satellites regularly monitor physical conditions on the sea's surface and at the interface of ocean and atmosphere.

Over the past decade, by initiating a series of important Earth observation programmes, Europe has become a major developer of technology for remote sensing. Of these Spot 1, launched in 1986, is the first European satellite designed for operational remote sensing of land masses. But it is not the first demonstration of Europe's technical capabilities in this area. As early as 1977, Meteosat 1, the first European meteorological satellite, started routine observations of the Earth's atmosphere and surface, thus inaugurating a service which has now become essential to weather stations in Europe and Africa, as well as to various agencies involved in oceanographic activities or global climatic research studies.

Experiments aboard the US Shuttle, such as the German MOMS (Modulator Opto-electronic Multispectral System) or the Metric camera tested by Spacelab 1 in 1983, have confirmed European technical expertise in high-resolution remote sensing. European ambitions in this area are further illustrated by the challenging development by ESA of ERS-1.

Current European programmes

In 1976 the common interest of ESA members was evident in their concern to gain access to data from the US Landsat satellite. ESA set up and consequently operated the Earthnet programme, consisting of receiving stations for Landsat data in Fucino (Italy) and Kiruna (Sweden), as well as for data from experimental US satellites (Seasat, Nimbus 7) in Oakhanger (UK) and Lannion (France).

Landsat data were used in many ways in Europe, from basic research to pollution monitoring, mapping, crop evaluation and so on. The potential economic and political relevance of these data led some ESA members, notably France and Sweden, to press for a European programme of remote-sensing satellites as soon as possible. The French proposed Spot, a system oriented towards high-resolution optical observations of the Earth's surface. Though some major countries, such as the Federal Republic of Germany, showed a lack of interest, Sweden and Belgium joined the French in their determination to go ahead with Spot 1.

After some soul-searching and redesign of its own strategy, ESA gradually became committed to an experimental mission, ERS-1, finally authorized in 1984. Meanwhile the French decided in October 1981 to start funding of Spot 2, identical to Spot 1, in order to operate Spot spacecraft on an operational and commercial basis. Sweden and Belgium again joined France and decided to participate in Spot Image, the entity set up to act as the commercial arm of the

Spot programme. The French followed with decisions to extend the series with Spots 3, 4 and 5.

Meteorological programmes

Since 1977 Meteosat satellites have provided satellite data for weather forecasting. The subsequent Meteorological Operational Programme of ESA provides further Meteosat satellites to give continuity of service to users until 1995. The new European Meteorological Satellite Organization, Eumetsat, stems directly from these programmes and is expected to cooperate closely with ESA on satellites, both polar and geostationary, for meteorology.

Spot programme

Spot 1 was put into orbit on 22 February 1986 by Ariane. Two more first-generation spacecraft, designated Spot 2 and Spot 3, have been ordered to provide continuity for the programme until the end of the 1980s. Spot 2 should be available for launch in mid-1987, and Spot 3 by August 1989.

Spot spacecraft carry two identical HRV sensors (high resolution visible), based on static solid-state arrays of detectors which operate in the visible and near infra-red part of the spectrum. The major features of Spot are the relatively high ground resolution of the imagery it will produce (ten metres in the panchromatic mode, twenty metres in the multispectral mode) and the ability of its sensors to point up to 27° East or West of the local vertical axis. (See Table 8.) This latter feature has the interesting potential of increasing the number of opportunities to obtain views of a given area. It also permits stereoscopic observations by combining views taken at different angles from the vertical, and therefore opens up the possibility of third dimension (or altitude) determination, an important requirement for cartographic applications.

The Spot programme was planned from the beginning as an operational and commercial system. CNES is in charge of spacecraft procurement, launch and operation, while Spot Image, a commercial corporation, is in charge of data distribution and all commercial relations with data users. Spot Image is developing a network of agents, distributors and subsidiaries to serve local markets.

ERS-1

ERS-1 is an experimental technology demonstration programme using active microwave, mostly aimed at oceanographic observations.

Land observations will also be carried out and will undoubtedly lead to major advances in mastering synthetic radar observation techniques. Its launch is planned for 1990 and it will have an expected life of 2–3 years. Canada has also joined this ESA programme.

ERS-1 will demonstrate in-orbit capabilities for all-weather radar observation, in both imaging (Synthetic Aperture Radar – SAR) and non-imaging modes (scatterometer, altimeter). It will operate at different frequencies and should constitute a major scientific and technological test of the operability of active microwave satellites for remote sensing.

The objectives of the ERS-1 programme are to define, develop and exploit remote-sensing data for coastal, ocean and ice applications, and to increase the scientific understanding of coastal zones and global ocean processes. Applications which will benefit from ERS-1 include off-shore oil activities, ship navigation, fisheries, oil pollution monitoring and such like.

Ground facilities and services

European industry has made great strides since the 1970s in developing ground facilities, such as receiving and processing equipment, digital image manipulation, film recorders and SAR processors. Whereas the first European stations for receiving Landsat data were for the most part made of US or Canadian equipment, European industry has built or is currently building five facilities outside Europe for receiving and processing Spot and Landsat data.

A similar trend can be seen further downstream in the chain from data acquisition to data interpretation. This area covers digital image analysis equipment and software as an aid to data interpretation, as well as associated services. European initiatives and developments in this field are numerous, with the result that European industry has become competitive in world markets.

With the preparation and launch of ERS-1 at the beginning of the next decade, further developments are expected in SAR data-processing and data interpretation and dissemination in near real time.* The requirements of the user communities in oceanographic and shipping will press ESA and its members to set up high-throughput, fast-turnaround facilities to process the enormous

*Converting data into useful pictures or maps is a complex and time-consuming process. The closer technology comes to instant or real-time processing, the more valuable the results, especially for military planning.

amount of data transmitted by ERS-1 and extract from them wave spectra, surface wind velocity and direction, ship positions and the like, all parameters whose usefulness is very short-lived.

Future European programmes

A second-generation Meteosat
This should provide data of improved quality as well as new data, such as profiles of temperature, water vapour and higher resolution data for regional applications. The launch is planned for late 1994 to ensure continuity with the present Meteosat Operational Programme, which runs until 1995. The operational system will be financed by Eumetsat, with government intervention in this field being limited to preoperational R&D.

Successor Spot satellites
Design work has now been completed on Spots 4 and 5. The preliminary industrial definition phase for the improved satellites began in early 1987. Production of Spot 4 is expected to begin in early 1988 for launch in mid-1992. Approval for Spot 5 is under consideration to allow its availability for launch in mid-1993.

Differences from the first-generation of Spot include the addition of a medium-infra-red band to the HRV instruments to measure vegetation humidity, and the installation of a new radiometer for large-scale monitoring of the Earth's vegetation and oceans. The new vegetation instrument, with five spectral bands, will be used for worldwide monitoring of crops and spontaneous vegetation, for forecasts of agricultural production, and for environmental studies. A secondary mission is the observation of oceans for scientific purposes or for specific applications such as fishing. The French also plan to develop a more efficient ground facility for image rectification with increased automation, so that more data can be processed at lower operational costs.

ERS-2
ESA plans ERS-2 for launch early in 1993. This would extend the ERS-1 mission, improve the benefit from the initial investment and provide users with 5–6 years of continuous data. The ERS-2 specifications are likely to be more or less identical to those of

ERS-1, in order to minimize modification costs, and to ensure continuity of data flows until the launch of the Columbus polar platform in 1995.

The European polar platform

The Columbus project, which includes the construction of a European polar platform, will permit a step-change in scope for remote sensing. Its increased power and load-bearing capabilities provide the opportunity to advance to the concept of an integrated multidisciplinary payload of instruments observing oceans, ice, land, meteorology and atmosphere, for both scientific and applied purposes. The project as a whole is discussed in more detail in Chapter 6. The programme requires integrated sets of instruments and should, with adjustment and replacement opportunities, permit long runs of observations.

Optimized versions of the instruments from ERS-1, together with new instruments, will allow the ocean and ice objectives of ERS-1 to continue to be addressed and to be improved upon. The major objectives for land observation would be to improve Europe's ability to manage its activities and resources in agriculture, forestry, land-use management, water resources, land surface processes, geology, cartography and environmental monitoring.

Similarly instruments on the platform would continue and improve upon the observations of the meteorological satellites currently in polar orbit. They would also provide regional and local data, profiles of atmospheric temperature and humidity, and research into and monitoring of atmospheric processes and constituents, particularly those which affect our climate.

The commercialization of remote sensing

So far this is the only sector, apart from space communications, to be exploited commercially, but still only to a limited extent. Many issues need to be addressed if successful commercial operations are to follow. The market for remote-sensing data is still difficult to predict in the short or long term. Its very diversity, as well as the need for continuing extensive investments, make it extremely difficult to estimate its future growth and potential size. Furthermore, political constraints on the distribution of high-resolution pictures of foreign countries may inhibit the viability of a remote-sensing system.

Uses

Users need a guarantee that satellite programmes for remote sensing will be able to provide a virtually continuous service over a long period of time. Otherwise they will be reluctant to rely on the data and to invest in the human and technical resources necessary to exploit them. Furthermore, the distribution of data to users has to be properly organized. Significant improvements are needed to meet the objectives of timeliness and reliability inherent in many applications of remote sensing. A sizeable 'value-added' industry is needed to extract the information needed by the users from the raw images provided by the satellites.

These points are interconnected. The market can develop to the point where it can sustain the cost of remote-sensing satellite systems only if the required continuity of service is guaranteed. However, governments are reluctant to subsidize such systems for too long, fearing that users may expect subsidies to continue indefinitely.

Some applications of remote sensing can be argued to be related to weather forecasting, traditionally accepted as a government responsibility, although some equate it rather with surveys of areas in which industry usually plays a direct role. This question can be debated at length, but existing institutional arrangements may condition decisions. Thus, weather forecasting as an accepted government domain already has a clearly identified customer for the collection by satellite of weather data; but there is not an established and structured market for surveying from space. The history of Landsat and early Spot operations demonstrate that the market for remote-sensing services is very diversified.

In spite of this situation, the United States and France have separately taken the decision to operate remote-sensing satellite systems on a more commercial basis. Their approaches, however, differ: the US government would like industry to take over and operate the whole system after a very short transition phase, whereas the French government has agreed to underwrite the initial capital costs of the satellites and has established Spot Image as a commercial corporation with data-marketing responsibility.

In both the French and American cases pricing policy is based on complete cost recovery and therefore should not differ significantly. However, the commercial risks involved in the business of remote sensing are very high, which explains the reluctance of US industry to make firm commitments to take over of the Landsat programme without a solid government guarantee to reduce those risks. The recent cancellation by the EOSAT corporation of the contract for Landsat 7 and 8 satellites, until it gets such a guarantee, aptly illustrates the limits of the US approach. The recent difficulties encountered in the US efforts to promote commercialization suggest that it is not yet possible to operate a self-sustaining and profitable satellite system. Interestingly, the Soviet Union has recently indicated a willingness to make available commercially images from their satellites to a resolution of 30 metres and perhaps subsequently 10 metres.

These issues have to be seen in the context of the market-place, and the market is as yet in its infancy. Applications of space imagery to geological exploration, for example, are already well developed and widely used by oil and mining industries. This is not the case for many mapping applications, which require high-resolution data and stereoscopic capabilities, or for crop conditions and pollution monitoring, which require frequent observations during the growing season. In addition, the operational use of such data requires technical equipment and training which are not usually available either in most government administrations or in private corporations (with the notable exception of mapping agencies).

The routine use of remote sensing will come about only if service industries bridge the gap between image producers and users. It is vital that a strong and efficient relationship develops between space imagery producers, distributors and the value-added image-processing industry. The development of the European market therefore requires producers who are responsive to market requirements.

(c) BASIC SPACE SCIENCE

For the purposes of convenience the following sections distinguish basic from applied science, even though the two shade into each other. Under basic science we include those activities which have an intrinsic intellectual value and which have been so important in stimulating European interest in space. The applied science section then focuses on activities with further, perhaps eventually commercial, applications.

Space flight has brought about major advances in several scientific disciplines, notably astronomy and astrophysics, solar physics, lunar and planetary science, and in the understanding of the sun/earth system, commonly referred to as solar terrestrial physics. Observations from space were bound to be highly significant for these areas of fundamental research, since the elimination of the blanketing by the Earth's atmosphere made the whole electromagnetic radiation spectrum accessible for the remote sensing of the physical and chemical properties of the sun, the stars and the galaxies.

Exploration of the Earth's radiation and magnetized plasma environment, and of the Moon, the planets and interplanetary space gave a totally new dimension to the study of the origin and evolution of our solar system. Recent spectacular, and extremely successful, examples of these endeavours were the encounter of the Nasa-Voyager space probe on its trajectory along the outer planets with Uranus and its moons in January 1986 and the encounter of ESA's Giotto probe with Halley's Comet in March 1986. Both missions have provided a wealth of new data which necessitate substantial revisions of our assumptions.

Traditional space science. The topics mentioned above belong to what one could nowadays call traditional space science. Scientists were the first users of the pioneering spacecraft and have ever since formed a large user group. The investigation of the origin and evolution of the universe still presents one of the largest scientific and intellectual challenges for mankind. Space-derived research has provided a quantum jump in the furthering of our knowledge, and space-borne instruments have opened up new wavelength windows on the universe in the infra-red, ultra-violet, X-ray and gamma-ray regimes. This has led to a drastic change in our perceptions, and the exploration of the universe from space will continue to play a dominant role in any further progress in these areas of basic science.

Microgravity science. Only a few years ago, the first space laboratory, Spacelab, brought a new category of experiments in fundamental science within reach: those which make use of the near absence of gravity within a space vehicle in continuous free-fall. This microgravity research entails the investigation of gravitational effects, or of effects that are normally completely masked by the influence of the Earth's gravity in a ground-based laboratory, in a number of physical, chemical or biological phenomena. Although very low gravity is the most obvious characteristic of an orbiter's space environment, other special environmental conditions can also be employed. These include energetic cosmic-ray and trapped particle fluxes, irradiation by the full solar spectrum and ultra-high vacuum conditions with practically unlimited pumping capacity.

Earth science from space. Observations from space provide a unique and powerful way of studying the Earth in a global sense. A coordinated programme, the International Geosphere Biosphere Programme (IGBP) is currently being formulated for a worldwide study of global change. Remote sensing of the Earth from space is potentially of great importance and will have a major impact on a number of established branches of fundamental science, namely, meteorology, oceanography, climatology and solid-earth sciences such as geokinematics and the study of the geopotential field.

Achievements and prospects

Traditional space science

In space astronomy, the era of exploratory survey missions is almost over. The latest, and remarkably successful, mission was the Iras satellite. A series of 'second-generation' projects have been carried out or are now in a preparatory stage. Europe has made a significant contribution in these areas, except in planetary surveys, which have been deemed too costly.

In the future the emphasis in astronomy will shift towards the development of the large semi-permanent observatories in space that are necessary for the next breakthroughs in astrophysics. The first such mission is Nasa's Hubble space telescope, scheduled for launch in 1988, with a small European contribution. In planetary science the next major step will be prolonged and in-depth exploration of topography, atmospheres and plasmaspheres with the aid of planetary orbiters and rendezvous of probes with 'primitive' bodies in our solar system (comets and asteroids). Such major scientific endeavours depend on international cooperation, in some cases on a worldwide scale. Smaller-scale missions are dedicated to specific scientific objectives and to the verification of new instrumental and operational concepts. A programme for traditional space science requires a mix of large- and small-scale projects.

The backbone of European space science over the past twenty years has been the mandatory science programme of ESA. Although much less ambitious than Nasa, ESA has built up an impressive series of 14 successful science missions in a row, without any failure. The scientific programme is so far the only mandatory ESA programme to which the member states contribute on the basis of their

GNP, and scientific merit has always been the paramount criterion. This has led to high-quality science products. In addition, national space science projects in some European countries, notably the United Kingdom, Germany, Sweden and the Netherlands, some associated with Nasa programmes, have further helped to build up Europe's reputation. A supranational effort is crucial to enabling Europe to take part in at least some of the main scientific space ventures of the future. The limited financial resources available within Europe and the necessity for a timely development of the enabling technologies demand a selective approach.

A long-term plan, covering at least two decades, was defined by a Survey Committee of leading European scientists in 1984. Horizon 2000 comprises four main elements, or 'cornerstones', for space science in Europe: two large-scale missions in X-ray and sub-millimetre astronomy; a coherent approach to solar terrestrial physics, comprising an upstream solar observatory and a multi-probe space plasma mission; and a cometary science project to return pristine material to earth. The plan is flexible enough to incorporate smaller-scale projects. Horizon 2000 meets the requirements of timely identification of areas of research in which Europe can claim a leading role or can make a substantial contribution vis-à-vis the other space powers. It possesses a high level of technological sophistication and thus acts as a catalyst for innovative technological developments in European industry. This technology-pull has been a unique feature of scientific satellites from the beginning and is likely so to remain, since it is the very nature of the scientific quest to strive for the nearly impossible. Horizon 2000 strikes a proper balance, albeit as a minimum programme, among the main areas of interest in the European scientific community. It does not exist in a vacuum but involves cooperation on a worldwide scale without any major loss of European autonomy and independence.

Apart from the science programme of ESA, national and multilateral projects can be made largely complementary and indeed can add scope and versatility. Moreover, a European science programme requires the national development of scientific instruments and their enabling technologies.

The programme for traditional space sciences is not contingent on the existence of a manned space station. But a space station could bring technical advantage and facilitate experimental techniques. Preliminary analysis of the potential of a station indicates that its usefulness as a payload carrier for large astronomy missions is limited by the disturbances it generates. It is technically useful as a service, assembly and staging base, although the cost-effectiveness is unclear, since a reliable cost estimate is not yet available. For many purposes a free-flying platform is a better solution.

Microgravity science

This branch of space science covers a large variety of research topics, including solidification and fluid physics, physical chemistry, biology and biotechnology, animal and human physiology, and medicine. Absence of gravity prevents quick segregation in heterogeneous mixtures, avoids gravity-induced convection arising from temperature or concentration gradients, eliminates hydrostatic pressure and provides the possibilities of containerless experiments. Some of these issues are considered in the section on applied science.

In the field of gravitational biology, the mechanisms of gravity sensing, the threshold of sensitivity and the mechanisms of response in living systems are not understood. The few experiments conducted in this field under microgravity indicate pronounced effects on replication and differentiation during culture owing to weightlessness. These results cannot be explained by the current biological theories and therefore constitute basic discoveries.

It is as yet premature to make well-founded statements about the potential breadth in scientific significance and rate of development of the microgravity sciences. Also, since the space-based research efforts are predominantly of an exploratory nature, basic and applied research are intertwined. Many of the current efforts can be considered to lie in the realm of basic science. Relevant microgravity investigations at present include tests of hypotheses which predict what should happen in the absence of a gravitational field, particularly in the material and fluid sciences; the detection of the effects resulting from the absence of a gravitational field and subsequent formulation of explanations, particularly in the life sciences; and the use of the results obtained under microgravity to improve and optimize the related research efforts in a terrestrial laboratory environment.

Earth science from space

Essential to the understanding of the Earth's history, behaviour and evolution is the observation and study of the interplay between its many different constituents, namely, oceans and ice, atmosphere, land and solid earth. Investigations of these phenomena over the past decades have delineated the profound effects, arising from interaction between these components, on the evolution of our planet. The exchange of energy and momentum between the land, the ice and oceans, and the atmosphere determine the thermodynamic equilibrium of the Earth. These interactions are controlled by many factors, such as the transport of heat by ocean currents, the distribution of atmospheric trace gases and pollutants, and the hydrological and biochemical cycles.

The significance of space-based observations lies in their global view of the planet and, most importantly, in the simultaneous measurements of the interaction between the different components of, for example, the climate system, from the same vantage points. An example is the measurement of the anomalies in the temperature of the sea's surface in the tropical Pacific, which have been linked with the observation of large-scale anomalies in the atmospheric circulation. Another example is the influence of major volcanic eruptions on the climate.

The effects of human activity on the behaviour of the Earth system are also rapidly becoming a major issue, and the potential changes induced by this activity need to be assessed. This can be done only if reliable models are available that describe the evolutionary behaviour of the global Earth system. A well-known example of human-induced interaction is the burning of fossil fuel at unprecedented rates, which injects enormous amounts of carbon dioxide into the atmosphere. A steady increase of carbon dioxide in the atmosphere can introduce global climatic changes as a result of the 'greenhouse' effect.

Although limited programmes for global observations of the Earth have already been carried out, the crucial elements for fundamental progress in Earth science will be continuity in observations and permanent monitoring of the Earth, both of which are essential for establishing a firm and coherent data base. In most cases it is necessary to have global coverage, which points to the need for space observations from polar orbits. A major problem is adequate sampling in space and in time. This certainly requires more than a single spacecraft. In other words, international coordination and collaboration are key elements.

An overview was prepared in May 1986 by an Earth System Science Committee, established by the Advisory Council of Nasa. This committee recommended that two programme paths for Earth system science be pursued in the coming decades. First, the study of the solid earth, including measurements of fundamental characteristics, would yield insights into long-term planetary evolution. This would encompass, among other things, investigations of plate tectonic motions, continental deformation and evolution, mantle structure, the Earth's gravity and magnetic fields. These studies would provide input also for the assessment of processes operating on a shorter time-scale as a result of coupling between atmosphere, hydrosphere, crust, mantle and core.

Second, the study of the fluid and biological Earth is a prerequisite for the understanding of global change into the next century. It involves the study both of processes directly relevant to global change (e.g., the physical climate system, the biogeochemical cycles and global moisture) and of the processes playing a significant indirect role, such as local weather patterns, sea-ice distribution and solar variations.

Up to now, Europe's role in this field of Earth system science has been very minor. Its main programmes in Earth observations have been mainly application-driven (e.g. Spot, ERS-1). A balanced European programme is feasible only in the context of a supranational commitment, in which ESA could be instrumental, and it depends on the availability of polar-orbiting platforms as the most advantageous and cost-effective means for global Earth observations in the 1990s. Multidisciplinary payloads provided by the United States and Europe can provide a coherent observational approach to the global study of oceans, ice, land and atmosphere, and need to be operated in a coordinated fashion. Using this opportunity as a stepping-stone, Europe could rapidly catch up with the developments in this field. Dedicated satellites in special orbits are still required for solid earth studies (crystal dynamics, gravity and magnetic field studies) and some atmospheric topics.

(d) APPLIED SPACE SCIENCE

State of the art

Microgravity conditions permit experiments or processes to be conducted in orbit which may subsequently have applications on

Earth. It is here that we encounter the ill-marked boundary between basic and applied science. For example, with regard to crystal growth, weightlessness permits scientists to grow large, near-perfect crystals. This is possible because, in the absence of gravity, little or no distortion of the crystal occurs and there is no contamination. The industrial applications of pure and near-perfect crystals are many. They are required in computers, lasers and numerous other optical and electronic devices. Microgravity conditions offer scientists an opportunity to investigate and improve methods for creating advanced metals, glass and ceramics. One benefit may be the development of lower-attenuation glass fibre for use in optical communication. Another important process that can be studied extensively in space is rapid cooling, which may give materials new properties and therefore contribute to advanced casting technology on Earth.

In the processing of biological material a widely used analytical technology is electrophoresis, in which a gel or another supporting medium is used to suppress convective flows. To apply this process to cells, cell components or other particles in the supporting medium must be eliminated, which becomes possible in microgravity. New medicines may then be produced.

There is, nevertheless, a wide gap between scientific exploration and possible industrial application. The first step in bridging this gap is experimentation and verification. This activity can be conducted on Earth, but it is expensive and available methods produce a very restricted time limit of from 1.7 to a maximum of 20 seconds. One alternative is to use sounding rockets, which launch a small payload into space. As the payload coasts upwards and falls back to Earth, a low gravity condition can be achieved for three to five minutes, permitting scientists to explore a wider range of phenomena. The Spar sounding rockets, for example, provided a large data base which served as the foundation of the microgravity programme aboard the Shuttle.

The second development stage requires the use of a space infrastructure and of the hardware necessary to conduct experiments of long duration. The third stage of the move to commercialization is demonstrating that the processing concept can work on a larger scale, is economically attractive and has a market. The routine production stage would normally follow, but at the moment it seems to be very far away. Currently the research stage is under way, but we have not yet reached the stage of commercial demonstration. The prospects of such commercial demonstration seem to depend primarily on the cost of space transportation and on the reliability of the space vehicles and all of the required infrastructures.

The Shuttle is at the moment the only vehicle in the West which enables experiments of long duration to be conducted in space. Experiments in microgravity using the Shuttle/Spacelab combination, which can accommodate a team of four research scientists, have so far yielded the best results. But there are three important limits. First, the movements of the members of the crew cause micro-accelerations which can compromise the results of certain experiments. Second, the length of the stay in space is about ten days, and many experiments require a significantly longer time. Third, the Shuttle does not produce enough energy for some applications of microgravity. Some of these inconveniences can be avoided by utilizing free-flying platforms, such as the Eureca, which will permit precise microgravity experiments to be conducted without disturbance.

Some American space companies are already thinking of beginning to market an industrial module which would be launched unmanned by the Shuttle. This launch could take place two or three years before the space station is ready and would carry a pressurized workroom in which astronauts from the space station could eventually work in a 'shirtsleeve' environment during the two or three days it would take to service the module.

The space structure which best provides for microgravity activities is, without a doubt, a manned space station.
This would permit experiments of a long duration to be performed directly by a team of scientists who would have the possibility of working in an environment with a microgravity level which is certainly of a higher quality than that of the Shuttle/Spacelab. The space station would also be the ideal structure within which routine production could be initiated. Nasa has begun a project to build a structure of this type which, if current plans and funding are maintained, should be operational some time in the 1990s.

The costs

By the end of the 1990s, if everything proceeds as planned, the Western space industry will have built all of the basic infrastructure necessary to begin the process of the industrialization of space. But what are the costs of the infrastructures, and will they be reliable? Today the cost of Shuttle transportation is very high. This sum must be reduced to approximately a tenth of its current level if it is going to be commercially feasible to produce new materials in space. In effect, the investments which industry must make in order to operate in space are too high at the moment, when compared to the amount industry spends on related research on Earth. A report from the American Office of Technology Assessment concluded that a commercial space venture would have to be assured of very high revenue before it became an attractive investment. Furthermore, there is always the risk that some activities which are now believed to be possible only in a space environment may one day become feasible on Earth at a much lower cost. A significant example of economic prospects discouraging the investor was Johnson & Johnson's cancellation of its proposed collaboration with McDonnell Douglas in the production of the first pharmaceutical products in space. Another factor of uncertainty is that the investors cannot count on a stable and predictable transport system.

In the wake of the Challenger tragedy, Nasa cancelled 15–18 Spacelab missions that were planned to fly during the next five years, greatly reducing the frequency of space experiment opportunities for scientists from the US, Europe and Japan. Nasa is likely to fly only three more Spacelab missions before the end of the decade, only one of which will use a pressurized module. These changes are likely to convince many scientists that they are better off flying their experiments on an unmanned launch vehicle. In discussions on the Hermes and the Columbus projects, many scientists argued strongly in favour of unmanned and cheaper projects, fearing that Columbus would absorb too much of the space budget.

Profitable products

Because of the high cost of space transportation, the conditions necessary for the commercialization of products and profitability are twofold: first, the product or the service rendered must have a sufficiently high intrinsic value; and, second, there needs to be a large enough market. As far as the first condition is concerned one must take into account the value-to-weight ratio. It is obvious that candidate materials for commercial manufacturing in space should be sufficiently light to minimize transportation costs, while being valuable enough to ensure that the market price offsets the costs attributable to transportation.

An example of such a product is pharmaceuticals, whose prices can be very high. The McDonnell Douglas and Johnson & Johnson pharmaceutical venture planned to use space processing for the production of new medicines and was expected to generate one billion dollars in annual sales by the early to mid-1990s. The venture focused on the use of the electrophoresis separation process in space to obtain mass quantities of a hormone. The Ortho pharmaceutical division of Johnson & Johnson expected to begin human patient testing in early 1985 having produced the material on Shuttle mission 41D. McDonnell Douglas astronautics, however, had encountered problems in making the product in the quantity required for clinical testing. At that point Ortho decided to cancel the agreement with McDonnell Douglas. This demonstrates that technical difficulties have not been successfully resolved, even in a type of microgravity activity which is more advanced than others.

So far only one product has been obtained, the monodispersed latex spheres. The spheres are used as tiny rulers to determine relative sizes of objects under microscopes and to calibrate filters, particle counters and porous membranes. Spheres smaller than one micron and up to three microns can be produced successfully in ground laboratories. Producing spheres of a uniform diameter larger than this is very difficult because of adverse gravitational effects. The spheres which have been produced in space by seeded emulsion polymerization are 10 microns (10 thousandths of a millimetre) in diameter. In spite of this good track record, the production of monodispersed latex spheres on the Shuttle stopped in 1985 because of lack of demand.

Another sector of interest to private companies is that of crystal growth. Microgravity Research Associates, Inc., together with Grumman Space, has planned to manufacture gallium arsenide semiconductor crystals in space. Gallium arsenide has properties far superior to silicon, which has been the basic electronic industry semiconductor material for several decades. New super-computers that will perform billions of computations every second, vital for strategic defence systems and advanced satellite communication systems, will have requirements that will surpass silicon technology and open the market for a new semiconductor material. Raw material on Earth is able to yield only 2 per cent of gallium arsenide as an end product, whereas in space it is possible to obtain a much purer product at 10 per cent. The cost of the production of four hundred chips on Earth with a yield of eight good gallium arsenide chips (2 per cent yield) is approximately $300. In space there is a yield of 10 per cent, so that from 400 chips we obtain 40 good chips. The processing costs are very large because of transportation prices. Even though the total cost is higher in space, the potential profits greatly exceed those available from earthbound production. But profits will flow only if there is a market ready to absorb the growing number of microchips.

Some marketing forecasts are decidedly optimistic about the future. The Centre for Space Policy, an American organization which specializes in estimating commercial or industrial opportunities in space, has predicted that by the year 2000 the annual revenue produced by the processing of material in microgravity will be roughly $41.5 billion. It is, however, difficult to make judgments from such assessments, because of the high number of variables involved and the lack of substantial reliability or applicability. Indeed, the commercialization of space may not occur until the twenty-first century, when cheaper transportation may be available.

Organization

In most of the industrialized world there is a marked and growing interest in microgravitational experimentation. These activities are mainly undertaken by governments and by space agencies. Only in the US has the private sector become involved to a relevant extent. Despite considerable promotion by Nasa of the potential commercial benefits from microgravity applications, the results of its own studies so far have not been particularly encouraging.

A European Low Gravity Research Association (ELGRA) has been set up with the sponsorship support of ESA and the Council of Europe in order to coordinate national activities in this sector. In 1982 ESA established the Microgravity Programme for basic research, with modest resources. The main programme involves the launching of sounding rockets for experimentation in fluid physics and material science. A series of experiments have been planned for execution on the Spacelab and Eureca. Within Europe the greatest interest in microgravity experimentation has been evident in Germany, where the Ministry of Scientific Research and Technology spent approximately $150 million between 1978 and 1985 on microgravity activities. The German programme for material processing is not purely scientific in orientation, but also seeks to encourage the industrial sector to explore the potential applications of space-processed metals, composite metals, chemicals and crystals. France is also carrying out a research programme to evaluate material processing applications. Some years ago, CNES designed a specialized automated 'manufacturing-in-space' system, called Solaris. The effort to promote interest in Solaris among other ESA member states has not been particularly successful. British interest in microgravity has been limited. As for the participation of private companies, a new organization called Intospace has been formed by the German MBB/ERNO and Italy's Aeritalia. This multinational marketing organization is designed to bring potential microgravity users together with the producers of European space hardware and systems. So far, however, the involvement of the private sector in Europe remains small.

The prospects for Europe

The industrialization of space, should it occur, would provide a number of important opportunities for those involved. The development of new materials, whose production is possible only under microgravity conditions, could prove to be an important factor, not only commercially but strategically as well (e.g., high-speed microchips for ultra-sophisticated computer systems). It is still far from certain, however, that microgravity activities can be made commercially sound, whatever the technical, economic and institutional framework. If the potential of these activities nevertheless remains attractive, international cooperation appears to be the best way to share the risks inherent in such an endeavour.

The most important decision currently outstanding is over the space station. Among the major divergences between Nasa and ESA, two are especially relevant to microgravity research: the technology transfer issue, and the right to the patents and exploitation of discoveries made through the space station. The German government wants to equip the permanent attached module as an advanced space laboratory for microgravity experiments. Nasa has not yet agreed, and its officials have discussed using it primarily for life sciences and related activities. The question of proprietary rights to exploit discoveries has not yet been resolved in negotiations. One of the objectives of Nasa has always been to channel funds and technical capabilities away from activities which are competitive or not compatible with US interests and to dominate as far as possible the definition of the missions of collaborative programmes .

To proceed with the immediate objective of establishing an independent European capability in the field of space industrialization would appear to be very difficult, unless Europe is ready to act on its own by developing its own infrastructure to which it can control access. Europe is technically capable of achieving such a goal, but the costs would be enormous and in microgravity the commercial prospects are unproven. But Europeans may have to consider whether to acquire the experience which would enable eventual commercial applications to be exploited.

(e) SECURITY

Current military uses of space

For the thirty years of the space age, military space activities have been divided into five traditional missions: communications, reconnaissance and surveillance, navigation, meteorology and geodesy. This section surveys these current space missions. The general characteristics of military satellites and their orbits are governed by basic technical considerations. These are set out in detail in the Annex and Figures 5, 6 and 7.

Communications

Most operational communications satellites (Comsats) today use ultra-high frequencies (UHF) and super-high frequencies (SHF), but extremely high frequency systems are under development. The move to higher frequencies for military satellite communication (Satcom) is motivated by two major factors. First, higher-frequency radio waves have a higher limit to their data-carrying capacity than lower frequencies. Second, transmitting antennas for higher frequencies can be made smaller without sacrificing performance, since the effectiveness of a transmitter dish is determined by the ratio of its size to the wavelength of the radio waves it is transmitting. There are three additional reasons for high frequencies (accompanied by wide bandwidths for some particular needs of Satcom). First, it is easier to protect higher-frequency links against countermeasures. Second, higher frequencies suffer less distortion in passing through an ionosphere disturbed by nuclear detonations. Third, although laser communication is also coming into use for satellite-to-satellite and satellite-to-aircraft links, ground-to-space laser communication links could obviously be frustrated by clouds. Covert 'low probability of intercept' (or LPI) communication, which does not betray the location of the transmitting ground terminal, is easier with wide bandwidths.

Military Comsats are deployed in a variety of orbits. Geostationary is high enough to allow widely separated ground stations to communicate through a single satellite, and a stationary satellite makes it easy for users to point their antennas. But the polar regions are invisible from geostationary equatorial antennas. The Soviet Union, which has many military installations at high latitudes, deploys Satcoms in Molniya orbits (see Annex). A communications satellite in LEO is visible at any given time from a relatively small 'footprint' of Earth below. Two terminals within the print can communicate directly, but widely separated users must store messages on board the satellite when it is overhead, ordering the satellite to 'dump' the message when it passes over the recipient. The Soviets deploy large numbers of such store-and-dump satellites.

Users of military Satcom fall into three categories: high-data-rate peacetime users, including intelligence and diplomatic terminals; tactical forces, which need moderate data-rates but worldwide coverage, small mobile terminals and resilience to disruption; and nuclear forces and their commanders, needing low data-rates but performance under severe stress.

Today's US Defense Satellite Communications System (DSCS), Fleet Satellite Communications (FleetSatCom) and Air Force Satellite Communications (AfSatCom) systems broadly represent this tripartite division. The DSCS provides the US government with worldwide (except the polar regions) high-data-rate voice and data communications.

The FleetSatCom (or FLTSAT) constellation consists of four synchronous satellites distributed about the equator in almost the same positions as the four DSCS spacecraft. The FLTSATs relay a fleetwide broadcast, transmitted to the satellites from ground stations, which almost all navy ships can receive. Surface ships normally transmit messages back to shore stations via the satellite link. Submarines, which can only receive and transmit at certain times, are provided with special facilities. All channels in the FLTSAT system operate in the military UHF band, except for an SHF uplink that carries the fleet broadcast up to the satellites.

The AfSatCom programme does not consist of a particular satellite constellation, but rather of communications packages on spacecraft sometimes designed for entirely other purposes. In addition to channels on the FLTSATs, the AfSatCom system includes communications packages on the DSCS spacecraft in Molniya orbits, and, in the future, on the 18 satellites of the Navstar Global Positioning System (GPS) constellation, providing worldwide coverage. Obviously, almost any military spacecraft would make a convenient 'host' for a small AfSatCom transponder, and it would be possible in principle for the Department of Defense to put simple transponders on civilian satellites as well. The array of Comsats available to the US military is completed by certain Nato systems and various experimental spacecraft.

The control of complex spacecraft that require frequent ground commands depends on a worldwide network of ground stations. Suitably located ground stations in areas resistant to political change, not to mention military conflict, are hard to provide. Direct satellite-to-satellite relay links avoid all these problems and obviate the need to store data on board, when the satellite is not in direct view of a control station. Nasa's Tracking and Data Relay Satellite System (TDRSS), consisting of a pair of spacecraft in synchronous orbit, will provide essentially uninterrupted relay between satellites at all altitudes with one ground station at White Sands, New Mexico.

Reconnaissance and surveillance

Electromagnetic radiation emitted or reflected from terrestrial objects can be detected from space in any of the three wavelength bands to which the atmosphere is transparent, i.e., the visible band, certain infra-red bands, and the microwave radio band. It follows that these are the bands used for military surveillance. In peacetime, these remote-sensing techniques are used for both tactical and strategic purposes. These include the collection of communications intelligence, the identification of the characteristics of new and deployed weapon-systems, and the gathering of tactical intelligence, such as tracking fleet movements, locating land unit headquarters and monitoring activities at air bases.

Although many of the same remote-sensing technologies apply to both tactical and strategic intelligence, there are three particular requirements for tactical missions. First, battlefield intelligence must be acquired, processed and disseminated rapidly, if it is to be useful. Therefore, the orbit and number of satellites should be chosen to allow continuous or very frequent survey of the battlefield: geostationary, Molniya or harmonic orbit, and a large number of satellites. Transmission time should be shortened. Data Relay Satellites and highly secure communications become very important. Second, tactical sensors of genuine military value must expect to come under attack, whereas peacetime intelligence collection is not likely to be directly impeded. So they must be well protected. Third, space-based sensors – a necessarily global capability – must compete in cost-effectiveness and survivability with other means of collecting tactical intelligence, such as aircraft and remotely piloted vehicles.

As for nuclear operations, space can be used to detect missile launches and nuclear detonations. Data that warn of missile launches permit the safe escape of bombers, tankers, cruise missile carriers, airborne command posts and, for launch under attack (LUA), ICBMs. Confirmation of detonations might also serve as a last check on an LUA decision. But the most important use of missile launch and nuclear detonation data would probably be to give decision-makers a clear assessment of what had happened – information crucial to responsible action and, under the likely chaotic circumstances, otherwise hard to come by.

Imagery

The resolution of a spaceborne optical camera is proportional to its altitude (See Table 8). Thus a photo-reconnaissance satellite orbiting at an altitude of 200 km and yielding imagery with one-foot resolution would at 5,000 km yield Landsat-type images that would be useful for forestry, but useless for most intelligence purposes. Photo-reconnaissance satellites are therefore confined to low earth orbit. Coverage at all latitudes requires polar orbits for these satellites. Furthermore, if the satellite has an on-orbit lifetime of longer than a month, it should make use of the sun-synchronous property of certain near-polar orbits, which maintain the same orientation with respect to the sun as the season change, always taking pictures at the same local time on the Earth below regardless of season. Low-altitude, polar, sun-synchronous orbit is therefore the home of long-life photo-reconnaissance satellites.

Infra-red cameras collect information about the surface temperature of objects on the Earth, potentially revealing features obscured at visible wavelengths. Radar images can be formed by illuminating the Earth with microwaves and processing the reflected signals. Active satellites provide night-time and all-weather imagery.

Signal detection

Satellites can also detect signals in the visible, infra-red and radio bands, including microwave pulses from the radars on a ship, telemetry from a cruise missile test vehicle, the visible flash of a nuclear detonation, or the infra-red plume of an ICBM launch. If the signal is sharply structured in time – like the flash of a nuclear burst or the pulses of a radar – the emitter's location can be deduced from the differences among the signal's arrival times at several well-separated satellites.

Intelligence satellites designed to operate at radio and radar frequencies are know as 'electronic ferrets' or 'electronic ears'. Their functions include eavesdropping on radio communications; locating sites with particular 'radio' characteristics; monitoring the intensity (or changes in intensity) of radio emissions (as an indication of new military activity); mapping the coverage and characteristics of radar installations (early warning; ground-to-air missile control); and intercepting telemetry transmissions from weapons under test.

One particular benefit of electronic reconnaissance is that it adds a dimension to photo-reconnaissance data. An electronic signature can convert a puzzling photographic 'blot' into a completely identified object. Ferret packages may be operated on their own or used 'piggyback' on a photo-reconnaissance satellite.

Orbits for signals detection should be chosen to provide continuous coverage of target areas, thereby preventing the opponent from concealing information by performing tests, sending messages, moving mobile radars or launching missiles during coverage gaps. Geostationary orbits offer continuous dwell over mid-latitudes; the US acknowledges that it stations warning satellites there. Long dwell times (and coverage of northern latitudes) are also possible from Molniya orbits; the Soviet Union deploys warning satellites in this way.

Continuous coverage by several widely separated satellites, permitting emitter location by the time-difference-of-arrival techniques, requires a 'birdcage' constellation; the US Nuclear Detection System (NDS) aboard the Navstar GPS satellite is in this kind of orbit.

Apart from geostationary missile warning and birdcage NDS deployment, the US has not revealed the locations of its other signal detection satellites. They could make use of all three possibilities: that is, geostationary, Molniya, and low and mid-altitude birdcage constellations.

Navigation

Accurate navigation is essential for supporting reconnaissance, weapon delivery (including sea-launched ballistic missiles), battle management, the precision emplacement of sensors and mines, and rendezvous. Terrestrial navigation systems have either restricted coverage or poor accuracy. In one satellite navigation method, used by the US Navy's Transit system and its Soviet equivalent, the user listens to how the received frequency of a radio signal changes as the transmitting satellite passes from horizon to horizon (Doppler shift). By knowing the satellite's orbit and the pattern of frequency change, the receiver can deduce its position on the Earth's surface. The need for global coverage makes polar orbits best for these satellites.

In a second navigation method, the user measures the arrival times of signals from several well-separated satellites, and then uses the inverse of the time-difference-of-arrival emitter location technique to deduce its position. The Navstar GPS provides just such a constellation.

Meteorology

Military operations, which are increasingly wide-ranging, require knowledge of the weather pattern in distant parts of the globe. The US Defense Meteorological Support Program (DMSP) satellites are in near-polar sun-synchronous orbits at 850 km. They orbit the earth 14 times per day at approximately 100 minutes per orbit. From an altitude of 850 km, the width of the swath of Earth visible below is almost 3,000 km – one fourteenth of the Earth's circumference. Thus, on its 28 passes over the equator each day, a DMSP satellite views nearly every point on the equator twice, once on an ascending pass and once on a descending pass. At other latitudes the swaths overlap. Data for military use can of course also be extracted from civilian weather satellites.

Geodesy

A precise knowledge of positions on Earth, and of Earth gravity on the various parts of a trajectory, is essential to the guidance and targeting of any vehicle that relies on inertial navigation, such as a submarine, aircraft or missile. Many of the well-publicized 'scientific successes' of space exploration and continuing work in that area are of military origin. For the development of ICBMs it was necessary to obtain precise measurements of the relative position of the planned missile launching points and the missile's targets. Such information could only be obtained from space. The accuracy of ballistic missile guidance systems was also critically dependent upon knowledge of the very small perturbations in the Earth's gravitational field. This too could only be explored effectively with the assistance of space-based measurements. Today, the development of effective cruise missile guidance systems also requires the use of space. Such guidance systems work on the principle of comparing a 'digitized map' of the Earth's surface held inside the missile with the picture of the ground being 'seen' by the missile's sensors as it flies to the target. The 'digitized map' has to be produced from space observations.

The basic operational mode of a geodetic satellite system consists of an ultra-precise measurement of orbital movements of special satellites. Since these satellites are designed to allow a precise prediction of their orbital parameters, any deviation from these values can be related to variations of the Earth's gravity. Conversely, knowledge of the orbital position of the satellite permits positioning in relation to it.

New military uses of space

Besides these traditional military uses of space, some more exotic ones are emerging.

Asat and Dsat

Asats (satellite attack systems) consist of capabilities to destroy, incapacitate or interfere with a satellite mission. They are a response to the growing military and political importance of satellites. They can be of many types ranging from ground-, air- or space-launched missiles, to lasers and high-power microwave transmitters. Studies have shown that means to destroy, incapacitate or interfere with satellite systems are so diverse, and so difficult to monitor, that it is difficult to imagine an effective and verifiable ban on Asat systems. Similarly, an attempt to segregate various kinds of satellite so that some of them will not be targets does not seem realistic, because it would significantly reduce the protection for the others. Last but not the least, some protection does exist for US and Soviet satellites that can be defined as 'national means of verification'.

The Soviet Asat system already has a ground-launched interceptor, which is sent into the same orbit as the target. This is attacked, after one or two orbits, by a radar-guided vehicle with an explosive warhead which is detonated when the two vehicles are close enough. The current Soviet system can reach satellites with altitudes below 2,000 km for polar orbits and slightly higher for orbits with lesser inclinations. Such a system has a built-in lack of flexibility, because co-orbital attack requires delay until the target orbital plane is over the launch site. The ability to strike a full constellation in a short time is limited. Furthermore the system's dependence on active radar makes it vulnerable to jamming.

The US system, still in development, is significantly different. An air-launched missile, using infra-red guidance, scores a direct hit on the target. This design allows much greater attack flexibility than the Soviet system, although its altitude capability is limited to around 1,000 km.

Dsat systems (satellite defence) are under discussion as a set of means and actions that could be used to defend satellites against Asat attacks, including electronic countermeasures. They include passive methods (orbital manoeuvres, decoys and hardening) and active methods (jamming, firing projectiles at the attacker, and so on).

Both Asat and Dsat systems rely heavily on a satellite tracking and identification network. This consists of radars and optical systems positioned around the world and linked to a central system for data management. Radars are used for trajectory and gross measurement of the objects in orbit, whereas optical telescopes, located in relatively cloud-free areas or airborne, provide identification. The comparison of orbital tracks with the predicted tracks of existing satellites provides real-time information on launches of new satellites or manoeuvres of existing ones. Such a system provides an early-warning function together with the information necessary, first, for the safe use of space, since it monitors all the debris that could be dangerous to manned or unmanned satellites, and, eventually, for the targeting of an Asat system or a Dsat system.

ABM (anti-ballistic missile) system

Since President Reagan's speech of March 1983, the world has realized that research had already been going on in the US and USSR to study the feasibility of using both space- and ground-based assets to build an ABM system. Any system of that kind would necessarily involve a detection and discrimination capability, ground- and/or space-based weapons and a battle management system. Space systems would be placed in different orbits: those directed at early warning and the tracking of a boosted vehicle would be in geostationary or Molniya orbits. Systems of discrimination and tracking at the ballistic phase would be on altitude orbits of 4,000–10,000 km in order to observe more closely the re-entry vehicles and the decoys, while being able to defend themselves against Asats. Space-based weapons would be in low altitude around 4,000 km, allowing a short time for the weapons to target, while being able to counter Asat attacks.

The implications for Europe

The profiles of national space policies set out earlier in this report drew attention to the considerable military space capabilities of the USA and the USSR, and to an emerging Chinese programme. Within Europe only France and the UK have any significant capabilities, both selectively focused on specific tasks. Italy is developing a military communications system. Europe, as such, has no collective system of its own, although members of Nato's integrated structure share in its collective satellite communications systems. Although the necessary technology and overall economic capability exist to go beyond this, Europe has been unable to resolve its military and political differences up to now, or to define specific and agreed collective requirements. However, some national requirements and interests can be identified which might form the elements of a collective approach.

Reconnaissance and surveillance satellites

French interest in an independent system is linked to nuclear deterrence (better knowledge of targets and ABM defences in the

Soviet Union, etc.), to the overseas operations (mainly in Africa) of the Force d'Action Rapide (FAR), and to a desire for independent intelligence sources to support French diplomacy. These concerns have been met by an optical reconnaissance satellite with a non-harmonic sun-synchronous orbit, in order to allow a periodic full scanning of the Earth, with non-real-time image availability.

The Federal Republic of Germany is interested in monitoring its eastern border zone, which is about 1,000 km North-South and 100 km East-West. This requires an all-weather capability to meet local conditions. A radar satellite has the advantage of overcoming the weather conditions, but it is relatively easy to foil when monitoring a land area, and in any case it would not meet the need for real-time imaging of a narrow area, such as the Central Front in Europe.

The UK has a keen interest in strategic information about targets for its nuclear forces and ABM defences, and in intelligence to support diplomacy. Under bilateral agreements the UK has access to wide-ranging US intelligence sources, which produce a quantity and quality of information which neither the UK alone nor Europe collectively could replace in the short-to-medium term. But British concerns are evolving and might begin to encompass a search for complementary capabilities within Europe. The Italians have a growing interest in access to a collective system, and the Spanish have signalled a similar concern; hence the decision by both to associate themselves with the French Helios programme.

Because military doctrines differ significantly among the four largest countries in Western Europe, it is difficult to reach a consensus on the military use of a single joint reconnaissance satellite. That situation is made even more complicated by domestic political issues: it would be difficult for example for the Federal Republic of Germany to be part of a satellite system that supported significantly British or French nuclear forces. Nevertheless, there may be scope for bilateral projects: the United Kingdom and France have nuclear forces with similar targeting and penetration problems. Similarly, Italian and French interests in the Mediterranean and in Africa could be rendered compatible.

A joint European reconnaissance system could have great political significance. It would enable Europe to monitor treaty compliance and crisis behaviour, and it would also give Europe a voice in any pertinent discussions on space law. European nations cannot expect to be involved in any East-West discussions, or to be signatories of a treaty, without an independent capability for monitoring treaty compliance. And, of course, such a reconnaissance system would permit crisis monitoring within Europe, without having to rely on indirectly received evidence. It should be underlined that for both uses – treaty compliance and crisis monitoring – every government would keep its independence of decision while benefiting from its shared source of information. Independent European analysis could well help rather than hinder transatlantic cooperation.

A reconnaissance system would also provide the military with data on the readiness, equipment and organization of enemy forces. Europeans would have the option to 'piggyback' an electric intelligence package.

Communication satellites

European cooperation would be simpler to set up in this field. By definition Comsats can be made transparent, so that every country could use its part of such a satellite. The service could include a worldwide coverage: one spot beam over Western Europe and one or two movable spot beams for overseas operations. Such a system would be a significant step forward in interoperability and could maintain each country's freedom by the allocation of separate but identical transponders. But the case for a European system depends on a judgment about what this would add to the mix of Nato and national systems already in operation or in prospect.

Nato's involvement in the field of satellite communications began in the mid-1960s with a technical group study of how a Comsat system might be used both to assist communication between the Nato Council and its member governments, and to provide means for exercising the commad and control of Nato ground, sea and air forces. In 1971, the North Atlantic Council agreed a policy statement which set out in definitive terms the role to be played by satellite communications in Nato. The implementation of this policy was entrusted to the Nato Integrated Communications Management Agency – now renamed the Nato Command Information Systems Agency (NACISA). The result of this policy today is the Nato Integrated Communications System (NICS), the third phase of which has recently been completed. It provides a sophisticated, flexible, survivable, high-capacity and complex network of ground- and space-based elements for mainly fixed telegraph and voice communication circuits, and it supplements appropriate national systems. It is interoperable with the US DSCS and used by the European members of Nato's integrated military structure.

When operated together with appropriate national space- and ground-based systems, NICS should allow the full military communication requirement within the North Atlantic Treaty area to be met. Technical and operational improvements, particularly in the field of signal-processing, are of course continuously under consideration. The basic principle, however, of a mix of national and Nato systems, including a diversity of space- and ground-based elements, provides a great deal of versatility and robustness. In a more narrowly defined European context, there is clearly room for cooperation between the two European powers which retain residual worldwide military commitments – France and the United Kingdom – in ensuring that there is not wasteful duplication in meeting their national global communication requirements. But there is also a case for wider European cooperation in ensuring that the architecture of the various developing military and civil communication systems, together with their system of operational management, provides the users with the maximum possible degree of operational flexibility for military use, particularly in the event of unforeseen circumstances. Such cooperation does exist but needs to be strengthened within the general framework of European military cooperative programmes. This European cooperation must take account of US national developments in military communication, but needs to recognize that some US national requirements may lead the US down paths which the Europeans may neither wish nor need to follow.

Asats

With the advent of Asats and space-based defence systems, the survivability of satellites is no longer assured. No protection exists today for European satellites, civilian or military. Two approaches to this problem can be envisaged. Europe could develop a full Asat system to deter any attack on its satellites by threatening to reply in kind. Such an approach would probably be effective but expensive. The other approach would involve strengthening space law to ensure the survivability of space-based assets. One proposal is to define 'keep-out zones' in space, which satellites from adversary countries would not penetrate except for a short period of time and with advance notice.

Uses

Without a military satellite of its own, Europe is not well placed to demand inclusion in any treaty to protect satellites or to define proprietary keep-out zones in space. Obviously such a move would imply the development of some limited Asat/Dsat capability.

Ground tracking and identification system

Whether or not Europeans decide to develop Asats or to participate in a treaty about the use of space, Europe will need a ground tracking and identification system in order to acquire an independent knowledge of satellites orbited by other nations, and to ensure that none of them is threatening European space assets. Such a facility is an obvious candidate for collaborative European development and operation.

6

MEANS

Access to space and the opportunities which are there to be seized demand a sophisticated technological and industrial infrastructure, as well as the capability to manage very large programmes coherently and cost-effectively. This chapter identifies the elements needed to give Europe the chance of turning its potential across the range of space applications into solid and sustained achievement.

Space transportation systems

The exploration and use of outer space require capabilities for lifting objects outside the Earth's atmosphere. Powerful launchers are required, whether to place heavy payloads into low orbits or to place smaller payloads into distant orbits, particularly the geostationary orbit located some 36,000 kilometres from the Earth's surface. Only a few countries – the Soviet Union, the United States, China, Japan and the European Space Agency – have met these requirements, though more have placed small payloads in low orbit. So far only the Soviet Union has achieved launches on a routine operational basis. (See Table 6.)

Launch vehicles have in the main been developed from ballistic missiles (the Indian and Japanese vehicles are exceptions), thus making the expendable launch vehicle (ELV) the primary means for obtaining access to outer space. The projected increase in the use of space led the Americans to perceive ELVs as inefficient, and encouraged them to develop the Shuttle as a more efficient and cost-effective system. The Shuttle is a partially re-usable and manned vehicle capable of carrying and returning cargo to and from space. The USSR continues to rely on ELVs, but it too is developing a shuttle system. There is also an increasing requirement for mobility within space, to reach satellites or platforms in close proximity to a large staging post such as a space station or shuttle. In the future an orbital transfer vehicle will be needed to move payloads from one orbit to another, well-separated orbit: for example, to bring a payload from GEO to a space station in LEO for refurbishment. Recently there has been growing interest in new launch systems, reflected in the rash of proposals for horizontal take-off and landing vehicles as possible replacements for the traditional vertically launched vehicle.

Shuttle

The first partially re-usable manned space vehicle was incorporated in the Space Transportation System introduced by the United States in 1981. This Shuttle system consists of an orbiter with three liquid-fuelled engines, two solid-rocket boosters and a large external fuel tank. Launch is achieved by the combined firing of the liquid-fuelled engines on the orbiter and the solid-rocket engines. The solid-rocket casings are parachuted back to Earth to be recovered and re-used, whereas the external tank is released just prior to orbital injection and burns up on re-entry into the atmosphere. The DC-9 orbiter carries a crew and payload of some 30 tons to LEO, and 15 tons can be returned to Earth; for higher orbits an additional stage is attached to the payload. Once in orbit, payloads can be lifted out or hauled into the cargo bay by a remote manipulator arm, currently built in Canada. A manned manoeuvring vehicle has been developed to allow access to satellites close to the Shuttle. The Shuttle is so far the only vehicle of its type.

In the event the Shuttle has not provided easy or low-cost access to space. It was envisaged in 1975 that by 1984 the price for a dedicated Shuttle launch would be $40 million, but in 1984 a Congressional estimate put the cost of a flight at some $375 million. Prices charged to customers have steadily increased from $80 million in 1984 to some $150 million (at 1984 values) in 1988. The American government has decided that prices will increase to reflect the real operating costs. The use of the Shuttle by Europeans for Spacelab, and eventually for transporting modules for the projected international space station, is expensive, even without the uncertainties resulting from the technical problems revealed by the Challenger tragedy. These factors are all the more significant because most of the American ELVs were phased out in the 1980s, when the Shuttle fleet became operational.

The main advantage of the Shuttle is its flexibility as a space system. This flexibility is based on its large payload capacity; its provision for the repair and maintenance of satellites and platforms in LEO; facilities for scientific and technological experiments over an eight-day period; the transport and assembly of space station modules; the servicing of a space station with crew and cargo; and the transport of cargo from space. Nevertheless, it should be noted that the Soviet Union has operated and routinely resupplied a space station for over a decade through the use of a heavy launcher (32 tonnes to LEO) with a space station as payload. This space station is, however, much smaller than that contemplated by the United States.

Ariane

Until Ariane was developed, launchers for heavy payloads, especially to GEO, were available only from the superpowers. This made Europe, and indeed other countries, highly dependent. Early European attempts to develop a launcher through ELDO were abortive, but reliance on the US subsequently proved unsatisfactory, for example in the difficulties over the launching of the Franco-German Symphonie satellite. The European decision to start the Ariane programme turned out to be a prudent one.

Ariane was developed at a cost of some 960 million EAU and had its first successful flight in 1979. It provided the Europeans with an independent launch capability, but early attention was also given to the scope for capturing a share of the market for launch services. Arianes 1, 2 and 3 are the original family; Ariane 4 has been up-rated from 1,750 kg to 4,300 kg for geostationary transfer orbit (GTO) in order to reduce the cost per kilogram in orbit. Since ESA could not direct its own commercial operation, Arianespace was established as a private concern, separate from ESA.

When Shuttle launches proved complex to manage and costly, and in the absence of other competing ELVs, Ariane became a successful commercial operation. It was capable of carrying two payloads on each flight at a cost for a dual launch of some $54 million in 1984 (cf. Shuttle costs of $80 million), and $75 million in

1985. An Ariane 4 launch, with a load factor of 100 per cent of the 4,200 kilograms, costs ESA customers around $95 million in 1987. The cost of a launch for other customers depends on market conditions. The international launch market is such that no launches are offered on the basis of full-cost recovery.

In 1985 the ESA Council decided that the Ariane 4 programme, though suited to current launch demands, would not be able to meet the future launch needs of Europeans and other users. They therefore approved an optional programme for the development of Ariane 5 as a heavy-lift vehicle (originally set at 15 tonnes to LEO), with estimated costs at the time of 2,600 million EAU. Ariane 5, which is to be ready by 1995,is intended to be 20 per cent more cost-effective than Ariane 4 for both LEO and GEO; capable of accommodating wide-diameter payloads comparable to the size of the cargo bay of the Shuttle; and more reliable. It is also to be man-rated, that is, capable of launching Hermes; and to carry modules that can be assembled into a space station.

Hermes

Several countries have plans to develop small manned orbiters, which would not have any role in the launch process, but would be attached as a payload to the launch vehicle. One such vehicle is currently under test in the Soviet Union and another – Hermes – was adopted by ESA in 1985 in a preparatory programme, and originates from a French proposal. The Hermes project's incorporation of the transport of both crew and cargo is potentially relevant for a future European space station. The capital costs of the project were estimated in 1985 by CNES at 2,000 million EAU, with annual operating costs of 220 million EAU for two flights per year. The 1987 ESA estimate is 4,534 million EAU for development, with operating costs of 276 million EAU. Hermes was originally judged to be admirably suited to the long-term ESA objectives of European autonomy. Recent assessments within ESA suggest, however, that the initial operational and budgetary forecasts were over-optimistic, and that the cargo-carrying capabilities of Hermes will be limited. Initially designed to weigh 16 tonnes with cargo, it will now weigh considerably more. Configured at 21 tonnes, Hermes could carry 3 astronauts and 3 tonnes of payload.

Ariane 5 and Hermes will be key and interdependent elements in the ESA plan to enhance European autonomy. A manned presence in space will require both. A European space station will require Ariane 5 and Hermes, with an appropriate cargo capacity. Both projects are technically and managerially demanding, as recent revisions of some of the design parameters and the timetable indicate.

Launch service

The provision of a launch vehicle is only one element of a launch service. Launch-pads must be built; special facilities provided for the integration of the payload and the launch vehicle; and equipment and personnel made available for tracking and control of the vehicle after launch. Ariane launches take place at the Centre Spatial Guyanais (CSG) in Kourou (French Guiana), close to the Equator and thus favourable for launch into GEO. The CSG was made available to ESA in 1979, after operating initially for the French national programme. ESA *and* its members are guaranteed access to the CSG facilities, and ESA has priority of use.

The main operations at the CSG are payload integration; launch facilities; and tracking and telemetry functions during launch. Two launch-pads are available, allowing a maximum of ten launches per year. The second launch-pad was completed in 1985 after four years' development and at a cost of 240 million EAU. It is the property of ESA and has been put at the disposal of Arianespace. Even these facilities for a rapid turn-around or launch-on-demand are limited, and any serious accident at the CSG would be a severe setback. It may be desirable to have additional facilities, but locations as suitable for GEO launches are not plentiful. One possibility would be the Italian site at San Marco, Kenya. Launches into polar orbit and orbits for many military payloads gain less advantage from an equatorial launch site, and alternatives may be available in Europe. In the future a European spaceplane might utilize conventional airfields and need less complex facilities.

The CSG tracking and telemetry facilities provide accurate orbit determination and allow launcher performance to be checked, using facilities in Brazil, Ivory Coast and the Nasa/DoD facility on Ascension Island. The status of the payload is monitored by ESOC at Darmstadt. ESOC's task is to monitor launch and early-orbit-phase operations so as to place the payload into the required orbit and perform in-flight commissioning and testing prior to putting the payload into service. For these purposes ESOC operates a dedicated ground system, including control and computing facilities and a terrestrial tracking network (Estrack) with stations located in Belgium, Spain, Germany, Gabon, Australia, Kenya, and French Guiana. ESOC is capable of controlling the complete range of spacecraft missions from near-earth orbit to deep space. Mission, spacecraft and ground station control are integrated into a single facility. However, this facility was not designed for manned missions, nor is it capable of tracking the full complement of objects in space.

Launch market

The demand for launch services is not precisely predictable. Forecasts have suggested a market of 120–160 payloads to the year 2000. But rapid developments in fibre optics complicate forecasts for communication satellites, and the expected demand for DBSs has been slow to materialize. The extent of the demand will also depend on other factors, such as the building of a space station, the promotion of other applications satellites and military activities.

A customer's choice of a launcher depends on price, availability, the perceived reliability of a launch vehicle and the insurance requirements. Offers from the Soviet Union and China raise further issues about technology transfers and foreign policy. The recent spate of launch failures has vastly increased insurance costs, to the extent that in some cases launch and insurance costs together reach the cost of the payload. Reliability and insurance arrangements could therefore be crucial factors in the future, if launcher supplies outstrip demand.

Ariane has been able to compete with the Shuttle because of its precision, its less complex procedure for planning and integration of payloads and its lower prices, together with an aggressive market strategy. However, there used to be no third competitor, a situation that has dramatically changed. New ELVs are under development in the United States; the USSR and China have entered the launch market; and Japan plans to produce a competitive launcher. Customers will no doubt be faced with an array of tempting terms for launch and insurance. In this milieu the competitiveness of European launchers will be of paramount importance, and the ability to attract foreign customers as well as to keep European loyalties will be sorely tested. To be cost-effective Ariane 5 almost certainly needs a significant export market and, on ESA estimates, a minimum of six launches per year.

New transportation vehicles

The forecast increase in space activities and long-term launch requirements, and the possible shift from partially to fully re-usable vehicles, present a case for new launchers to LEO (and perhaps from LEO to GEO) with a very fast turn-around time and little complexity in the preparation of payloads. Such vehicles might combine a lower cost per kilogram in orbit; a potential for cargo recovery; and manned capability. Several designs are under investigation in France, Germany, Japan, the United States and the United Kingdom. Preliminary work has commenced in Europe on Sänger, a German project, and Hotol, a British project. These vehicles are intended to be re-usable; to operate like an air liner; to contain no expendable parts (apart from fuel); and to be capable of extended operation with minimal ground support and facilities. They should be able to take off from a conventional runway.

Such vehicles would have many advantages, apart from the projected low cost of access to outer space. They would allow routine access into and out of the atmosphere, and thus rapidity and flexibility for low orbit missions. Some people have gone further and suggested that they may also provide hypersonic air services between far-flung destinations on Earth. For military purposes there is the possibility of reconnaissance and other missions. In the United States the spaceplane is viewed as potentially offering both low-cost access to space and a superior vehicle for all long-haul transportation.

No design for a spaceplane can be achieved without major advances in the propulsion systems, materials structures, computer hardware and software, and other technologies. These developments straddle traditional disciplines and require new skills, since such vehicles are neither aircraft nor spacecraft. Their development will be technically demanding, and initial costs may be very high, perhaps 4–6 billion EAU over 10 years; but the potential long-term benefits are too significant to ignore. Hermes will provide some knowledge on hypersonic technology, but it will not help with the horizontal take-off and air-breathing engine technologies that are essential to the next generation of launchers.

Satellites

Increased interest in the potential of space has stimulated the development of large numbers of different types of satellite. Since the first satellite was launched in 1957, over 3,500 have been placed in orbit. Technological developments, as well as experience in their use, have generated great variations in the complexity and size of satellites, not least in those developed by European industry.

The necessarily long lead-times mean that many satellites that will be deployed over the next decade are already in the design or production phase. For some satellite systems there will be just minor changes to versions already developed. For other applications new types of satellite will be introduced, particularly for mobile communications. Of great importance will be the development of a data-relay satellite system (ideally three satellites) capable of relaying data in a permanent or semi-permanent mode to and from stations, vehicles and platforms orbiting at altitudes below GEO. This will allow continuous contact with a space station or manned vehicle and allow data from Earth observation satellites to be received on a more timely and secure basis.

The costs of a data-relay system for Europe are a matter for speculation. ESA's estimate is 700 million EAU for DRS-1 at an operational minimum of two satellites for 1996–7. However, European industry's estimate is in the order of 900 million EAU. It is not yet clear that the traffic demand would warrant a wholly owned European system. A possible option, which might be more cost-effective, would be to join Japan and the United States in developing a global system.

For some operations it is more cost-effective to use a single large satellite or platform – instead of a number of smaller satellites – which might also carry multi-mission platforms. Such a platform can, for example, combine into a single polar orbiting platform, with sensors associated with different small Earth observation satellites. Eurospace has costed this option at some 700 million EAU for a ten-year operating lifetime, with servicing by an orbiter such as the Shuttle or Hermes. There are many technical and economic advantages to be derived from such a polar platform.

Over the next decade some communications satellites will be larger and more complex and have significant increases in radiated power, thus allowing smaller receiving antennas to be used. Olympus will be the first such European satellite (built at a cost of some 600 million EAU) and will reach much of Europe with its powerful beam, thereby making possible a truly pan-European television service. Intelsat 6 and Eutelsat 2 satellites will follow this trend. Smaller antennas and lower costs should stimulate demand for further satellite communication services.

ESA has allocated some 1,660 million EAU over the next decade towards research and development in advanced techniques to keep Europe in the forefront of this technology. Expected developments include on-board switching and satellite clusters with optical links integrating the fixed, broadcast and mobile services. However, not all of these developments will necessarily have commercial application. The emphasis is to be focused on the needs of the users through improved integration of space and terrestrial systems. It should also be noted that subsequent generations of Comsats may be smaller and less complex.

Satellites for security purposes would use many of the same technologies, as is the case with those few military satellites already developed in Europe. However, any moves to enhance European capabilities in this domain would require further investment in technologies and systems to perform precise security tasks, for example with photographic, microwave and electronic reconnaissance satellites. For many of these technologies Europe is currently dependent on US suppliers.

Space station

A space station is a durable infrastructure providing power, storage, docking capabilities and, if manned, life support, as well as residential and working space. This can be achieved in several different ways. The USSR has operated space stations with human crews for over a decade.

A space station allows sophisticated experiments in life and material sciences; storage facilities for fuel and other supplies; repair and maintenance of satellites and platforms; and more efficient staging of voyages to higher orbits and to the Moon, planets and deep space. In conjunction with other existing and projected facilities, such as a cargo-recovery vehicle or a data-relay satellite, new space services become possible.

As indicated in an earlier section of this report, space stations can be operated by various combinations of robotics and human crew. The necessary range of automated systems is not yet available, so at least initially some type of human presence is required. And for some tasks human crews are essential, albeit costly.

The decision to acquire new space assets of this kind represents a quantum jump within a space programme and necessarily involves the commitment of major resources, technological as well as financial. It would also have a big impact on the overall direction of future space activities. Europe has responded favourably to the American invitation to participate in an international space station, along with Japan. Indeed, Europe has already acquired some relevant experience with Spacelab. This pressurized module was an ESA programme designed, funded and developed as a European manned element in the American STS programme at a cost of 1,000 million EAU over a period of ten years. Spacelab's performance and its use as a tool for experimentation in a variety of science and technology applications have been demonstrated in the three flights of the first module since 1983. Spacelab is totally dependent on the Shuttle for transport to and from space. The second module, delivered in 1985, has not yet flown. In September 1986 Nasa announced that over the next ten years there would be only three, not fifteen, Spacelab flights, a great disappointment to the Europeans. This decision resulted from changing American priorities in favour of military missions, and clearly demonstrated the penalties of dependence on the US.

Experience with Spacelab, though limited, has shown the usefulness of deploying an unmanned platform for certain types of experimentation. This in turn has spawned the Eureca free-flier programme, whereby a platform carrying experiments can be placed into orbit and, after a period of about six months, recovered and returned to Earth. It is then refurbished and re-used. Eureca is dependent on the Shuttle for launch and retrieval services, a dependence which Hermes would remove only to the extent that it can take over Eureca's functions.

The Columbus programme is directed at European participation in the US-initiated international space station, but would pave the way for an independent European capability. (See Figure 8.) The structure of Columbus draws on Spacelab, but is not yet fully established. The ESA ministerial meeting in 1985 approved the outline programme and set a budget of 2,600 million EAU up to 1995, including a three-year period of operation. A new estimate, based on industrial studies, now puts the development of the Columbus programme at 3,611 million EAU, at 1986 exchange rates. The elements of this programme are a pressurized attached module; a man-tended free flyer (MTFF); a polar platform; and an enhanced Eureca. The name Columbus was chosen at a time of optimistic assessments that construction in space would start in 1992, the quincentenary of Columbus' voyage.

Quite what the European components and functions within this international space station will be depends on the outcome of negotiations currently under way. ESA and its members are keen to operate a space laboratory, whether attached to the space station or in a free-flying mode, for the conduct of material and life science experiments; and to include a European polar platform with a co-orbiting unmanned platform (the enhanced Eureca). Negotiations are in train with the United States over the jurisdiction, ownership and control of the space station. Until these issues have been resolved, the level and status of European involvement is uncertain, since European participation will need to reflect European interests. In so large a venture asymmetric financial contributions are bound to generate keen bargaining over the relative shares of material benefits and control.

Europe does not currently have the all-round capability to develop an independent space station, nor will it have such a capability for some time to come, without Ariane 5. Europe lacks a launcher capable of carrying space station modules; a manned vehicle to assist in construction; supporting automata; and a cargo-recovery system. As long as these are not available, the Europeans will be dependent on the United States. The compensating benefits would be of two kinds. First, the Europeans would acquire experience and know-how which could then be drawn on if the decision were taken to develop a European space station. Second, there would be European access to facilities for the conduct of life and material sciences experiments, as well as the use of the programme for various other applications. These policy issues are discussed in Chapter 8.

A space station is expensive. The original American estimate of about $8 billion is already much too low; it will cost a good deal more. Apart from the considerable capital costs, the operating costs will be very high. The additional costs of the Shuttle might be as much as $350 million per flight, if American policy to recover a return closer to the full costs is implemented.

Means

The investment costs will be incurred well before any economic returns can be reaped, and will have to be borne by governments. But early involvement of users in space station activities may be a prerequisite for balancing the infrastructure with the capacity to use it. This may require the inclusion of special provisions oriented towards the private sector.

People and robotics

In planning a new generation of space programmes, Europeans must address the question of what exactly is involved in seeking autonomy in manned space flight. The answer is crucial in determining the level and distribution of resources, and also in setting the performance and safety standards for space systems. Europe has already developed substantial know-how in manned space flight through collaboration with Nasa in developing Spacelab, still the only manned payload for the US Shuttle. The development of Spacelab has helped to build up a European ground infrastructure to train European astronauts and to control missions. The first European missions, involving European astronauts – still in cooperation with Nasa – have shown the public, politicians and policy-makers that manned space flight is a key element of the space capabilities of a nation or, in our case, of Europe.

But it is not only that people in space have a symbolic and a political value; human crews can perform operational tasks, and thus provide an important capability that is necessary for the effective utilization of space. As other sections in this report show, space provides a large potential both for the application of many technologies and for scientific research. The opportunities to exploit this potential strongly depend on two factors which will drive the concepts of future space systems: first, the reduction of transportation costs and, second, the reduction of operational costs.

One way to achieve the first is to acquire fully re-usable manned and unmanned transportation systems. Achieving the second requires the availability of in-orbit facilities for repair, maintenance, servicing and payload exchange, as well as in-orbit construction. Such operations require an infrastructure that can carry permanent or visiting human crews. The economic viability of both will largely depend on an optimal sharing of operational tasks between people and robots. (See Table 9.)

The capabilities of people and robotics

In industrial operations on Earth, there is a developing trend towards replacing humans with robotic systems whenever the latter can perform tasks more economically or to a better quality. Such tasks consist of simple generic manipulation in a predefined sequence. A robotic system is able to adapt, even improve on, the abilities of a worker doing repetitive tasks, while more delicate tasks, such as reasoning, planning and trouble-shooting, are performed by people. The crucial advantages of people are their flexibility in adapting to new tasks or unknown situations, their intelligence and innovative capability in solving very different problems, their mobility and excellent perception, and their wide-ranging physical dexterity. Humans are capable of integrating new and diverse information so as to arrive at new or alternative solutions to any problems which arise. A routine operational sequence may be projected in detail, or the result of scientific experiments may be expected in advance, but not all occurrences, opportunities and even findings can be anticipated.

Hence people in space can recognize and analyse problems as they arise and, if necessary, redefine an operational procedure or experiment to obtain a controlled influence on the results. A human in space may perform a wide variety of tasks with very different objectives; a mission specialist trained to perform or operate specific experiments may also be able to perform maintenance, repair or assembly tasks. Any limitations do not depend on human characteristics in general, but flow from the individual's personal training and assigned task allocation. People operating in space can interact and discuss with the scientists or mission specialists on ground, and thus do more than act as a routine operator, who can be replaced effectively by a robotic system. Tasks which often change their sequence or character or are hard to predefine are an obvious domain for human crew.

Another major advantage is that people have a unique capability to perform diagnoses, to learn, and to acquire and interpret multisensory information. People can recognize complex patterns and extract relevant information from scenes observed from new viewing angles or distorted by a noisy background. They generally also have the capability to assess the likely overall results of an operation. This unique feature makes it impossible to dispense with people in operating space systems, if we take into account the current state-of-the-art of robotic systems.

But of course human crews have some disadvantages and impose constraints which have to be taken fully into account. In short, these comprise the medical health issues (radiation, microgravity, environment); limited physical strength; the psychological dimension (fatigue, stress, attention, motivation); safety considerations; and, of course, the costs.

As for robotic systems, three major categories are relevant: teleoperated, hybrid and autonomous systems. These three categories represent an increasing order of versatility, capability and independence.

Teleoperator systems, such as the Shuttle RMS, represent the current state-of-the-art in robotics for space applications. They are usually multi-jointed, single-arm manipulator systems. The operation of these teleoperator systems hardly depends at all on the 'man-in-the-loop' for real-time analysis of the non-structured environment, decision-making and control of the manipulator. Teleoperator systems today have limited capabilities with regard to dexterity, versatility, perception and diagnosis systems. However, the enhancement of the perception systems and feedback of information to the human operator, as well as of the dexterity of the manipulator arm, will give the ground operator the feeling of being present at the worksite. This enhancement of the teleoperation mode is called telepresence.

Hybrid robot systems will provide increased autonomy wherever simple and repetitive subtasks can be preprogrammed and assigned to automatic operation. This will relieve the human operator of routine tasks so that he or she can concentrate on high-level tasks. The operator still has to control the total procedure and give high-level commands to the robotic systems, using an advanced teleoperator work-station as the interface.

The operation of fully autonomous robots will not require permanent supervision by a human operator. Autonomous robots will be able to adapt to a wide variety of tasks, since they will have at their disposal a broad range of embedded extendable expert systems and methods for artificial intelligence (AI) systems with which they will diagnose, decide and optimize operations in unfamiliar situations.

The problem today is the limited autonomy of robotic systems. Those we have are capable of reacting to an individual situation which is well predefined, but cannot adapt independently to unknown events. In this case a human operator has to analyse the new situation at the remote work-station, which is equipped with a multiplicity of information systems such as graphic and data displays. Robotic systems in the future will have at their disposal learning capabilities which will support the extension of a knowledge and diagnosis system.

A solution to the problem of the unknown will be the extensive use of data bases for operational planning and of more sophisticated computer data on the space elements. The use of these categories of data bases will enhance robots' capabilities and will provide an important basis for achieving systems autonomy. Robots will be self-adapting to unknown environments and eventually perform complex tasks, once they have the means for scene analysis, reasoning and mission planning.

Teleoperation will be one valuable means of performing operations in orbit, though the human factors in telerobotics operation have to be extensively investigated. Efforts are already under way to improve information feedback systems, such as predictive stereoscopic and graphic displays at the work-station, and simulators for planning and comparison with real operations in orbit. It will not be possible to separate completely the 'man-in-the-loop' from the teleoperator system (that is, with a human operator on the ground and the teleoperator on an unmanned space system), mainly because the transfer of information between space and ground is subject to delays.

Allocating tasks to people and robotics

In determining the allocation of tasks between people and robotics in the future we have to take the following into account: technological developments in robotics; robotic capabilities; the complex character of the tasks; the orbit location of the worksite; the availability of manned transport and launch vehicles; and the in-orbit infrastructure. People will also be involved in in-orbit operations if teleoperation modes are applied. It is not possible to identify quantitatively absolute figures to govern the choice, but we can make general statements about the allocation of tasks between people and robotics depending on operational modes, trends in the development of robotic technology, task categories and different orbits.

Table 9 illustrates the possible distribution of tasks between people and robotics. The involvement of people will shift from manual, planning, control and supervision tasks to supervision only, if robots can be developed as autonomous systems. Supervision may not even be needed continuously, in that routine tasks may be performed by robots with their own supervision systems.

Towards a European strategy

This analysis assumes that robotics will make remarkable progress in the future, especially in the area of artificial intelligence. In an optimistic scenario, it may be possible within three decades to have available autonomous robotic systems which could take over a number of tasks currently performed by human crews during both intravehicular activities (IVA) and extravehicular activities (EVA). But it has to be emphasized that people will remain of paramount importance to in-orbit operations. The efficiency of future space systems will depend on a close cooperation between people and task-oriented robotic systems, whether with people acting as decision-makers, supervisors or teleoperators, or with a human manual worker who is cooperatively supported by robotic systems.

In other words, if Europe is to engage in all those activities which are necessary for the effective use of space, it requires the skill and the special capabilities of people in space. Manned space flight is an essential part of an autonomy strategy. Without it, Europeans will always have just a fragment of the autonomous capabilities of the large space nations such as the USA and USSR. Moreover, Japan and China have already decided to develop in the long term their own manned space-flight capabilities.

ESA's long-term plan envisages two systems that will lead to autonomous manned space flight. One is Hermes, which can carry three or four astronauts to low earth orbits in order to join up with the international space station or one of its European elements, especially the MTFF. The MTFF is the second key element for achieving autonomy in orbit infrastructure, because it provides the working base for the astronauts to perform their tasks. In its first configuration the MTFF is not a permanently manned facility; it will be visited by humans only during specific servicing periods. This development strategy leaves open a European decision later on (in ten years) as to whether or not a permanently manned facility will be needed, since the technology of robotics may have developed so rapidly that the transfer of many more tasks to robots will make a permanent human role in space unnecessary. This development strategy is quite different from those of the USA and the USSR, and offers a unique way to develop European autonomy.

Ground segment

The development of the ground segment is crucial to any operations in space. Within the past decade, a number of large receiving stations have been established for Earth observation, originally under the European Earthnet programme for the reception of Landsat data. Earthnet acquires, pre-processes and distributes satellite remote-sensing images and data to receiving stations in other member states. Similar facilities have been established for other types of Earth observation satellite. To benefit from their own satellites, Europeans have had to conclude agreements with some foreign countries for the reception of data at ground stations within the latter's territory. The development of a Data Relay Satellite system would remove the need for foreign receiving stations and allow data to be received in a more timely and secure manner.

Antennas are used to receive communications from satellites. Developments in satellite technology have permitted the receiving antennas to become progressively smaller. These antennas can be located on the premises to which communications are to be sent, thus in principle reducing costs, since they obviate the need for land lines. This, coupled with the increasing range of communication services, has prompted a great increase in the ground segment market, so that it is now much larger in value terms than the space segment. For the period 1985–95 ESA estimates the total ground segment market to be in the region of 15,000–18,000 million EAU.

However, the price, character and performance of European designs have not been competitive with those from Japan and the United States. A very large part of the ground segment market in Europe has been captured by these two countries. The limitations of European industry in this sector stem partly from the fragmentation of the European market into many small markets separated by artificial barriers. The removal of these barriers, the further development of common specifications and standards and, most importantly, the opening up of public procurement would allow a single European market to be developed with the attendant benefits of large unit production. Some of these questions have long been addressed in the various ITU and satellite consortia and fora. More recently the European Commission has been instrumental in the promotion of common technical standards for the telecommunications sector, as well as in the development of an enlarged common European market.

It does not, however, follow that European industry will necessarily capture a larger share of the market either in Europe or globally. The performance capabilities of ground segment equipment manufactured by Europeans could certainly be improved, especially in relation to microelectronics and information technology. But over-sophisticated equipment is not profitable either; so a balance is required. The creation of a space infrastructure in communications is only one part of a long chain in which the multiplier effect is very significant. It would be ironical if Europe were to be independent in the construction of satellites and their launching, but did not have the independent means to use them.

The industrial infrastructure

Over a period of some 25 years or so Europe has acquired a solid industrial base in the space sector. Over 200 firms now have a primary interest in producing space systems and subsystems, with some hundreds more providing other subsystems and components. It has been estimated, for example, that 800 firms might eventually contribute to the Hermes programme. The major firms involved have solid foundations in the aeronautical and defence sectors – both traditional European strengths – and many have a role as national industrial champions, either directly through public ownership or indirectly through government contracts.

There is also an important and growing secondary industry, including the suppliers of equipment to end users; industries which are actual or potential users of the space segment; and commercial services which increasingly make use of space capabilities. ESA's activities are not the central focus for these groups, since their involvement is either with the organizations which operate space-based technologies (national PTTs, the international consortia – Intelsat, Eutelsat, Inmarsat, and so on) or with direct commercial and financial activities which happen to use space technologies.

European expenditure has been modest but also relatively cost-effective. No entirely satisfactory yardsticks exist for evaluating the return on investment, but all studies suggest that space R&D has had positive productivity effects and brought substantial benefits to a number of industries. The growing maturity of the European space industry is reflected in the increasing share of the commercial market (and not just government and ESA contracts) in the turnover of the industry, from an estimated 3 per cent in 1977 to 15 per cent in 1983. A 30 per cent share is forecast for 1988. Much of this has been due to launch contracts for Ariane and to earnings from communications satellites involving consortia with American manufacturers.

The future of these markets is not secure, given the intense, and subsidized, competition from launchers and rapid developments in fibre optic technology. Some markets have been slow to materialize: DBS and remote sensing are both examples. The ground segment market is subject to fierce competition, with the US and Japan taking 80 per cent of the international market. In addition, the European economies have won a return in the form of positive inflows on the balance of payments. Eurospace estimated these at 285 million EAU per annum over the period 1980–4, together with the tax revenue from the producers.

The personnel directly employed in the European space industry totalled some 21,000 in 1983, with a further 10,000 in institutions, consultancies and universities. Another 18,000 or so are employed in secondary industries associated with space technologies. The space industry employs highly qualified personnel; the proportion of engineers and university graduates is greater than 40 per cent of the total. The rate of growth in employment was some 7 per cent per annum between 1977 and 1983, but this is now tailing off. Indeed, delays in current ESA contracts are making Eurospace very anxious that the level of the workforce may drop if programmes are not brought forward.

The increased budgets agreed by ESA in 1985 had led analysts to suggest that there would eventually be a demand for 44,000 personnel in primary institutions and industries, and 29,000 in the secondary industries. Europe has many fewer space specialists than the United States; Eurospace suggests as few as 12 per cent of the US total. Surveys of European industry also indicate that there are serious shortages of some specialists. Expensive training schemes have had to be developed so that qualified engineers can acquire the further five years' training necessary.

Europe has not yet developed a space industry policy, although ESA has recognized that its programmes both depend and have an impact on the space industry. ESA's mandatory programmes helped to consolidate three industrial consortia – Mesh, Cosmos and Star – which emerged in the early 1960s as groupings of companies in the aerospace sector from different European countries. These consortia continue to function in contracting for ESA's scientific programmes, but the patterns of tendering for ESA's optional programmes, for international contracts and within individual European countries are quite different. The strengths and weaknesses of this pattern of industrial collaboration are discussed in Chapter 7.

ESA's second main impact on the industry has been through the rule of 'fair industrial returns' or 'juste retour', aimed at ensuring an equitable spread of national industrial participation across a range of ESA programmes. The results are set out in Table 5. Originally conceived as a necessary and valuable tool for distributing industrial benefits among ESA's members, this system has come under some criticism, and its consequences are also discussed in Chapter 7.

The third strand of ESA's approach has been to hand on the responsibility for the commercialization of space technologies. ESA has defined its role as providing the initial impetus from its R&D, which is then picked up by commercially oriented ventures, and either marketed as a particular product, for example by Arianespace, or operated as a specific service, for example by Eumetsat.

Finally we should note that Eurospace, the Paris-based representative grouping of European space companies, has played to some effect the two roles typical of comparable industrial organizations. It has built bridges among the companies and with ESA in order to channel views and test opinion; and it has developed considerable technical expertise and evaluative skills which help to ensure that industrial considerations are taken into account in ESA's planning and management processes.

Space law

Outer space is an international commons outside the sovereign control of any state. The laws determining its use are developed primarily in the UN by Copuos, which has a current membership of over 50 states. Within Copuos the status of ESA is restricted to that of observer, but ESA members are full and effective members of the Committee. Decisions in Copuos are made by consensus rather than by majority vote.

The principal instrument of space law is the Outer Space Treaty of 1967. This sets out the general principles for the use of outer space, but does not deal with specific applications of space technologies, most of which were not developed when the principles of the treaty were first agreed. The treaty stipulates that outer space, unlike air space, is not subject to national appropriation and is free for use by all states in accordance with international law. Three additional agreements cover the return and rescue of astronauts and objects launched into space, the liability for damage caused by space objects, and the registration of objects launched into space. This last agreement stipulates that states must register information on the space objects for which they are responsible, including their general function and orbital parameters. However, there have been many instances when the information provided by the launching state has been unsatisfactory with regard to the function of the space object, especially in the case of military satellites.

The Outer Space Treaty prohibits the stationing of nuclear weapons and other weapons of 'mass destruction' in space, but does not discuss the militarization of space or consider other types of weapon. Military activities have been carried out by some states and considered acceptable in so far as they have not involved the use of any form of weapon and do not breach international agreements on non-proliferation. They are therefore regarded as a 'peaceful' use of outer space. Some of these activities, such as the use of 'national technical means' (NTM) of verification, have been accorded a special status in bilateral arrangements between the superpowers. In 1981 and 1983 the Soviet Union proposed draft treaties prohibiting the stationing of all weapons in outer space, but these have met with several obstacles, among which is the problem of verification. A treaty of this kind would probably be multilateral, but Europe has not had an important role in the discussions to date.

It has been less easy to secure international agreement to regulate those activities that have significant economic and national security linkages. The European countries have refused to support the UN General Assembly resolution requiring the prior consent of the receiving state as a precondition for the use of direct broadcasting satellites. Europeans have considered this to be in violation of the principle of the free flow of information. There are also great variations among national laws pertaining to matters such as advertising and decency, and even in a European context some measure of harmonization and agreement is now considered necessary. Accordingly efforts to achieve this have been undertaken by a number of European institutions, including the European Community, the European Broadcasting Union and the Council of Europe. So far progress has been slow.

The expected increased participation of the private sector in space activities may raise some difficulties. The existing rules state that the activities of non-governmental entities require authorization and continuing supervision by the appropriate party to the treaty. Hence private unregulated activity in space is prohibited, and some measure of licensing and government regulation remains necessary. Three problems arise. First, where space activities involve non-governmental agencies from more than one country, some transnational regulation would be necessary. Second, the move towards privatization, so far more advanced in the UK than any in other European country, makes it harder to maintain regulation based on designated signatories who may have to choose whether and how to license a competitor. Third, the deregulation increasingly practised to enhance competition and entrepreneurship is a powerful constraint on international regulation.

The development of a multinational space station and the resulting space commerce raise legal issues not yet fully provided for by existing treaties. In the case of a national space station, it will be important to establish whether it would be desirable or viable to apply terrestrial laws to space activities or to introduce the required modifications and innovations necessary to ensure that individuals living and working in space are adequately protected. In a multinational venture such as the US suggestion for an international space station, there is the complication of the determination of jurisdiction or the right of a state to prescribe and enforce its rules of law. This is necessary to resolve such issues as intellectual property rights, product liability and export law. The question of jurisdiction will be determined by the terms of the relevant space station agreement, and there have been some divergences on this matter in negotiations between the United States and Europe. The ESA Council of Ministers declared in 1985 that a fundamental objective of European participation in the space station would be responsibility for the design, development and management of those modules built by Europe. On the other hand, the United States considers that its technological and economic leadership in the venture requires that it should have jurisdiction.

ESA members, in espousing the goal of autonomy, have embarked on a new phase of large, complex and expensive projects which may also be flanked by a military space programme. This new phase of activities will of necessity entail managerial and organizational systems that will be of much greater complexity than anything previously experienced. The international context, both political and economic, is also evolving rapidly. New international regimes will be necessary, and Europeans will need to consider carefully how to maximize their influence on them.

Part III
TOWARDS A COMMON SPACE POLICY

7

THE BALANCE-SHEET

Western Europe has managed, in a relatively short time and with limited means, to achieve a number of impressive results in space technology and the space industry. It ranks as the third presence in space. However, despite these notable achievements, essentially in the field of space science and in the area of launchers, European space efforts fall short of a truly effective joint policy. Nor are they commensurate with the kinds of technological, industrial and strategic stakes involved in space activities or with the human and economic potential available in Europe. This calls for a closer look at European strengths and weaknesses in this area, given that Europe is not held back by scientific obstacles or intrinsic inferiority. The problems are those of purpose and of the best use of resources.

In financial, technological, industrial and human terms, Western Europe's resources are on the same scale as those of the two largest space powers: the United States and the Soviet Union. And yet the gap between Europe's space programme and those of the two space giants remains considerable. Moreover, Europe is under increasing challenge from Japan. Significantly, the gap between European and American investments in space (around 1.8 billion EAU and nearly 20 billion EAU respectively in 1985) gives a ratio of less than 1 to 10.

Yet with this modest investment Europe has managed to master the key space technologies, and now possesses the modern industrial infrastructures required to design, construct and exploit most space systems. In addition, over the years European firms have gradually acquired substantial experience of working together. Organizational as well as personal relations have been built between many companies throughout Europe in the course of carrying out joint programmes. All this has contributed to the emergence of a vigorous European 'space lobby' at the industrial level.

ESA's contribution to the development of a collective European space capability has been fundamental. A unique example of a successful multinational space organization, ESA has provided the infrastructure for space collaboration among European states. It has fostered the emergence of a European 'space spirit' within the space community. Six European astronauts have been into space: two from France, three from Germany and one from the Netherlands. By formulating options jointly and exchanging knowledge and experience, the space science community in Europe has generated a momentum capable of persuading governments to adopt programmes at the European level. And by supervising the implementation of these programmes, ESA has also developed impressive managerial skills within a relatively efficient and integrated framework: it awards between 500 and 600 contracts per year to industrial firms in ESA countries and distributes the technological risks and benefits among member states.

Meanwhile, Europe's human resources engaged in the space industry and R&D are also impressive. European manpower is certainly comparable in quality to that of the other space countries. However, Europe still lags behind the US and the USSR in the overall number of qualified personnel. In the United States, there are some 250,000 people employed in space activities, whereas Europe, with around 30,000, has only about 12 per cent of the US number in spite of having a much larger total population.

Fragmentation persists

Europe's relative success in space from a limited financial and human base is, however, no cause for complacency. The resource shortage, compared with the dominant space powers, is in itself a fundamental weakness which has major implications in the long run for Europe's technological and economic strength, as well as for its security. Financial allocations, in a time of heavily constrained budgets and dependence on limited public support, tend to promote 'national' ventures and 'national' perceptions of what should be done in space. Differences in the absolute size of national space budgets are striking. These budgets are dispersed between various ministries and agencies, with military and civilian expenditures usually kept separate, and each with their own contracting terms. The unpredictability and slowness of commercial returns over the medium to long term makes it hard as yet to attract high levels of investment from private financial and industrial sources.

But this basic handicap is further aggravated by the persistent fragmentation of effort at both the political and the industrial level. The result is that Western Europe's stretched resources are deployed in a far from optimal fashion. While other space powers, such as the United States, the Soviet Union or Japan, have founded their space policies on an essentially national basis, European space activities have been carried out within a double framework. Most European countries have been involved in collaborative space endeavours, first in ELDO/ESRO, then, since the mid-1970s, within ESA, whose famous initial 'package deal' was most persuasive. Simultaneously, however, some European states have built up national space programmes to serve their specific national interests. ESA's programmes depend on the national facilities of its members. Decisions do not rest on a single and coherent approach within a supranational framework. European achievements in space are therefore the result of a compromise, not always an easy one, between national and collaborative options.

Disparities within Europe

A further complication is the unavoidable disparity between the larger and the smaller European countries. The former (i.e., France, the Federal Republic of Germany, Italy and the United Kingdom), with larger financial and economic capabilities, have chosen to develop their own national programmes, to an extent independently from the ESA framework. In each case attitudes towards space have been shaped by specific national and foreign policy ambitions or constraints. France, for example, has sought greater autonomy, while Germany and the UK have sought to maintain a close relationship with their American ally as a key dimension in their defence policies.

The smaller European partners may have an important capability in space exploration as well as an impressive industrial infrastructure (as in the Netherlands or Belgium), but they lack the budgetary and institutional means to succeed individually. With the sole exception of Sweden, they have therefore concentrated the bulk of their space effort on the collaborative option, by participating actively in ESA-sponsored programmes and by developing niche production. The result is a spectrum in which the national space expenditures of some countries, such as France, Sweden or, until recently, West Germany, are greater than their contributions to ESA, while Ireland has no national space programme at all. (See Figure 2.)

Burden-sharing through ESA

These important discrepancies have meant that ESA's attempts to harmonize national programmes and to guarantee as much equality as possible between member states have not always been successful. Two novel rules were developed within ESA to prevent this gap between the leading European countries and their less well-endowed partners from growing: namely, the distinction between mandatory and optional programmes, and the principle of fair industrial returns on investment. Despite these rules the gap, if anything, has actually widened.

Member states contribute to ESA's mandatory activities on the basis of their average GNP calculated over the past three years. This ensures both the stability of this part of the budget and a parity of decision for member states as regards the mandatory scientific projects. But many of ESA's programmes are optional: the countries participate and contribute according to their policy or industrial interests and their national priorities. Significantly, the major achievements of ESA in the field of launchers and operational systems belong to this category of 'à la carte' programmes. The mandatory programmes now account for only some 15 per cent of ESA's work.

These mechanisms have produced a situation in which the weightier members (in terms of industrial and technological capabilities) become the obvious 'project leaders' of crucial, but optional, programmes, thereby further increasing discrepancies among European countries. Indeed, the four major contributors to ESA (France, West Germany, Italy and the United Kingdom) together account for around 80 per cent of ESA's budget, while some of the other partners have difficulties on occasion in honouring their limited budgetary commitments. (See Figure 3.)

Fair returns

This feature should have been offset by the ESA rule of fair returns, whereby a country's percentage share of the total value of contracts placed under the mandatory programmes should correspond to its percentage financial contribution to a given programme. (See Table 5.) The industrial 'return coefficient', both as an average and individually, has succeeded in spreading the benefits of ESA contracts. But this outcome has been reached only by difficult and time-consuming negotiations that continue to prevent the smooth and harmonious integration of European and national space endeavours.

The fair returns policy is open to other criticisms. First, there is some inflexibility in that fair returns are calculated for each phase of a project rather than spread across projects, in spite of ESA's endeavours to compensate across 'packages'. Second, the cushion of guaranteed contract shares to some industrial participants removes some of the pressures for rationalization of the European space industry. Third, in practice the fair returns to smaller countries tend to be awarded through contracts to their smaller companies. Although this helps to encourage effective niche suppliers in those countries, it impedes access to contracts for smaller hi-tech companies as they emerge in the larger countries. The new CoSpace consortium is an attempt to remedy this.

A fourth problem relates to economies of scale. The smaller companies which can benefit in smaller countries are in practice more often suppliers of components than producers of subsystems, because the value of the latter's share (perhaps 15 per cent of a total contract) would generally exceed the fair return. In the Netherlands Philips has virtually left the space segment because there is not enough headroom for both it and Fokker, which draws in 60 to 70 per cent of the Dutch industrial return. All this impedes economies of scale for small companies. Finally, the large companies from the large countries to an extent win disproportionately high industrial returns as prime contractors, and the truth of the matter is that companies are sometimes chosen for political rather than commercial reasons, or because it is 'their turn'.

ESA's approach to contracts is a further complication. In its attempts to keep down the costs of already expensive projects, ESA is highly, but sometimes too, demanding in squeezing the profit margins of contractors. In their concern for value for money and exacting standards, ESA managers sometimes set very onerous and overspecific conditions and take too long to make decisions. The costs of tendering alone are sometimes a disincentive to bids. And of course the multilingual and multinational complications remain real. Similar problems arise with tendering for other international contracts, since the prolonged bidding procedures take place for a small number of expensive contracts in a context in which competition is politically constrained.

Public attitudes

Europe also suffers from persistent problems arising from differing perceptions and misunderstandings among ESA partners. For example, while some countries tend to criticize France for its overly 'nationalistic' approach to joint European programmes, the dominant perception in France is that many successful European space activities would not have been possible without the French contribution and drive.

More importantly, perhaps, the fragmentation of European space activities in a politically fragmented Europe is reflected in the absence of broad public support for a European presence in space. This sometimes shades into a wider public disenchantment with high-technology ventures, both military (SDI) and civilian (nuclear power). The result has been a vicious circle. On the one hand, neither the limited and fragmented national programmes nor the equally limited ESA projects have been able to generate on a European scale the kind of public support necessary to convince both tax payers and legislative authorities to increase their investment in space. And yet without the new resources required to produce a qualitative jump in European ambitions, the various programmes are likely to stagnate at their present levels, and thus fail to produce the kind of highly symbolic and ambitious project which could unite the Europeans in a common vision. There has after all been no European equivalent to the Apollo programme. This handicap is especially relevant at a moment when the Europeans face important decisions about two large new ventures: Hermes and the space station.

Consequently, where public support does exist, for example in France, it is still largely associated with mainly national activities and goals (including in the political-strategic area) rather than with 'Europe' as a whole. This further reinforces the fragmented character of the various national activities. Meanwhile, elsewhere in Europe, public opinion tends to view space as an esoteric field which should be left to the superpowers or in which, at best, Europe should simply follow in their shadow.

Civilian and security options

Political and industrial fragmentation has been most obvious in the use of space for defence and security. Here there has been, thus far, virtually no European cooperation, except partially through Nato. ESA has limited itself to the 'peaceful uses of space', an important touchstone for its neutral members, which do not belong to Nato. France does not currently belong to Nato's integrated military structure and therefore is not tied into the Nato system of communications satellites. Furthermore, bilateral collaboration may be a counter-attraction, as in the US-UK case, which removes what might otherwise be a powerful impetus for European collaboration.

Bilateral cooperation in Europe itself is not always easy. The failure of the proposal for a joint Franco-German surveillance satellite is a case in point. There were differences over the technologies to be chosen, over the distribution of costs and production shares, and over the operational nature of the satellite. The Germans had the impression that they were being asked to pay for a 'precooked' French project. The French had the impression that they were dealing with German self-imposed (or US-imposed) political limitations. Although such disputes between France and Germany do not necessarily prejudice European cooperation as a whole, they certainly complicate it.

This fragmentation has major implications, not just for Europe's long-term security interests, but also for the overall technological strength and volume of European space efforts. For it should be kept in mind that in the US and the USSR, military applications have acted as a primary driving force for civilian space programmes. By contrast, Europe has not been able to generate the civilian benefits that often result from military R&D and production in terms of dedicated budgets, dual technologies in which the R&D costs can be spread, the volume of activity and some economies of scale. Only in France has the spin-off been significant. Yet this outcome is paradoxical in so far as many of the European space firms are also defence contractors, and would presumably welcome the opportunity to foster more dual use with military space requirements.

Strategic doctrines

For the superpowers, the essential importance of a space-based dimension to their respective force structures flows from their strategic nuclear confrontation and military capabilities on a global scale. In Europe only France and, to a lesser extent, the UK have significant military space programmes. But there is as yet no explicit strategic consensus upon which the common requirements and precise mission for an integrated military space component could be built. The military doctrines and strategic postures of the different European countries remain divergent, largely because of the wider geostrategic considerations inherent in the political-strategic order in Europe since World War II. This situation is unlikely to be radically modified overnight.

Significantly, although Europe's two nuclear powers – France and the UK – have each developed independent military programmes for space communications (the British Skynet and the French Syracuse), they have done so quite separately. In 1993 France will launch a military reconnaissance satellite, Helios, but again after failing to convince West Germany to join that venture. Although Italy and Spain recently decided to play a modest part in it, the project still falls short of being genuinely multilateral. The essentially national character of these military activities has aggravated the fragmentation of European space efforts, especially given the extremely high cost of military programmes. In the case of France, for instance, the rapidly developing military effort is expected to produce a driving impetus for space programmes as a whole, but it will necessarily put a strain on other commitments. By 1995 France's military space budget might exceed its annual contribution to ESA.

Pressures for change

Nevertheless, it is clear that in military space activities autonomy, however it is defined, is not an achievable goal for individual European countries, because the costs are prohibitive. For both financial and strategic reasons, European collaboration in military space activities could play an important role in strengthening Europe's global defence capability and in diminishing its dependence on the USA. Europe's existing space assets have the potential for security applications on a collective basis. There is already some dual use, in that some European civilian satellites have capacity reserved for military use. Technologies either achieved or under development provide Europeans with the means to operate a reconnaissance system for arms control and verification and crisis-monitoring, or to support collectively the deployment of military forces. The industrial infrastructure could fairly easily adapt. The political context is evolving fast and could well promote acceptance of shared European needs to be pursued through Europe's different institutional frameworks.

European strengths and weaknesses

If we take a closer look at what Europe has so far achieved in space, a mixed picture emerges, revealing different strengths and weaknesses from sector to sector. In space science and launchers, Europe's successes are remarkable, but the results are less satisfactory across the range of applications satellites and even worse in the ground segment.

Science

ESA's scientific programme has been one of the foundations of European cooperation in space. The development of scientific missions was the initial goal of ESA and was based on a wide consensus among the member states that has not so far been questioned. ESA's scientific budget, linked to GNP shares, has been very stable in comparison with both other European space programmes and the cognate investments of other space powers, notably the United States.

As a result, Europe has been able to achieve and maintain a real strength in science on a world scale, with an eminent scientific community and industrial capabilities for the development of highly sophisticated platforms and instruments. Between 1968 and 1986, 15 scientific spacecraft were launched, ranging from simple payloads for studies of the magnetosphere to ESA's X-ray observatory satellite, Exosat. The recent success of Giotto demonstrated that European space science now stands at the forefront of scientific research. Moreover, European involvement in projects such as the joint ESA/Nasa Spacelab and Hubble space telescope, and French collaboration with Soviet programmes, has enhanced European capabilities and expertise.

These achievements stem from Europe's success in allocating financial and human resources prudently, but also selectively, to key projects of considerable scientific yield, a potentially crucial point for the future of the European space programme as a whole. Clearly, however, the scientific challenge is by no means over. The very successes of the past have widened our research horizons. These now require a broader range of resources, including substantial new flight opportunities. The trend of space research is towards the use of large multi-purpose facilities that will be able to supply a permanent flow of scientific data. All this adds up to major new challenges over the next twenty years for European space cooperation, challenges as great, if not greater, than those made during the pioneering years after 1964.

Launchers

In the field of launchers, Europe's achievements are also quite remarkable in terms of both production and commercialization. A relative latecomer to the launcher market, Europe has succeeded, thanks to the development of the Ariane family of ELVs, in moving closer to the two major space powers. Since the 1973 decision to initiate the Ariane programme, Europe has had the means to launch its own satellites. Despite a series of technical mishaps and delays as well as a limited production capability, Ariane now represents about 45 per cent of the world market in conventional launchers, and its launch schedule is booked up for some time ahead. It should, however, be noted that Europe cannot yet launch on demand for unexpected and time-constrained needs. The success of Ariane is linked to two positive factors. First, the European launcher, with a geographically advantageous launch site, became operational in December 1979, when there was a growing demand for launches, particularly in GTO, and especially for telecommunications satellites. Second, the European partners were able to set up Arianespace, an organization that was well adapted to market conditions.

Success is linked to the very specific nature of the launchers: the high costs and crucial requirement for technological standardization make a collaborative venture absolutely vital. This is not the case for satellites: once a country has reached a certain level of technological and industrial development, the natural step is often to go for indigenous production.

However, the decision to concentrate all the efforts on ELVs left major gaps elsewhere. For manned space flight in particular, without a re-usable orbiter Europe was completely dependent on the US. Until the Challenger accident, the operation of Spacelab was possible only through cooperation with Nasa and the Shuttle. The programme has since then been brutally stopped. But it is not just a question of maintaining the current generation of launchers. Europeans have also prudently to look towards the next generation, which will be necessary early in the next century. In the medium term a small launcher could be useful, such as the Marianne proposal by Eurospace.

Commercial pressures

Results are less satisfactory when it comes to applications satellites. As regards their production and commercialization, Europe today faces two major problems. First, it has to preserve its own home market in the face of competition from other satellite producers (for the moment, essentially the major US firms such as Hughes, RCA or Ford Aerospace). In this context, the recent Luxembourg purchase of Astra and the decision of the new British Satellite Broadcasting Consortium to accept Hughes's tender, both US-made telecommunication satellites, are setbacks. Second, Western Europe's position in external markets is still very weak: Arabsat has been a rare example of European success in marketing space products on a worldwide scale. European firms may win part or all of a major contract outside Europe perhaps just once every eighteen months.

Europe's fragility in the face of growing international competition is even more evident in the ground segment, a market of possibly greater commercial importance which also happens to be far more open than that of operational space systems. The US and Japan account for some 80 per cent of the ground segment market. Since all producers have to respect strict technical standards set up by international space-user organizations, such as Inmarsat or Intelsat, competition in this field is essentially a matter of price, financial arrangements and delivery schedules – decidedly weak points in European performance. This has already become clear in the field of large telecommunication ground stations, in which Europe has to contend with the growing capability of Japan, the world's first producer of ground receiving stations and switching systems. One telling example is the sale in 1983 by Japan's Nippon Electric to the Arab League of a ground station for the Arabsat satellite, despite a tender from Aérospatiale, the satellite's prime contractor.

To make things worse, key elements of the ground station market are only just beginning to be developed in Europe, although the imminent stimulus provided by DBS to new markets for antennas is likely to result in increasing competition between European and non-European producers. Finally, as regards ground stations for receiving remote-sensing data, Europe again lags behind – in this case, Canada and Japan.

Industrial collaboration

These weaknesses in the satellite and ground sectors are linked to a series of structural problems that affect the European space industry, as well as to the specific character of the European market for space products. For a long time, the European industrial landscape in the field of space was dominated by three consortia – Cosmos, Mesh and Star. However, this initial structure has changed significantly. The scientific programmes in which the consortia still operate have decreased in number, and other ESA programmes – in both space transportation systems and applications satellites – have been supplied by differently composed industrial groupings.

In the 1970s new European groupings were formed notably to develop Spacelab and the Meteosat satellites. Another telling example was the group established to construct L-Sat/Olympus. This group came into existence as a response by British and Italian companies to the German and French decision to build TDF-1 and TV-Sat independently (Germany and France having judged ESA's telecommunication satellite programme, adopted in 1979, to be too slow-moving). New groupings are currently being formed to prepare the Columbus and Hermes programmes.

Europe's growing involvement in the field of operational systems resulted at the industrial level in increased competition among aerospace firms for the contracts awarded within ESA's optional programmes, and in the establishment of *ad hoc* groupings precisely to supply these new programmes. In tendering for international contracts, European firms have formed alliances with American companies: BAe and Thomson CF with Hughes, Aérospatiale and MBB with Ford, Matra with TRW. These alliances have contributed to a sharpened commercial edge to the European firms. In thc meantime, national procurement has been substantially supplied by national contractors, sometimes, as in Italy, organized into a national consortium. Many of the major European companies are substantially publicly owned. Only in France has competition between two prime contractors been a regular feature.

Finally, the national mergers between firms belonging to different consortia have undermined the original framework of the European space industry. At the same time, the decrease in the number and change in character of programmes adopted by ESA, together with the increase in their technological complexity and overall costs, have contributed to the breaking up of the 'historical' consortia. This picture sharply contrasts with the US industry, where the consortia are occasional, project-related and based on commercial and technical criteria.

Today, Europe's industrial scene is characterized by a growing disparity between European aerospace firms: the most competitive among them (Aérospatiale and Matra in France, MBB-Erno and Dornier in West Germany, British Aerospace and Marconi in the United Kingdom, Aeritalia and Selenia-Spazio in Italy) tend to dominate the market and become obvious 'prime contractors'. More and more frequently, these companies form joint ventures, such as Eurosatellite or Satcom International, generally with the aim of exploiting together a commercial market, and in recognition of the importance of entrenching habits of collaboration.

Obstacles to rationalization

One result of this pattern is that the division of tasks and responsibilities among European industrial firms has proved rather rigid. Space platforms are developed by companies belonging to the aeronautics sector (such as Aérospatiale, MBB and BAe), and their payloads are generally produced by firms in the telecommunications and electronics sectors (such as Alcatel-Thomson-Espace, AEG, ANT and Marconi). In the United States, the same firm is responsible for the whole product. Consequently in Europe the prime contractor usually manufactures only 30 per cent of the product value, whereas in the US the share is in the region of 80 per cent. It is, however, often the case that the US supplier subcontracts, but on a commercial basis and not by international negotiation as in Europe. Thus in America a highly concentrated national industry benefits from a large, open national market, whereas in Europe there are too many overlaps between too many national competitive units; insufficient industrial linkages take place across national borders; and the national markets are smaller and still fragmented. As a result, European firms find it more and more difficult to maintain a sufficient 'critical mass' (in terms of financial capacity, employment and R&D), in comparison with an increasingly concentrated US aerospace industry. However, the state of the US industry is changing: the combination of delays in launch opportunities and a saturation of the satellite market has produced overcapacity. US firms are consequently chasing international contracts even harder and sometimes with terms which European firms cannot reasonably match.

These problems are aggravated by a business planning cycle which is vulnerable to significant fluctuations that are linked to long lead-times and poor correlation between political decisions and industrial capacities. For example, since the costs of satellite systems have increased more rapidly than space budgets, the number of programmes has declined and the industrial firms have had to contend with an unhealthy succession of fat and lean years. Thus, today's apparent overcapacity in the European space industry might actually be a problem of underbudgeting: the big programmes decided by ESA in the mid-1970s are reaching completion, and governments have not agreed new programmes over the past few years that reach the level of industrial capacity. This situation may, however, improve in the future as the new round of programmes (Columbus, Hermes, Ariane 5) is adopted at ESA's ministerial conference scheduled for late 1987. Even so, however important these programmes may be, they account for perhaps only a quarter of the combined effort of European firms.

Market conditions

On the purchasing side, the structure of national monopolies still largely dominates the organization of European markets for satellites (especially in telecommunications). The PTTs are the key players in the relevant procurement and operating organizations, both international and European – Intelsat, Inmarsat and Eutelsat. In West Germany, for example, resistance to the loosening of the Bundespost's monopoly runs deep, although it is recognized that new services will have to be provided at internationally competitive rates. A recent comparison, published in the report of a government-appointed commission, showed that tariffs for a few services may be as much as 15 times as expensive as their equivalents in Britain or the US. That this is so even in a relatively liberal country demonstrates the scale of the European problem.

European industries have not yet benefited from the deregulation process that took place in the United States in the early 1980s, although the UK is now beginning to follow suit. But firms are still dependent upon public purchasing by the various national agencies as the driving force for the development of space products and services in Europe. In short, what Europe lacks above all is the momentum of a truly open market: the weakness of private-sector demand and the persistence of national monopolies contribute to the continuing fragmentation of the European space industry.

Deregulation is of crucial importance. The early American decision to opt for an 'open skies' policy dramatically reinforced the buoyancy and sharp competition which have characterized the American market and given US firms such an edge internationally. The Americans are now taking this a stage further by possibly allowing private business networks to compete with Intelsat. Hence deregulation also carries risks for Europe. Open tendering means contracts go to the company or consortium which can offer the most attractive terms.

As the use of space technology spreads from the more esoteric to the more routine applications, more of the purchasers will be operating agencies or companies for which space *per se* will be irrelevant but the commercial product and its apparent profitability will be vital. Space technologies have to compete with alternative technologies. In the telecommunications sector the evolution of fibre optics has fundamentally changed the questions and the relative costings for the purchaser. Moreover, traffic by cable is much more amenable to national control than traffic by satellite, even though satellite links offer immense advantages for communications to remote areas, over great distances and to mobile ground stations. The various issues at stake also raise difficult choices about the respective character of satellites and ground stations: for instance, is it better to opt for more complex satellites and simpler ground systems or vice versa; and should one go for niche markets or for higher volume and more diverse markets? The sophistication of engineering and the tight specifications needed for ground stations that serve scientific satellites may be too complex and too costly for routine telecommunications.

The opening up of the European space market, a necessary condition for future competitiveness, is by no means an easy task. ESA's contribution in this area is minimal, since it is fundamentally a product-development and R&D organization with little competence in the operation of space systems and their commercial uses. It has no regulatory teeth or enforcement capabilities.

Meanwhile, the impact of the European Community on the space sector has so far remained very limited. Its rules on opening up public procurement have been slow to bite in the telecommunications sector. The various satellite consortia have done some of the standard-setting, but more remains to be done. The EC has now begun to move in the field of direct broadcasting, but only just.

The interconnections between adjacent technologies have hardly been addressed on a concerted European basis. There is no collective stimulus to users and potential users of some space technologies. Remote sensing is a good case in point: large mass markets remain a very distant prospect, but specialized trans-European markets might be viable, especially perhaps for those applications which are relevant to European public policies. Crop-forecasting is an obvious example.

The dynamics of adjustment

But the situation is rapidly evolving. The drive within the EC to create a single internal market by 1992 is serious and has now been reinforced by the Single European Act, which is relevant not just to the EC but to its other European associates, which include the other ESA members. The EC is involved in many relevant areas of technology, as its burgeoning R&D programmes make clear, and may well serve as a vehicle for enmeshing space technologies in a wider process of technological modernization, a dimension which goes beyond the functionally specific remit of ESA. The EC has user needs of its own and to an extent can promote them in its developing country associates. And the EC has the European Parliament, which can be a vehicle for promoting a wider political awareness of the relevance of space, as its recent debates on the 1987 Toksvig report illustrate. The important question which follows is how far ESA and the EC will succeed in establishing complementary and mutually supportive roles.

It is clear that overall Europe's industrial infrastructure and performance in the space segment could be improved, and indeed must be improved if Europe is to sustain a larger space programme and withstand international competition. Firms, governments and institutions have a common interest in this, though inherited patterns and divergences of interest will remain limitations. Some are more interested in the less exotic and potentially more profitable areas, while others see their best returns from large projects with a long life-cycle. In neither field do the conditions in Europe currently exist for effective competition. In the contemporary climate it will be difficult to persuade politicians and public opinion across Europe to support larger public investment in space, unless the economic and technological returns are demonstrably positive.

It would be unrealistic to expect a wholesale restructuring and rationalization of the European space industry. But there is a visible and increasing willingness on the part of the firms to improve the industrial infrastructure on a sustained and trans-European basis. The most likely prospect is a continuation of the regrouping of alliances among European firms in response to contract opportunities, an adjustment process which could result in a smaller number of larger firms with a network of smaller companies clustered around them. But this will not happen unless both governments and the various European organizations help to nudge that process along. And a balance remains to be struck between the development of large public programmes and the opening up of markets for products and services derived from space technologies.

Larger programmes and big projects also demand a different management style from that appropriate to the majority of ESA programmes so far, with the key exception of Ariane. Managerial difficulties within Nasa provide a salutary warning. More than one management model is possible, from the hegemonic to the collective. Hegemonic methods, however valuable, are probably not appropriate within Europe. Some lessons can be drawn from other areas of technology, both civilian and military, especially in the aviation sector. Collective management can be made to work, but it requires more than loosely-knit consortia.

As for the link with adjacent technologies, many European collaborative initiatives are now under way through Eureka, the EC's R&D programmes (notably, for space purposes, Esprit, Race, Brite and Star) and the IEPG. The optimization of resources and technology transfers within Europe are unlikely to be achieved unless bridges are built to produce effective cross-disciplinary technologies and greater competitiveness.

US-European relations

Lastly, Europe's relationship with the US raises difficult policy issues across the range of space applications. Many ESA missions have depended on Shuttle launches, with the current result that the science programme, for instance, is now way behind schedule. Europe remains dependent upon the United States in certain important technologies: above all, its dependence in the field of electronic components for space systems is becoming alarming, as a recent IEPG report argued in the context of European defence industries in general. Overall, about 10 per cent of components used are US-sourced and not currently capable of European substitution – for example, high-quality magnetic tape recorders. But on some projects the imbalance is disturbing: ERS-1 is 60 per cent sourced from Europe, 30 per cent from the US, and 10 per cent from Japan; and on Rosat, more than 50 per cent of the inputs are from the US. No doubt the relevant technologies could be produced indigenously, but not necessarily at competitive prices.

The development of the US military space effort has had positive effects on the technological level of the US civilian programmes, whereas in Europe there are still practically no such spin-offs to the civilian space sector, and the European electronics industry anyway has many other problems. So far, however, Europe has not been technologically dependent upon Japan on a major scale, even though there is growing European-Japanese competition in an increasing number of space-oriented activities.

On the other hand, the European space industry has had access to important international markets through consortia with US firms. To take one example, British Aerospace does more business with Hughes than it does for ESA, a commercial response to the level and character of demand in the market-place.

This is not to say that the European space industry's experience of collaboration with the US has always been happy. Several European firms have found the conditions attached to collaboration onerous. US federal regulations impede the access of European firms to the US market. And not far down the road lie important issues about technology transfer which will arise the moment the first decision is made, if it is, to accept a Soviet (or even Chinese) launch of a European or part-European satellite. The attractive terms being offered for Proton may prove irresistible. The foreign policy dimension is of crucial importance.

But the most sensitive issues in US-European civilian space relations arise from the big projects in which Europeans have joined, or contemplated joining, Nasa. ESA took part in Spacelab with a degree of real enthusiasm. It provided a welcome opportunity to take part in a manned programme with important scientific opportunities. ESA could perhaps have negotiated more advantageous terms, but it could not overcome dependence on American decision-making over which Spacelab missions flew and when. This negative experience has bred intense European caution over the international space station. The Americans see the station as an extension of their national territory and thus governed by their own laws. For example, they consider that all inventions aboard the space station will be American. Similarly, they have proposed a rather unequal sharing of the workload: experiments on material in microgravity will be conducted in American modules; the European laboratory is to deal more with life sciences, which are far less attractive commercially. A request by the DoD that the space station be utilized mainly for military programmes and functions is also posing a grave political problem for ESA, which is limited by its charter to activities of a 'peaceful nature'. Even so, it may be a valuable means for Europe to develop an indigenous space station at a later date. Everything hangs on whether terms can be negotiated which will permit European interests to be adequately safeguarded.

Conclusion

Since European space activities remain characterized by the coexistence of national and collaborative programmes, Western Europe's space policy reflects a complex game of competition and cooperation among member states. For reasons of national pride and economic interest (space programmes help maintain a country's industrial capacities and are expected to have a positive impact on employment), European governments are reluctant to move squarely towards a fully integrated space policy, even though the size of investments and of the programmes ahead require such a qualitative jump to be made urgently. In the meantime, national agencies (for example, in the case of PTT administrations) still behave as competitors, and the virtual absence of transnational actors favours the persistence of individual interests. In the field of defence and security Europe still lacks the political cohesion and the common definition of requirements needed to promote a collective view of where space technologies can enhance security goals.

All in all, despite remarkable achievements over recent decades, Europe is still far from having a true *policy* of its own in space. Persistent protectionist tendencies, a fragmented industrial base, differing security perspectives and the lack of wide popular support constitute major obstacles to the much-needed unification of European space efforts in both the civilian and the military sectors. Yet recognition of these constraints and weaknesses is beginning to be reflected in a debate which is both more realistic and more open. The key question is whether this can produce the redefinitions of policy goals and the capabilities necessary to guarantee an effective future in space for Europe.

8

PRIORITIES

There are no insurmountable obstacles to maintaining and improving Europe's performance in space. Nothing is needed that cannot be built upon existing assets. Some courage, coupled with a limited increase in space budgets and a greater awareness of common aims, should suffice. But the time for decision is now: international competition is growing, and the European space industries and scientists need time and money to set the necessary programmes in motion. Different perceptions and priorities, in the various European nations, are part of the problem. Should the required qualitative jump in space activities be made only by some European countries, leaving the others behind, the outcome would probably not improve Europe's international competitiveness. These differences should not, however, overshadow the importance and the degree of consensus already in place. The biggest threat to Europe's future role in space comes from its cumbersome and slow procedures for decision-making and from the delays in coordinating and integrating the various European activities.

The limitations

The importance of the national and European programmes described earlier in this report cannot be denied. However, they will not easily guarantee Europe's autonomy in space, the sustained competitiveness of its space industries or its position as the 'third' space power.

External constraints

Some uncertainties derive from the past experience of cooperation between ESA and Nasa. Many European programmes have been based on cooperation with the United States. Should this cooperation become more difficult, a complete re-examination of the European programmes would be unavoidable. The lessons from the past give ground for future concern. The negative experience of Spacelab is a case in point and may have held Europeans back from exploiting the potential of applied science in space.

In general, the major space projects have been decided by the major space powers, the USA prominent among them. The US-European experience has been one in which the Europeans have only been able to reject or accept participation in US-defined and US-led projects, and never the other way round. Even good European ideas have sometimes had to be implemented as American-led projects, with European involvement only later. France, to a lesser extent Britain, and now also ESA, are enlarging their options by cooperating with the USSR; Japan and China, too, are already attractive collaborators. But the USA is, and will remain, Europe's main partner for many years to come.

Internal constraints

Problems within Europe also have negative repercussions on European space policy as a whole. The level of funding, in particular, is crucial. ESA members have accepted a significant increase in the ESA budget, while at the same time expanding their national space budgets. But the new programmes, with their ambitious aims, require far larger funding. By mid-1987, calculations of the total cost of programmes such as the Ariane 5 launcher, the Columbus space station and the Hermes shuttle had grown from 7.2 to 11.3–12 billion EAU. The total cost of these big programmes was thus broadly comparable with the US share of the international space station's total cost, then estimated at about $12 billion.

Financial allocations in a time of heavily constrained budgets and lukewarm public support increase the temptation to identify 'national' ventures based on 'national' perceptions of what should be done. Differences between the various national space budgets in absolute terms, the fragmentation of those budgets between various ministries, the division between military and civilian expenditures, as well as the division between European and national allocations, have to be taken into account as added weaknesses. If any major European country were to drop back from its commitments to ESA, as has been contemplated in Britain, the new ESA package would be put in jeopardy. The other members would then have to face the choice between increasing their own financial contributions and scrapping or delaying parts of the programme.

Many of these problems could be avoided if there was a direct source of secure funding for the European space programme. For example, a levy based on national wealth, building on the existing arrangements for funding ESA's mandatory programmes, would enlarge the resource base available for optional programmes.

Pressures of time

The timing of the proposed European programmes is crucial. It is clear that some projects will not become fully operational until well into the 1990s, by which time other space powers will have significantly progressed in competitive areas, such as space transportation, electronic intelligence and telecommunications, ground stations and cheap expendable space launchers. Therefore Europe cannot risk major delays.

Industrial performance

A further constraint is Europe's limited industrial and technological capacity in some space activities. European dependence on certain American and Japanese technology is undeniable, as is the relative difference in scale between European and American space manpower.

In short, the biggest risk for Europe is that of remaining the junior partner of the two existing major space powers (and particularly the USA), while losing ground to Japan. The objective of European 'autonomy' identified by ESA would then become unachievable and Europe's competitiveness would decline.

The challenges

Two challenges are particularly relevant, and whether or not they are met depends on how far West European governments are prepared to go in taking effective independent action. Telecommunications and security are the two key fields in which Europe must develop independent capabilities if it is to achieve real autonomy in space. And so far Europe has not found a satisfactory way of dealing with them on a collective and multilateral basis.

Telecommunications

It is in the field of telecommunications that space has had its greatest economic success and has attracted the largest number of 'private' users. But Europe still lags behind the US and Japan, because a combination of industrial inefficiencies, national protectionism and the monopolistic structures of most PTTs continues to impede European effectiveness. American industry has thus been able to capture much of the world market, notwithstanding the existence of important European technological capabilities and significant ESA and national programmes. An energetic European policy must therefore include not just high-quality hardware and software, but a deregulation of all the PTTs and the elimination of protectionism in this field. This in turn should be part of the implementation of the Single European Act and the drive to complete the European internal market by 1992.

Security

The second challenge is security, for which there is no multilateral European space policy. European space requirements that are not dealt with nationally are generally handled through Nato or in direct cooperation with the USA. Even bilateral European projects in this field, so far generally inspired by France (which is outside Nato's military framework), have not been easily agreed.

There is, however, a growing awareness of the utility of space for security purposes. Although defence budgets, too, are heavily constrained and cannot easily be stretched to increase spending on space activities, there is an obvious advantage in planning, wherever possible, for the closer integration of military and civilian ventures. Already some of Europe's civilian satellites have a built-in capacity reserved for military use. The increasing 'need to know' and the necessity of having better, more secure C3I systems are recognized by European military establishments.

Satellite reconnaissance stands out as a crucial resource for the future international and diplomatic role of Europe: without satellites the Europeans will lack a very important element of independent appraisal of developments in many regional theatres, for example, Libya and Afghanistan. The resolution of images required for such a purpose is higher than the 10–20 metres offered by existing commercial optical satellites such as Landsat, though it does not need to be as sharply defined as the 1–2 metres used for specific, tactical military purposes. (See Table 8.)

Already Landsat and Spot images have been very useful for assessing such events as Chernobyl; the Iran-Iraq war; the purported construction of a SAM SA-5 Soviet site in Libya; the launch test of Soviet SLBMs beneath the ice of the Arctic Ocean near Wrangel Island; the purported construction of new Soviet air and naval bases in the Kola peninsula; the establishment of an alleged new Soviet base of mobile intermediate-range ballistic missiles (SS-20s) near Kirov; the Soviet space shuttle facilities at Tyuratam; and the secret Iraqi facility near Samarra, allegedly producing chemical weapons. In January 1986 it was also reported that Landsat imagery was being used to monitor Soviet military activities in the Far East.

Arms control may now stand a better chance than it has for some while, and as a result increased attention will be paid to the balance of conventional forces. The Atlantic Alliance will need enhanced capabilities for observation from space, both to verify arms control agreements and to ascertain whether the military situation is stable. The European members of Nato could contribute to an equitable sharing of burdens within the Alliance by creating an observation capability of their own. A West European system, adapted to the special circumstances of the European continent, would guarantee a desirable degree of redundancy within the Alliance and could also strengthen the strategic consensus between the United States and its European allies. Both domestic public opinion and other interested parties could be reassured that such a West European observation system would serve no other purpose than to ensure compliance with agreed understandings and to give support for world stability.

Although ESA cannot deal with military programmes, a case could be made for involving it in arms control programmmes of a clearly peaceful nature, and in other systems of communication and warning. More specific military programmes, directly related to battle management and crisis management, could be dealt with through other European institutions, such as WEU, or specific multilateral agreements. So, too, could systems for communications and electronic intelligence, though they also have a utility for arms control purposes.

Industrial competitiveness

No long-term solution to Europe's future in space can be found without a more competitive European space industry. National divisions (evident in the telecommunications field), the quicker pace of some competitors (such as Japan), the absence of a large military sector (in contrast to the American and Soviet cases) and the comparative economic advantage of less developed countries (such as China) represent tremendous challenges. Europe should first of all tackle its internal industrial problems.

Lack of standardization and economies of scale are problems for the entire European space industry, particularly in the critical field of space launchers. The whole of the Ariane programme has produced a very large number of single 'prototypes', without yet developing sufficient capacity to permit launches 'upon request'.

The economic and industrial importance of the 'ground segment' of space operations should not be underestimated. In commercial terms ground stations and infrastructure are crucial, and it is here that Japanese and American competition is toughest. More open public markets are needed, especially in the PTT field, as well as common European standards. In specific fields such as telecommunications, links should be encouraged with other European ventures in high technology.

It should be possible to develop cost-effective collaboration which would both stimulate competition and avoid wasteful duplication. But it would be wrong for Europe to rely on a single launcher system, a single engine construction site and a single launch site. European needs are not necessarily best served by monopolistic suppliers of space systems: some intra-European competition is desirable in each main area of application. Many national programmes were justified in the past by the absence of a European alternative. The problem now is to avoid the persistence of purely nationalistic approaches and protectionist measures, since these increase the user's costs and progressively push the European industry out of the market. Only if these conditions are fulfilled can there be an effective industrial base in Europe.

It is painfully clear, at this point, just how far Europe still is from a common and coherent space policy. What Europe needs is a common approach to the exploitation of space. The basic elements of that policy are set out below.

Autonomy: a new policy approach

European governments must accept the fact that Europe's future role and influence in world politics, in global markets and in the cultural life of the next century may largely depend on its capacity and willingness to explore and use space, to develop the necessary technology and to build up the required industrial infrastructure. At the same time governments and the public need to recognize space as an important instrument in achieving the political (including security), economic and cultural integration of Western Europe, as part of the process of developing a stronger European identity. These considerations all point to the overall political objective, already identified by the 1985 ESA ministerial meeting, of acquiring a European identity and autonomy in space. This autonomy, in the European context, can be defined as 'a capability to reach, to operate in and to return from space, and to do so, not on sufferance of friend or foe, but according to its own perception of what is to the common good'.

Autonomy thus connotes the ability to shape a future space programme independently of others. It presupposes that Europe should have space transportation and recovery systems for crews and materials; a space infrastructure consisting of scientific satellites and observatories, space-based remote-sensing systems and telecommunications networks, manned or man-tended space platforms, and a satellite-based capacity for the verification of arms control agreements and the observation of relevant military developments; and the necessary ground installations to establish, maintain and use these systems. Europe would also need to deploy both people and robotics in space in complementary roles.

There is, however, an important difference between autonomy and autarky. Europe is not, and cannot be, autarkic in space, whereas European autonomy, even if based on a large utilization of foreign components and technologies, is politically and technically compatible both with Europe's industrial capacities and with its relations with the USA and Japan (and with the USSR). Moreover, Europe has a vested interest in creating an open and interdependent space market in which national preferences are not allowed to dominate international cooperation, and from which protectionist barriers have been removed. Our objective is to build a competitive and autonomous Europe in space, not to help segment space within national boundaries. On the contrary, Europe should also work towards establishing an international space order.

Components of autonomy

The first and most evident priority lies in the field of *launch capabilities*. Important economic and industrial interests in this field should be encouraged. Europe must free itself from its current dependence on the USA for the costs, timetable and completion of important European space projects.

The development of new launchers with greater lift capability (Ariane 5), of partially recoverable systems (Hermes), and possibly of new, fully retrievable ones, should increase the competitiveness of European space transportation, especially if a greater degree of 'industrialization' in the production and operation of launchers is established. More satellites and more activity in space will require more, and more efficient, launchers: Ariane 5 and Hermes, based on mature, traditional transportation systems, are already in sight. Industrialized launchers, routinely produced and operated, should be developed from the Ariane family. Fully retrievable and reliable launchers are rather further off, but remain the best hope for cheaper and wider access to space.

The present scale of European space operations does not seem to require new launch sites immediately. The industrialization of traditional launchers could, however, require an additional European launching site in the not too distant future. Should a new generation of launchers be developed, the problem would probably be simplified, since these 'space planes' might use traditional air bases, with few modifications.

The choice of whether or not to send human crews into space has, at some point, to be openly confronted. Although no robot can substitute for the versatility of a human being, people cannot be regarded as expendable. The cost of making a system 'safe' for humans, as opposed to making it 'reliable', is huge. Europe should aim at complementarity between robots and humans, and avoid concentrating on humans for the sake of it or to create an image of prowess.

Another key priority for any European space policy will remain the construction and utilization of *various kinds of satellite*. In this sector European technology is often very competitive; a high degree of autonomy can be reached, and a greater European presence in some fields (such as telecommunications, Earth observation) would have the beneficial effect of increasing international competition. Developments in direct broadcasting and TV distribution satellites, the growth of new services in education and training, and the prospect of the deregulation of telecommunications all point to the need for a determined European effort. The fact that telecommunications in space have already reached a degree of self-financing is a positive element.

In the field of Earth observation, the dual use of satellites for military and civilian purposes (both long-term evaluation and management), the forecast growth of a private market for satellite images, and the value of remote sensing in such fields as meteorology and environmental protection, all serve to emphasize the increasing importance of space systems. Similar conclusions can be drawn from recent developments in the field of communications (for maritime, aeronautical and land-based mobile vehicles), search and rescue, and monitoring systems such as Elint.

A third priority for European space policy is the *space station*. In the short term the likelihood is that this will be achieved by collaboration rather than independently. Cooperation with the USA and Japan through the Columbus project of ESA opens up bright prospects. This should allow Europe to develop capabilities which will increase its autonomy and identity in space. A higher degree of European control over the costs and operation of the space station, and especially of the European platforms, is an absolute requirement for European participation in the overall project.

Finally, the importance of the *ground segment* of space operations must be reiterated as a precondition of achieving effectiveness on a European scale.

Security

Europe has no collective space capabilites for security purposes. It is clear from the points made earlier in this report that it is in the security field that the challenge to Europe is the greatest. Like it or not, the militarization of space is already well advanced, as the sections on the USA and USSR show. Space has become a major arena of security. Even if Europe had all the means at its disposal, it should not opt for a massive and aggressive military presence in space. On the contrary, it should do its utmost to support the ABM

Treaty and to uphold all the legal instruments available to prevent the further exploitation of space for military purposes.

However, this does not mean that Europe should remain idle in this area; it must look to its own long-term security and be able to influence the international space order. Europe should therefore equip itself with the following minimum requirements: (1) autonomous capabilities for observation, reconnaissance and verification of arms control agreements, based on optical and radar satellite systems; (2) closer intra-European military communications systems, building on what is already available; (3) shared intelligence-gathering, to give European nations a capability for early warning and crisis management; (4) the means to protect European assets in space; and (5) studies of the option to deter attacks on European assets.

Obviously, to carry out such a programme would require an appropriate balance between the Nato framework and the existing programmes being pursued in Europe. It would, however, be a mistake to wait for a joint European military and defence authority to be created. Pragmatic bilateral or trilateral channels, as well as such institutions as the Eurogroup, the IEPG, WEU and EPC, have roles to play in developing a collective approach. The development of systems capable of performing the relevant civilian and security-oriented operations should also be encouraged within the ESA framework, where the existing technical infrastructure could be legitimately drawn on for peaceful security purposes. Reconnaissance satellites are a prime example of where ESA could make a contribution. An evolution of this kind would allow individual European nations to make differentiated use of the various systems available.

The legal order

One key feature of Europe's autonomy and identity in space will be its ability to develop a credible and influential role in worldwide space cooperation, both in specific space projects and through its capacity to envisage a new space order.

The problem of establishing a new space law, and new legal instruments (besides those already in force), capable of dealing with the many security and economic problems related to the human presence in, and utilization of, space, is becoming urgent. The interests of both producers and users must be taken into account, as must the legitimate concerns of the international community as a whole. Indeed, a greater awareness of the international interests of all mankind might help to defuse Soviet and American inclinations to engage in superpower confrontation in space. The establishment of a new space regime should be made a priority subject of EPC, so that Europe can become a major actor in this arena.

Institutional requirements

The future of Europe in space has to be built on the existing reality. European space activities are generally carried out by the various national agencies or ministries, since it is they that take the relevant budgetary decisions past institutional and political obstacles, lobby for greater space budgets, gather public support and identify economic interests and technical capabilites.

One necessary step forward, therefore, is a more strategic coordination of space activities and policy at the national level, to overcome at least some of the problems deriving from intra-governmental conflicts. One means to this end is the use of national space agencies, which may tend to promote particular national interests, but have the advantage of providing a single and coherent interlocutor.

Other institutions, such as the EC and WEU, have existing and potential competences related to space policy. ESA has a wider European membership than the EC or WEU, or even Nato. No real conflict need develop among these organizations. On the contrary, at least as far as the EC and WEU are concerned, there is a strong case for putting their political weight behind the general thrust of ESA's programme. The European Parliament, for example, has endorsed the necessity of increasing ESA's budget, as a way of helping Europe towards the goal of autonomy.

But there is a need for democratic scrutiny and control over the substantial expenditures needed for space, and for public debate on the strategic and security dimensions. Hence we welcome the general thrust of the debates in the European Parliament on this subject. But we deplore the Parliament's very limited impact on the actual decisions taken. Currently European space policy decisions are shaped by the demands, interests and the almost exclusive expertise of the specialized space policy community (the manufacturers, the military, the bureaucrats and the scientists, at both national and European levels). Space is more important than those interests.

The general goal of a strong autonomous European presence and identity in space will not be attained without greater public support.

Furthermore, a greater degree of democratic control and public awareness of the choices involved in European space policy is not merely desirable; it will become a necessity as that policy reaches a higher degree of importance and maturity.

9

RECOMMENDATIONS: TOWARDS EUROPEAN AUTONOMY

The analysis of Europe's needs and achievements in space clearly shows that a significant increase in European activities in this field is an urgent priority. The indispensable basis for European action consists of implementing ESA's Programme of Action. All governments and political forces in the participating countries should give full support to the ESA programme as the precondition for achieving Europe's stated goal of autonomy in space.

I Space transportation systems

1 European autonomy in space can be achieved only if Europe has the capability to place objects in space and to launch and recover manned spacecraft. The present policy of developing Ariane 5 should therefore be continued. The goal should be to turn Ariane 4 and Ariane 5 into cost-effective launchers for Europe that will allow routine production, and so be competitive in the market. This is all the more pressing now that Soviet and Chinese launches are being offered at attractively low prices. The present policy of developing Hermes should also be continued.

2 The next generation of fully retrievable space transportation systems, required early in the next century, should be embarked on within ESA, in order to give Europe a wider option for operational access to space. New and re-usable hypersonic systems which have air-breathing as well as traditional rocket propulsion systems, and which can start and land horizontally on conventional airfields (like the Hotol and Sänger projects), should

be pursued in the study phase today as competitive programmes, with the aim of then selecting one project for joint European development in the future.

3 Europe must ensure that it preserves the present launch site at Kourou. An expanded programme may require additional launch facilities.

II Satellites

1 Europe needs to have its own eyes and ears in space. It must make wider use of space-based systems for telecommunications, observation, navigation and meteorology, and in doing so strengthen Europe's technological base and its competitiveness in the world market.

2 Europe should develop and operate a joint reconnaissance satellite system in low earth orbit for safeguarding Europe's political and diplomatic interests, as well as for arms control verification, the monitoring of military movements and crisis control. This system should be based on existing remote-sensing techniques (optical and microwave) and could be complemented by a jointly operated electronic intelligence-gathering package.

3 Europe should lose no time in developing a data relay satellite (DRS) system for multiple purposes, and examine whether an international DRS system between Europe, the US and Japan would meet European needs.

4 European cooperation in using communications satellites for security purposes should be strengthened. Collective planning of the architecture of national systems, both civil and military, and of their management would be a valuable step towards achieving interoperability and operational flexibility.

5 In order to protect European satellites against potential Asat threats, Europe should actively prepare for and participate in negotiations on a code of conduct for the peaceful uses of space, based on a regime of confidence-building measures and rules for space use. At the same time, Europe should study means of protection for its satellites in the context of the right of self-defence.

III Space platforms and stations

1 Europe's long-term goal should be to establish a space station of its own to provide a central element of its space activities and a highly visible symbol of Europe's capabilities and common will.

2 Europe's immediate and medium-term goal should be to participate through the Columbus programme in the development and operation of the space station proposed by the US and augmented by a man-tended free-flying module. This should be done as a step towards establishing an autonomous European space station at a later stage. However, that participation should be contingent on European conditions for using Columbus being met: namely, genuine partnership in the management, utilization and financing of the station. If the negotiations fail to create the preconditions for such a partnership, Europe should build, launch and operate a space station of its own.

3 Europe should continue to develop partially or fully automated space platforms for scientific and industrial purposes.

IV Ground infrastructure

1 The ground segment for the operation and use of space systems must be planned and built as an integral part of Europe's space activities.

2 Nationally operated ground facilities should be incorporated more closely in a European system. To this end, European countries should agree on common technical standards for the ground segment. The export of ground stations should be strengthened through the production of standardized and low-cost systems.

V Industrial effectiveness

1 Governments and the European organizations should set the framework for achieving the competitive and sophisticated industrial infrastructure that is required to underpin a more vigorous European space policy in both civilian and security uses of space.

2 It is vital to ensure that collaborative programmes are developed with a steady rhythm, to provide steady business expansion and to raise the level of skilled employment.

3 Greater economic efficiency should be sought by urgent efforts to improve the competitiveness of the space sector and to liberalize markets. Europe must deregulate, unify its internal market, europeanize public procurement, consolidate viable transnational consortia of space companies, supply a larger share of the home

market for the ground segment, and improve its capacity to bid successfully for international contracts in both the space and ground segments. The implementation of the Single European Act is an essential precondition.

4 Further efforts are needed to promote technological innovation across the sectors relevant to both the production and the application of space technologies. This should be accompanied by measures to stimulate user demand on a European basis.

VI Space science and research

Europe should maintain a strong scientific programme, in order to improve its position in the exploration of the universe and in the widening of our horizons. European scientists should take a broader share of space science activities than in the past and be encouraged to give greater emphasis to the application of science to practical use. The ESA programme 'Horizon 2000' points in the right direction and deserves full support by the governments.

VII Institutional framework

1 ESA should be supported and strengthened as the centre of European space policy. Its successful mixture of mandatory and à la carte programmes and its focus on peaceful purposes should be maintained. Over time it should consolidate its role as the central policy-making body for Europe in the civilian field and represent Western Europe vis-à-vis the rest of the world.

2 Europe should continue to develop and operate jointly European applications satellites through specialized European organizations such as Eutelsat and Eumetsat.

3 The European Community should play a greater role in setting common norms and standards for the European market and in helping to develop relevant technologies.

4 The development of European security collaboration in space should proceed in spite of the current institutional inadequacies. But it will eventually require an appropriate institutional framework, building on the existing potential of WEU, EPC, IEPG and the Eurogroup. A strengthened WEU, for example, could become

the European institutional framework for reconnaissance satellite systems and eventually electronic intelligence satellite systems.

VIII And beyond ...

An effort should be made within ESA to study European space policy up to the middle of the next century. The study should in particular cover the possibility of using a European space station for the exploration of, and the establishment of bases on, the Moon and planets of the solar system, either in a European context or in cooperation with non-European countries.

To achieve autonomy, a substantial increase in funding has to be decided on and a fairer distribution of the costs among our countries has to be accepted. Even though it is notoriously difficult in this field to pin down exact and reliable figures, it should be realized that the implementation of the recommendations outlined above would require approximately the doubling, in volume terms, of the current level of annual European expenditure up to the year 2000. This would provide the resources for our recommended ESA plan and a new European security programme, as well as enabling national expenditure to continue. This figure of on average about 4.5 billion EAU per year over 13 years, while considerable in absolute terms, would represent only about one-fifth of current US spending. A reasonably modest price to pay for European autonomy in space.

TABLES AND FIGURES

Table 1 European members of relevant international organizations

	ESA	Eutel-sat	Eumet-sat	EC	WEU	Nato (Pol.)	Nato (Mil.)	Intel-sat	Inmar-sat
Austria	×	×	×	–	–	–	–	×	–
Belgium	×	×	×	×	×	×	×	×	×
Cyprus	–	×	–	–	–	–	–	×	–
Denmark	×	×	×	×	–	×	×	×	×
Finland	–	×	×	–	–	–	–	×	×
France	×	×	×	×	×	×	×	×	×
W. Germany	×	×	×	×	×	×	×	×	×
Greece	–	×	×	×	–	×	–	×	×
Iceland	–	×	–	–	–	–	–	×	–
Ireland	×	×	×	×	–	–	–	×	–
Italy	×	×	×	×	×	×	×	×	×
Liechtenstein	–	×	–	–	–	–	–	×	–
Luxembourg	–	×	–	×	×	×	×	×	–
Malta	–	×	–	–	–	–	–	–	–
Monaco	–	×	–	–	–	–	–	×	–
Netherlands	×	×	×	×	×	×	×	×	×
Norway	×	×	×	–	–	×	×	×	×
Portugal	–	×	×	×	–	×	×	×	×
San Marino	–	×	–	–	–	–	–	–	–
Spain	×	×	×	×	–	×	–	×	×
Sweden	×	×	×	–	–	–	–	×	×
Switzerland	×	×	×	–	–	–	–	×	–
Turkey	–	×	×	–	–	×	×	×	–
UK	×	×	×	×	×	×	×	×	×
Vatican City	–	×	–	–	–	–	–	×	–
Yugoslavia	–	×	–	–	–	–	–	×	–

Table 2 European collaborative programmes for satellites and vehicles (a broad selection)

Name	Date	Function	Sponsors
Heos 1	1968	Scientific/magnetic fields	ESRO/ESA
Heos A2	1972	Scientific probe	ESRO/ESA
TD-1A	1972	Scientific/radiation	ESRO/ESA
ESRO-4	1972	Scientific/particles	ESRO/ESA
Symphonie A/B	1974-5	Experimental Comsats	Franco-German with Belgium
Cos B	1975	Scientific/cosmic rays	ESRO/ESA
ISEE	1977	Earth-sun probe	Nasa/ESA cooperative
Meteosat 1/2	1977	Weather satellites	ESA/Eumetsat
IUE	1978	Ultra-violet	Nasa/ESA/UK
Geos 2	1978	Scientific geomagnetic	ESA mandatory
OTS-2	1978	Pre-operational Comsat	ESA
Marots	1978	Maritime communications	ESA
Marecs A/B2	1981–4	Maritime communications	ESA, lease to Inmarsat
Spacelab	1983	Manned scientific module	ESA/Nasa using Shuttle
ECS-1-5	1983–5	Comsats	ESA, lease to Eutelsat
Giotto	1985	Halley's Comet intercept	ESA mandatory
Exosat	1986	X-ray observation	ESA mandatory
Meteosat PL	1987	Weather satellite	Eumetsat
TDF-1/TV-Sat 1/2	1987–9	TV satellites	Franco-German
Inmarsat 2	1988	Maritime communications	Inmarsat, procured in Europe
Olympus (L-Sat)	1988	TV satellite	ESA
Hipparcos	1988	Scientific/astronomical	ESA mandatory
TDF-2	1988–9	DBS	Franco-German
Meteosat 3	1988-90	Weather satellite	ESA/Eutelsat
Eureca	1988–9	Retrievable carrier	ESA
ERS-1	1989	Ocean sensing	ESA
Ulysses	1989	Scientific/astronomical	ESA mandatory
Iso	1992	Scientific/astronomical	ESA mandatory
Columbus	late 1990s	Space station module	ESA
Hermes	late 1990s	Manned spaceplane	CNES/ESA study

Table 3 European national satellites

Country/Name	Date	Function/remarks
France		
Asterix	1965	First French satellite
Diapason	1966	Scientific
FR-1	1966	Scientific
Diadème 1	1967	Scientific
Diadème 2	1967	Scientific
Wika/Mika	1970	Scientific (first foreign payload)
Peole	1970	Experimental Meteosat
Eole	1971	Experimental Meteosat
Tournesol 1	1971	Scientific
D-2A	1973	Scientific (launch failure)
Starelette	1975	Research vehicle
Castor/Pollux	1975	Twin scientific launch
Aura	1975	Scientific
Télécom	1985	Civil/military Comsat
Spot 1/2	1986–9	Remote sensing
W. Germany		
Azur	1969	Scientific
Dial	1970	Scientific
Avos 1	1972	Scientific
Avos 2	1974	Scientific
Helios 1	1974	Solar probe
Helios 2	1976	Solar probe
SPAS–01	1983	Shuttle payload
IPS	1985	Shuttle payload
Rosat	1987	X-ray telescope (launch deferred)
Italy		
San Marco 1	1964	Scientific
San Marco 2	1967	Scientific
San Marco 3	1971	Scientific
San Marco 4	1974	Scientific
Sirio	1977	Experimental Comsat

(*continued*)

Table 3 (*concluded*)

Country/Name	Date	Function/remarks
Netherlands		
ANS-1	1974	Scientific
Iras	1983	Scientific (with US and UK)
Spain		
Intasat 1	1974	Experimental
Sweden		
Viking	1985	Experimental
Tele-X	1987	DBS, launch pending (based on TDF)
UK		
Ariel 1	1962	Scientific
Ariel 2	1964	Scientific
Ariel 3	1967	Scientific (first *all*-British satellite)
Ariel 4	1971	Scientific
Ariel 5	1974	Scientific
Ariel 6	1979	Scientific
X-4 (Miranda)	1974	Experimental
UoSat 1	1981	Experimental (ham radio)
UoSat 2	1984	Experimental (ham radio)
UKS	1984	Scientific
Skynet 1	1969	Military Comsat
Skynet 1B	1970	Military Comsat (failed in orbit)
Skynet 2A	1974	Military Comsat (launch failure)
Skynet 2B	1974	Military Comsat
Skynet 4	1987	Military Comsat (launch pending)

Table 4 Profile of expenditure for ESA programmes 1987–2000 (in millions of EAU at 1986 value)

Programmes	1987	1988	1989	1990	1991	1992	1993	1994	1995	1996	1997	1998	1999	2000	Total
General budget	124.7	120.0	126.0	138.1	148.0	161.0	167.0	170.0	170.0	170.0	170.0	170.0	170.0	170.0	2174.8
Associated to general budget	66.1	55.0	52.6	57.7	62.9	67.1	67.1	68.1	68.4	69.7	70.8	71.8	72.9	74.0	924.2
Science	170.3	176.8	185.3	196.7	206.4	216.7	227.9	232.0	232.0	232.0	232.0	232.0	232.0	232.0	3004.1
Other technology programmes	6.7	4.4	12.8	21.0	23.9	27.3	30.3	32.4	32.4	32.4	32.4	32.4	32.4	32.4	353.2
Earth observation	174.1	197.9	228.3	224.2	234.9	239.7	237.2	236.9	230.7	252.3	255.9	251.4	261.1	267.4	3292.0
Microgravity	13.0	46.9	78.1	110.0	110.0	110.0	110.0	120.0	130.0	142.0	142.0	142.0	142.0	142.0	1538.0
Telecommunications	289.6	248.5	215.5	192.6	228.7	244.1	235.7	267.8	293.0	319.0	339.0	275.7	222.9	247.9	3620.0
Space station and platform	195.3	228.1	312.5	363.4	469.0	478.0	513.0	525.0	555.0	600.0	678.0	681.0	890.0	877.0	7365.3
Space transportation systems	446.4	516.4	669.6	833.0	867.0	980.0	1074.0	1010.0	952.0	810.0	707.0	772.0	599.0	575.0	10811.4
Grand total	1486.2	1594.0	1880.7	2136.7	2350.8	2523.9	2662.2	2662.2	2663.5	2627.4	2627.1	2628.3	2622.3	2617.7	33083.0

Source: European Space Agency, 1987.

Table 5 ESA: industrial returns, from 1 January 1972 to 31 December 1984 (in thousands of EAU).

Country	Unweighted amount[a]	Weighted amount[b]	Ideal amount[c]	Surplus/ deficit[d]	Return coefficient
Austria	14866	11597	12481	– 884	0.93
Belgium	174609	163451	175390	– 11939	0.93
Denmark	76696	62237	60551	+ 1786	1.03
France	1665255	1384526	1353034	+ 31492	1.02
W. Germany	1204736	1118191	1085754	+ 32437	1.03
Ireland	4772	3933	2716	+ 1217	1.45
Italy	532404	497197	519210	– 22013	0.96
Netherlands	201864	142776	145633	– 2857	0.98
Norway	7130	7053	7741	– 688	0.91
Spain	120608	101257	120572	– 19315	0.84
Sweden	92219	82672	89215	– 6543	0.93
Switzerland	87123	78684	84303	– 5619	0.93
UK	658996	592787	574134	+ 18653	1.03
Canada[c]	55435	55435	61712	– 6277	0.90
Total	4896713	4301896			

Source: European Space Agency, ESA/IPC (85)4, 15 January 1985.
[a] I.e., actual amount spent.
[b] Amount after correction, to take account of distortions.
[c] ESA's notional calculation of what ought to be the case.
[d] Difference between weighted amount and ideal amount.
[c] Associate member.

Table 6 Launchers: a historical survey by area

Area/name	Date	Performance	Remarks
Europe			
Europa	1962–71		ELDO experiment, failed
Black Arrow	1969–71		UK, discontinued
Diamant	1970–3	0.23 tonnes LEO	France, pioneering venture
Ariane 1	1979–86	1.7 tonnes GTO 0.95 tonnes GEO	ESA series
Ariane 2	1984	2–2.5 tonnes GTO	
Ariane 3	1984	2.5 tonnes GTO	
Ariane 4	1986	1.9–4.2 tonnes GTO	6 versions
Ariane 5		15 tonnes LEO 8 tonnes GTO	Preparatory study
USA			
Atlas family	1958–87	3.8–5.9 tonnes LEO 1.2–2.2 tonnes GEO	Includes Agena and Centaur
Delta	1960–87	1.8 tonnes LEO 1.25 tonnes GEO	Future use under review
Saturn family	1963–75		Heaviest US launcher, discontinued at end of Apollo/Spacelab missions
Saturn 5		152 tonnes LEO 53 tonnes lunar mission	
STS (Shuttle)	1981–	29.4 tonnes LEO 14.5 tonnes polar orbit 1.99 tonnes GEO	Launches halted in 1986 by Challenger accident
Thor	1959–76	0.43 tonnes LEO	10 in store, ex-IRBM*
Titan	1959–	15.87 tonnes LEO 3.1 tonnes planetary missions	Main military ELV
USSR			
A1/A2	1959–60	5–7.5 tonnes LEO	B1 and C1 used for less heavy LEO missions
D (Proton)	1967–	20+ tonnes LEO 2.5 tonnes GEO	
F1/F2	1970s	4–5.5 tonnes	
Energiya	1987	100–250 tonnes LEO	Under test

(*continued*)

Table 6 (*concluded*)

Area/name	Date	Performance	Remarks
Japan			
M Series	1970	0.7 tonnes LEO	
N Series	1978	0.35 tonnes GEO	
H Series	1986	0.55–2 tonnes GEO	
China			
CZ-1	1970–1	0.3 tonnes LEO	Long March One
FB-1	1975–83	1.2 tonnes LEO	
CZ-2/3	1984–	1.4–2.2 tonnes LEO 1.3 tonnes GTO	Long March Two
India			
SLV-3	1980	0.05 tonnes LEO	
ASLV	1985–	0.15 tonnes LEO	Under development
PSLV	Early 1990s	1 tonne polar orbit	Under development

* Intermediate-Range Ballistic Missile.

Table 7 Unmanned US missions: approximate numbers launched by December 1986

Military		Civilian	
Photo-reconnaissance	255	Communications	62
Electronic surveillance	99	Astrophysics	76
Nuclear explosion detection	22	Land remote sensing	5
Early warning	51	Ocean remote sensing	1
Geodesy	19	Environmental and meteorological	65
Meteorology	80		
Communications	138	Geodesy	10
Navigation	39	Planetary	27
Ocean reconnaissance	31		
Total	734	Total	246
	Combined total	980	

Table 8 Image resolution for Earth observation (in metres)

Objects	Detection	Identification	Precise identification	Description
Bridges	6	5	2	1
Radars	3	1	0.3	0.15
Telecommunication relay	3	1	0.3	0.15
Troops	20	2	1	0.3
Airfields	6	5	2	0.3
Artillery battery	1.5	0.5	0.3	0.2
Airplane	5	1.5	1	0.2
Control centre	3	1.5	1	0.2
Tactical missile sites	3	1.5	0.5	0.3
Ships	8	5	0.6	0.3
Vehicles	1.5	0.6	0.3	0.1
Fields	10	6	1	0.1
Ports	30	15	6	3
Railroad system	30	15	6	1.5
Urban zone	50	30	3	3
Submarine on surface	30	6	1.5	1

Note: the smaller the figure, the better the quality of the picture.

Table 9 People and robotics: operational modes and task allocation

Operation mode and basis	EVA-IVA	Teleoperation/ telepresence	Teleoperated hybrid system	Autonomous system
Task	People	Manipulator	Manipulator/ robotics	Robotics
Task and situation analysis and reasoning	People	People	People/robotics	Robotics
Strategy and path planning	People	People	People/robotics	Robotics
Procedure control	People	People	People/robotics	Robotics
Monitoring and control of robotics system	N/A	People	People/robotics	Robotics
Object handling	People	Robotics	Robotics	Robotics
Supervision of operational sequence	N/A	People	People	People/robotics
Contingency support	People	People	People	People/robotics

Figure 1 **Expenditure on space** International comparisons 1985

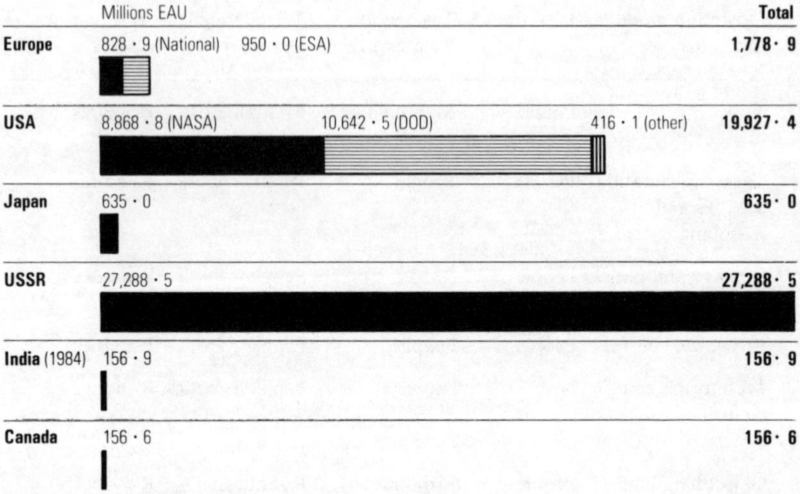

NB Figures for China not available
European figures do not include military expenditure
No account is taken of commercial expenditure

Figure 2 **European expenditure on space in 1985**

Millions EAU	National programme	ESA contributions	**Total**
Austria	0 · 3	2 · 5	**2· 8**
Belgium	4 · 4	37 · 0	**41· 4**
Denmark	2 · 5	13 · 0	**15· 5**
France	437 · 5	255 · 2	**692· 7**
Federal Republic of Germany	186 · 3	175 · 0	**361· 3**
Ireland	0 · 0	1 · 5	**1· 5**
Italy	81 · 7	105 · 0	**186· 7**
Netherlands	9 · 5	32 · 0	**41· 5**
Norway	3 · 1	3 · 0	**6· 1**
Spain	2 · 5	21 · 0	**23· 5**
Sweden	64 · 0	19 · 0	**83· 0**
Switzerland	1 · 3	17 · 0	**18· 3**
United Kingdom	36 · 0	129 · 3	**165· 3**

NB These figures cover civil expenditure direct by governments and exclude both explicit defence and commercial expenditures.

Figure 3 **Member States' contributions to ESA programmes 1986**

Shown in Million Accounting Units and as percentage shares

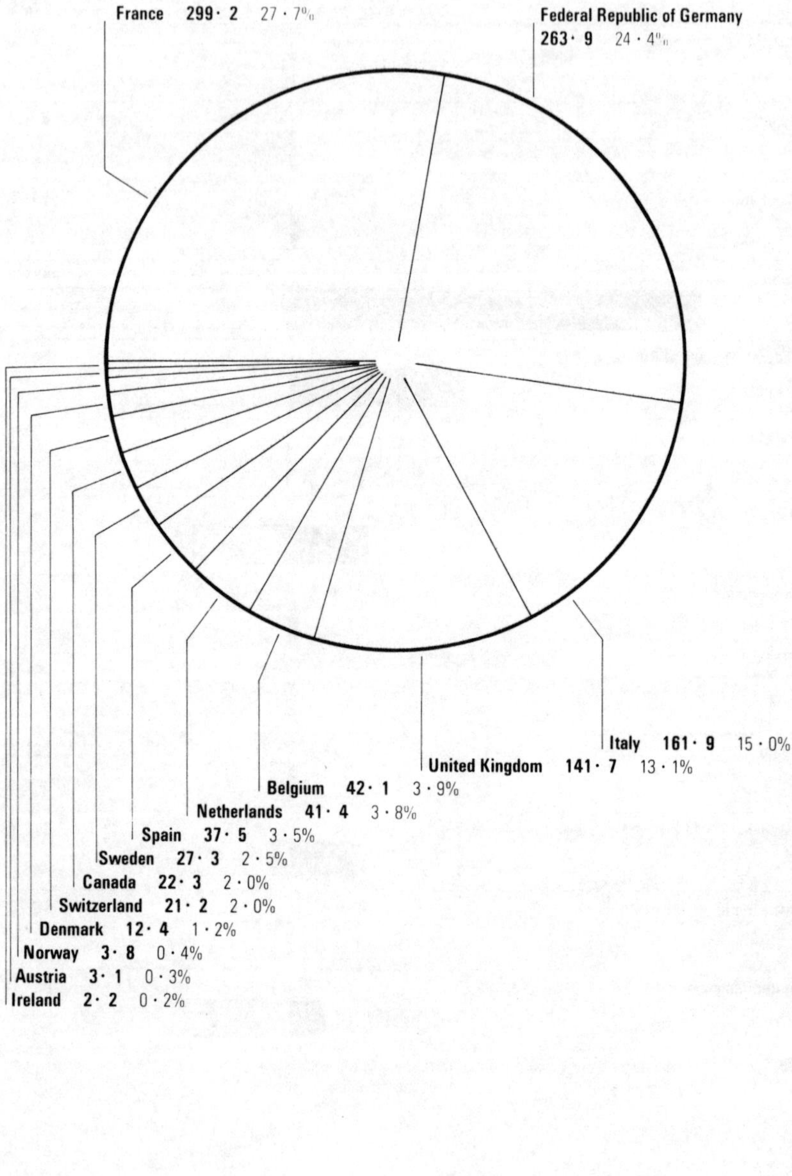

Figure 4 **Ariane 4** National workshares

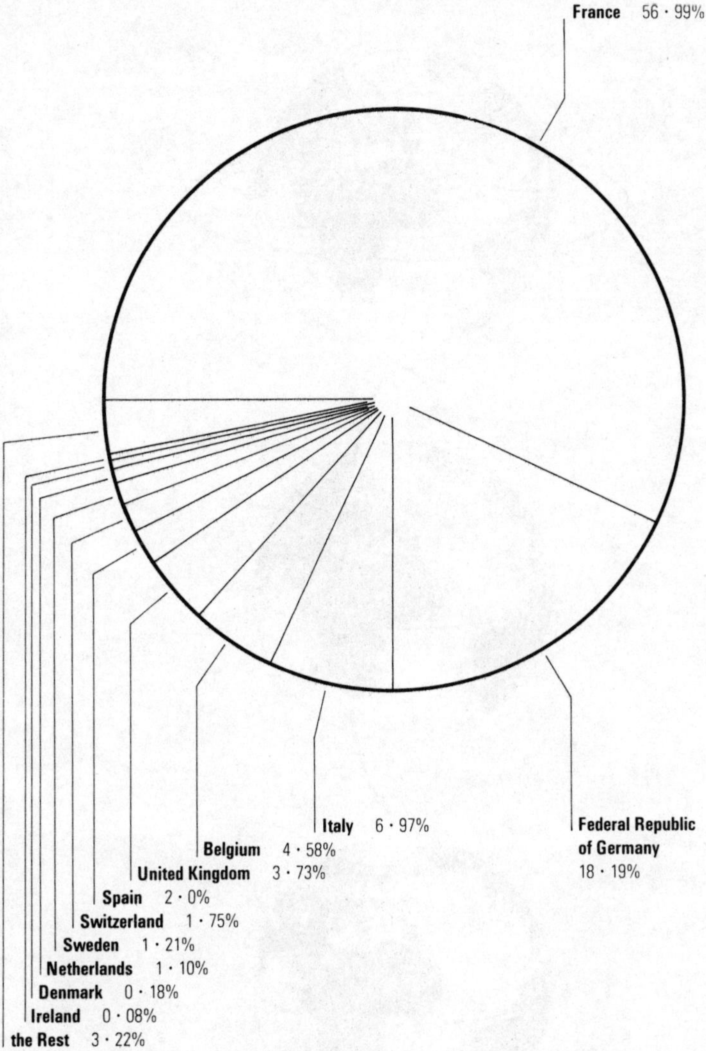

source **Flight International** 4 May 1985

Figure 5 **Orbit positions**

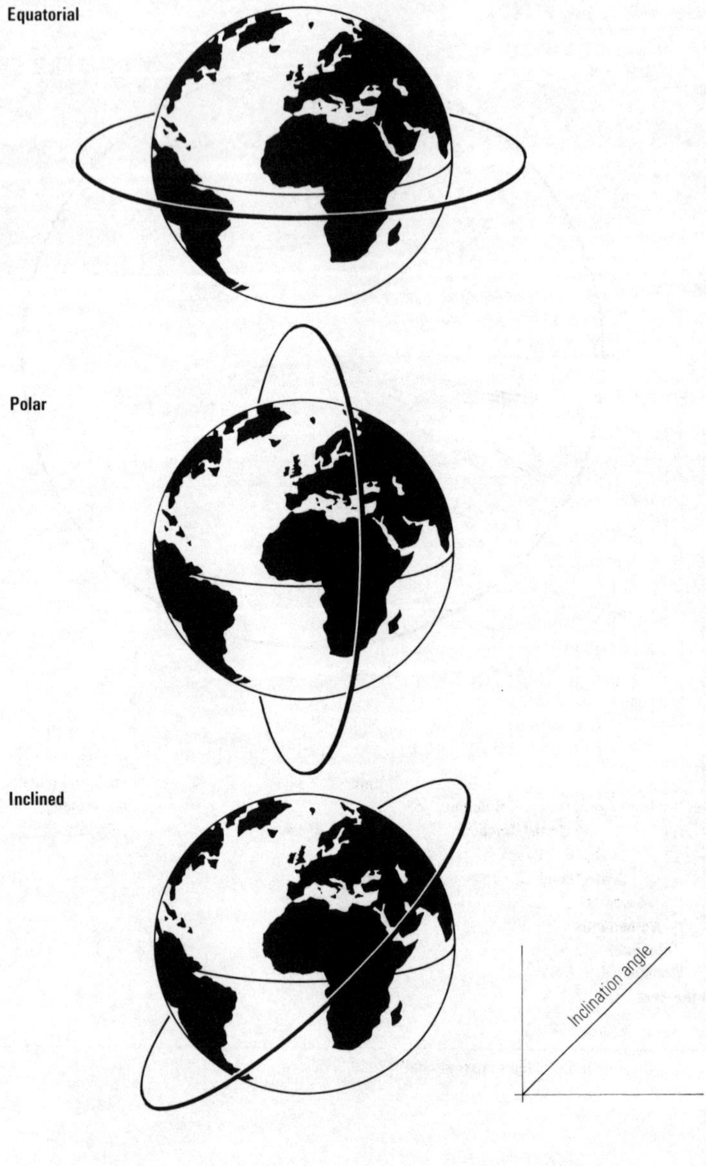

Figure 6 **Major orbital types**

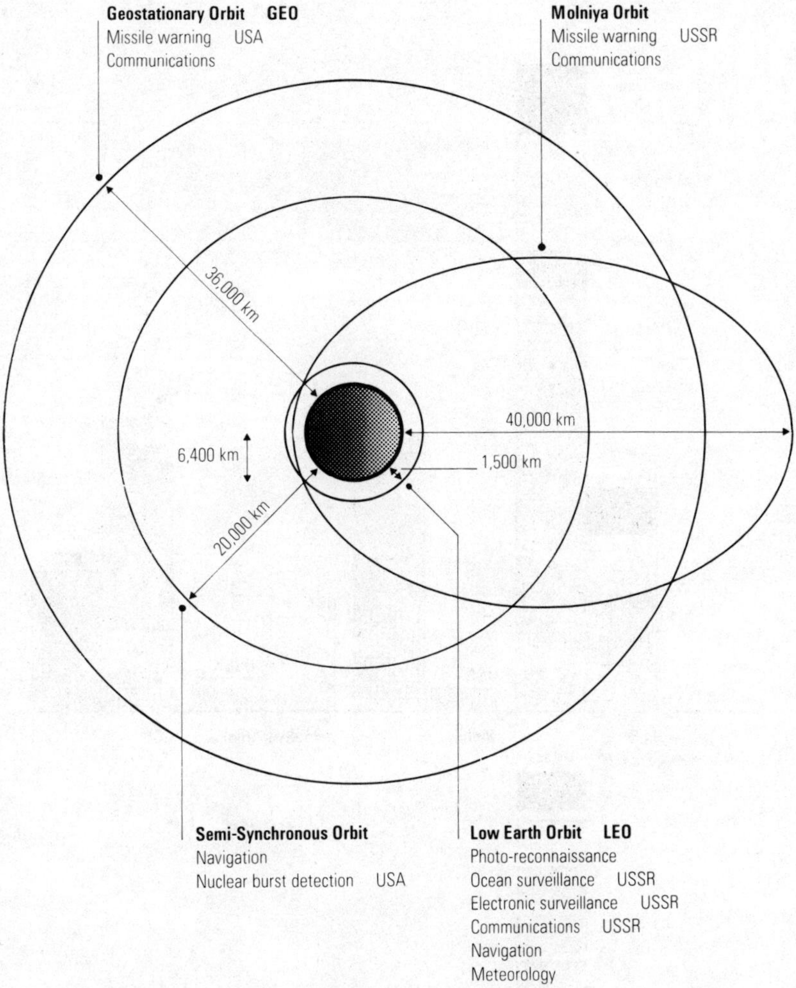

Figure 7 **Typical distribution of military satellites** deployed by the USA and the USSR at any given time

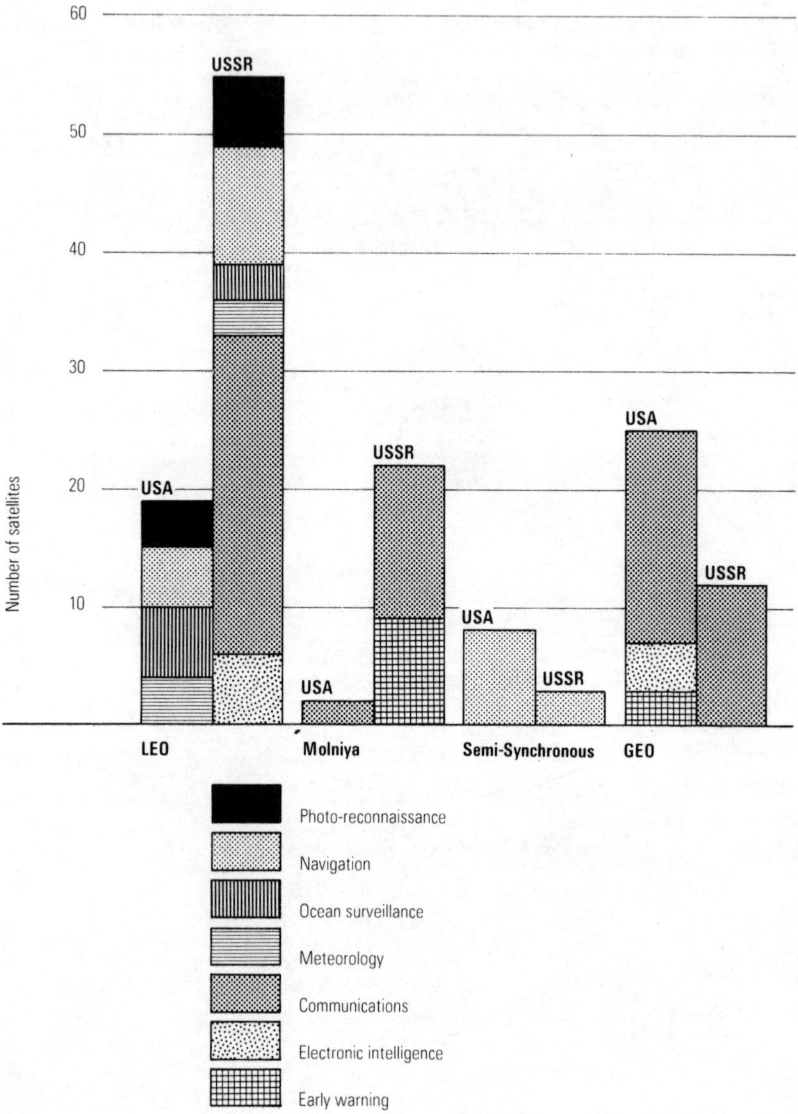

source **Scientific American** June 1984

Figure 8 **International space station**

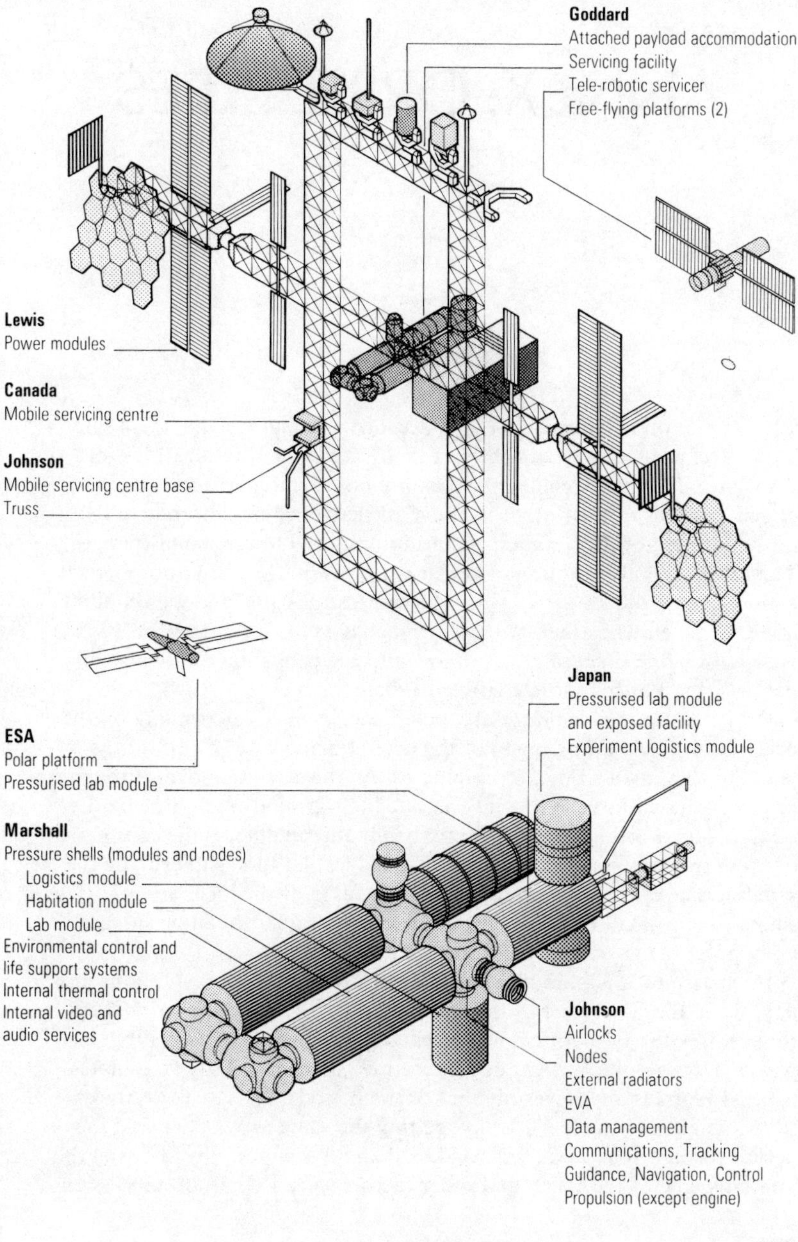

ANNEX: ORBIT TYPES

Orbital motion is independent of spacecraft size and weight, so all satellites travel around a given orbit in exactly the same way regardless of their nature. Orbits are either circles or ellipses. Circular orbits are described in terms of their altitude above the earth; elliptical orbits in terms of the altitude of their highest point (apogee) and lowest point (perigee). The lowest possible orbit – a circular orbit just above the atmosphere at an altitude of 200 km or so – has the shortest possible period, about 90 minutes. Satellites in high-altitude orbits have most of a hemisphere in view at any one time; satellites in low-altitude orbits have only a small patch of the Earth's surface visible to them.

Orbits are termed 'equatorial', 'polar', or 'inclined', depending on the orientation (inclination angle) of the orbital plane with respect to the plane of the Earth's Equator (Figure 6). As the satellite moves through its orbit, the Earth turns below it. The satellite's ground track, tracing the point on the Earth's surface directly below the satellite, is the composite of these two independent motions. The highest latitude surveyed by the satellite is equal to the chosen inclination angle. If the inclination angle is small, only a small band above and below the Equator can be surveyed. Only polar orbits overfly the Earth's entire surface.

If we assume a 90-minute orbit, the satellite makes 16 revolutions in a day, while the Earth turns just once beneath it. The ground track therefore crosses the Equator on 16 descending passes and 16 ascending passes, retracing itself every day. Since the period is exactly 90 minutes (i.e., 24 hours is an integer number of the period), the satellite retraces the same ground track every day, getting the same view. Changing the period to slightly more or slightly less than 90 minutes – by allowing the uneven shape of the Earth and other factors to cause natural precession –

would cause the entire ground track pattern to shift slightly sideways every day rather than reproduce itself exactly. The shifting pattern would pass over the entire Earth between 60° North and 60° South, allowing this whole region to be surveyed in detail.

As a consequence, if 24 hours is an integer number of orbital periods, two opportunities will exist every day to overfly a given location on the Earth, but a large portion of the Earth will never be overflown. Such an orbit is called harmonic. If 24 hours is not an integer number of orbital periods, it will be possible to scan the entire Earth over several days, but a given location will be overflown only every few weeks. Thus, harmonic orbits are best suited to monitor closely a crisis, or a sensitive area, whereas non-harmonic orbits are best suited to strategic surveillance.

Just five orbits contain all satellites (four of them are illustrated in Figure 7).

Low earth orbit (*LEO*) is the term applied to the region below an altitude of about 5,000 km. Orbits there have periods ranging from 90 minutes to a few hours. LEO is used for Earth observation.

Geostationary orbit (*GEO*) is circular orbit with an altitude of 35,700 km (6 Earth radii), and has a period of 24 hours. A satellite in eastward equatorial orbit at this altitude therefore remains over the same point on the Equator as both it and the Earth go around. The satellite has line-of-sight contact with more than 80 per cent of the hemisphere beneath it. To all observers who can see the satellite, it remains in the same position in the sky at all times. Several satellites spaced evenly around the Equator give coverage of the whole world except the polar regions. Most US, and an increasing number of Soviet, communications satellites are in geostationary orbit.

Satellite orbits may be sun-synchronous at different altitudes. Inclined to the Equator, they always follow the sun and thus the light, a feature which considerably simplifies satellite design.

Molniya orbits partially achieve the useful property of GEO and are named after the family of Soviet communications satellites that makes use of them. The orbit is highly elliptical – 40,000 km apogee and 500 km perigee – and has a 12-hour period. Because a satellite in Molniya orbit travels very slowly near its apogee and very rapidly at perigee, it spends more than 11 or even 12 hours on one side of the Earth. The orbit is inclined, so the satellite dwells high over the northern hemisphere, a convenient position for communications with the USSR. The particular inclination angle of 63° is chosen because at this angle the orbit is stable and does not drift under the influence of the Earth's equatorial bulge.

Semi-synchronous orbit is a circular orbit at an altitude of 20,000 km, and has a 12-hour period. This general altitude band is referred to as semi-synchronous orbit, although it is sometimes defined as ranging all the way from LEO to GEO.

Super-synchronous orbit. Orbits between GEO and the Moon are thinly populated today, but offer vast reaches for stationing future military satellites. The US deployed a constellation of nuclear burst detection satellites in 11-day orbits half-way to the Moon in the 1960s.

(NB. *Geostationary transfer orbit* is a temporary, highly elliptical orbit that satellites are put into before, for example, being placed more permanently in GEO or being despatched on interplanetary travel.)

GLOSSARY

ABM: Anti-Ballistic Missile.
AGV: Avion à Grande Vitesse, a hypersonic aircraft under study in France.
AI: Artificial Intelligence.
AM-136: Italian military telecommunications satellite programme.
ANS: Dutch astronomy satellite.
Anik: The Canadian network of communications satellites for domestic use.
Apollo: American lunar exploration programme, 1961–70.
Arabsat: Arab Satellite Communication Organization. Consists of 22 Arab PTTs and provides telecommunications and television services to the Middle East and North Africa from two satellites. The first was launched by Ariane in 1985.
Argos: French-US venture on platform location and data collection.
Ariane: ESA's launch programme, begun in 1973, and the first to be operated commercially. After development by ESA, manufacture and launch responsibility passed to Arianespace. Arianes 2 and 3 are improved versions of Ariane 1, and differ only in that Ariane 3 has two strap-on solid-rocket boosters. Ariane 4 is designed to carry heavier payloads. Ariane 5, at the design stage, will be able to carry even larger spacecraft, such as Hermes.
Arianespace: A commercial company formed to manufacture and launch Ariane. Its shareholders are 36 European aerospace and electronics firms, European banks and CNES. France is the major shareholder, with an approximately 60 per cent stake split between CNES and industry.
Ariel: Early British series of scientific satellites.

Asat: Anti-satellite system.
ASEAN: Association of South-East Asian Nations.
Asterix: First French satellite, launched 26 November 1965.
AT&T: American Telephone and Telegraph.
Atlas: Expendable launch vehicle developed for Nasa by General Dynamics from early US ICBM.
BAe: British Aerospace.
BNSC: British National Space Centre, established in 1985 to coordinate British space activities.
Brasilsat: Brazil's first communications satellites, for television, telephone, telex and data links.
Brite: Basic Research in Industrial Technologies for Europe (EC programme).
C3I: Command, Control, Communications and Intelligence.
CMEA: Council for Mutual Economic Assistance, also known as Comecon.
CNES: Centre National d'Etudes Spatiales, the French national space agency, founded in 1962.
CNR: National Research Centre (Italy).
Columbus: European programme to be operated as part of the international space station.
Comsat: a generic term for communications satellites, but also refers specifically to the Communications Satellite Corporation, established in 1962 by the Federal Communications Satellite Act to operate and supply satellite services commercially in the US. It is the US signatory to Intelsat. Its main satellites are in the Comstar series.
Copuos: Committee on the Peaceful Uses of Outer Space, established by the UN General Assembly initially as an *ad hoc* committee in 1958, following the launch of Sputnik; now a permanent committee with over 40 members, operating by consensus rules. Copuos has adopted four international agreements: the Treaty on the Use of Outer Space; the Agreement on the Rescue of Astronauts; the Convention on International Liability; and the Convention on Registration of Objects. A fifth agreement, on the Activities of States on the Moon, has not been ratified by all the members.
Cos B: ESA scientific satellite, 1975 (gamma-ray astronomy).
Cosmos: European industrial consortium of MBB, Marconi, Selenia and SNIAS.
Cospace: European industrial consortium to promote access for small firms to space contracts.
Cospar: (Soviet) Committee of Space Research.
Cospas/Sarsat: A search and rescue system using spacecraft, originally developed by the US, Canada and France (parties to Sarsat – Search and

Rescue Satellite-Aided Tracking) and the USSR (with the Cospas programme). The system became operational in July 1985.

CSG: Centre Spatial Guyanais, ESA launch site at Kourou in French Guiana.

DBS: Direct Broadcasting Satellite, capable of broadcasting TV and radio directly to ground-based stations.

Delta: Three-stage expendable launch vehicle developed for Nasa by McDonnell Douglas.

DFS/Kopernikus: Satellite developed for the German Federal Post Office by a consortium led by Siemens (including AEG, Standard Elektrik Lorenz and MBB).

DFVLR: Deutsche Forschungs-und Versuchsanstalt für Luft-und Raumfahrt, German Aerospace Research Establishment, founded in 1968.

Diamant A: Early French launcher.

DMSP: Defense Meteorological Support Program (US).

DoD: US Department of Defense.

DRS: Data Relay Satellite.

Dsat: Satellite defence system.

DSCS: Defense Satellite Communications System, the US DoD communications network of two spacecraft (often referred to as 'discus') at each of four geostationary locations.

Earlybird: US telecommunications satellite, launched in 1965, renamed Intelsat 1.

Earthnet: The European network for the acquisition, archiving and distribution of remote-sensing data, based at Esrin in Frascati. The network distributes information collected from the US Landsat. Dissemination services are handled nationally.

EAU: European Accounting Unit used by ESA and derived since 1979 from the ECU, the currency unit of the EC. Calculations of the rate are fixed for the current calendar year on the basis of the currency conversion rates for the previous year. Adjustments are then made retroactively to the contributions of member states to cover the difference between the fixed and the market rates. The EAU is close but not identical to the ECU in value.

EBU: European Broadcasting Union.

EC: European Community.

ECS: European Communications Satellites, developed for regional telecommunications within Europe under the direction of ESA and operated by Eutelsat. The first satellite was launched in 1983.

EIRP: Equivalent Isotropic Radiated Power.

ELDO: European Launcher Development Organization, created in 1962 with seven members to develop the European launcher.

ELGRA: European Low Gravity Research Association.

Elint: Electronic Intelligence, usually for military use.
ELV: Expendable Launch Vehicle, such as Ariane or Delta, which can be used once and then disintegrates on re-entry.
Energiya: New Soviet launcher.
EOSAT: American ocean surveillance satellite, based on Landsat.
EORSAT: Elint Ocean Reconnaissance Satellite.
EPC: European Political Cooperation.
ERS: ESA's Earth-Resources Satellite. ERS-1 is designed for ocean surveillance (wind and wave) as the forerunner for operational remote-sensing spacecraft in the next decade, using Synthetic Aperture Radar.
ESA: European Space Agency, formed in 1975 as the successor to ESRO and ELDO 'to provide for and promote, for exclusively peaceful purposes, cooperation between European states in space research and technology and their space applications'. Based in Paris, it has 13 member states, and Canada and Finland as associates.
Esdac: European Space Data Centre, based in West Germany.
ESOC: European Space Operations Centre, which runs ESA's satellite operations and the corresponding ground facilities and communications networks. Located in Darmstadt, West Germany, with a staff of over 200, its network includes a central control centre as well as telemetry, tracking and control facilities at various international ground stations, principally in Belgium, Spain, French Guiana, Australia and Gabon.
Esprit: European Strategic Programme for Research and Development in Information Technology (EC).
Esrin: European Space Research Institute, based in Frascati near Rome, created out of the never-completed Research Institute, operates ESA's Information Retrieval Service and the Earthnet distribution system for remote-sensing material.
ESRO : European Space Research Organization, set up to coordinate the scientific efforts of its members through shared satellites and ground facilities, and absorbed by ESA in 1975.
Estec: European Space Research and Technology Centre, based at Noordwijk, in the Netherlands.
ETS: Engineering Test Satellites, developed by Nasda. The first of the ETS series was launched in 1975 and the fifth, ETS-5, was launched in 1987.
Eumetsat: European Organization for the Exploitation of Meteorological Satellites, established in 1983 to coordinate and develop European interests in satellite systems for weather forecasting, in association with ESA and using Meteosats 1 and 2.
Eureca: European Retrievable Carrier, a re-usable payload carrier, designed by ESA to stay in orbit for up to six months. Eureca is part of

ESA's follow-on programme to Spacelab. The first flight, to include an inter-orbit communications package as a forerunner of a data relay system, was to have been launched by the Shuttle in 1987.
Eureka: European Research Coordinating Agency.
Europa: Unsuccessful ELDO programme for launchers.
Eurosatellite: Joint venture by AEG, Aérospatiale, Alcatel-Thomson, ETCA and MBB in the field of DBS.
Eurospace: Organization of European space companies, based in Paris.
Eurostar: Franco-British telecommunications platform, developed by Satcom International, a joint venture by British Aerospace and Matra.
Eutelsat: European Telecommunications Satellite Organization, based in Paris and with 26 members; the PTTs are designated by governments as signatories on an investment share basis. It provides telephone, telegram and telex communications throughout Europe, together with television and radio transmissions. It currently uses ECS leased from ESA and is procuring a second generation system.
EVA: Extravehicular Activities.
Exosat: European X-ray Observatory Satellite, launched in 1983 to examine the location, form and characteristics of X-ray stars as part of the ESA Space Science Programme. Results are received at the Villafranca Earth station in Spain.
Ferrets: Military intelligence satellites using electronics to pick up signals on radio and radar frequencies.
FGMDSS: Future Global Maritime Distress and Safety System, an automated communication system for deployment in 1991–7.
FOBS: Fractional Orbital Bombardment System.
FRG: Federal Republic of Germany.
GEO: Geostationary (or geosynchronous) Orbit (see Annex).
Geodesy: Study of Earth's gravity and magnetic field.
Geos: European scientific space probe.
Giotto: Named after the Italian Renaissance painter who included Halley's Comet in a fresco ('The Nativity'), this ESA spacecraft was launched in July 1985 by Ariane under the ESA Space Science Programme. It passed within 500 km of the nucleus of Halley's Comet in March 1986. The encounter lasted four hours, during which time Giotto took one picture every four seconds.
Glasnost: Soviet domestic mechanism equivalent to liberal democracy.
Glonass: Soviet navigation satellite system equivalent to Navstar.
GOES: Geostationary Orbit Operational Environment Satellite, a series of meteorological satellites operated by the NOAA in conjunction with its Tiros polar-orbiting spacecraft.
GPS: Global Positioning System.
GTO: Geostationary Transfer Orbit (see Annex).

H-1 and H-2: Expendable launch vehicles being developed in Japan by Nasda, with the H-2 planned to place heavier satellites in orbit than the H-1.
Helios: Proposed French optical reconnaissance satellite system, expected to be operational in 1993.
Heos: European scientific satellite, to investigate interplanetary magnetic field.
Hermes: A mini-shuttle spaceplane proposed by France, to be launched by Ariane 5, designed to undertake manned operations in orbit for up to one month. The preliminary studies and contracts were approved by ESA in November 1986.
Hipparcos: An ESA space science craft, designed to measure the positions, parallaxes and motions of about 100,000 stars and due to be launched by Ariane in 1988.
HM-60: Large European cryogenic engine to be developed for Ariane 5.
Horizon 2000: Long-term European space science programme proposed by European scientists' panel in 1984. It envisages four 'cornerstone' elements: X-ray research, solar observation, plasma research and a cometary science project.
Hotol: Horizontal Take-Off and Landing Vehicle. Proposed British concept for a re-usable and man-rated spaceplane. It would make a novel combination of new air-breathing and conventional rocket technology in a single-stage vehicle.
HRV: High Resolution Visible.
Hubble Space Telescope : This will be the largest instrument ever orbited when it is launched by Nasa. It is to be maintained in orbit for 15 years, though instruments will be replaced. Among ESA's contributions are the Paint Object Camera and the solar arrays.
ICAO: International Civil Aeronautical Association.
ICBM: Inter-Continental Ballistic Missile.
Ice : International Cometary Explorer. Formerly the third International Sun-Earth Explorer, Nasa's ISEE-3 was redesignated Ice. It conducted the first close encounter with a comet in the autumn of 1985.
ICR: (Dutch) Interdepartmental Committee on Space Research and Technology.
IEPG: Independent European Programme Group, a European grouping for collective defence procurement.
IGBP: International Geosphere Biosphere Programme, an international study of global change.
IMO: International Maritime Organization.
Inmarsat: International Maritime Satellite Organization, based in London and operating since 1982; offers a system of satellites to provide telephone, telex, data and facsimile, distress and safety communications

services, for the shipping and offshore industries. It leases the Marecs A and B2 spacecraft from ESA, maritime communications subsystems from Intelsat, and three Marisat satellites from Comet. It is now procuring a second generation of spacecraft of its own. Inmarsat is financed by the designated signatories – the PTTs – of the 48 member countries, which include the USSR and other Eastern bloc countries, each with an investment share based on usage.

Insat: The Indian series of satellites, which combine telecommunications, direct broadcasting by satellite and weather-imaging in a single geostationary orbit craft. The operational network is for two satellites, which were built by Ford Aerospace. Insat 1A was abandoned five months after launch.

Intelsat: International Telecommunications Satellite Organization, based in Washington since 1965; owns and operates over two-thirds of the satellites used for international communications. With some 112 member countries, Intelsat operates as a financial cooperative, with investment shares linked to usage.

Intospace: European consortium led by MBB and Aeritalia to encourage private investment and commercial development in microgravity.

Intersputnik: CMEA/Soviet allied communications satellite system.

Iras: Infra-red Astronomical Satellite, Dutch-built for US-Dutch-UK research programme. Launched 1983.

Iris: Italian Research Interim Stage, a propulsion project for launchers.

IRS: Indian satellite for Earth observation.

ISAS: (Japanese) Institute for Space and Astronautical Sciences.

ISEE: International Sun-Earth Explorer.

ISMA: French-proposed international satellite monitoring agency.

Iso: Infra-red Space Observatory, an ESA space science project, intended to complement Nasa's space telescope, and to offer high sensitivity observing facilities for a large region of the electromagnetic spectrum. The satellite will be launched by Ariane in the late 1980s.

ISRO: Indian Space Research Organization.

Italsat: Italian telecommunications satellite, scheduled for launch in 1988.

ITU: International Telecommunications Union.

IUE: International Ultraviolet Explorer, a joint project between the SERC, ESA and Nasa, launched in 1978 to provide an observatory facility for ultraviolet spectrometry.

IVA: Intravehicular Activities.

Kopernikus: See under DFS/Kopernikus.

Kosmos: Soviet satellite series begun in 1962. Kosmos is a blanket term for Soviet military applications.

Landsat: The most important range of remote-sensing spacecraft so far, this American series flies in near-polar orbit at an altitude of about 900 km. The primary use of the five spacecraft is US crop forecasting and land surveying. A significant secondary purpose has been to provide data to numerous other countries, a number of which operate Landsat ground stations.

LDC: Less Developed Country.

LEO: Low Earth Orbit (see Annex).

Long March 2 and 3: Chinese ELVs.

LPI: Low Probability of Intercept.

L-Sat: Renamed Olympus.

LUA: Launch Under Attack.

Marecs: Maritime telecommunications satellites developed by ESA. Marecs A was launched in 1981 and Marecs B2 in 1984. These satellites are leased to Inmarsat.

Marisat: Maritime communications satellites developed for Comsat by Hughes.

Marots: ESA's Maritime Orbital Test Satellite.

Mesh: European industrial consortium of Aeritalia, British Aerospace, Erno, Matra and Saab.

Meteosat 1 and 2: Meteorological satellites developed by ESA and launched in 1977 and 1981. The Meteosat operational programme began in 1983.

Milstar: Military Strategic-Tactical and Relay.

Mir: Soviet space station.

Molniya: Elliptical orbit developed by USSR (see Annex).

MOU: Memorandum of Understanding.

MTFF: Man-Tended Free Flyer, proposed element of Columbus.

N-1 and 2: Expendable launch vehicles developed in Japan for Nasda by Mitsubishi Heavy Industries.

NACISA: Nato Command Informations System Agency, replacing NICSMA.

Nasa: National Aeronautics and Space Administration of the USA, founded in 1958.

Nasda: National Space Development Agency of Japan, founded in 1969.

Nato: North Atlantic Treaty Organization.

Nato III: The current generation of Nato communications satellites, which provide secure links within Europe and to the USA. The operational network comprises four satellites, one of which is a spare. The satellites are interoperable with the US DSCS series. Skynets built for Nato will provide the next series.

Navsat: Developed to provide precise positional updates to Polaris submarines. (Navsat is also the name of an ESA navigational satellite proposal under consideration.)

Navstar: A series of US DoD spacecraft using the Global Positioning System to provide continuous, worldwide and 3-D position, velocity and time checks. The full service is due to become available in 1989, when the planned constellation of 18 spacecraft is complete. The system will also accommodate an unlimited number of civilian users, who will receive slightly less accurate information.
NDS: Nuclear Detection System (US).
NICS: Nato Integrated Communications System.
NICSMA: Nato Integrated Communications Management Agency, now replaced by NACISA.
Nimbus: US remote-sensing system.
NIVR: Nederlands Instituut voor Vliegtuigontwikkeling en Ruimtevaart, Dutch Agency for Aerospace Programmes.
NOAA: National Oceanic and Atmospheric Administration (US).
NRL: National Aerospace Laboratories (Dutch).
NTM: National Technical Means (of verification of arms control agreements).
Olympus: Formerly L-Sat, this will be a new generation of large European communications satellites, developed by ESA. Olympus 1 will be a test satellite, combining DBS, business telecommunications and experimental communications payloads. One of the two DBS transponders on Olympus 1 will be available to the EBU for a three-year experimental period; the other DBS transponder will be used by the Italian broadcasting organization, RAI.
Orion: The first of six American communications companies to file applications with the Federal Communications Commission for the operation of private transatlantic telecommunications satellite systems in competition with Intelsat.
OTS: Orbital Test Satellite, developed by ESA for telecommunications experiments within Europe; launched in 1978.
Palapa B: The second generation of telecommunications spacecraft built for Indonesia. After the successful launch on the Shuttle of Palapa B1, the second craft was placed in the wrong orbit and was one of two satellites successfully rescued by the Shuttle in November 1984 at the instigation of the mainly London-based insurance underwriters. It is now up for resale.
Phobos: Soviet interplanetary space probe.
Progress: Soviet automatic cargo spacecraft.
Proton: Expendable Soviet launch vehicle.
PSN: National Space Plan (Italy).
PTT: Post, Telegraph and Telephone organization.
Race: Research and Development in Advanced Communications Technologies for Europe (EC programme).

Radarsat: A Canadian satellite which will fly in LEO to monitor Arctic ice conditions and Canadian renewable resources such as crops and forests. The first spacecraft was due to be launched in December 1990 on the Shuttle.
Rorsat: Radar-equipped ocean surveillance satellite.
Rosat: A German-sponsored scientitic satellite with some US and UK involvement. It was due to be launched on the Shuttle in 1987, and is intended to complete an all-sky survey of X-ray sources as well as a detailed observation of specific sources.
Sänger: Proposed German concept for a re-usable and man-rated space-plane, taking off and landing horizontally. One stage of it would use novel air-breathing technology, and the other (conventional) rocket propulsion.
Salyut: Soviet space station.
Samro: French Satellite Militaire de Reconnaissance Optique (now overtaken by Helios).
SAR: Synthetic Aperture Radar.
Sarsat: See Cospas-Sarsat.
Satcom: Term for military communications satellites.
Saturn: American heavy-lift launcher.
Sax: Italian scientific satellite.
Scout: US ELV used for small payloads.
SDI: Strategic Defense Initiative. US programme to explore a range of strategic missile defence technologies, many of which will depend upon deployment in space.
SEA: Single European Act.
Semyorka: Soviet launcher.
SERC: Science and Engineering Research Council (UK).
SES: Société Européenne des Satellites, DBS consortium.
SHF: Super-high Frequency.
Sicral: Italian military communications satellite system, to be used also by police and civil defence agencies, under AM-136 programme.
Sirio: Italian telecommunications satellite, launched 1977.
Skylab: US space laboratory, placed in orbit in the early 1970s and now disintegrated, originally the Apollo Applications Program.
Skynet: British military satellites for strategic and tactical communications between ships and both fixed and mobile land terminals. Skynet 4A was due to have been launched by the Shuttle in 1986. A further two spacecraft in the same series (Skynets 4B and 4C) are being built for launch by the end of 1988.
SLBM: Submarine-Launched Ballistic Missile.
SLV: Indian launcher.
SL-X-16: New Soviet launcher.

Soyuz: Series of manned Soviet space vehicles.

Spacelab: A re-usable and modular manned laboratory designed to fly in the cargo bay of the Shuttle, as ESA's contribution to Nasa's STS. By the terms of the agreement negotiated with Nasa by ESA, Spacelab became the property of Nasa after its first flight.

Space station: This will provide for a permanent, manned presence in space. Economic summit countries and ESA members were invited by President Reagan during his 1984 State of the Union address to make it an 'international' project. In January 1985 ESA agreed to take part in the initial design stage (Phase B) through the Columbus module, and associated free-flying, unmanned platform. Negotiations for intergovernmental agreements between America and participating countries began in September 1985. Both Canada and Japan are also taking part in Phase B studies.

SPAS: Shuttle Pallet Satellite, built in Germany.

Spot: Satellite pour l'Observation de la Terre, French-built for Earth observation. Spot 1 was launched in 1986 and three more are planned over 10 years.

Spot Image: The company, backed by the French government, with small contributions from Belgium and Sweden, which aims to make remote sensing from Spot a commercial enterprise.

SRON: Space Research Organization of the Netherlands (Stichting Ruimte Onderzoek Nederlands).

SSBN: Ballistic-missile-carrying nuclear submarine.

Star: European industrial consortium of British Aerospace, Dornier, SEP and Thomson.

STAR: Special Telecommunications Action for Regional Development (EC programme).

STS: Space Transportation System, better known as the Shuttle, is Nasa's largest programme since the Moon landings. It was conceived in 1972 as a replacement for all existing expendable civilian and military boosters. The first Shuttle flight, the orbiter Challenger, took place in April 1981. The loss of Challenger in January 1986 has grounded the Shuttle until at least mid-1988.

STW-2: Chinese telecommunications satellite.

Symphonie A and B: Early Franco-German communications satellites, launched in 1974 and 1975.

Syracuse: French military communications network, in operation since 1984.

TDF-1 and TDF-2: The French version of the Franco-German DBS craft, being built by Eurosatellite, a consortium of Aérospatiale, MBB, AEG Telefunken, Thomson, and ETCA of Belgium. TDF-1 was due to be launched on Ariane in 1986, and be operational from the start of 1987

with four channels. A second craft has been ordered to provide a fifth channel and a back-up service. A French commercial company, Télévision par Satellites, will lease the satellites from the French government.
Télécom 1: France's first domestic telecommunications satellites; the series provides business data and videoconferencing facilities in France, long-distance telecommunications to its overseas departments, and military communications.
Titan: Expendable launch vehicle developed for US Department of Defense by Martin Marietta.
Topex-Poseidon: French-US project on oceanographic satellites.
Transpace Carriers, Inc. (TCI): Started in 1982 by a group of Nasa technicians to exploit the US administration's decision to allow private companies to launch rockets, the organization secured the agreement of McDonnell Douglas to build the Delta rocket for it.
TSS: Tethered Satellite System, developed in Italy with US.
TV-Sat: Designed as the German equivalent to TDF-1, was due for launch in May 1986, but disagreements between the Federal and Land governments have delayed its development.
UHF: Ultra-high Frequency.
Ulysses: A joint ESA space science/Nasa project (formerly the International Solar Polar Mission) to investigate solar wind, the sun/wind interface, solar and galactic cosmic rays, cosmic dust and other phenomena. Ulysses was to have been launched by the Shuttle in 1986. After a 14-month journey it would be within Jupiter's gravitational field, and the entire mission would last five years.
Vega: Soviet interplanetary space probes.
VSAT: Very Small Aperture Terminal, a small antenna-sized satellite signal reception system.
WARC: World Administrative Radio Conference.
WEU: Western European Union.
WMO: World Meteorological Organization.
X-30: US National Aerospace Plane, a proposed hypersonic re-usable and single-stage vehicle, under study since 1986.
Zircon: Proposed British Elint satellite.

Foreword

In the mid-Seventies FAO initiated and coordinated the International Friesian Strain Comparison Trial. It was one of the most extensive cattle experiments of all times, which continued for more than a decade. During the first three years, 80 000 doses of semen from ten strains of the best Black and White cattle populations in the world were used on 30 000 cows located on 70 state farms in Poland. The semen used came from Canada, Denmark, the Federal Republic of Germany, Israel, the Netherlands, New Zealand, Sweden, the United Kingdom and the United States of America. Polish semen was also included in the experiment.

Virtually all the economically important features of this cattle breed were studied during the experiment. The results clearly show which strains excel in different characteristics including milk yield, butterfat and protein production and percentage, beef performance, meat quality, body size, feed conversion, health and reproduction.

It is noteworthy that no individual strain proved to be uniquely superior or inferior in all characteristics; each tested strain showed its strong and its weak features. Governments and breeders around the world will find the results of this endeavour invaluable for making their choice of the available Black and White strains for the improvement of their indigenous cattle populations.

Various results of the experiment have already received global appreciation in many scientific papers and in the discussions at scientific conferences on animal production. This book, however, brings together for the first time all the findings, including those not published previously. The opportunity has also been taken to include additional information. The history of the different Friesian strains in the ten countries which participated in the trial, as well as the current breeding strategy, have been reviewed. An assessment of crossbreeding studies using Black and White cattle in most countries of the world has been given. The book indicates how the results of this international experiment can be used in the developing countries in programmes for the improvement of their indigenous breeds.

I hope that the cooperation and friendship established between scientists and breeders during this experiment will continue in the future and will result in strengthened collaboration. I am pleased to know that several of the participating countries have set up further joint research work as a result of the FAO experiment in Poland. FAO will continue to encourage this trend and, whenever possible, will make such important results available internationally for use in both developing and developed countries.

Edouard Saouma
Director-General

Errata

M-22
ISBN 92-5-102593-2

Page 7 Table 1.4. *Second column of figures to be read under* "1985"

Page 23 Caption should read:

"Patra", the highest yielding cow from the intensive-feeding trial — and F_1 from Canadian sire Roybrook Starlite 24106-4-7. Patra produced 15 014 kg milk (3.49 percent fat) in her fourth lactation — a total of 523.4 kg butterfat (Agric. Univ. Exp. Sta. Brwinów).

Page 31 Table 2.3. *Footnote should read:*

1 kg of concentrates was equal to 1.05 oat units and had 136 g of digestible crude protein.

Pages 61-5, 69, 70, 72, 73

Tables 2.20 - 2.29. *For* effects', *read* effects[1]

Page 74 Table 2.30. *Subheading under* "Net energy intake" *and* "Total gain" *should read:*

F_1 s.e. R_1 s.e. F_1 s.e. R_1 s.e.

Last two lines should read:

Total	1 135	6	1 126	7	628	3	601	4	34.7	0.2	32.4	0.3	386	1	350	1
Significant differences*	A-K,D,E,F,H,G,B C,I-D,K				K-A,I,C,B,F,H D-A,I,C,B,F G-A				A-K,D,F,H,B K-C,G,I,B,H,E,F,D							

Page 75 Table 2.31. *The* "Significant differences" *should read:*

E-A,D,C,G,I — K-G,F,E,H,C,I,D,A

H-A,D

F-A

Page 79 Table 2.33. "Significant differences" *under* "Lean: bone, F_1" *should read:*

B-C,A,G,D,I,K

E-C,A,G,D,I,K

H-C,A,G

F-C

The second column of "Significant differences" *should be under* "Lean: bone, R_1"

Page 80 Table 2.34. *Both columns of* "Significant differences" *should be under* "F_1".

Page 94 Table 2.37. "Significant differences" *should be under* "Liveweight at 10 days after calving"

Page 95 Table 2.38. "Significant differences" *should be under* "Height at withers 200 days after calving"

Page 100 Table 2.39. *For* effects', *read* effects[1]

Page 101 Table 2.40. Heading, line 1. *For* effect, *read* effects[1]

At bottom of Table. *Delete* 2.65, 31.4, 2.05, 24.3

Footnote. *For* [2]See, *read* [1-2]See

Page 104 Second caption, last line. *For* 4.0%, *read* 4.40%

Page 105 First caption. *For* 4.16 fat, *read* 4.16% fat

Second caption. *Should read* Sire from Israel. Average milk yield 5 487 kg, 3.89% fat.

Page 106 Second caption. *For* 8 364 kg, *read* 4 833 kg

Page 109 Table 2.41. ''Significant differences'' *should be under* ''Height at withers''

Page 114 Table 2.46. ''Significant differences, protein'', last line. *For* E to H *read* E to H

Page 124 Table 2.48. *For* effects', *read* effects[1]

Page 125 Table 2.49 Heading, line 1. *For* effects', *read* effects[1]

Column 5, line 6. *For* 7.1, *read* -7.1

Column 5, line 7. *For* 8.4, *read* -8.4

Page 134 Table 2.53 *Last two lines of footnote should read:*
* significant influence of strain ($P < 0.05$)
** highly significant influence of strain ($P < 0.01$)

Page 140 Table 2.60. Column 10, lines 6,7,8. *For* 3.19, 0.55 *and* 1.37, *read* 13.19, 10.55 *and* 11.37

Page 154 Table 2.69. Table heading, line 3, columns 6 - 10. *For* **w,s,w,s,w,** *read* **s,w,s,w,s.**

Page 158 Table 2.72. Column 1, line 5. *This should read* UK

Page 251 Reference 13. (Dohy. J. and Szüllo. B.) *For* 1984, *read* 1974

Page 296 Table 5.15. Column 3, line 4. *For* 2 06, *read* 2 065

Page 306 Table 5.21. Lines 6 and 7. *Should read:*

Milk production per lactation (kg) (estimated)	1 491	± (16)	184	2458	± (22)	154	[1]2695	± (64)	122

Page 318 Table 5.27. Heading. *For* 1985, *read* 1984

Page 341 This photograph should be substituted for the one on page 344

Page 344 This photograph should be substituted for the one on page 341

Page 375 Table 5.59. (continued). Notes. *For* (P(((((0.05) *read* ($P < 0.05$)

Page 376 This photograph should be substituted for the one on page 382

Page 382 This photograph should be substituted for the one on page 376

Page 424 Paragraph 1, line 3. *For* Merket guinea, *read* merker and guinea

Page 441 First reference. *For* Chaundhry, *read* Chaudhry

Page 442 Reference 5. *For* g-14, *read* G-14

Page 392 and 393 Table 5.70. *Insert brackets and close text as indicated:*

Colombia	Turipana	Costeno con cuernos Holstein	896 (652)	2 000 (197)		1 832 (35)	1 950 Rubio (1976) (162)
	El Nus Milk with calves	Blanco Orejinegro Holstein		1 039 (31)			Botero (1976)
Cuba	Grazing + concentrates	Holstein Zebu	500	1 811 (1 300)	3 100 (200)	3 339 (65)	3 554 Prada (1978) (high)
Venezuela	Farm 1	Holstein Native					
	Farm 2	Holstein Native					
	Farm 3	Holstein Native					
	Part-stabled Highland Region (n = 382)	Native Holstein					

Pages 394 and 395 Table 5.71. *Insert brackets and close text as indicated:*

Colombia	Turipana	Costeno con	C.I.	453 (3 554)	426 (47)		505 Rubio (1976) (14)
		cuernos Holstein	S.C.	1.7 (475)	1.9 (90)		3.0 (216) Huertas (1972)
Colombia	El Nus Grazing	Blanco Orejinegro Holstein	C.I.				
Cuba	Grazing + concentrates	Holstein Zebu	C.I.				
Venezuela (1977)	Lake Maracaibo	Criollo B. Swiss					
	Region (n = 1 349)	Holstein Zebu				(Holstein)	419 ± 4.42
Part-stabled Highland		Native Holstein					
Region (n = 576)		Native Holstein					

Contents

Authors' Preface

The primary justification for this book is to present and assess the results of a ten-year international comparison of the performance of ten of the principal strains of Friesian and Holstein-Friesian cattle worldwide, a comparison which was carried out in Poland under the sponsorship of the Food and Agriculture Organization of the United Nations. Much has already been written about the different stages of this experiment in scientific papers and popular articles, but this is the first time that the manifold results, which represent a milestone in sustained international cooperation in animal science, have been brought together. They are presented in a way which it is hoped will appeal both to the general and the scientific reader. However, the intention of the book is wider. Because of the worldwide implications of the results for livestock improvement, the results have been placed in the wider context of cattle production and experimentation in both the developed and developing world - with particular reference, of course, to Friesian and Holstein-Friesian cattle and their extensive use in crossbreeding. The book therefore starts with an overview and develops its theme country by country.

The experiment in Poland would not have been possible without the willing cooperation and collaboration of the participating countries. Acknowledgment of specific individuals and the role of FAO is made in the appropriate chapter. The work for the Friesian comparisons and analysis of the results would also not have been possible without the unstinting help of many colleagues. The authors wish to thank:

- All members of the Technical Advisory Committee (TAC) for their valuable advice on the experiments, formulation and evaluation of the results, and the three successive chairmen of TAC-FAO staff members-Drs. I.L. Mason, C.G. Hickman and J. Hodges;
- Friesian breeders' associations and AI organizations from the participating countries for the bull semen donations;
- Staff on the state farms, the regional breeding offices and AI centres in Poland which were included in the project;
- Prof. Dr. A. Zarnecki and Dr. M. Lukaszewicz for the statistical evaluation of the results of the field project and intensive trial, respectively; and
- Dr. G. Kalinowska for supervising the participating farms and collection of data for the field trial.

Similarly, information on the distribution, development and performance of the Friesian breed and its crosses in the large number of countries covered, in both practical and experimental situations, owes much to those who responded to our enquiries and appropriate reference is made in the text and tables. In this regard, the help is particularly acknowledged of Drs. R.M. Acharya, V. Buvanendran, A.W. Qureshi and L. Vaccaro who revised the text and supplied additional data. A number of colleagues have helped more particularly in the preparation of the manuscript: Dr. C.B. Coulson, Mr. T. Vejaratpimol, Ms. N. Minghi and Mr. G. Beccaloni who drew the figures.

The authors are grateful to Elsevier Science Publishers B.V., Amsterdam, Netherlands for their agreement to using data from published material.

Finally, the authors are especially indebted to Dr. G. Wiener who, as technical editor, substantially improved the text.

The Authors

Dr. H.A. Jasiorowski is Professor of Animal Science at Warsaw Agriculture University and, during 1969-75 and from 1984 to the present, Director of the Animal Production and Health Division, FAO.

Dr. M. Stolzman was formerly with the Central Animal Breeding Office, Warsaw.

Dr. Z. Reklewski is Professor of Animal Science and Director of the Institute of Genetics and Animal Breeding of the Polish Academy of Sciences, Mrokow, Warsaw.

1. The World Cattle Population and its Development - An Overview

1.1 WORLD CATTLE POPULATION

Black and White Friesian cattle are the most numerous and productive among the three hundred or more cattle breeds in the world today. Artificial selection has been successfully applied to endow the breed with, on average, the highest genetic merit in production traits. The breed has achieved the biggest expansion in numbers of any breed in the present century and its distribution and importance is worldwide. There are very few countries where it is not now present as a pure breed or used for crossbreeding.

From beginnings in Asia, cattle spread from early centres of domestication in the river valleys of the Indus, the Euphrates and Tigris and the Nile to many other areas of Asia, Africa and Europe (Friend, 1978) (Figure 1.1a). Further large migrations of cattle took place from the sixteenth century onward to all other parts of the world-then largely from Europe but also of zebu cattle from India to America and Australia (Friend, 1978) (Figure 1.1b) leading to the present-day distribution of cattle as shown in Figure 1.2.

At present there are more than 1.2 billion cattle and 130 million buffaloes in the world (FAO statistics) dispersed and utilized all round the world under different economic, social and ecological conditions. More than 300 breeds of cattle live and produce under various climatic conditions from the Equator to the Poles.

Table 1.1 gives data on cattle numbers in different parts of the world and their increase during the last 25 years. It is interesting to note that the growth rate in cattle numbers in some developing regions was similar to the growth rate in the human populations.

Two thirds of the world's cattle population are in the developing countries. As ruminants, they utilize mainly natural roughage and crop residues and by-products thus competing either not at all or only minimally for feed resources which can be used directly for the human population. This can be seen from the figures given below Table 1.1 on p. 4, which show the percentages of metabolizable energy (ME) made available to ruminants from different sources (Winrock, 1978).

Fig. 1.1a Early cattle migrations

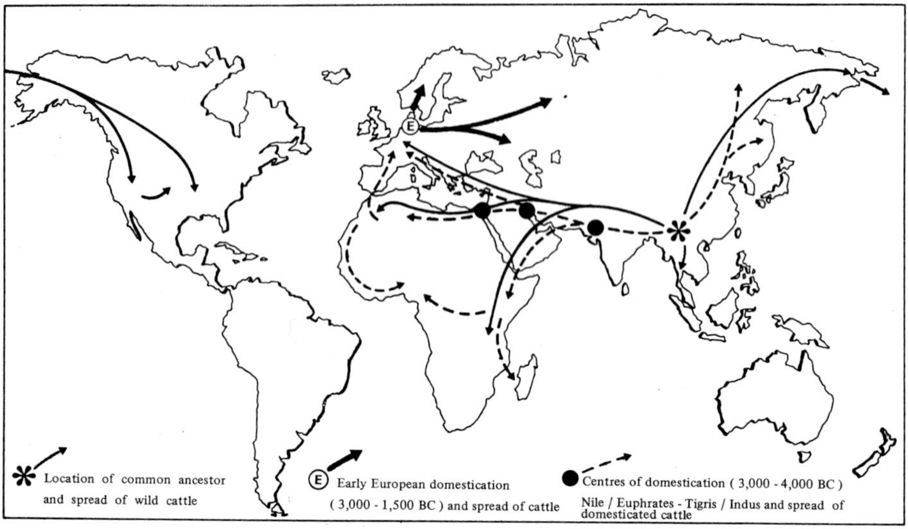

Fig. 1.1b Cattle migrations after the sixteenth century

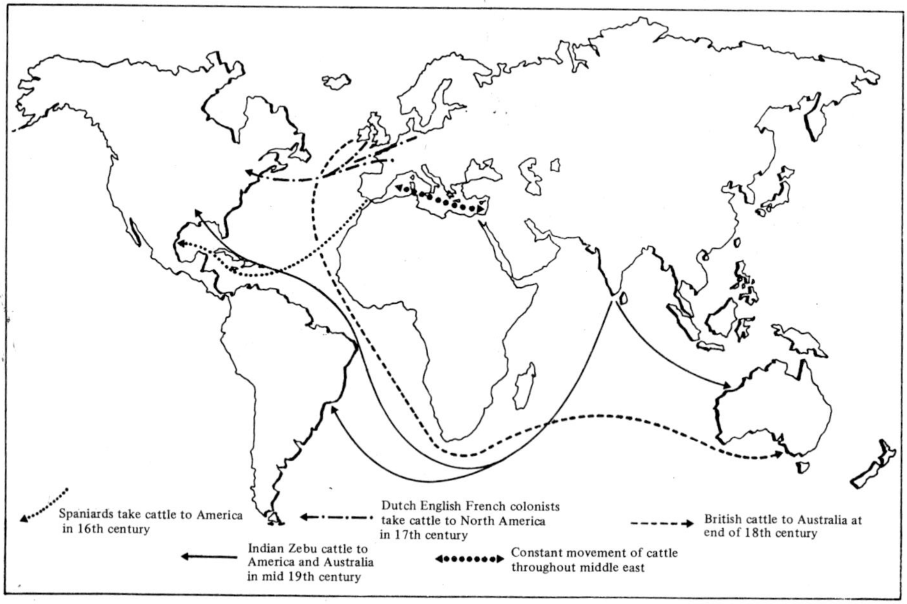

1.2 OVERALL PRODUCTION OF MILK AND BEEF

The main products from cattle are milk and meat, though by no means the only ones. These account for 58 percent of total animal protein available for human consumption.

Table 1.2 shows the increase of milk and beef production in different parts of the world during the last 25 years. This shows that meat and milk production has risen continuously in all countries, but per head of the human population the picture is quite different.

Table 1.3 shows milk and beef production per head of human population in 1961 and 1985 in different regions of the world. The decrease in milk production per head of human population in North America and Oceania is due to changes in consumption patterns and economic factors. Quite different reasons apply in developing countries of Africa and the Near East where low levels of production persist in the face of increasing human population. A similar picture can be seen for meat production per head of human population. During the last 25 years there was practically no increase in developing countries and in fact a decline in Africa and Latin America.

Fig. 1.2. Present-day distribution of cattle

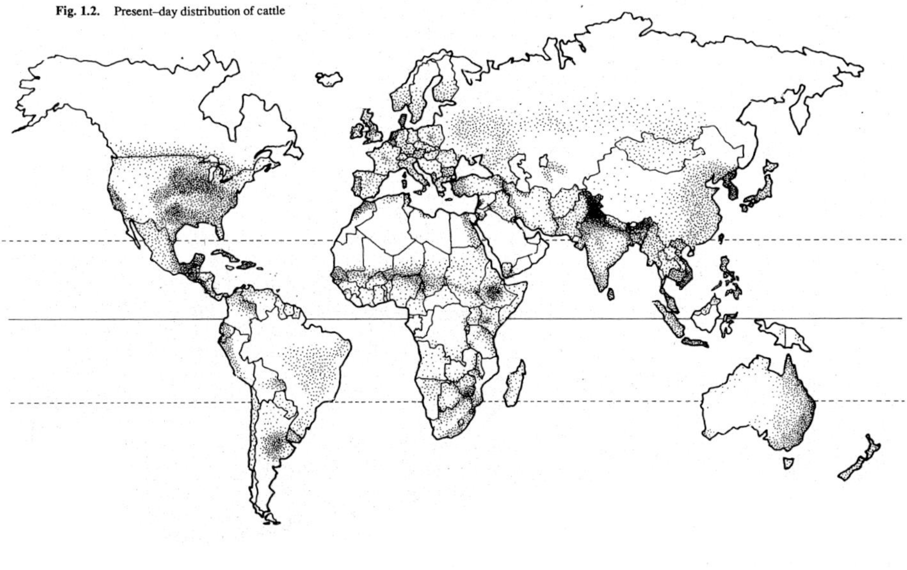

Table 1.1 **Total cattle and human population in millions** (from Agrostat Information System, FAO)

	Cattle			Humans		
	1961	1985	Growth rate	1961	1985	Growth rate
World	952	1 268	+ 1.2	3 075	4 842	+ 1.9
North America	110	122	+ 0.4	202	264	+ 1.1
Western Europe	90	99	+ 0.4	330	377	+ 0.6
Eastern Europe	26	34	+ 1.1	99	114	+ 0.6
USSR	76	121	+ 2.0	218	278	+ 1.0
Oceania	24	31	+ 1.1	13	19	+ 1.6
Others	16	18	+ 0.5	116	158	+ 1.3
Total developed	342	425	+ 0.9	978	1 210	+ 0.9
Africa	105	138	+ 1.1	227	449	+ 2.9
Latin America	179	318	+ 2.4	223	406	+ 2.5
Near East	35	56	2.0	126	240	+ 2.5
Far East	235	271	+ 0.6	787	1 379	+ 2.4
Asian CPEs	7	8	+ 0.6	63	108	+ 2.3
China	49	51	+ 0.2	668	1 044	+ 1.9
Others	—	1		3	6	+ 2.9
Total developing	610	843	+ 1.4	2 097	3 632	+ 2.3

Sources	Percentage of metabolizable energy
Grazing area	42
Areas with non-agricultural functions (roadsides, forests, etc.)	11
Cropland	21
Crop residues	24
Cereals	0.5
Oil seeds	0.5
Agro-industrial by-products	1

1.3 OTHER OUTPUTS FROM CATTLE

Products other than meat and milk are important, even of primary importance, in many countries.

For example, on a world scale hides produced by cattle and buffalo amount to 6.5 million tons per year (FAO, 1985), and 44 percent of this comes from

Table 1.2 **Total production from cattle of meat (beef/veal) and milk** (from Agrostat Information System, FAO)

	Beef/Veal			Milk from Cows		
	1961	1985	Annual growth index	1961	1985	Annual growth index
	(milion metric tons)		*(%)*	*(milion metric tons)*		*(%)*
World	27.4	46.3	2.2	316.5	455.6	1.5
North America	8.1	11.9	1.6	65.3	73.0	0.5
Western Europe	5.8	8.8	1.8	106.3	141.9	1.2
Eastern Europe	1.1	2.0	2.5	27.6	41.2	1.7
USSR	2.9	7.5	4.0	61.6	98.8	2.0
Oceania	0.9	1.8	2.9	11.5	14.0	0.8
Other developed	0.6	1.2	2.9	5.0	10.9	3.3
Total developed	19.4	33.2	2.3	277.3	379.8	1.3
Africa	1.3	2.0	1.8	3.7	6.1	2.1
Latin America	5.4	8.4	1.9	18.2	35.2	2.8
Near East	0.5	1.1	3.3	5.0	9.5	2.7
Far East	0.6	1.1	2.6	11.5	22.3	2.8
Asian CPEs	0.2	0.2	0	0.2	0.4	2.9
China	—	0.3	—	0.6	2.3	5.8
Other developing	—	—	—	—	—	—
Total developing	8.0	13.1	(2.1)	39.2	75.8	2.8

developing countries. While the main value of hides is to produce leather, they are used for food in many developing countries in Africa and Asia (Cockrill, 1974).

Cattle dung plays an important role in many parts of the world. About 40 percent of the farmers of the world depend wholly or partly on animal wastes to fertilize soil. In many developing countries cattle dung is used for fuel. In some parts of India with no firewood, per capita use of dried cattle dung for fuel is one ton per year. In its absence US$9 billion annually would be needed to pay for other sources of energy for cooking, heating etc. (Winrock, 1978).

Cattle and buffalo have a large role as draught animals. In Africa, animals provide 82 percent of total power used in agriculture; in the Far East 99 percent, in the Near East 88 percent and in Latin America 75 percent of which cattle and buffalo are a very important component. In Asia, for example, less than 20 percent of the cattle population are females, because males are kept mainly for power. Without animal power, India would need about US$1 billion per year to cover additional expenses for fuel to allow land cultivation and the operation of irrigation pumps.

Table 1.3 **Average production, per person, of meat (beef/veal) and milk from cattle** (Agrostat Information System, FAO)

	Beef/Veal			Milk		
	1961 (*kg*)	1985 (*kg*)	Annual growth index (*%*)	1961 (*kg*)	1985 (*kg*)	Annual growth index (*%*)
World	9	10	+0.4	103	94	-0.4
North America	40	45	+0.5	324	276	-0.7
Western Europe	17	23	+1.3	322	376	+0.6
Eastern Europe	11	18	+2.1	232	251	+0.3
USSR	13	27	+3.1	282	356	+1.0
Oceania	68	92	+1.3	886	740	-0.8
Others	5	7	+1.4	43	68	+1.9
Total developed	61	85	+1.4	284	314	+0.4
Africa	6	4	-1.7	17	14	-0.8
Latin America	24	21	-0.6	82	87	+0.2
Near East	4	5	+0.9	40	39	-0.1
Far East	1	1	0	15	16	+0.3
Asian CPEs	—	—	—	—	—	—
China	—	—	—	1	2	+3.0
Others	2	3	+1.7	9	6	-1.7
Total developing	4	4	0	19	21	+0.4

In many African countries animals and especially cattle are kept primarily for social status. For example, nomadic tribes of the Sahelian belt keep large herds not so much for marketing as for social prestige. Often the reason for hoarding animals in this part of the world is to achieve greater security. Another reason for keeping cattle may be religion, as for example in India, but in other countries also, cattle still play a part in sacrificial ceremonies.

Another purpose of cattle raising is for bull fighting. In Spain and Latin America a special breed is maintained to supply bulls for the corrida. In Indonesia and Malaysia fights of bull swamp buffaloes are very popular. In these countries the breeding of cattle and buffalo to obtain fighting bulls can be quite a lucrative business.

Cattle and buffalo racing may be less popular now, but not so long ago in India cattle cart racing was a substantial sporting event. Buffalo racing with riders is a great sport in Indonesia, Sabah, Malaysia and the Philippines (Cockrill, 1974).

Table 1.4 **Production of meat (beef and veal) and milk from cattle** (from Agrostat Information System, FAO)

	Meat			Milk		
	1961	1985	Annual growth index	1961	1985	Annual growth index
	(kg per animal)[1]			(kg per animal)[1]		
World	29	36	+0.9	1 738	2 058	+0.7
North America	74	98	+1.2	3 230	5 693	+2.4
Western Europe	64	89	+1.4	2 786	3 822	+1.3
Eastern Europe	42	59	+1.4	2 057	2 974	+1.5
USSR	38	62	+2.1	1 768	2 271	+1.0
Oceania	37	57	+1.8	2 224	3 492	+1.9
Other Developed	34	64	+2.7	3 047	4 268	+1.4
Total developed	56	78	+1.4	2 444	3 337	+1.3
Africa	13	14	+0.3	301	351	+0.6
Latin America	30	26	-0.6	892	968	+0.3
Near East	15	20	+1.2	566	659	+0.6
Far East	3	4	+1.2	440	582	+1.2
Asian CPEs	17	24	+1.4	540	548	+0.1
China	1	6	+7.8	2 556	2 583	+0.1
Others	16	14	-0.6	1 012	950	-0.3
Total developing	13	16	+0.9	572	702	+0.9

[1] Meat (beef and veal) from total production divided by total number of cattle irrespective of type, age or use. Milk from total production divided by total number of cows kept from milk production.

Sometimes cattle play an important role in exhibitions and shows, e.g. in the famous rodeos in USA and Coleaderos de Toros in Latin America.

It is important to bear in mind all the various uses of cattle and their products particularly when constructing breeding plans and the introduction of exotic breeds of cattle to the developing countries.

1.4 PRODUCTIVITY OF CATTLE

Table 1.4 presents the average annual milk and meat output from cattle on a per animal basis in the years 1961 and 1985 but derived in the case of meat from the total out-turn in relation to all cattle irrespective of type, age or use, and in the case of milk from the total production in relation to the number of cows kept for milk production. In general, the results show that while a rapid increase in cattle productivity was observed in developed countries there was practically no increase in cattle productivity in developing regions over this 25-year period particularly in view of the low levels from which any changes are measured. The meat output per head of cattle in Latin America even declined and in Africa remained practically at the same level.

Particularly in relation to meat production, basing the calculation of output on these official statistics can lead to serious over-simplification and inaccuracy since there is no uniform connection across all regions between meat output per animal and the total numbers which include all the animals used or merely kept for other purposes. This, as well as age at slaughter of those cattle killed for meat, may well differ from region to region. In the case of milk production what is included under milking cows may also differ between regions and will include different proportions of single, dual or triple-purpose animals. Yet provided that the basis of comparison over the period from 1961 to 1985 has not changed drastically within a region, the conclusion seems to be that what little increase in total output there has been in developing regions over those years is more likely to have arisen from an increase in cattle numbers than from any improvement in output per animal. Comparison of Tables 1.1 and 1.4 supports this view. But for the increase in cattle numbers, undernourishment of the human population in these regions might have been exacerbated. However, it is doubtful whether further increases in cattle numbers, even if attainable, are a sensible way of maintaining the level of meat and milk consumption by the human population. Overstocking with associated overgrazing is already causing serious problems in many parts of the developing world. Increases in production will have to be sought by improving output per animal. Lack of progress in this respect in the developing world contrasts with substantial progress in developed countries.

Human welfare in economic, social and nutritional terms is the basic motivation for improvements in cattle productivity. The average yield per cow per year is now as high as 7 000 kg milk and 250 kg meat in some countries and of course much higher in some herds. This remarkable advance has been achieved through improved methods of selection, a wide exchange of genetic material (including importation of frozen semen), increased crossbreeding including the wider use of large sire breeds, and, of course, through changes in management practices and intensification of feeding especially with the use of large amounts of grain and other concentrates. The real genetic improvement achieved in only relatively few countries has, however, spread to other developed countries through importation of animals or semen.

Unfortunately, developing countries are lagging behind in this progress with serious economic and social consequences. On the one hand, good progress in cattle production in the developed countries has created serious surpluses,

Tablc 1.5 **Intervention stocks of dairy products** (from GATT, 1985)

	EEC		USA		
	Butter	Milk powder	Butter	Milk powder	Cheese
	(thousand metric tons)		(thousand metric tons)		
1975	140.3	350.0	6.6	97.5	—
1980	370.7	201.1	69.2	177.5	1.3
1985	957.3	635.9	117.7	531.0	281.6

Table 1.6 **Total imports in developing regions of cows' milk and milk products** (from FAO, 1984)

	Milk (fresh/con./dried)		Cheese (soft and hard)		Butter	
	1961	1984	1961	1984	1961	1984
	(thousand metric tons)					
Africa	397	1 390	29	27	21	88
Latin America	1 212	3 534	19	45	14	53
Near East	203	3 080	11	280	6	168
Far East	1 329	3 454	3	12	8	42
Asian CPEs	73	532	—	1	—	3
China	—	46	—	—	—	4
Others	26	73	1	4	2	6
Total developing	3 149	12 109	63	369	51	364

mainly in the EEC countries and in the USA. As can be seen from Table 1.5, the surpluses of dairy products have grown very rapidly. On the other hand, many developing countries, unable to meet the demands for milk with national production, turn to imports. Table 1.6 shows how rapidly the import of milk and dairy products has grown in developing countries. This in itself would not be unreasonable if it were not for the growing debts of the developing countries and the poor prospects they have of earning sufficient income from their own exports to pay these debts.

For the developing countries there is therefore no other solution but to improve the productivity of their cattle. Indeed, many attempts in this direction are being undertaken. Fascination with the rapid progress in developed countries is leading many developing countries to attempt to follow a similar pattern of intensification. Semen of the same best bulls is found in the USA and Europe as well as in Africa, Asia and Latin America. Similar selection methods

Fig. 1.3 Genesis of livestock systems (*from Humphrey, 1980*)

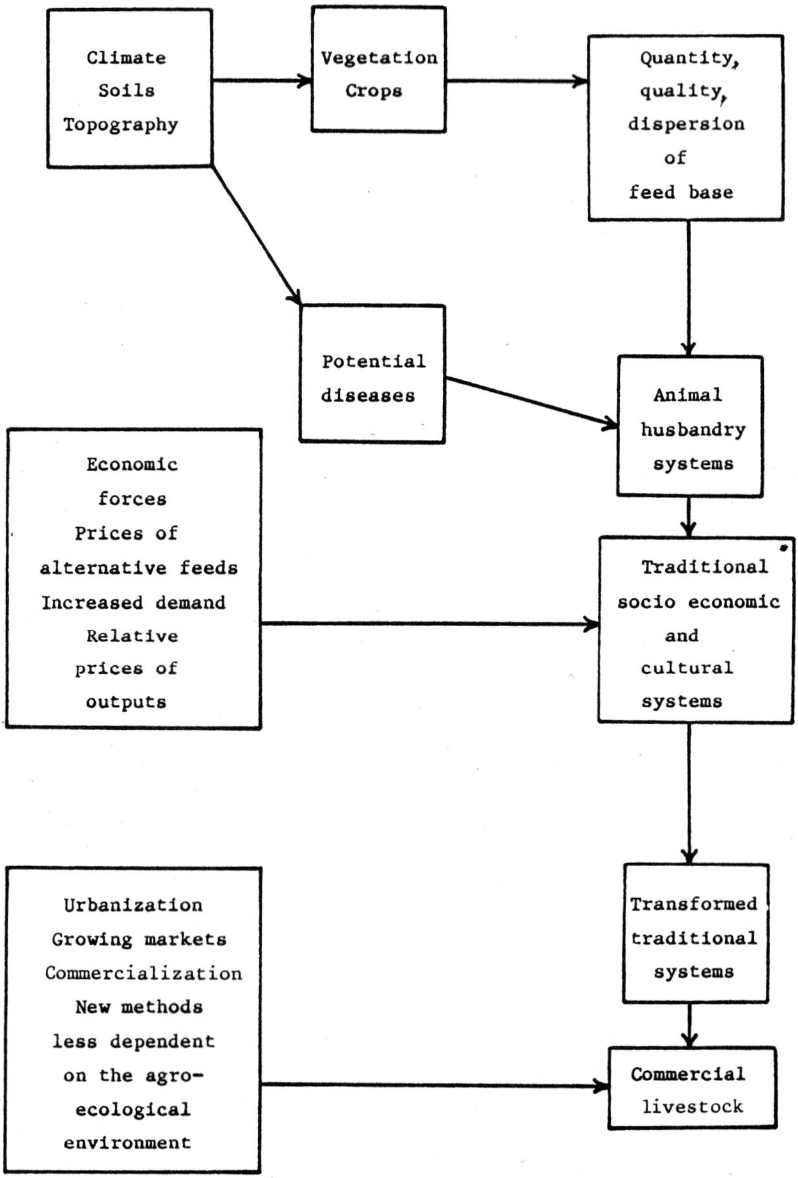

and technologies are introduced wholesale around the world, often without taking into account the stage of development, resources and climate of a given country. Results are often disappointing, because many natural constraints which limit the productivity of cattle in developing countries have been neglected.

Constraints may be of different types operating separately or together (Figure 1.3). Ecological constraints such as the quality of land and type of climate including soil fertility, rainfall, temperature and plant growing season play important roles. Biological constraints, like feed and water supply, are of primary importance. Animal diseases may strongly influence cattle production. Trypanosomiasis transmitted by the tsetse fly serves as perhaps the best example.The availability of productive genotypes may be a lesser constraint than in the past with the possibility of semen shipments and outcrossing provided that other biological factors allow the potentially more highly productive animals to survive. Socio-economic constraints are rather complex, but include factors such as supply-demand relationship, input-output prices, and availability of capital.

1.5 CATTLE PRODUCTION SYSTEMS

Livestock production systems are numerous, as can be seen from Figure 1.4. Cattle are raised under almost all of these systems. The classification of cattle production systems depends on the criteria used. For example, classified from the standpoint of different inputs required by different types of agricultural production, the cattle production systems will divide into:

- *Land intensive* (amount of land usage per animal or unit product is high)
- *Labour intensive* (high labour usage per animal or unit product)
- *Capital intensive* (high amount of capital investment per animal or unit product)

Differences in the intensity of land utilization are well illustrated by the production of only 54 kg animal protein per 100 ha in Africa, 186 kg in Australia and 3 800 kg in Europe (Jasiorowski, 1983).

On a labour intensity classification, the Far East produces 7 kg animal protein per person active in agriculture whilst in Australia this is as high as 1 625 kg.

Capital intensity is at its greatest when large herds are totally housed and all feed (concentrates and roughage) brought to the cattle and with high mechanization leading to low labour input (e.g. dairy herds in California).

Fig. 1.4 **World livestock production systems** (from Humphrey, 1980)

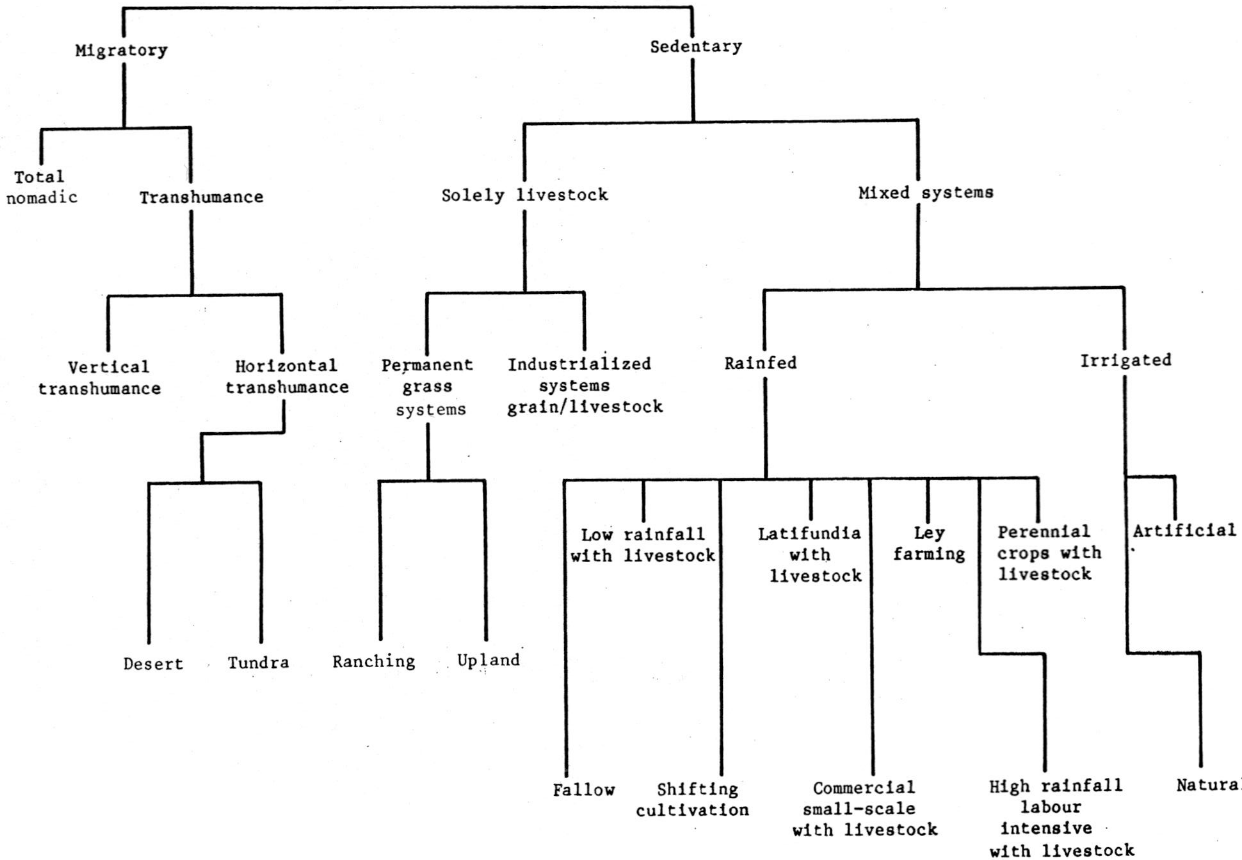

Although each system in the above classification has one dominant feature it is clear that several factors play some role. For example, low density of cattle in Australia (high land intensity) is nonetheless supported by substantial capital input for water supply systems, fencing and even the use of helicopters for herd management. Medium-size family dairy farms in Europe are both labour and capital intensive, or at least both factors play significant roles. By contrast, land and labour are the major components of cattle production in Mongolia. Herding takes place on large steppes by whole families who obtain the milk. In winter the cattle have to be helped to find feed under the heavy snow.

Taking into consideration all ecological, biological and socio-economic constraints, almost countless cattle production systems can be distinguished. For practical reasons we will group them here into two main categories, migratory and sedentary systems, and will describe only the most important of them.

1.5.1 Migratory systems

In some regions of the world the system of pastoral migration of cattle plays a very important role. The migratory systems can be divided further into nomadic and transhumant. In Africa, nomadism and transhumance combined utilize 35 percent of the grazing land (Raun *et al.*, 1982).

a) The nomadic system

This is probably the oldest way of using animals. It consists of constant migration of whole tribes and families together with their herds, usually of cattle, goats, sheep and camels. The direction of the migrations is dictated by seasonal changes of grass growth and accessibility to water. Pastoral nomadism prevails in dry, thinly populated areas of about 1-10 people per 1 km^2, as for example in Sahelian Africa.

The nomadic system is still the main form of animal utilization in semi-arid areas, which in practice are not suited for settlement. These areas are often uninhabited for most of the year, and might otherwise be left unused. Nomadism may be considered, therefore, a highly developed method of fitting into the ecosystem, in which the nomad does not attempt to alter the environment but adapts his way of life to it (Chalmers, 1976).

The nomadic system of animal production exists in many countries of Africa, in the Near East and in West Asia. Tuaregs in Niger and other parts of the Sahel travel with their herds from the borders of tsetse fly territory up to the

Saharian outskirts following the rainfall. A good example of nomadism is provided by the excellent cattle breeders in East Africa, the Masai tribes, whose diet consists almost exclusively of milk, cattle blood and meat. Nomadic herds on the move, with their baggage animals loaded with all the belongings and a rider, average 4-5 km per hour and cover 16-24 km daily. They travel about 50 days in a year and cover about 650 to 800 km (Chalmers, 1976).

Prospects for an increase of nomadic herd output are rather poor. There is a possibility that in future the output of animals in this system may even decrease. In the meantime, however, the number of animals possessed by nomadic herdsmen increases due to the nomads' fear of a reduction in the supply of milk and to the fact that the size of the herd is related to the prestige of the whole family. However, the nomads, who are known as excellent breeders, are not concerned with proper soil utilization. As a result, excessive plant devastation takes place, which, in turn, has caused deterioration of, for example, Sahel and Sudan zones as well as the spreading of desert areas.

b) Transhumant migration system

The transhumant (vertical or horizontal) method of cattle utilization makes use of sometimes distant pastures during the rainy season and of fodder and post-harvest remains in dry periods.

In summer this system also makes use of mountain pastures, which are inaccessible at other seasons of the year.This system of production can be found not only in Africa and Asia, but also in some regions of southern Europe, the Asian republics of the Soviet Union and Mongolia. Modern technology, however, is being introduced even to this system of pastoral production, such as the use of lorries or trains in some developed countries to carry the animals to far-away pastures.

In the transhumant as in the nomadic system, practically no marketable milk is produced, but sometimes cheese is produced and marketed. In neither system, very little, if any, crossbreeding with the temperate breeds is practised.

1.5.2 Sedentary systems of cattle production

Population density, a shrinking availability of land and the emergence of markets are the main reasons for settlement. The existing sedentary systems of cattle raising in the world differ widely from one another, depending on soil and climatic conditions and the level of development of individual countries. Sedentary systems come under several headings:

a) Shifting cultivation

In west and southwest Asia, East Africa, and South America the system is based on the cycle of cultivation and periodic abandonment of land. The result is the growth of natural vegetation in the abandoned areas (weeds, grass), which is used for cattle grazing.

In some regions of West Africa, as for example in the Gambia, this system is used to breed trypanotolerant cattle, the N'dama. Here the cattle graze freely around villages during the dry season and are herded by herdsmen during the rainy season when plots of land around villages are cultivated. The calves are kept close to the villages, which in turn makes the cows come once a day to be milked.

In regions of dry savannah in Africa, large cattle herds are kept under this system. The alternation of land cultivation and grazing is also typical of cattle production in Ethiopia, Tanzania and Burundi.

b) Cattle in low rainfall areas

This system is quite frequent in the sub-humid tropics where land is under permanent cultivation. Under this system cattle are fed chiefly waste fodder and what remains in the fields after harvest. Sometimes they are grazed on fallow land. Communal schemes are often adopted to allow all cattle in the village to graze in all fields after they have been cleared of crops. In this system, African farmers who keep no cattle themselves allow nomads to bring in their herds so that the animals may graze in their fields and manure them at the same time. Under this system marketing of milk and beef is often less developed. Cattle, together with buffaloes, are intensively used for draught power.

In general, the productivity of cattle and buffalo kept under the subsistence production system is low.

c) Cattle production under perennial crop systems

This system plays a certain role in tropical regions of West and East Africa, east Asia and South America mainly for beef production. Waste products are very important here. Sugarcane plantations, for instance, provide cattle feed in the form of sugarcane tops, molasses and fibrous remains after the sugar extraction process. Pineapple plantations, cleared of fruit, also provide feed for cattle as well as plantations of agave, plants which are used for the production of sisal fibre. The grazing of cattle under coconut, oil palm and rubber trees is gaining popularity.

d) Ranching

This system, found in different climatic conditions, is a commercial form of pastoralism adapted to the conditions of urbanization. Popular in regions with poor soil and little rain, this extensive system is used chiefly for beef production.

It can be found in the steppe regions of Central Asia, in Central America, and also in Africa; but the best examples of this form of production are the pasturing of cattle in Australia, in the prairies of the USA and in the steppes of South America. The shortage of water and the low quality of land require large inputs per animal, while high labour costs encourage investment in permanent fences, watering places, roads, etc. In some cases helicopters are used to watch the animals.

e) Cattle production system based on crop rotation

In this system, cattle are fed on fodder grown in fields or allowed to graze pastures during the summer. This system is predominant in a temperate climate, but appears also sporadically in the tropics.

In the developed free market economy, family commercial farms with little or no hired labour dominate the dairy and beef industry. Here, the concentration of land and cattle has recently taken place leading to bigger farms and herds, and to their specialization (Gravert, 1982).

In the developed free market economy saturation of the market with dairy products and a build-up of surpluses have led to rather low prices for milk. Farm income has been compensated for by high milk yields which in turn have been possible only with intensive concentrate feeding.

Due to high labour costs and the scarcity of land, cattle production in the developed world is becoming more and more capital-intensive.

f) Large-scale cattle units

This very capital-intensive system can be found in both Americas, in Europe and recently in the Middle East. In their most developed forms, the operations are independent of the farming of land and all inputs, including feedstuffs, are purchased.

Large dairy farms in California provide a good example, where the owner of a herd of 2 000 cattle may have less than 20 hectares of land and buys all fodder from outside, including maize for silage, hay and straw. An extreme example is the large fattening farm in Colorado, which fattens some 100 000 cattle annually.

Large cattle units are frequent in the USSR and East European countries, where they are organized on state or collective farms. For dairy cattle, the size of a unit is usually 600 - 1 200 cows, with some herds up to 2 000 cows. In the German Democratic Republic (GDR) young stock are kept in separate units with a capacity for heifers of 1 000 - 2 000 head and for calves from 400 to 1 000 head. In the USSR, Czechoslovakia, Bulgaria, Romania and Hungary large cattle production units are integrated with large-scale farming, but in GDR they are separate economic enterprises obtaining feedstuffs on contracts from farms specializing in plant production (Jasiorowski, in press).

Large-scale production units are sometimes criticized as being detrimental to the environment and too capital- and grain-intensive. Low fertility of cows and high mortality of calves are frequently noted under this system. While this system of cattle production may sometimes be advantageous in developed countries where capital is cheap and labour expensive, it is misguided to recommend it for the developing countries.

1.5.3 Distribution

The distribution of cattle and buffalo among different farming systems is as follows (Payne, 1976): nomadism, 5 percent; transhumance, 10 percent; ranching, 29 percent; mixed farming with crops, 55 percent; specialized and intensive farms, 1 percent.

1.6 CONCLUSIONS

Cattle production in general and milk production in particular are offered the challenge of a wide variety of ecological niches, management systems and economic opportunities. The realization of improved productivity, whether of meat, milk, or a combination of these and other uses of cattle, will depend on a better understanding of the basis of present variation in production levels, including its genetic and environmental components. This in turn will affect future prospects for change and in particular the direction of change. It is in this context that the comparison of the major strains of Black and White cattle in Poland has to be seen and, it is to be hoped, will make a contribution toward decision-making in cattle production both in developed and developing countries.

Succeeding chapters will present the results of the strain comparisons in Poland followed by a survey of the development and breeding programmes with Black and White cattle in the countries contributing the strains. Further chapters will deal in a similar way with the distribution of this breed in other developed countries and, importantly, in the developing world where it has a particular role in crossbreeding. In view of the widespread use of American Holstein-Friesians in the European Friesian population a chapter will review the evidence of this to date. A final chapter will discuss some of the general issues arising from the results and surveys.

References

CHALMERS, A.W. Advantages and disadvantages of nomadism with particular reference to the Republic
1976 of Sudan. *In* Smith A.J., ed. Beef cattle production in developing countries. Centre for Tropical Veterinary Medicine, Edinburgh. pp. 388-397.

COCKRILL, W.R. *The husbandry and health of the domestic buffalo.* Rome, FAO.
1974

FAO. *Production yearbook.* Vol. 38, Rome, FAO.
1984

FAO. *Production yearbook.* Vol. 39. Rome, FAO.
1985

FRIEND, J.B. *Cattle of the world.* Poole, Dorset, UK, Blandford Press.
1978

GATT. *The world market for dairy products.* Geneva, GATT.
1985

GRAVERT, H.O. *Family farming milk production trends.* International IDF-EAAP Seminar on The Dairy
1982 Cow of the Future. Warsaw, 11-13 August 1982.

HUMPHREY, J.H. *The classification of world livestock systems.* AGA/MISC/80/3. Rome, FAO.
1980

JASIOROWSKI, H.A. *World animal production: present situation.* OIE Symposium on Anabolics in
1983 Animal Production. Paris, OIE.

JASIOROWSKI, H.A. *In* Cattle production. Cattle production systems in practice. Elsevier Science
1987 Publishers, Amsterdam (in press).

PAYNE, W.J.A. Systems of beef production in developing countries. *In* Smith, A.J., ed. Beef cattle
1976 production in developing countries. Centre for Tropical Veterinary Medicine. Edinburgh, UK. pp 118-131.

RAUN, N.S., HART, R.D., DE BOER, J., FITZHUGH, H.A. & YOUNG, K. *Livestock program priorities*
1982 *and strategy.* Morrilton, Arkansas, USA, Winrock International Livestock Research and Training Center.

ROUSE, J.E. *World cattle. Vol. I. Cattle of Europe, South America and New Zealand.* University of
1972 Oklahoma Press, Norman.

WINROCK INTERNATIONAL. *The role of ruminants in support of man.* Morrilton, Arkansas, USA,
1978 Winrock International Livestock Research and Training Center.

2. International comparisons of ten strains of Friesian cattle

2.1 INTRODUCTION

The project entitled ‘‘Testing of ten different strains of Friesian cattle’’, popularly known as ‘‘The Polish Friesian project’’, was the largest of its kind in the world. Initiated by the Food and Agriculture Organization of the United Nations, the project was aimed at comparing the genetic value for milk and meat production of ten Friesian strains. It was carried out under field conditions at Polish state-owned commercial agricultural enterprises as well as at research centres. The experimental part was started in 1974 and completed in early 1984 (Fig. 2.1). According to an internationally agreed procedure, semen from young unproven AI bulls selected at random was to be used to inseminate Polish Friesian cows for production of F_1 and backcross (R_1) progeny. Apart from the field experiments it was decided to make more detailed observations of milk production and fattening in conditions of intensive feeding. The Breeding and AI Organizations from the participating countries donated semen and provided technical and scientific advice through the Technical Advisory Committee (TAC) established by FAO. The international coordination of the project was done by FAO.

Ten countries took part in the project-six European and four from outside Europe; these were: Denmark (DK), the Federal Republic of Germany (FRG), the Netherlands (NL), United Kingdom (UK), Sweden (S), Canada (CND), the United States of America (USA), Israel (IL) and New Zealand (NZ). The tenth country was Poland (PL), which provided a control group. The Technical Advisory Committee (TAC) met 12 times (from 1973 to 1984), usually coinciding with meetings of the European Association for Animal Production (EAAP). The members of the TAC were:

DK - Prof. Dr. A. Neimann-Sørensen
FRG - Prof. Dr. H.O. Gravert
NL - Prof. Dr. Ir. R.D. Politiek
UK - Dr. L.K. O’Connor
S - Dr. B. Lindhe and Dr. J.A. Eriksson
CND - Prof. E.B. Burnside

USA - Dr J.M. Smith, Prof. A.E. Freeman and Dr. H.D. Norman
NZ - Dr. H.J. Clifford and Dr. B.W. Wickham
IL - Dr. R. Bar-Anan
FAO - Mr. I.L. Mason, Dr. C. Hickman, Dr. J. Rendel and Dr. J. Hodges
PL - Prof. Dr. H. A. Jasiorowski, initiator of the project and principal investigator of the intensive-feeding dairy trial; Prof. Dr. Z. Reklewski, principal investigator of the intensive-feeding beef trial; Dr. Maria Stolzman, principal investigator of the field trial; Prof. Dr. A. Zarnecki, statistical evaluation of the field trial.

The Polish research institutes involved were: the Institute of Zootechnics in Cracow (for the field trial), the Institute for Genetics and Animal Breeding at Jastrzebiec (for the intensive-feeding beef and dairy trials), the Agricultural University in Warsaw (for part of the intensive-feeding dairy trial), the Central Animal Breeding Office in Warsaw, the staff of Animal Breeding Offices in Gdansk, Bydgoszcz and Olsztyn, as well as of four Artificial Breeding Stations (Sztum, Olsztyn, Myslecinek and Srawecinek). Dr. G. Kalinowska, research assistant from the Institute of Zootechnics, supervised the farms and collected data.

The field trial was largely financed by the Institute of Zootechnics. Approximately 15 percent of the total cost was covered by USDA PL-480 and M. Curie-Sklodowska funds. The intensive-feeding trials were financed by the Institute for Genetics and Animal Breeding at Jastrzebiec.

Fig. 2.1 Time scale of different aspects of the strain comparisons

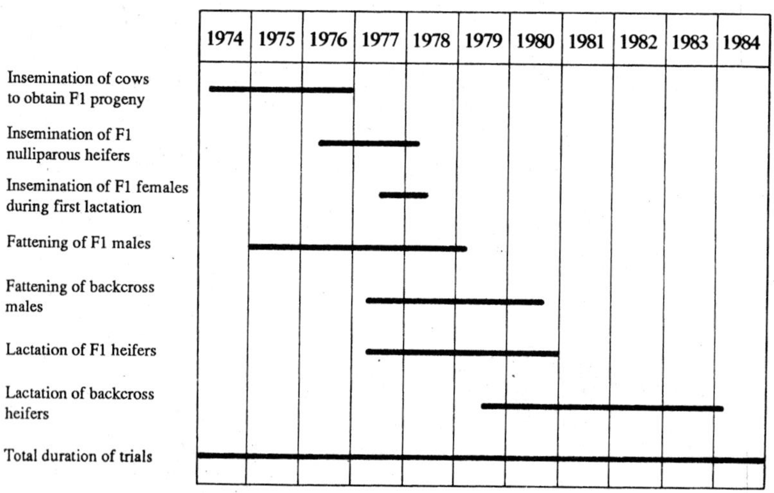

«Patra», the highest yielding cow from the intesive-feeding trial — and F_1 from Canadian sire Roybook Starlife 24106-4-7. Patra produced 15 014 kg milk (3.49 percent fat) in her fourth lactation — a total of 523.4 kg butterfat (Agric. Univ. Exp. Sta., Brinóv).

2.2 MANAGEMENT AND FEEDING SYSTEM OF ANIMALS ON THE POLISH AGRICULTURAL ENTERPRISES PARTICIPATING IN THE TRIALS

The 20 state-owned enterprises were located in the two northern provinces — Elblag (80 percent of animals) and Bydgoszcz (20 percent of animals) (Fig. 2.2). They were of different size with from 1 to 7 cow herds per enterprise — a total of 71 herds — and a corresponding number of young stock rearing and fattening units. Individual herd size averaged 180 cows, totalling 12 752 Friesian cows in 1973 (Table 2.1). According to official milk recording results in 1973 the

Fig. 2.2 Geographical location of the strain comparisons

average milk yield of enterprises varied from 3 052 kg to 4 085 kg, and fat content from 3.69 to 4.09 percent. One enterprise, Trampowo, was later excluded from the project and two others, Susz and Waplewo, were included in 1976.

All farms applied a similar system of cattle management and feeding and the experimental animals received the treatment which is normal on these enterprises. The system was as follows:

Calves stay for their first 7-10 days in cow barns and are then transferred to rearing units (3-5 cow barns support 1-2 rearing units) where they are kept until 6 months of age. Females then go to young stock barns (1-2 rearing units supporting 2-3 stock units) until 15 months of age. For mating, females are usually transferred to another barn where they stay until the seventh or eighth month of pregnancy. At this stage of pregnancy, heifers are distributed among cow herds as replacements. Usually, heifers are kept as replacements within an enterprise but sometimes they are transferred between enterprises. Male calves remain entire (uncastrated) throughout. Males older than 6 months of age are

Table 2.1 **Numbers of herds and cows and milk records (in 1973) of state-owned agricultural enterprises involved in the Friesian project**

Enterprise	No. of herds	No. of cows	Milk (kg)	Fat (%)
Lisewo	3	757	3 201	3.84
Szymankowo	3	763	3 265	3.76
Miloradz	4	954	3 052	3.81
Stogi	5	905	3 471	3.81
Lasowize	4	852	3 267	3.76
Nowy Staw	3	658	3 159	3.76
Krzyzanow	3	1 042	3 228	3.84
Lipowo	3	758	3 099	3.69
Zielonki	4	568	3 174	3.87
Trampowo	1	105	3 089	3.98
Kadyny	2	317	3 984	3.99
Rzeczna	4	558	3 733	3.88
Ulnowo	4	520	3 658	3.93
Szymbark	3	345	4 039	3.89
Baloszyce	3	342	3 714	3.93
Tegowo	4	542	3 445	3.83
Grodziec	3	284	3 644	4.09
Piaski-Kobylniki	7	1 400	3 639	3.77
Sobiejuchy	4	459	4 085	4.01
Chwaliszewo	4	623	3 558	3.88
Total	71	12 752		

kept in fattening units (loose housing) and transferred for a finishing period to special barns, where they are generally tied up. In this period, bulls that are sold alive for export undergo veterinary tests. All cattle are routinely examined for tuberculosis and brucellosis.

Cows spend about five months at pasture, from mid-May to mid-October. In late summer and autumn, the pasture is supplemented by cut green fodder, or silage and some hay. In winter, cows receive roughage in the form of silage (from maize, grass or sugarbeet leaves) plus 3-5 kg of hay and some straw. Fodder beet is given in winter, and by-products, including sugarbeet pulp (provided dry or as silage) and sometimes brewer's grains, are also available. Concentrates (about 12 percent crude protein) are given all the year round according to milk yield. The average use of concentrates is 0.25 kg per kg of milk (600-1 000 kg per year). Cows are usually tied, but some large barns for 500-600 cows have loose housing with a milking parlour. Cows are milked twice a day, but some farms milk three times daily during the first 100 days of lactation.

Calves are born throughout the year, with a small winter peak. Calves receive colostrum, milk (150-200 kg) and milk replacer (30-40 kg). They are also given hay and concentrates and seasonal roughage. Heifers are put to pasture during the summer but spend seven months of the year indoors. They are given green cut fodder (in late summer and autumn), hay (1-2 kg), some straw, dried sugarbeet pulp, and concentrates (up to 2 kg per day per animal). Pregnant heifers receive supplementary feeding consisting of hay and concentrates (1-1.5 kg).

Bulls are fed indoors on cut green fodder or silage, grass meal, straw and concentrates (1-1.5 kg). During the finishing period animals receive a supplement of concentrates (2 kg per day). Bulls are fattened, usually to a liveweight of 450 kg.

Polish Friesian cows on farms included in the project were of a dual-purpose type with well-developed musculature. Genetically they were closely related to the Dutch strain, being daughters of Dutch sires or of sires with a high proportion of such blood. In the northern part of Poland some bulls were also imported from the Federal Republic of Germany and their sons were used. All dairy herds were milk recorded (monthly recording by official milk recording assistants) and all cows and heifers were inseminated artificially.

2.3 EXPERIMENTAL PROCEDURE

2.3.1 Field trial

The comparison of strains was made by the use of first cross (F_1) progeny from Polish Friesian cows inseminated with semen of bulls from the participating

F_1 cows from the international experiment on one of the participating state farms in Poland

countries. This was followed by backcrosses to bulls of the paternal strain or to Polish bulls (Fig. 2.3). All progeny were reared and bulls (entire males) were fattened to a liveweight of 450 kg. Heifers stayed in the test until the end of the first lactation except that the farms were allowed to cull those producing less than 10 kg of milk daily up to the third month of lactation. The following data were recorded:

Foundation dams

- Identification of farm and herd where inseminated, dates of services and bulls used.

Progeny generations (F_1 and backcrosses)

- Identification of sire and dam, birth date and weight, sex.
- Causes of death or disposal of animals during period of project.
- Monthly weights up to one year of age for males and up to 18 months old for females.
- Height at withers, chest girth, at 12 months of age for males and at 12 and 18 months for females.
- Blood type for a sample of the progeny.

Female progeny only

- Age at first service and first calving.
- Dates of inseminations and identities of bulls used.
- Date of calving, calving difficulties, liveweight after calving (within 24 hours).
- 305 days' first lactation milk yield.
- Fat and protein percentage and first lactation length.
- Rate of milking and udder measurements (for a sample of female progeny only).

The timetable of the experimental part of the project is shown in Fig. 2.1.

Statistical procedure for the field trial

The statistical models used for the evaluation of the results were discussed and agreed within the Technical Advisory Committee. The calculation of the results of the field test was based on linear mixed models which are given in some detail by Stolzman *et al.* (1987). The factors included in the models, are summarized here:

All traits: strain, herd, year and season (all as fixed effects), sire within strain (random effect) and additionally:

Growth traits: a regression on the number of days between date of birth and date of observation.

First lactation traits: a regression on age of the cow at first calving. *Speed of milking and udder measurements:* time of milking (e.g. morning), a regression on the number of days between date of calving and date of observation, a regression on the quantity of milk at a milking, and further, for calculations on differences between teat placement areas before and after milking, a regression of teat placement areas after milking was fitted.

2.3.2. Intensive-feeding trial

a) Beef trial

The bull calves for the intensive-feeding trial were those born in the field test and purchased for the purpose from the state farms involved in the Friesian project. Completing the F_1 groups took approximately three months in the autumn of both 1976 and 1977. Half of each genetic group was fattened to 450 kg at slaughter and the remaining half to 550 kg. The animals were randomly distributed between the slaughter weight groups. It was intended to have 28 animals

Fig. 2.3 Mating system to produce animals for the strain comparisons

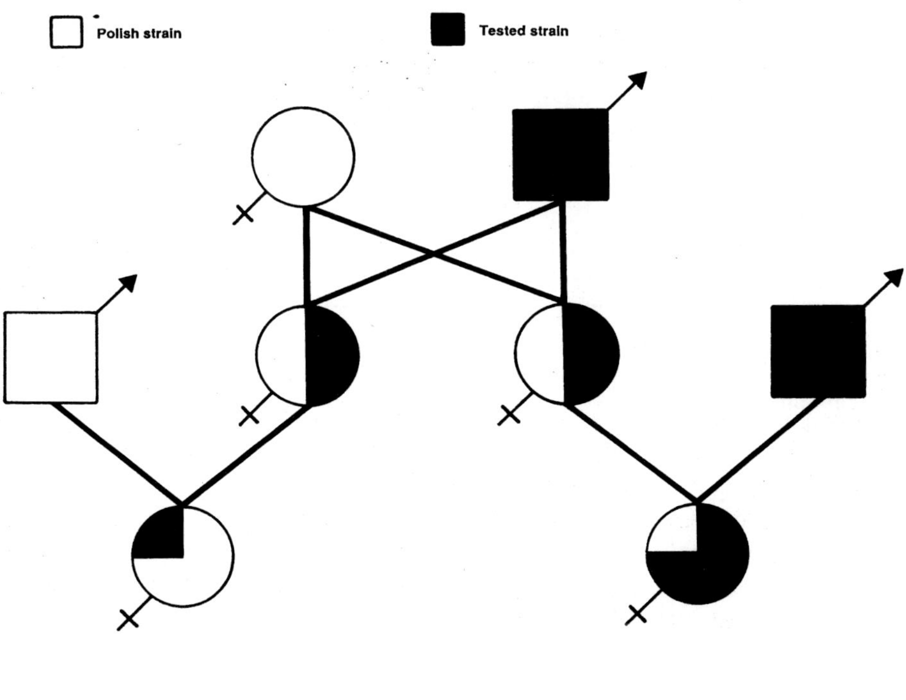

per strain (i.e. 14 per slaughter weight group). However, due to leg and respiratory problems, four bulls did not finish the fattening period — one each from USA and Danish crosses and two from UK. The New Zealand strain was represented by 30 animals and the Canadian and Polish strains by 29 animals each. The Polish animals were all purebred.

The backcross (R_1) generation bull calves for the intensive-feeding trial were similarly obtained in the autumn in 1978. There were 14 backcross calves per strain from each of Canada, Israel, the Netherlands, New Zealand, UK and the USA, 15 from Germany, Poland and Sweden, 17 from Denmark. The Polish strain was, again, represented only by purebred animals. All the animals finished the fattening period which, in the case of R_1, was terminated at 450 kg.

For both the F_1 and R_1 crosses, the offspring were sired by at least seven sires per strain. The age of calves at purchase was approximately eight weeks.

Feeding. During the initial period (56-100 days) calves were given milk substitute, commercial concentrate for calves, and hay. From the start of the test the bulls were given two daily feeds of complete concentrate mixtures and 1 kg of hay. The composition and nutritive value of the mixtures are shown in Table 2.2.

The first mixture was used from the beginning of the test (Day 110) until the animals reached 250 kg, the second mixture was used there after. The feeding regimen employed in the test allowed daily weight gains of up to 1 100 g over the entire fattening period. The nutrient requirements were determined according to Nutrient Requirements of Beef Cattle, National Academy of Sciences, Washington, D.C. (1970). The daily quantities of feed given were determined individually on a body weight basis and adjusted every two weeks. Feed not eaten was weighed back twice a week.

Table 2.2 **Composition of complete concentrate mixtures for intensive feeding trial** (from Reklewski *et al.*, 1985)

Components	**Concentrate I** *(%)*	**Concentrate II** *(%)*
Ground grain	70	48
Commercial concentrate for calves	28	20
Sugarbeet pulp (dried)	—	10
Dried grass meal	—	20
Mineral components	2	2
Per kg concentrate:		
Dry matter	87	88
Digestible protein	12.8	11.0
Net energy for gain MJ	5.1	4.9

Slaughter. Prior to slaughter the animals were fasted for 24 h with access to water. After slaughter the carcass weight was determined. The right side of the carcass was dissected and contents of valuable cuts (chuck back, ribs, best ribs, rump, round, shoulder) (Weniger *et al.*, 1963), lean, bone and fat trim were measured. In addition, the weight of subcutaneous fat was determined separately. Beef performance evaluation was based on indicators characterizing both the growth rate and lean content in the carcass. The two criteria used were:

1. Mean daily lean gain $= \dfrac{\text{carcass weight} \times \text{percent lean}}{\text{days from birth to slaughter}}$

2. Net energy consumption per kg lean gain $= \dfrac{\text{total energy consumption}}{\text{weight of lean in carcass}}$

b) Females for lactation performance

F_1 generation. 400 pregnant F_1 heifers were randomly selected from the state farms included in the field part of the project in the Elblag and Bydgoszcz districts. Each strain was represented by at least 14 sires. The parentage of the animals was checked by blood typing. The test was performed in two series in the years 1978 and 1979. Half the heifers were tested at the Institute for Genetics and Animal Breeding in Jastrzebiec and the other half at the Agricultural University of Warsaw in Brwinow. There were thus 100 F_1 heifers at each location in each year.

Two months before calving, the females were put on slats in two sheds (without bedding) with facilities for individual feeding.

Table 2.3 **Composition of the concentrates**[1] (from Jasiorowski *et al.*, 1983)

	%
Crushed barley	35
Wheat bran	20
Crushed corn	20
Soya bean oil meal	10
Rape seed oil meal	10
Yeasts	1
Chalk	2
Mineral Mixture	1
Salt	1

[1] 1 kg of concentrates was equal to 1.05 oat units and had 136 kg of digestible crude protein

Feeding. The heifers during the test were fed hay, dried beet pulp, maize silage and concentrates. The feeding system was the same during the two successive years.

The composition of the feed is shown in Table 2.3.

The basic ration per heifer per day given the whole year round was 15 kg corn silage, 6 kg meadow-hay and 2 kg dried beet pulp. During lactation, concentrates were offered according to milk yield and fat percentage. For each kg of milk the amounts were:

Fat (%) of milk	**Concentrate (kg)**
3.5 or less	0.40
3.6-4.0	0.44
4.1-4.5	0.48
4.6-5.0	0.52
5.1-5.5	0.56

In addition, during the first month of lactation, the cows were given an extra 1 kg of concentrates.

Milk production was recorded weekly, and the amount of concentrates was adjusted accordingly. The fat, protein and total solids content was estimated in milk samples. Ease of milking was examined at the beginning of the lactation period, using Alfa-Laval equipment.

The heifers were weighed on the 10th day after calving, and again at 305 days of lactation. Body parts were measured after 200 days of lactation. The health of the animals was recorded.

Backcross (R_1) generation. In total, 220 R_1 heifers were randomly chosen and purchased in 1978 or 1979 from the farms involved in the project. Backcrosses were only to the "imported" strains and the Polish strain was represented by purebred animals. They were randomly distributed between the same two sites (Jastrzebiec and Brwinow) as the F_1 generation and procedure and feeding were the same. (Backcrosses to the Polish strain also became available but could not be included in the main results.)

c) Statistical procedure for intensive-feeding trial

Statistical analysis took the form of least squares [for details see Jasiorowski *et al.* (1983) and Jasiorowski *et al.* (1986)] for milking traits with factors for the effects of strain, herd (where located), year and season of mating and a regression on age at calving. For liveweights and body dimensions factors were included for strain, herd (where located), year and season of birth and a regression on age at calving.

However, the factors for year and season of birth on liveweight and body measurements, and the effect of year and season of conception on dairy traits were excluded from the final model having been found on preliminary analysis not to be significant. Ease of milking was recorded in Brwinow only (8-12 first calvers per strain). The least squares model for this included factors for the effects of the daily milk yield and of the number of days between calving and date of observation.

Statistical analysis of the slaughter data used the method of least squares with the model described in more detail by Reklewski *et al.* (1985) including factors for the effects (in the F_1 generation) of strain, slaughter weight class and an interaction of strain with weight class; and in the R_1 generation strain alone (there was only one slaughter weight).

The significance of differences between individual strains was determined by the Duncan test as modified by Kramer (1957).

Estimates of additive and non-additive (heterosis) effects of strain were calculated. It was assumed that all the heterosis and half of a strain additive value were present in the F_1 generation and 50 percent of heterosis and 75 percent additive effects in the backcross. If f is the difference between the average for F_1 and the average of the Polish strain; and r is the respective difference in the backcross generation, the additive value of a strain equals $2r$-f and heterosis $h = 1.5f$-r.

2.3.3 Sire information and insemination results

Two batches of young unproven AI bulls were randomly selected in the nine cooperating countries (outside Poland) and semen from them was sent to Poland in 1974 (first batch) and in 1975 (second batch). The first batch comprised semen from 168 bulls and the second batch from 180 bulls. In addition there was semen from 40 randomly chosen young Polish bulls. Each strain was represented by between 37 and 40 bulls.

All the bulls used in the project were later progeny-tested in their respective countries of origin. Available information on the average breeding value of bulls used in the project and of all contemporary young AI bulls sampled in a given country, as provided by the cooperating countries, is shown in Table 2.4. It appears that the breeding value of Canadian bull semen sent to Poland may have been slightly better than the average of all their contemporary bulls, while Israeli bulls used in Poland had a slightly lower average breeding value than their contemporaries. After long discussions within the TAC it was decided that no correction factor for the difference in the breeding value of bulls used in Poland

and of their contemporaries should be introduced into the statistical model because of differences among countries in the size of their AI bull populations, and in the methods of calculating breeding values.

Using four-generation pedigrees of the 388 bulls used as sires, the relationship among bulls within and between strains and the inbreeding coefficient within strains were calculated according to the covariance matrix approach (Table 2.5). These bulls were born in the period from 1970 to 1973. At that time no relationship between North American Holstein-Friesians and bulls of the Polish, Danish, Dutch and New Zealand strains was found. A small relationship existed between Canadian Holsteins and the British and Swedish Friesians. Among European strains, the German Friesians were most related to North American Holstein-Friesians. Dutch Friesians were related to all the European strains. Israeli cattle showed a strong influence of the USA Holsteins and to a lesser extent of the Canadian ones. The New Zealand strain was not related to any other strain included in the project.

The highest inbreeding coefficient was found in the Canadian and Dutch group of bulls. The UK bulls were not inbred at all.

Table 2.6 shows the number of bulls and semen doses delivered to Poland, the use made of the semen and the AI results. From 348 bulls 83 570 semen doses were delivered to Poland. Five different forms of semen packaging were used — pellets, mini and midi straws, ampoules and mini tubes. All semen was deposited at the AI centre at Sztum (province Elblag) which served as a semen bank and distributor among 3 other AI centres. Inseminators received 50 doses of semen (10 doses of semen from one bull of each of 5 strains). When this consignment had been used up the next was composed of 50 doses from 5 bulls belonging to the remaining 5 strains. In each herd, the cows were inseminated at random. Such a system of semen distribution ensured the use of bulls of each strain on each farm.

Insemination of cows (for second or later calving only) to produce the F_1 progeny was started on 15 March 1974 in 70 dairy herds, and ended on 31 December 1976.

Pregnancy diagnosis was based on veterinary examination. In total, 33 699 cows were inseminated at least once. The pregnancy rate over all inseminations was 78.7 per hundred inseminations. The number of inseminations per pregnancy (AI index) to produce the F_1 generation was 2.38 (Table 2.6). The largest number was required for semen from Polish bulls but this may have arisen because semen from these bulls was used after July 1975 (following a TAC decision) on all the difficult breeders (i.e. cows inseminated more than two times). The F_1 generation was produced from 89.2 percent of semen doses from non-Polish bulls of the first batch and from 34.7 percent of the second batch of semen.

Table 2.4 **Average breeding value of bulls (other than Polish) used in the Polish Friesian project and of all contemporary bulls sampled in the country of origin (information provided by countries participating in the project)** (from Jasiorowski *et al.*,, in press)

Country	Method of sire evaluation	Traits	Breeding value of bulls used in Poland: Batch I bulls: No.of bulls	Av.	Batch II bulls: No.of bulls	Av.	All bulls: No.of bulls	Av.	Corresponding average for all progeny tested: Contemporaries of batch I: No.of bulls	Av.	Contemporaries of batch II: No.of bulls	Av.	All bulls: No.of bulls	Average
USA	MCC	PD milk kg					37	238					1 391	233
		PD fat kg						5.3						5.2
CND	BLUP	Milk BCA	17	+5.3	19	+5.3							63	4.7+5.1
		Fat BCA		+5.0		+5.6								3.3+4.8 in 1972
													82	3.0+3.9 entered 2.4+4.1 in 1973
													87	3.5+4.2 entered 3.0+5.1 in 1974
DK	HMC	R-figure	20	99.45	20	99.50	40	99	278	99	207	98	103	97 sampled 1977
													198	98 " 1978
													184	99 " 1979
UK	BLUP	ICC milk kg	19	+147	20	+58		+102						
		ICC fat kg		+7.1		+3.5	39	+5.3					256	+110.5 +5.5
S	BLUP	BV for FCM	18	98,39	22	98.23	40			97.14		97.20		
FRG	HMC and	BV Milk kg					38	+104						Av. for all
	BLUP	BV fat kg						0						breeding values in a year of sampl-
														ing assumed to be zero
NL	BLUP	BV milk kg	20	-1	20	+26								
		BV fat +												
		prot.per day		+2		+12								"
IL	MCC	PT 1974												
		Milk kg					38	124					92	153
		Fat kg						5.4						5.7
NZ	MCC	Breeding												
		Index	20	114	20	107			91	111	89	112		
		Milk		120		114				118		118		
		Fat												

Table 2.5 **Average relationship (as percentage) among bulls within strains (on diagonal) and between strains (above diagonal) and average inbreeding coefficient (as percentage) within strains (lower diagonal)** (from Stolzman *et al.*, in press)

	Paternal strain origin									
	USA	**CND**	**DK**	**UK**	**S**	**FRG**	**NL**	**IL**	**NZ**	**PL**
USA	3.288 0.380	0.761	0.000	0.003	0.000	0.446	0.000	1.554	0.000	0.000
CND		7.319 1.921	0.000	0.025	0.045	2.328	0.000	0.149	0.000	0.000
DK			4.099 0.160	0.000	0.013	0.149	0.228	0.000	0.000	0.141
UK				1.187 0.000	0.000	0.008	0.000	0.001	0.000	0.000
S					2.325 0.781	0.008	0.069	0.004	0.000	0.036
FRG						1.924 0.082	0.326	0.143	0.000	0.039
NL							4.660 1.074	0.000	0.000	0.197
IL								3.878 0.185	0.000	0.000
NZ									2.597 0.449	0.000
PL										1.014 0.664

Table 2.6 **Semen received and used in the project and results of artificial insemination, by strain**

Semen received	Use of semen to produce:											AI results			
Paternal strain origin	No. of semen doses			No. of bulls			F_1 generation			Back-cross gener.		F_1 generation[1]		Back-cross generation	
	I batch	II batch	Total	I batch	II batch	Total	I batch	II batch	Total	II batch	Total	No. of pregn. cows	AI index[2]	No. of pregn. cows	AI index[2]
USA	3 825	5 000	8 825	17	20	37	3 518	1 648	5 166	769	5 935	2 207	2.34	353	2.18
Canada	3 825	5 000	8 825	17	20	37	3 464	1742	5206	778	5984	2318	2.24	411	1.89
Denmark	4 275	5 000	9 275	19	20	39	4 050	1 714	5 764	826	6 590	2 459	2.34	414	2.00
United Kingdom	4 275	5 000	9 275	19	20	39	3 815	1 700	5 515	745	6 260	2 338	2.36	365	2.04
Sweden	4 320	5 500	9 820	18	22	40	4 040	1 765	5 805	852	6 657	2 394	2.43	423	2.01
Fed. Rep. of Germany	4 500	4 050	8 550	20	18	38	3 745	1 675	5 420	812	6 232	2 541	2.13	405	2.00
Netherlands	4 500	5 000	9 500	20	20	40	3 939	1 700	5 639	696	6 335	2 443	2.31	357	1.95
Israel	5 000	5 000	10 000	18	20	38	4 315	1 767	6 082	938	7 020	2 473	2.46	383	2.45
New Zealand	4 500	5 000	9 500	20	20	40	3 905	1 740	5 645	904	6 549	2 466	2.29	394	2.29
	39 020	44 550	83 570	168	180	348	34 791	15 451	50 242	7 320	57 562	21 639	2.32	3 505	2.09
Poland				20	20	40	7 986	4 981	12 967	5 608	18 575	4 885	2.67	2 524	2.22
				188	200	388	42 777	20 432	63 209	12 928	76 137	26 524	2.38	6 029	2.14

[1] For second and later calving only
[2] No. of inseminations per pregnancy

The insemination of F_1 heifers (first calving) and F_1 cows (after first calving) to produce the backcross generation took place from May 1976 to the end of March 1978. Two types of backcrosses were produced — F_1 females mated to bulls from each paternal strain (giving progeny with 75 percent blood of each of these strains) and to Polish bulls (giving progeny with 75 percent Polish blood) as seen in Fig. 2.3. Only semen from bulls belonging to the second batch was used to produce backcrosses. The heifers were inseminated for the first time at the age of at least 15 months and 360 kg of liveweight. In total 7 326 F_1 females were inseminated (of which 5 700 were heifers for their first calving). The pregnancy rate was 82.3 per 100 inseminations. Each pregnancy required on average 2.14 inseminations to produce the R_1 generation (Table 2.6).

To increase the number of backcross heifers, 474 such animals born in 1979 (after the end of the originally intended period) were included as pregnant animals. For this reason the experiment ran one year longer than intended. The inseminations of F_1 and R_1 females took place on 34 units for raising heifers — six of these units were very large, accommodating between 650 and 850 animals each.

In total 69 percent of semen doses sent to Poland by the participating countries was used to produce the F_1 and backcross progeny (Table 2.6). The best farms included in the project used the rest of the semen.

Table 2.7 **Involuntary losses of F_1 males from birth to 12 months of age, by strain**

Paternal strain origin	Percentage of losses per age class				Total number lost
	Birth to 30 days	31 to 90 days	91 to 180 days	181 to 365 days	
USA	41	32	11	16	195
Poland	41	25	17	17	419
Canada	36	26	19	19	170
Denmark	35	27	22	16	248
United Kingdom	36	28	17	19	198
Sweden	41	22	15	22	175
Federal Republic of Germany	40	23	20	17	231
Netherlands	38	23	17	22	246
Israel	40	24	20	16	221
New Zealand	38	27	19	16	209
Total	894	591	411	416	2 312
	39%	25%	18%	18%	100%

2.4 RESULTS

2.4.1 Losses of animals (field trial)

a) F_1 generation

From 1974 to 1977 a total of 24 846 calves were born alive and survived at least 24 hours. Ninety-four percent of diagnosed pregnancies resulted in live calves indicating that abortions, stillbirths and prenatal mortality accounted for 6 percent. The proportion of male calves at birth was 51.6 percent. All calves were to be reared without any selection. Later losses resulted from animals dying or being culled by the veterinarians for serious illnesses, accidents and major defects or failure of prescribed disease tests. In the F_1 generation the involuntary losses amounted to 5 538 animals.

In males, the rate of loss in the rearing and fattening period up to 12 months of age was 18 percent. A breakdown by age at time of loss is given in Table 2.7. The highest rate of loss was in the first month of age — 39 percent of all losses. The main reasons for death or for culling by the veterinarians were respiratory diseases, diarrhoea and cachexia caused by these diseases.

Losses of F_1 heifers (Table 2.8) up to one year of age (15.1 percent) were lower than in males; 24 percent of all those lost died in the first month of life; above 18 months of age and before first calving, 1 285 F_1 heifers were lost (10.7 percent of all heifers born alive and 39.8 percent of the losses of F_1 females).

There were no major differences among strains either in the proportion of each sex lost-in relation to the numbers born — or in the age distribution of losses.

In Table 2.9 the reasons are given for culling F_1 heifers over 18 months old. Fifty-one percent were culled for infertility and 35.6 percent because of general underdevelopment.

b) Backcross generation (R_1)

In the years 1977-78 (see Table 2.6) 5 353 backcross calves were born out of the 6 029 pregnant F_1 females (89 percent). The sex ratio was 50.5 percent males and 49.5 percent females. Apart from abortions, stillbirths and prenatal mortality some pregnancies resulted from inseminations with semen other than from a bull of a designated strain. Such progeny were eliminated from the experiment. Total involuntary losses in the backcross generation were 1 629 animals.

Twenty-nine percent of the males did not finish the experimental period, 11 percent higher than in F_1 bulls. Details are shown in Table 2.10. Proportions lost in each age group were similar to the F_1 generation.

Table 2.8 **Involuntary losses of F_1 females from birth to 18 months of age, by strain**

Paternal strain origin	Percentage of losses per age class						Total number lost
	Birth to 30 days	31 to 90 days	91 to 180 days	181 to 365 days	366 to 540 days	above 540 days	
USA	26	19	8	7	5	35	255
Poland	28	16	5	5	4	42	591
Canada	23	18	12	6	5	36	261
Denmark	27	16	10	7	3	37	323
United Kingdom	25	15	8	8	4	39	309
Sweden	24	18	7	5	5	41	279
Fed. Rep. of Germany	25	18	8	6	3	40	315
Netherlands	24	14	12	6	3	41	306
Israel	16	18	10	9	3	44	292
New Zealand	23	16	10	7	4	40	295
Total	790	543	280	205	12	1 285	3 226
	24%	17%	9%	6%	4%	40%	100%

Table 2.9 **Number of F_1 heifers culled for different reasons at over 18 months of age and before calving, by strain**

Reasons for culling	Paternal strain origin										Total
	USA	CND	DK	UK	S	FRG	NL	IL	NZ	100% Poland	
Infections	3	3	5	3	4	1	1	0	0	2	22
Injuries, foreign bodies, bloat	0	1	1	0	3	2	1	0	1	1	10
Cachexia	26	27	42	44	28	38	52	52	43	100	452
Sexual organs underdeveloped	12	8	5	9	8	10	11	9	11	16	99
Infertility	39	41	56	49	61	51	45	54	52	108	556
Abortion	2	4	3	2	0	7	8	2	2	5	35
Non-identified reasons	8	9	7	15	11	15	9	12	8	17	111
Total	90	93	119	122	115	124	127	129	117	249	1 285

Losses of backcross females up to one year of age were 21 percent of those born alive — higher than in the case of F_1 females. As shown in Table 2.11, 26 percent of all R_1 heifers which died did so in the first month of life. From 18 months to first calving, 273 R_1 heifers were culled-10.3 percent of all females born alive and 32.0 percent of all heifer losses in this generation (Table 2.12). Culling for infertility, underdevelopment of sexual organs before first service, or failure to conceive accounted for 54 percent of all heifers lost. Cachexia was also a major reason for culling R_1 heifers above 18 months old.

As in the F_1 generation, the results of the backcross males and females provide no evidence that losses were significantly higher or lower than average in any strain.

2.4.2 Blood typing of progeny in the field trial

In the years 1975-79 the total number of F_1 and backcross progeny blood typed was 16 096, i.e. 53 percent of all animals born alive in both generations. Nearly all female progeny were blood typed, but only 20 percent of males.

Table 2.10 **Involuntary losses of backcross males by age classes and strain**

Paternal strain origin	Percentage of losses per age class				Total number lost
	Birth to 30 days	31 to 90 days	91 to 180 days	181 to 365 days	
USA	50	18	14	18	50
Canada	33	30	17	20	54
Denmark	41	25	18	16	51
United Kingdom	41	30	9	20	44
Sweden	37	31	13	19	54
Fed. Rep. of Germany	40	25	19	16	48
Netherlands	55	19	19	7	53
Israel	37	35	11	17	54
New Zealand	39	27	11	23	44
Poland (R_1)	38	35	15	12	223
Pure Polish	41	27	10	22	101
Total - number	312	227	109	128	776
- percent	40	29	14	17	100

Table 2.11 **Involuntary losses of backcross females by age classes and strain**

Paternal strain origin	Percentage of losses per age class						Total number lost
	Birth to 30 days	31 to 90 days	91 to 180 days	181 to 365 days	366 to 540 days	above 540 days	
USA	7	34	10	7	7	35	41
Canada	22	37	10	12	2	17	51
Fed. Rep. of Germany	34	15	10	7	0	34	59
United Kingdom	23	20	14	0	4	39	44
Sweden	18	14	5	2	0	61	44
Denmark	22	22	11	7	4	35	55
Netherlands	35	27	0	11	0	27	44
Israel	21	23	21	8	2	25	39
New Zealand	29	17	11	6	2	35	63
Poland (R_1)	26	27	79	5	3	30	292
Pure Polish	33	19	10	8	2	28	121
Total - number	220	204	79	55	22	273	853
- percent	26	24	9	6	3	32	100

Table 2.12 **Number of backcross heifers culled for different reasons at an age above 18 months and before calving, by strain**

Reason for culling	Paternal strain origin											Total
	USA	CND	DK	UK	S	D	NL	IL	NZ	PL (R_1)	100% PL	
Injuries, foreign bodies, bloat	0	0	1	0	0	1	0	1	2	1	0	6
Cachexia	4	4	7	7	12	8	4	3	7	39	11	106
Sexual organs underdeveloped	0	1	1	0	3	0	1	0	3	1	3	13
Infertility	10	3	8	8	12	10	7	6	9	45	15	133
Abortion	0	0	0	1	0	0	0	0	0	1	1	3
Non-identified reasons	0	1	3	1	0	0	0	0	1	2	4	12
Total	14	9	20	17	27	19	12	10	22	89	34	273

The error rate in parentage detected as a result of blood typing was 4.8 percent arising from the large number of bulls used in the project and a procedure of semen distribution in which each cow returning to service was mated to a different bull. Nonetheless this rate is regarded as relatively low. Numbers excluded are shown in Table 2.13.

2.4.3 Liveweight, daily gains and body measurements of F_1 and R_1 males and females in the field trial

The numbers of animals, overall averages and standard deviations are given in Tables 2.14 to 2.19. The results of strain comparisons are shown in Figures 2.4 to 2.20. Each figure shows both generations (F_1 and R_1) and the ten strains. For the Polish strain, instead of F_1 values the results are for pure Polish Friesians and for the R_1 generation values are for pure Polish and for backcrosses to Polish bulls. Actual values from the linear mixed model solutions with standard errors are presented in Tables 2.20 to 2.29 along with approximate values for least significant differences between strains.

a) Liveweight and daily gains

The overall average weight of F_1 male calves was 37.3 kg at birth and 290 kg at one year of age (Table 2.14). Females of the F_1 generation weighed 35.5 kg at birth, 272 kg at 12 months and 369 kg at 18 months (Table 2.15). For both sexes and all weights, variation among strains was highly significant (P   0.01) with the Holstein strains — USA, Canadian and Israeli — generally above average in weight and animals of the Dutch and Polish strains usually the lightest (Figs. 2.4 to

Table 2.13 **Number of progeny excluded from the experiment after blood typing due to errors in recorded parentage**

Generation	Males	Females	Total
First crosses (F_1)	171	376	547
Backcrosses (R_1)	43	190	233
Total	214	566	780

2.8). From birth to 12 months old, the average daily gain of F_1 males was 711 g (Table 2.14) whilst females from birth gained 682 g daily to 12 months and 622 g daily to 18 months (Table 2.15). Strain variation was again highly significant ($P < 0.01$) with, not unexpectedly, the heavier (Holstein) strains making the greater gains — this is not, however, uniformly the case since some strains, for example the Israeli, started at no more than an average birth weight but became relatively heavy later in life (Figs. 2.9 to 2.14).

Table 2.14 **Liveweights and daily gains of F_1 males (overall mean and s.d.) for ten strains in the field trial** (after Stolzman *et al.*, in press)

	Liveweights			Daily gains		
Paternal strain origin	**No.**	**Birth** *(kg)*	**12 m** *(kg)*	**No.**	**6 to 12 m** *(g)*	**Birth to 12 m** *(g)*
USA	747	38.6 5.5	294.2 40.7	746	723 172	719 110
CND	829	38.4 5.5	293.5 39.6	829	712 172	717 111
DK	802	37.8 5.3	287.2 40.4	801	703 172	703 112
UK	760	37.1 5.0	289.6 40.1	760	708 169	710 111
S	786	37.3 5.3	291.3 39.6	786	714 167	713 111
FRG	812	37.8 5.2	289.3 41.0	812	710 172	708 114
NL	756	36.8 4.8	283.0 40.9	756	683 170	692 115
IL	782	37.4 5.7	293.5 41.3	782	709 174	721 115
NZ	762	36.9 5.4	290.8 39.9	762	718 175	714 109
Pure Polish	1 516	36.2 5.1	289.5 40.3	1 514	708 172	711 111
Total	8 552	37.3 5.3	290.1 40.5	8 548	709 172	711 112

For the R_1 generation, males weighed 34.5 kg at birth and 273 kg at 12 months, gaining 672 g daily over that period (Table 2.16). For females, correspon-ding weights were 32.8 kg at birth, 266 kg at 12 months and 359 kg at 18 months with daily gains from birth being 643 g and 598 g respectively (Table 2.17). As before, females weighed and gained less than males. Strain rankings were not dissimilar for the R_1 and F_1 generations but, perhaps because of the much smaller number of animals involved in the backcrosses, variation in daily gain was

Table 2.15 **Liveweights and daily gains of F_1 females (overall mean and s.d.) for ten strains in the field trial** (after Stolzman *et al.*, in press)

Paternal strain origin	No.	Liveweights: Birth (kg)	Liveweights: 12 m of age (kg)	Liveweights: 18 m of age (kg)	No.	Daily gains: 6 to 12 m (g)	Daily gains: 12 to 18 m (g)	Daily gains: Birth to 12 m (g)	Daily gains: Birth to 18 m (g)
USA	592	36.5	272.9	372.8	569	666	539	670	625
		4.9	31.0	38.1		122	147	84	68
CND	640	36.9	277.5	375.6	618	679	530	684	631
		5.0	29.6	38.0		124	146	83	68
DK	664	35.9	270.9	370.2	643	663	539	670	625
		4.5	29.3	36.1		122	143	82	67
UK	663	35.1	267.4	363.2	652	652	519	661	612
		4.3	29.2	37.0		122	145	82	68
S	692	35.3	270.8	368.3	679	651	532	668	621
		4.7	30.2	36.9		135	137	84	67
FRG	677	35.9	275.2	369.6	667	677	513	681	623
		4.8	29.0	37.1		119	137	79	67
NL	616	34.6	265.7	361.5	603	645	520	656	609
		4.4	30.7	36.9		128	136	85	67
IL	664	35.6	275.9	372.9	654	666	530	682	630
		4.9	28.9	36.1		124	143	81	66
NZ	698	35.1	272.5	368.5	688	665	524	673	622
		4.7	27.9	32.4		118	137	75	59
Pure Polish	1 127	34.6	272.8	367.1	1 105	667	515	674	620
		4.4	29.3	36.0		131	146	83	66
Total	7 033	35.5	272.2	368.8	6 878	664	525	672	622
		4.7	29.7	36.6		126	142	82	66

not significant for males but reached statistical significance for females ($P < 0.05 - 0.01$, see Tables 2.20 to 2.25). In the R_1 generation, as for the F_1, Holstein types were the heaviest with German Friesians following closely later in life. The other Friesian strains, especially the Polish and Dutch, were lighter. However, the relative positions of the strains varied somewhat from age to age as seen from Figs. 2.4 to 2.14.

Table 2.16 **Liveweights and daily gains of backcross (R_1) males (overall mean and s.d.) for ten strains in the field trial** (after Stolzman *et al.*, in press)

Paternal strain origin	No.	Liveweight birth *(kg)*	Liveweight 12 m *(kg)*	Daily gain 6 to 12 m *(g)*	Daily gain birth to 1 *(g)*
USA	90	35.5 4.6	270.8 42.4	662 185	665 117
CND	99	35.5 4.2	276.3 43.1	679 173	679 119
DK	98	35.4 5.2	273.1 46.9	654 198	668 132
UK	101	34.0 4.7	267.4 41.4	652 191	658 116
S	101	35.4 4.7	272.6 39.8	649 165	670 113
FRG	100	34.0 3.9	281.9 42.0	706 191	701 120
NL	95	34.4 4.3	264.6 48.4	635 226	649 138
IL	112	35.3 4.9	289.4 45.9	679 199	715 128
NZ	92	35.2 4.1	272.6 40.3	672 177	673 116
PL (R_1)	534	33.9 4.4	272.2 42.2	663 174	670 118
Pure Polish	196	34.0 4.2	269.7 45.4	634 193	659 126
Total	1 618	34.5 4.5	273.2 43.7	661 186	672 122

Table 2.17 **Liveweights and daily gains of backcross (R_1) females (overall mean and s.d.) for ten strains in the field trial** (after Stolzman *et al.*, in press)

Paternal strain origin	No.	Liveweight: birth *(kg)*	Liveweight: 12 m *(kg)*	Liveweight: 18 m *(kg)*	Daily gains: 6 to 12 m *(g)*	Daily gains: 12 to 18 m *(g)*	Daily gains: birth to 12 m *(g)*	Daily gains: birth to 18 m *(g)*
USA	91	32.9	265.5	360.2	620	515	644	600
		3.4	30.6	42.6	149	178	84	77
Canada	132	33.0	267.5	366.0	634	540	645	611
		4.0	30.2	45.9	126	189	82	85
Denmark	108	33.2	260.5	356.9	604	527	628	594
		4.0	27.5	43.7	125	160	74	77
UK	103	32.7	264.8	348.3	619	455	638	577
		3.9	29.4	44.3	135	171	81	81
Sweden	141	33.3	269.2	365.1	636	524	649	607
		3.8	31.4	45.3	136	162	86	83
Federal Republic of Germany	135	32.9	268.1	361.4	623	511	649	602
		4.2	27.6	43.1	132	160	77	79
Netherlands	98	32.5	262.7	351.6	624	485	637	586
		3.5	32.5	42.4	134	150	89	80
Israel	113	33.2	275.6	378.5	646	562	668	633
		4.2	30.7	44.7	138	186	86	84
New Zealand	160	33.3	266.8	362.7	636	524	645	604
		3.9	28.2	40.0	124	156	76	72
Poland (R_1)	461	32.6	262.4	351.0	598	484	632	583
		4.4	28.6	38.0	133	156	77	69
Pure Polish	254	32.5	269.3	362.3	629	508	652	604
		4.5	28.9	41.1	122	167	78	75
Total	1796	32.8	266.1	359.1	620	512	643	598
		4.1	29.6	42.6	132	167	81	78

Table 2.18 **Body measurements of F_1 males and females (overall mean and s.d.) for ten strains in the field trial** (after Stolzman *et al.*, in press)

		F_1 males			F_1 females			
		12 m of age			12 m of age		18 m of age	
Paternal strain origin	**No. of animals**	**Height at withers** *(cm)*	**Chest girth** *(cm)*	**No. of animals**	**Height at withers** *(cm)*	**Chest girth** *(cm)*	**Height at withers** *(cm)*	**Chest girth** *(cm)*
USA	756	112.8	155.3	539	112.2	154.3	120.5	169.0
		5.1	8.9		4.6	7.1	5.3	8.5
CND	824	112.7	154.6	589	112.3	154.2	121.0	169.5
		5.1	9.2		4.3	6.6	4.7	7.9
DK	804	111.5	154.5	613	110.4	154.0	119.0	169.5
		5.5	9.3		4.6	6.7	5.0	7.8
UK	770	111.9	154.1	590	110.5	153.4	118.6	168.1
		5.3	9.0		4.3	6.3	4.3	7.6
S	804	111.9	154.4	637	110.6	153.4	119.0	168.3
		5.1	9.3		3.9	6.8	4.9	7.1
FRG	807	111.9	154.2	600	111.5	154.3	119.7	169.1
		4.6	8.7		4.9	6.1	4.7	7.6
NL	744	111.7	153.4	566	110.8	153.9	118.6	168.3
		5.9	9.1		3.9	6.8	4.6	7.7
IL	785	112.2	155.2	625	111.9	154.0	120.1	168.6
		5.2	9.6		4.0	7.6	4.9	7.6
NZ	771	112.1	155.1	663	111.0	153.2	119.5	168.3
		5.3	8.8		4.1	6.9	5.3	8.0
Pure	1 574	110.9	153.8	1 045	109.8	153.6	117.9	167.6
Polish		5.4	9.1		4.1	6.6	5.1	7.8
Total	8 639	111.9	154.4	6 467	111.0	153.8	119.3	168.6
		5.3	9.1		4.3	6.8	5.0	7.8

Table 2.19 **Body measurements of backcross males and females (overall mean and s.d.) for ten strains in the field trial** (after Stolzman *et al.*, in press)

		R_1 males			R_1 females			
		12 m of age			12 m of age		18 m of age	
Paternal strain origin	**No. of animals**	**Height at withers** *(cm)*	**Chest girth** *(cm)*	**No. of animals**	**Height at withers** *(cm)*	**Chest girth** *(cm)*	**Height at withers** *(cm)*	**Chest girth** *(cm)*
USA	67	112.3	155.0	72	111.0	151.5	120.5	163.3
		5.4	6.2		4.1	5.7	4.2	6.1
CND	83	111.9	154.1	108	110.4	152.9	120.8	165.7
		4.9	7.5		4.6	5.6	4.7	6.3
DK	86	112.0	155.8	97	110.2	150.5	119.2	164.8
		4.8	7.0		3.6	5.8	4.7	7.2
UK	89	111.5	153.3	75	109.5	152.8	118.6	165.4
		4.8	8.2		4.2	5.7	4.5	7.1
S	87	111.6	155.9	113	110.4	153.0	120.0	164.7
		5.1	8.5		4.5	5.3	4.4	7.5
FRG	86	113.0	154.7	102	109.7	152.2	119.8	165.4
		4.9	7.1		4.9	5.8	4.9	7.3
NL	85	111.0	153.9	89	108.6	152.4	118.6	164.3
		4.3	7.4		4.2	5.7	4.2	7.2
IL	99	111.9	155.2	98	110.9	152.2	121.3	166.6
		5.4	8.6		4.0	5.4	5.2	7.9
NZ	82	111.4	154.4	145	109.7	152.1	119.3	165.0
		4.7	6.7		3.6	5.4	4.9	7.5
PL (R_1)	484	111.6	152.1	378	109.3	152.9	118.3	166.2
		4.6	9.4		3.9	5.9	3.8	6.5
Pure Polish	170	110.9	154.0	220	109.1	152.1	118.5	166.2
		5.5	8.1		4.9	5.7	5.4	7.5
Total	1 418	111.6	153.7	1 497	109.7	152.3	119.2	165.5
		4.9	8.4		4.3	5.7	4.7	7.2

Fig. 2.4 Birthweight of males in the field trial showing each strain as a percentage deviation from the overall mean (approx. s.e. $F_1$0.5 $R_1$1.1)

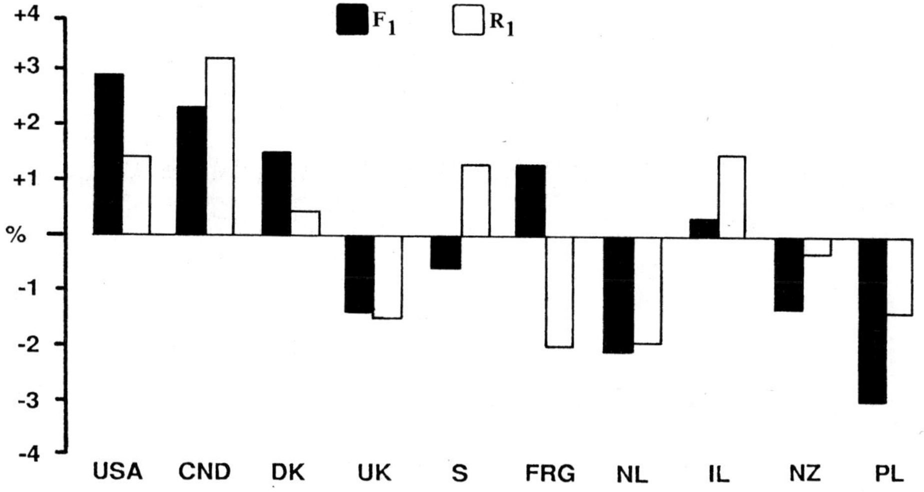

Fig. 2.5 Birthweight of females in the field trial showing each strain as a percentage deviation from the overall mean (approx. s.e. $F_1$0.5 $R_1$1.0)

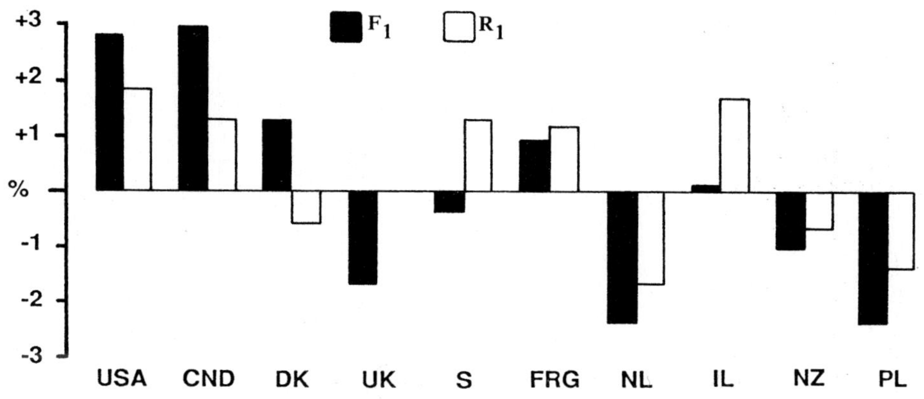

Fig. 2.6 Liveweight of 12–month–old males in the field trial showing each strain as a percentage deviation from the overall mean (approx. s.e. $F_1$0.4 $R_1$0.5)

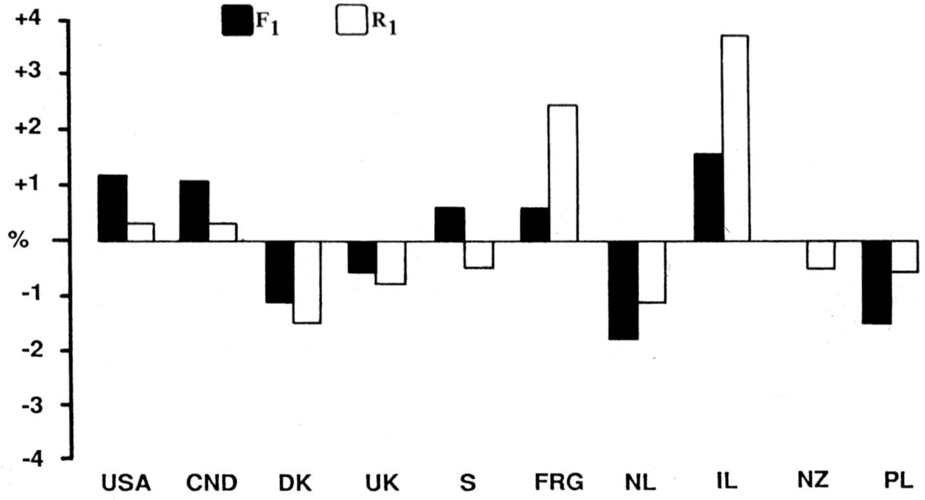

Fig. 2.7 Liveweight of 12–month–old females in the field trial showing each strain as a percentage deviation from the overall mean (approx. s.e. $F_1$0.3 $R_1$0.5)

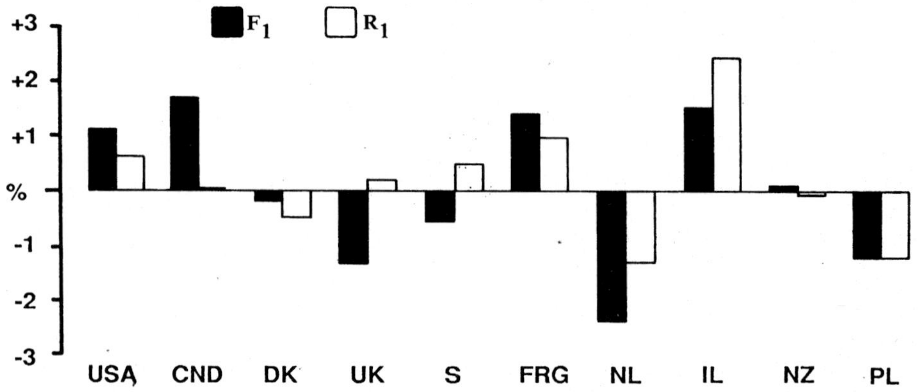

Fig. 2.8 Liveweight of 18–month–old females in the field trial showing each strain as a percentage deviation from the overall mean (approx. s.e. $F_1$0.3 $R_1$0.4)

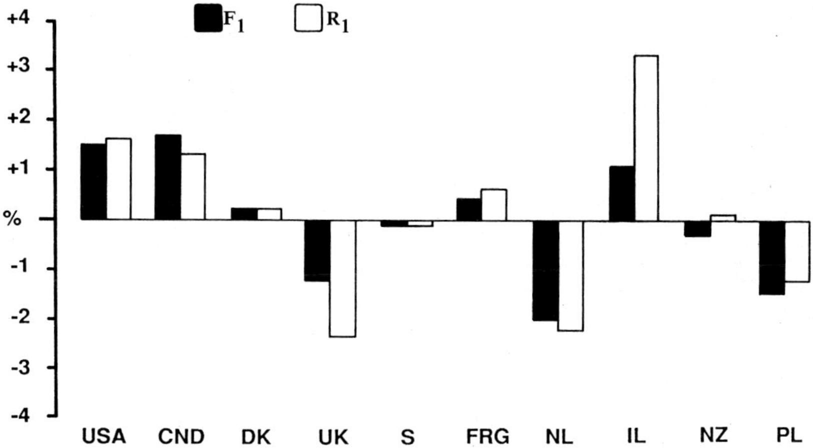

Fig. 2.9 Daily gain in weight of males from 6 to 12 months old in the field trial showing each strain as a percentage deviation from the overall mean (approx s.e. $F_1$0.7 $R_1$2.3)

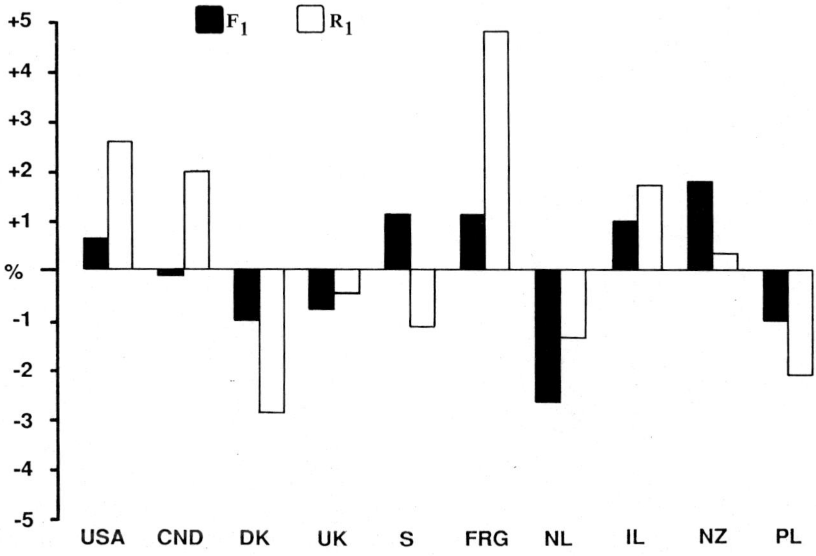

Fig. 2.10 Daily gain in weight of females from 6 to 12 months old in the field trial showing each strain as a percentage deviation from the overall mean (approx. s.e. $F_1$0.7 $R_1$1.9)

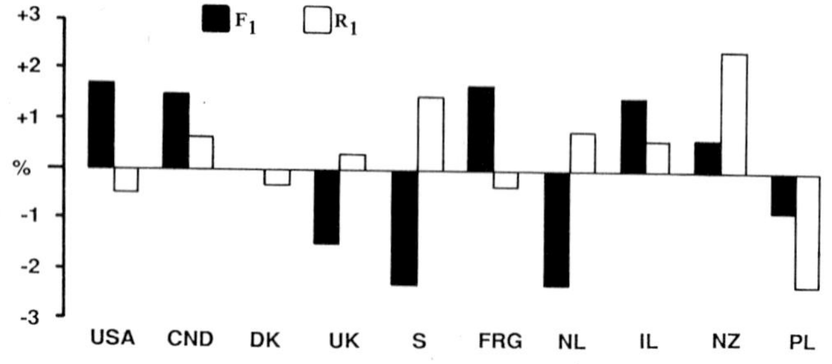

Fig. 2.11 Daily gain in weight of males from birth to 12 months old in the field trial showing each strain as a percentage deviation from the overall mean (approx. s.e $F_1$0.5 $R_1$1.4)

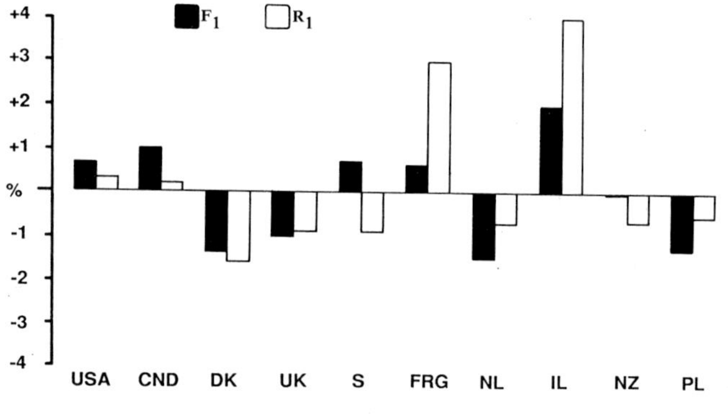

Fig. 2.12 Daily gain in weight of females from birth to 12 months old in the field trial showing each strain as a percentage deviation from the overall mean (approx. s.e. $F_1$0.4 $R_1$1.1)

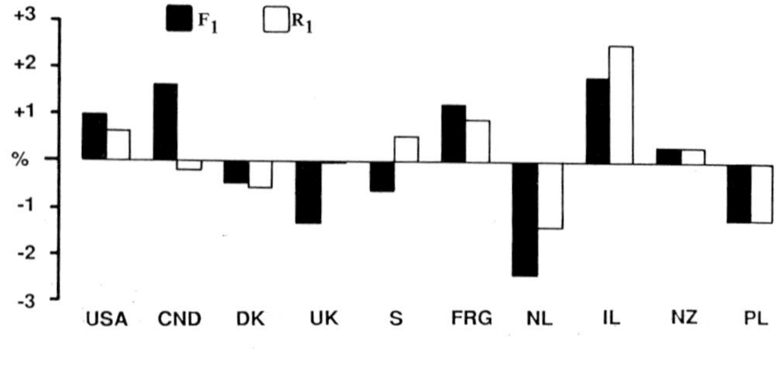

Fig. 2.13 Daily gain in weight of females from 12 to 18 months old in the field trial showing each strain as a percentage deviation from the overall mean (approx. s.e. $F_1$0.9 $R_1$2.5)

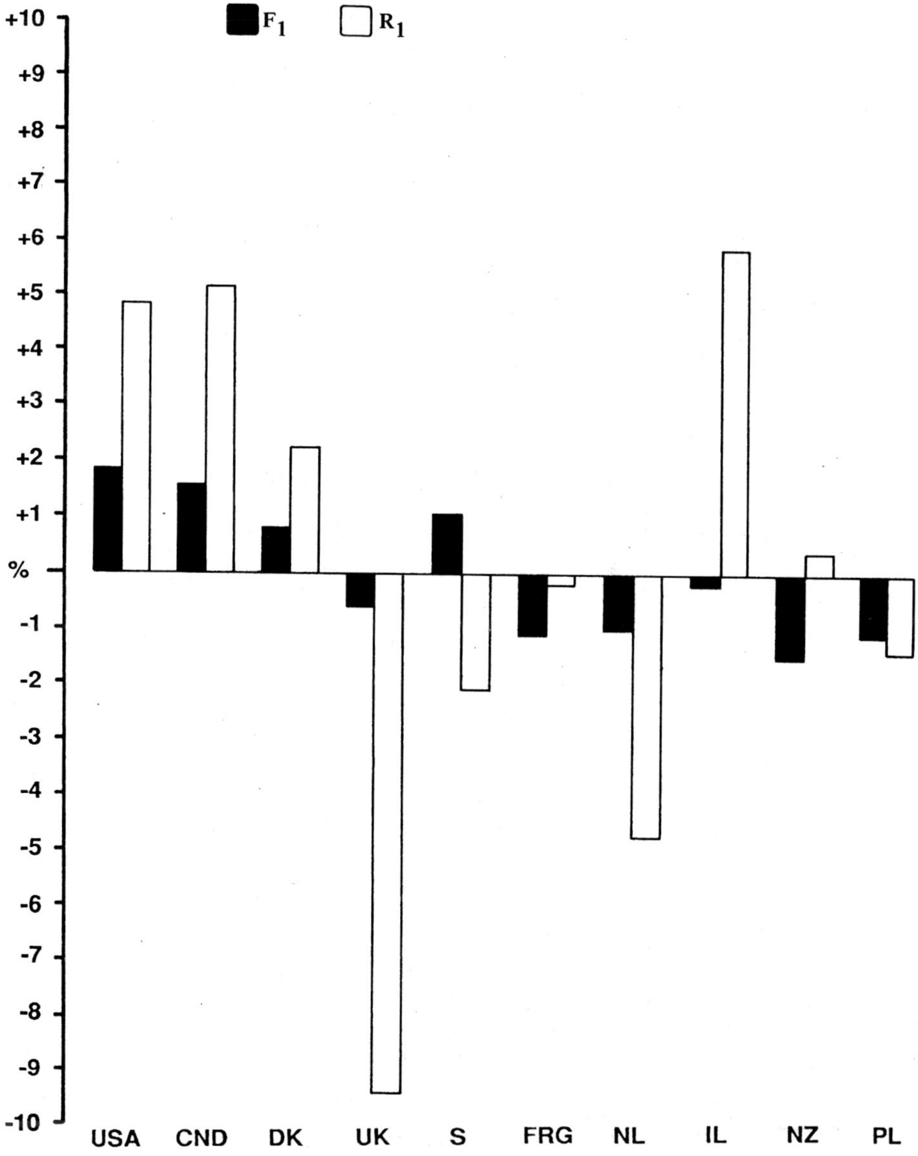

Fig. 2.14 Daily gain in weight of females from birth to 18 months old in the field trial showing each strain as a percentage deviation from the overall mean (approx. s.e. F_1 0.3 R_1 1.0)

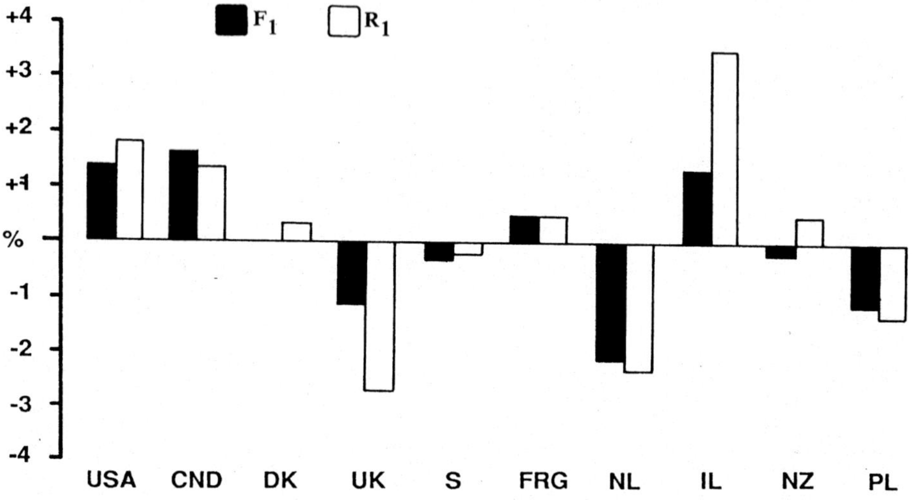

Heterosis estimates. Heterosis was estimated for liveweight and weight gains. The effects, in general, were smaller than the standard errors attaching to the estimates. Only in the case of birth weight were estimates of the effect of heterosis higher. Those involving crosses with strains from the USA, Canada, Denmark and the Federal Republic of Germany were of the order of 5 to 6 percent($\pm$3.1 percent) of the Polish mean for female calves, and 5 to 9 percent ($\pm$3.5 percent) for male calves of these strains and those of the United Kingdom and the Netherlands.

b) Body dimensions

The effect of strain on height at withers of 12-month-old bulls shown in Tables 2.18 and 2.19 was highly significant only in the F_1 progeny. As seen from Figure 2.15 Holsteins from the USA, Canada and Israel were the tallest. Swedish and New Zealand bulls were slightly above the average. For heifers (Tables 2.18 and 2.19) the influence of strain was found to be important both in the F_1 generation ($P < 0.01$) and in the backcrosses ($P < 0.05$) (see Tables 2.26 to 2.29). The ranking of strains at 12 months of age was broadly similar in both generations with the three Holstein strains and the German strain most often

Table 2.20 **Liveweight (kg) and daily gain (g) of first cross (F_1) males in the field trial — the effects of paternal strains as deviations (±s.e.) from the mean** (after Stolzman *et al.*, in press)

Paternal strain origin	Liveweight				Daily gain			
	At birth**		12 m**		6-12m		Birth-12 m**	
	Effect	s.e.	Effect	s.e.	Effect	s.e.	Effect	s.e.
USA	1.12	0.18	3.38	1.22	4	5.5	4	3.4
Canada	0.85	0.17	3.09	1.16	-1	5.2	7	3.3
Denmark	0.58	0.18	-3.08	1.20	-7	5.5	-10	3.4
United Kingdom	-0.51	0.18	-1.88	1.19	-6	5.4	-4	3.3
Sweden	-0.22	0.17	1.53	1.19	8	5.4	5	3.3
Fed. Rep. of Germany	0.50	0.17	1.78	1.18	8	5.3	4	3.3
Netherlands	-0.81	0.18	-5.20	1.20	-19	5.4	-11	3.4
Israel	0.13	0.18	4.52	1.23	7	5.6	14	3.5
New Zealand	-0.50	0.18	0.09	1.20	13	5.5	0	3.4
100% Polish	-1.14	0.14	-4.23	0.96	-7	4.3	-9	2.7

Significance of variation among strains: ** ($P < 0.01$)
[1] Estimated by using linear mixed model.
Approximate least significant differences[2]

$P < 0.01$	0.67	4.19		13
$P < 0.05$	0.52	3.24		10

[2] In view of the large number of potential comparisons between any two strains, the least significant differences shown at the 0.01 level of probability are more appropriately regarded as those which would be equalled or exceeded by chance (between two strains) less than 1 time in 20.

above average (Figs. 2.15 and 2.16). At 18 months of age the tallest strains with the greatest height at withers were again the USA, Canadian, Israeli and German in both F_1 and R_1 heifers. Among the European strains the pure Polish and F_1 Dutch females were the smallest (Fig. 2.17).

For chest girth there was an influence of strain only for animals of the F_1 generation (Tables 2.26 to 2.29); however, the differences among strains were small (Figs. 2.18, 2.19, 2.20). In heifers, the ranking of strains was the same at both ages. In general, Holstein strains were above average in girth just as they had been in height.

Fig. 2.15 Height at withers of 12–month–old males in the field trial showing each strain as a percentage deviation from the overall mean (approx. s.e. $F_1$0.2 $R_1$0.4)

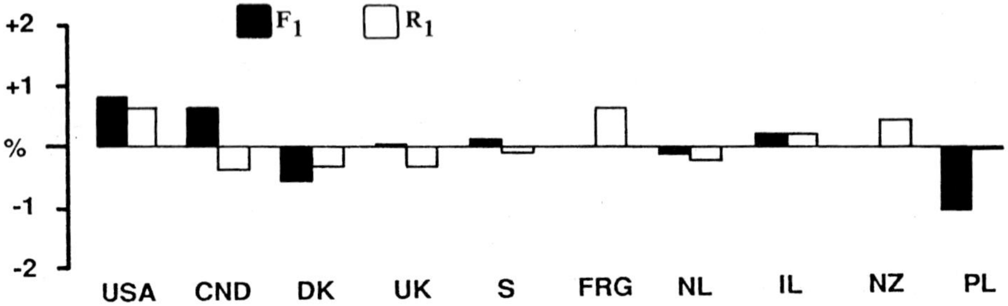

Fig. 2.16 Height at withers of 12–month–old females in the field trial showing each strain as a percentage deviation from the overall mean (approx. s.e. $F_1$0.1 $R_1$0.3)

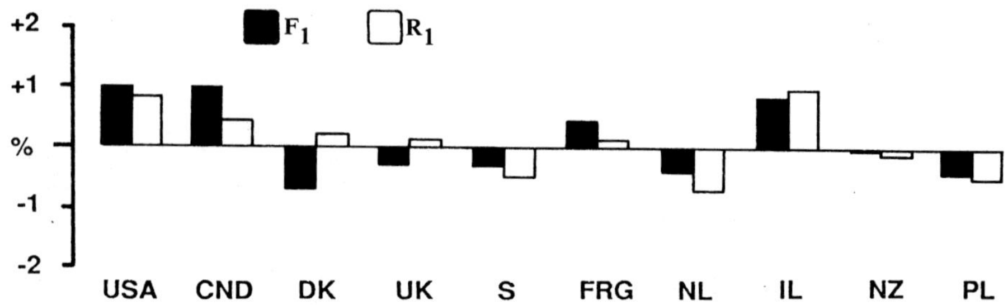

Fig. 2.17 Height at withers of 18–month–old females in the field trial showing each strain as a percentage deviation from the overall mean (approx. s.e. $F_1$0.1 $R_1$0.3)

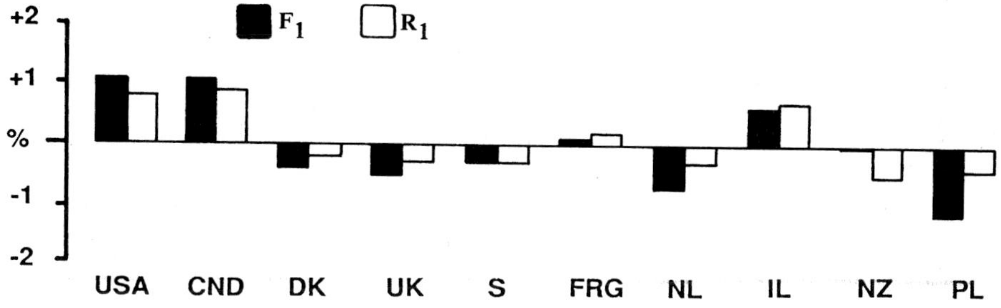

Fig. 2.18 Chest girth of 12–month–old males in the field trial showing each strain as a percentage deviation from the overall mean (approx. s.e. $F_1$0.2 $R_1$0.4)

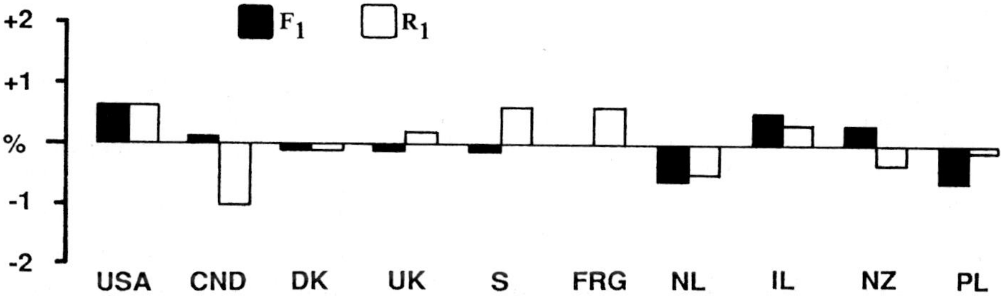

Fig. 2.19 Chest girth of 12–month–old females in the field trial showing each strain as a percentage deviation from the overall mean (approx. s.e. $F_1$0.2 $R_1$0.3)

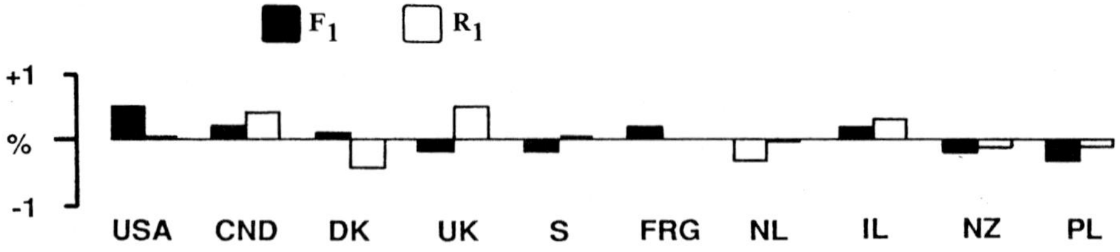

Fig. 2.20 Chest girth of 18–month–old females in the field trial showing each strain as a percentage deviation from the overall mean (approx. s.e. $F_1$0.1 $R_1$0.3)

2.4.4 Results on liveweight, feed intake and carcass characteristics of F_1 and R_1 bulls in the intensive-feeding beef performance trial

In order to evaluate whether strains would differ in the efficiency of growth and in carcass characteristics at different slaughter weights, two slaughter weights were chosen for the F_1 animals in the intensive trial — 450 kg and 550 kg. There were, as expected, clear average differences between these two groups and, in particular, at the higher slaughter weight the animals were fatter (19.4 percent vs. 17 percent), had a higher dressing percentage (58.5 vs. 56.9) and a larger MLD area (79.8 cm² vs. 70.0 cm²) and correspondingly lesser proportions of lean meat and valuable cuts. Efficiency of growth, in terms of energy and protein consumed per unit of gain in liveweight, was lower for animals taken to the higher slaughter weight but the average daily gain of lean weight did not differ between the two slaughter groups. However, there was no overall evidence that the strains differed relative to each other at these two slaughter weights, as judged by the absence of a strain × treatment interaction, although there were some specific differences which, in view of the relatively small number per class (approximately 14), are difficult to interpret. Accordingly, since the main purpose of the trial is to estimate the importance of strain variations, the results from the two slaughter groups have been pooled in order to test strain differences with the larger total numbers then involved. Backcross (R_1) animals were slaughtered at 450 kg alone. It is recognized that this reduces the validity of the F_1 vs. R_1 comparison for the purpose of estimating heterosis. For that reason, separate results will be quoted where appropriate. Further details of the results from the two slaughter ages are given by Reklewski *et al.* (1985).

Table 2.21 **Liveweight (kg) and daily gain (g) of backcross (R_1) males in the field trial — the effects' of paternal strains as deviations (±s.e.) from the mean** (after Stolzman *et al.*, in press)

Paternal strain origin	Liveweight				Daily gain			
	At birth		12 m		6-12m		Birth-12 m	
	Effect	s.e.	Effect	s.e.	Effect	s.e.	Effect	s.e.
USA	0.49	0.44	0.88	1.38	17	17.5	2	10.9
Canada	1.12	0.41	0.88	1.35	13	16.3	1	10.2
Denmark	0.13	0.42	-4.07	1.36	-19	16.5	-11	10.3
United Kingdom	-0.51	0.43	-2.09	1.36	-3	17.1	-6	10.6
Sweden	0.70	0.41	-1.40	1.35	-8	16.3	-6	10.1
Fed. Rep. of Germany	-0.68	0.41	6.76	1.35	32	16.3	20	10.1
Netherlands	-0.65	0.44	-3.03	1.37	-9	17.3	-5	10.8
Israel	0.51	0.41	10.06	1.35	11	16.4	27	10.2
New Zealand	-0.11	0.41	-1.31	1.35	2	15.8	-4	9.8
Poland (R_1)	-0.45	0.30	-1.58	1.25	-14	11.7	-4	7.3
100% Polish	-0.54	0.31	-5.11	1.26	-22	12.3	-14	7.7

[1] See footnote Table 2.20

Table 2.22 **Liveweight (kg) of first cross (F_1) females in the field trial — the effects' of paternal strains as deviations (±s.e.) from the mean** (after Stolzman *et al.*, in press)

Paternal strain origin	At birth**		12 m**		18 m**	
	Effect	s.e.	Effect	s.e.	Effect	s.e.
USA	0.92	0.18	3.25	1.10	5.67	1.22
Canada	1.06	0.17	4.60	0.99	6.25	1.20
Denmark	0.45	0.17	-0.52	0.77	0.83	1.20
United Kingdom	-0.59	0.17	-3.58	0.96	-4.33	1.15
Sweden	-0.17	0.17	-1.55	0.95	-0.35	1.15
Fed. Rep. of Germany	0.31	0.17	3.11	0.94	1.64	1.14
Netherlands	-0.84	0.17	-6.50	0.98	-7.30	1.18
Israel	0.04	0.17	4.11	0.99	4.04	1.19
New Zealand	-0.37	0.16	0.37	0.88	-1.12	1.12
100% Polish	-0.83	0.14	-3.30	0.80	-5.35	0.98

Significance of variation among strains: ** ($P < 0.01$)
Approximate least significant differences[2]

	At birth	12 m	18 m
$P < 0.01$	0.65	3.64	4.48
$P < 0.05$	0.50	2.82	3.46

[1] [2]See footnotes Table 2.20

The results of the strain comparison are shown in Figures 2.21 to 2.33. The results for the backcross generation are based on only approximately half the numbers in the F_1 generation. In general this has affected quite markedly the statistical significance of strain differences. Least squares solutions are given in Tables 2.30 to 2.34.

a) Weight gains and energy intake

Mean *daily gain* from 110 days of life to the final body weight (450 or 550 kg for F_1 and 450 kg for R_1) was 1 135 g in F_1 (1 127 g to 450 kg and 1 144 g to 550 kg) and 1 126 g in R_1 animals. Differences among strains were significant in the daily gain for the F_1 bulls only (Table 2.30) though not separately

Table 2.23 **Daily gain (g) of first cross (F_1) females in the field trial — the effects' of paternal strains as deviations (±s.e.) from the mean** (after Stolzman *et al.*, in press)

Paternal strain origin	**6 to 12 m****		**12 to 18 m**		**Birth to 12 m****		**Birth to 18 m****	
	Effect	**s.e.**	**Effect**	**s.e.**	**Effect**	**s.e.**	**Effect**	**s.e.**
USA	11	4.7	10	5.0	7	3.0	8	2.3
Canada	10	4.6	9	4.9	11	2.9	10	2.2
Denmark	0	4.7	4	5.0	-3	2.9	0	2.3
United Kingdom	-10	4.4	-3	4.7	-9	2.8	-7	2.1
Sweden	-15	4.4	6	4.7	-4	2.8	-1	2.1
Fed. Rep. of Germany	11	4.4	-6	4.7	8	2.7	3	2.1
Netherlands	-15	4.5	-5	4.8	-16	2.7	-13	2.1
Israel	9	4.6	-1	4.9	12	2.9	8	2.2
New Zealand	4	4.3	-8	4.6	2	2.5	-1	2.0
100% Polish	-5	3.8	-6	4.0	-8	2.3	-7	1.7

Significance of variation among strains: ** ($P < 0.01$)
Approximate least significant differences[2]

$P < 0.01$	17				11		8	
$P < 0.05$	13				8		6	

[1] [2]See footnotes Table 2.20

at the two slaughter weights. The highest daily gains were achieved by the Holstein strains (from USA, Canada and Israel). The German F_1 and the Polish bulls were also above the average. There were differences in ranking of strains between the R_1 and F_1 generation (Fig. 2.21). Thus, Swedish and Danish backcross males were above average while in F_1 they were below; conversely, Canadian and Polish males which were ranked below the average for R_1 had been above average in the F_1 generation.

The mean *net energy intake* (MJ per kg gain) was 34.7 in the F_1 male progeny (Table 2.30). The fast growing paternal strains had the better feed efficiency with the USA strain having the lowest intake and the New Zealand strain the highest (Fig. 2.22), though for the New Zealand strain efficiency was about average for the group slaughtered at 550 kg (feed efficiency differed little for this strain between the two slaughter weight groups whereas for all other strains

Table 2.24 **Liveweight (kg) of backcross (R_1) females in the field trial — the effects' of paternal strains as deviations (±s.e.) from the mean** (after Stolzman *et al.*, in press)

Paternal strain origin	**At birth***		**12 m**		**18 m****	
	Effect	**s.e.**	**Effect**	**s.e.**	**Effect**	**s.e.**
USA	0.59	0.39	1.58	1.28	5.90	1.36
Canada	0.44	0.34	0.10	1.24	4.53	1.31
Denmark	-0.21	0.37	-1.40	1.27	0.86	1.34
United Kingdom	0.01	0.37	0.42	1.26	-8.34	1.33
Sweden	0.44	0.33	1.29	1.24	-0.40	1.30
Fed. Rep. of Germany	0.38	0.31	2.56	1.23	2.31	1.29
Netherlands	-0.55	0.38	-3.42	1.28	-7.64	1.35
Israel	0.56	0.36	6.52	1.26	11.88	1.33
New Zealand	-0.23	0.29	-0.04	1.24	0.19	1.27
Poland (R_1)	-0.38	0.29	-3.31	1.21	-4.31	1.26
100% Polish	-1.05	0.28	-4.29	1.20	-4.98	1.26

Significance of variation among strains: * ($P < 0.05$)
** ($P < 0.01$)

Approximate least significant differences[2]

	At birth	12 m	18 m
$P < 0.01$	1.27		5.07
$P < 0.05$	0.98		3.92

[1] [2]See footnotes Table 2.20

it declined markedly at the higher weight). In the R_1 group average intake was slightly lower at 32.4 MJ per kg gain probably because the R_1 bulls were slaughtered earlier, i.e. all at 450 kg of liveweight. (The lower proportion of Polish blood in this generation relative to the F_1 would have worked, if anything, in the opposite direction.) In the R_1 generation the differences among strains were not significant.

The mean *daily net gain* (Fig. 2.23), showing the rate at which the carcass weight was reached, follows daily gain fairly closely in respect of strain differences. Differences among strains were significant only in the F_1 generation (but only on the pooled slaughter results) (mean 628 g, and 617 and 640 respectively at the two slaughter weights) (Table 2.30). The daily net gain was

Table 2.25 **Daily gain (g) of backcross (R_1) females in the field trial — the effects' of paternal strains as deviations ($\pm$s.e.) from the mean** (after Stolzman *et al.*, in press)

Paternal strain origin	6 to 12 m		12 to 18 m*		Birth to 12 m		Birth to 18 m**	
	Effect	s.e.	Effect	s.e.	Effect	s.e.	Effect	s.e.
USA	-3	13.5	25	14.8	4	7.8	11	6.6
Canada	4	11.8	27	12.9	-1	6.8	8	5.8
Denmark	-2	13.0	12	14.2	-3	7.5	2	6.4
United Kingdom	2	12.6	-49	13.9	0	7.3	-16	6.2
Sweden	9	11.5	-11	12.6	3	6.7	-2	5.7
Fed. Rep. of Germany	-2	10.8	-1	11.8	6	6.2	3	5.3
Netherlands	5	13.4	-24	14.7	-9	7.8	-14	6.6
Israel	4	12.5	30	13.7	16	7.3	21	6.2
New Zealand	15	10.0	2	10.9	2	5.8	3	4.9
Poland (R_1)	-14	10.1	-7	11.0	-8	5.8	-8	5.0
100% Polish	-18	10.0	-4	11.0	-10	5.8	-8	4.9

Significance of variation among strains: * ($P < 0.05$)
** ($P < 0.01$)

Approximate least significant differences[2]

$P < 0.01$			48				22	
$P < 0.05$			37				17	

[1] [2]See footnotes Table 2.20

significantly lower for the New Zealand strain (in both generations) and the Danish and German strains in the F_1 were particularly below average and the USA, Israeli and Swedish strains most clearly above (in the F_1). The mean in the R_1 generation was 601 g.

b) Carcass results

For *dressing percentage* the fitted mean for the F_1 generation was 57.7 percent (56.9 at 450 kg and 58.5 at 550 kg) and slightly lower at 56.7 for the R_1 (Table 2.31). Only some of the strains in the F_1 generation differed significantly

Fig. 2.21 Mean daily gain in weight from 110 days to slaughter of males in the intensive–feeding trial showing each strain as a percentage deviation from the least squares mean (approx. s.e. $F_1$1.7 $R_1$2.0)

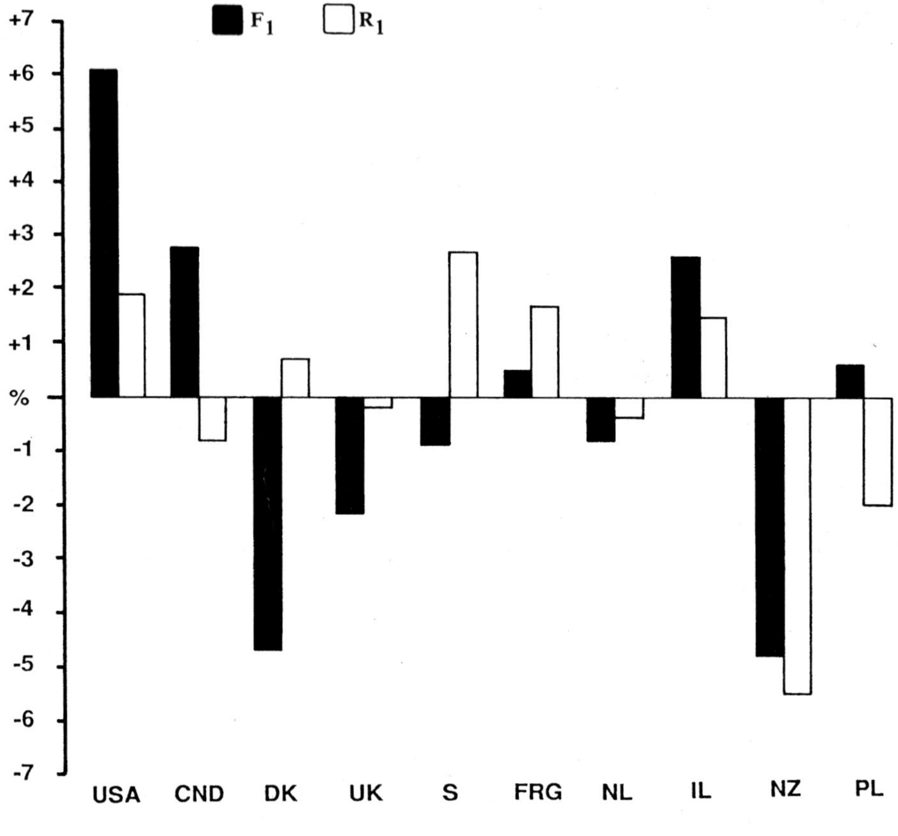

from others (specifically because of differences at 550 kg slaughter) with crosses from British, Dutch and Swedish Friesians having the best and those from USA Holsteins the poorest dressing percentage (Fig. 2.24). Differences in the backcross generation were not significant but most of the strains deviated from average in the same direction in both generations (Fig. 2.24).

The differences among strains in the proportion of valuable cuts in the carcass were small (Table 2.31 and Fig. 2.25). For the F_1 progeny the mean was 62.4 percent (62.8 and 62.0 at 450 and 550 kg slaughter respectively) with carcasses of the New Zealand strain having significantly the lowest proportion (Fig. 2.25). A slightly higher proportion of valuable cuts (63.2 percent) was found in the R_1 generation. Although strain differences appeared to be greater in the backcross generation with Swedish and German having the highest proportion of valuable cuts and Canadian and New Zealand the lowest, none of the differences were significant.

The carcass dissection results are presented in Table 2.32. The muscles are the most valuable part of the carcass. The mean *lean content* in the carcasses of F_1 bulls was 64.6 percent (65.3 and 63.8 respectively at the two slaughter weights). An above-average proportion of lean was found in carcasses of the Swedish, USA, Polish and British strains. Variation among strains (Fig. 2.26) was relatively small and statistically not significant.

A slightly lower proportion of lean in the carcass was found in R_1 bulls (64.1 percent) at equivalent end points of slaughter at 450 kg perhaps reflecting

Fig. 2.22 Net energy intake (MJ/kg gain) from 110 days to slaughter of males in the intensive–feeding trial showing each strain as a percentage deviation from the least squares mean (approx. s.e. $F_1$1.7 $R_1$2.2)

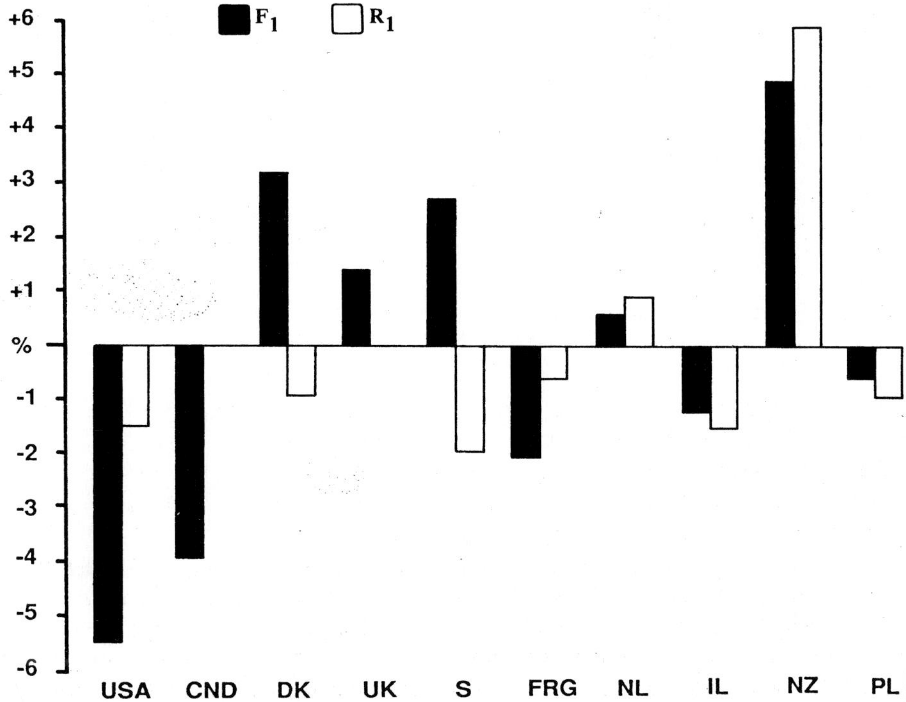

Fig. 2.23 Mean daily net gain in weight from 110 days to slaughter of males in the intensive–feeding trial showing each strain as a percentage deviation from the least squares mean (approx. s.e. F_1 1.6 R_1 2.0)

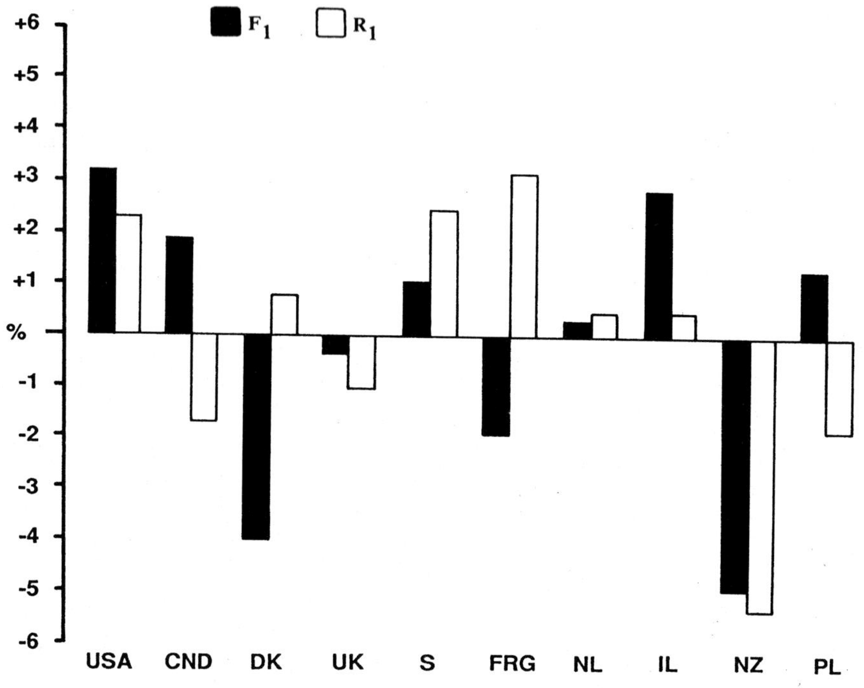

Fig. 2.24 Dressing percentage of males in the intensive–feeding trial showing each strain as a percentage deviation from the least squares mean (approx. s.e. F_1 0.5 R_1 0.7)

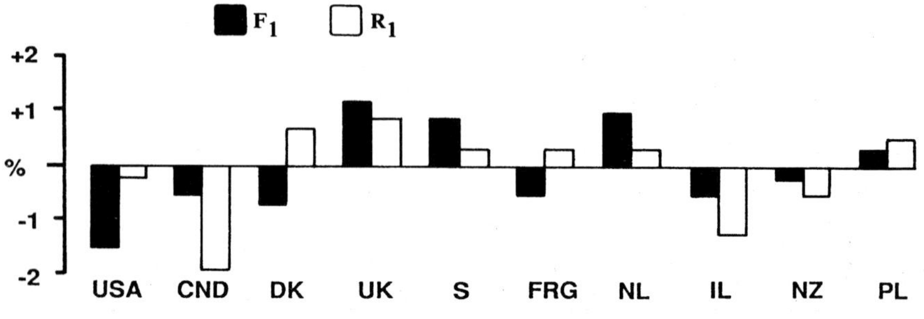

Fig. 2.25 Proportion of valuable cuts in the carcasses of males in the intensive–feeding trial showing each strain as a percentage deviation from the least squares mean (approx. s.e. $F_1$0.3 $R_1$0.5)

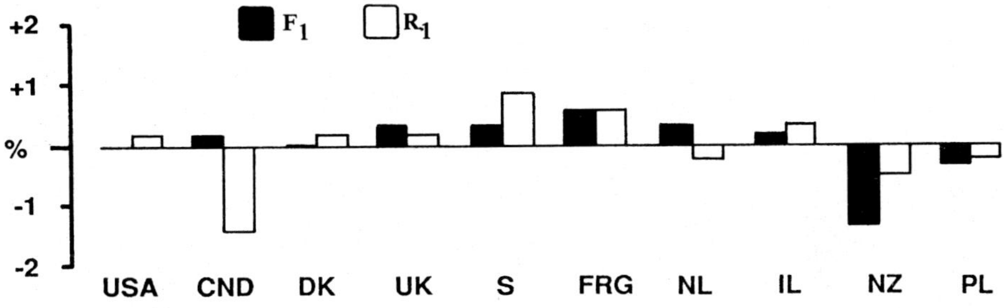

Table 2.26 **Height at withers (cm) and chest girth (cm) of first cross (F_1) males at 12 months old in the field trial — the effects' of paternal strains as deviations (±s.e.) from the mean** (after Stolzman *et al.*, in press)

Paternal strain origin	Height at withers**		Chest girth**	
	Effect	s.e.	Effect	s.e.
USA	0.85	0.19	0.90	0.33
Canada	0.62	0.17	0.10	0.30
Denmark	-0.65	0.19	-0.17	0.33
United Kingdom	0.00	0.19	-0.16	0.33
Sweden	0.12	0.19	-0.13	0.32
Fed. Rep. of Germany	0.00	0.19	0.01	0.32
Netherlands	-0.07	0.19	-0.86	0.33
Israel	0.24	0.20	0.76	0.33
New Zealand	0.04	0.19	0.54	0.32
100% Polish	-1.14	0.15	-0.97	0.25

Significance of variation among strains: ** ($P < 0.01$)
Approximate least significant differences[2]

	Height at withers	Chest girth
$P < 0.01$	0.72	1.22
$P < 0.05$	0.55	0.94

[1] [2]See footnotes Table 2.20

Fig. 2.26 Proportion of lean in the carcasses of males in the intensive–feeding trial showing each strain as a percentage deviation from the least squares mean
(approx. s.e. $F_1$0.8 $R_1$1.1)

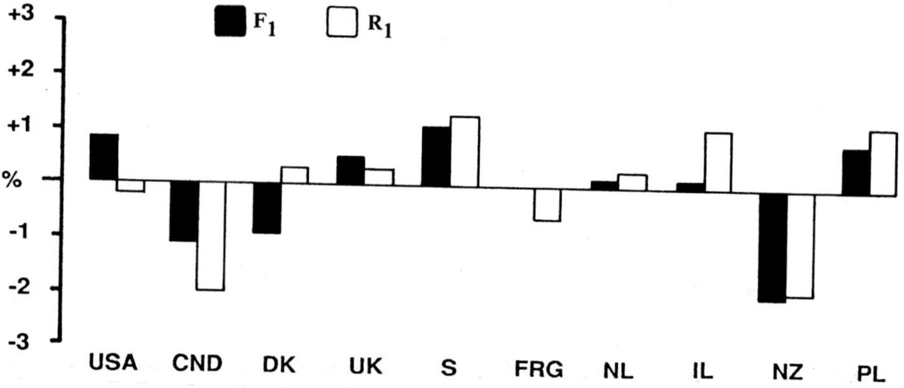

Table 2.27 **Height at withers (cm) and chest girth (cm) of backcross (R_1) males at 12 months old in the field trial — the effects' of paternal strains as deviations (± s.e.) from the mean** (after Stolzman *et al.*, in press)

Paternal strain origin	Height at withers		Chest girth	
	Effect	s.e.	Effect	s.e.
USA	0.71	0.49	0.90	0.76
Canada	-0.44	0.42	-1.58	0.67
Denmark	-0.22	0.43	-0.12	0.67
United Kingdom	-0.30	0.43	0.24	0.67
Sweden	-0.05	0.42	0.95	0.65
Fed. Rep. of Germany	0.63	0.42	0.94	0.66
Netherlands	-0.23	0.43	-0.74	0.68
Israel	0.18	0.42	0.52	0.65
New Zealand	0.42	0.42	-0.39	0.65
Poland	-0.02	0.30	-0.09	0.47
100% Polish	-0.68	0.32	-0.63	0.50

[1] See footnote Table 2.20

Fig. 2.27 Proportion of fat in the carcasses of males in the intensive–feeding trial showing each strain as a percentage deviation from the least squares mean (approx. s.e. F_1 3.3 R_1 4.0)

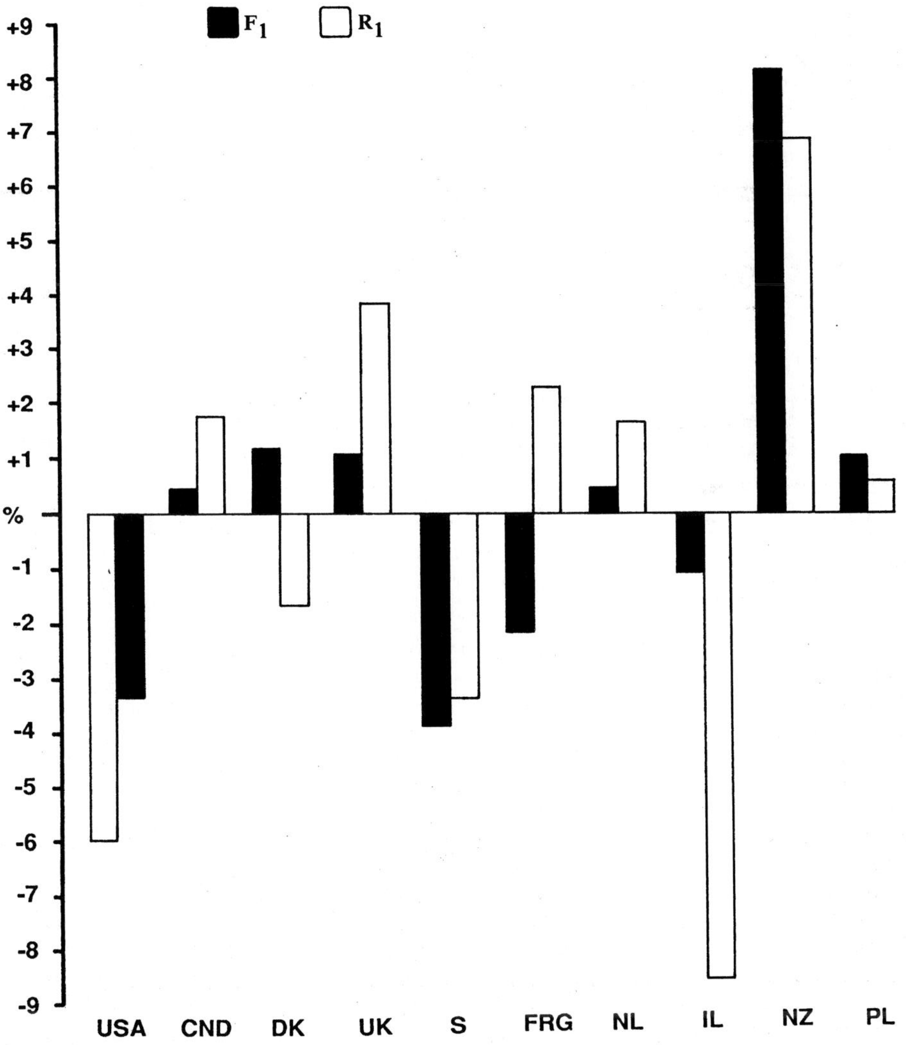

Table 2.28 **Height at withers (cm) and chest girth (cm) of first cross (F_1) females in the field trial — the effects' of paternal strains as deviations (± s.e.) from the mean** (after Stolzman *et al.*, in press)

Paternal strain origin	12 months				18 months			
	Height at withers**		Chest girth*		Height at withers**		Chest girth**	
	Effect	s.e.	Effect	s.e.	Effect	s.e.	Effect	s.e.
USA	1.07	0.17	0.76	0.27	1.37	0.19	0.53	0.27
Canada	1.06	0.17	0.33	0.26	1.29	0.18	0.56	0.26
Denmark	-0.75	0.17	0.19	0.26	-0.51	0.17	0.38	0.24
United Kingdom	-0.36	0.17	-0.35	0.26	-0.59	0.18	-0.37	0.26
Sweden	0.37	0.14	-0.31	0.22	-0.29	0.17	-0.30	0.25
Fed. Rep. of Germany	0.45	0.17	0.36	0.26	0.14	0.18	0.17	0.26
Netherlands	-0.44	0.17	-0.40	0.26	-0.83	0.18	-0.67	0.26
Israel	0.85	0.17	0.27	0.26	0.76	0.18	0.47	0.26
New Zealand	-0.01	0.14	-0.37	0.22	0.05	0.16	-0.15	0.23
100% Polish	-1.50	0.14	-0.47	0.22	-1.38	0.15	-0.64	0.22

Significance of variation among strains: * ($P < 0.05$)
** ($P < 0.01$)

Approximate least significant differences[2]

$P < 0.01$	0.63	0.97	0.68	0.98
$P < 0.05$	0.48	0.75	0.52	0.75

[1] [2] See footnotes Table 2.20

Table 2.29 **Height at withers (cm) and chest girth (cm) of backcross (R_1) females in the field trial — the effects' of paternal strains as deviations (±s.e.) from the mean** (after Stolzman *et al.*, in press)

Paternal strain origin	12 months				18 months			
	Height at withers*		Chest girth		Height at withers*		Chest girth	
	Effect	s.e.	Effect	s.e.	Effect	s.e.	Effect	s.e.
USA	0.90	0.47	0.07	0.59	0.93	0.49	0.06	0.66
Canada	0.41	0.40	0.62	0.50	1.12	0.42	0.44	0.57
Denmark	0.24	0.42	-0.61	0.53	-0.23	0.44	0.08	0.60
United Kingdom	0.15	0.45	0.81	0.57	-0.33	0.48	0.58	0.64
Sweden	-0.51	0.39	0.05	0.50	-0.38	0.41	-0.77	0.56
Fed. Rep. of Germany	0.14	0.38	-0.02	0.48	0.28	0.40	0.21	0.55
Netherlands	-0.79	0.42	-0.03	0.53	-0.31	0.45	-0.23	0.60
Israel	1.07	0.42	0.40	0.53	0.79	0.44	0.34	0.59
New Zealand	-0.06	0.32	-0.18	0.40	-0.58	0.34	-0.54	0.46
Poland (R_1)	-0.59	0.33	-0.11	0.41	-0.47	0.34	0.01	0.47
100% Polish	-0.97	0.32	-1.00	0.41	-0.81	0.34	-0.19	0.46

Significance of variation among strains: * ($P < 0.05$)

Approximate least significant differences[2]

$P < 0.01$	1.46				1.54			
$P < 0.05$	1.13				1.19			

[1] [2]See footnotes Table 2.20

Table 2.30 **Least squares mean daily gains (±s.e.) from 110 days to slaughter and net energy intakes per kg gain of males of ten different strains at different slaughter weights* (intensive trial)** (from Reklewski *et al.*, 1985)

Paternal strain origin		Mean daily gain (g)				Mean daily net gain (g)				Net energy intake (MJ/kg gain)				Total gain (kg)			
		450 + 550 kg		450 kg		450 + 550 kg		450 kg		450 + 550 kg		450 kg		450 + 550 kg		450 kg	
		F_1	s.e.	R_1	s.e.	F_1	s.e.	R_1	s.e.	F_1	s.e.	R_1	.e.	F_1	s.e.	R_1	.e.
USA	(A)	1 204	13	1 147	20	648	10	515	7	32.8	0.6	31.9	0.5	381	2	345	4
Canada	(C)	1 167	19	1 117	19	640	9	591	11	33.7	0.6	32.4	0.4	388	2	353	3
Denmark	(D)	1 086	18	1 134	30	603	9	606	14	35.8	0.5	32.1	0.8	387	2	350	4
Sweden	(F)	1 125	19	1 156	26	635	5	616	13	35.3	0.6	31.8	0.8	384	2	352	3
Fed. Rep. of Germany	(G)	1 141	19	1 145	24	616	9	620	12	34.0	0.6	32.2	0.6	387	2	346	4
Netherlands	(H)	1 126	19	1 121	18	630	10	604	14	34.9	0.6	32.7	0.5	392	2	359	3
Israel	(I)	1 164	19	1 143	21	646	10	604	9	34.3	0.6	31.9	0.6	378	2	345	4
New Zealand	(K)	1 080	18	1 064	19	597	9	569	10	36.5	0.5	34.3	0.7	391	2	350	3
Poland (pure)	(B)	1 142	18	1 103	16	636	9	590	9	34.5	0.5	32.1	0.6	385	2	351	4
Total		1 135	6	1 126	7	628	3	601	4	34.7	0.2	32.4	0.3	386	1	350	1
Significant differences*		A-K,D,E,F,H,G,B				K-A,I,C,B,F,H				A-K,D,F,H,B							
		C,I-D,K				D-A,I,C,B,F				K-C,G,I,B,H,E,F,D							
						G-A											

* Differences between strains significant at: P ≤ 0.05 — single underlining
P ≤ 0.01 — double underlining

Table 2.31 **Least squares means of post-slaughter traits of males of ten different strains at different slaughter weights· (intensive trial)** (from Reklewski *et al.*, 1985)

Paternal strain origin		**Dressing percentage**				**Valuable cuts** *(%)*			
		450 + 550 kg		**450 kg**		**450 + 550 kg**		**450 kg**	
		F_1	s.e.	R_1	s.e.	F_1	s.e.	R_1	s.e.
USA	(A)	56.9	0.3	56.6	0.4	62.4	0.2	63.6	0.2
Canada	(C)	57.4	0.3	55.6	0.5	62.5	0.2	62.3	0.3
Denmark	(D)	57.3	0.3	57.1	0.4	62.4	0.2	63.3	0.2
United Kingdom	(E)	58.4	0.3	57.2	0.4	62.6	0.2	63.3	0.2
Sweden	(F)	58.2	0.3	56.9	0.4	62.6	0.2	63.8	0.2
Fed. Rep. of Germany	(G)	57.4	0.3	56.9	0.3	62.8	0.2	63.6	0.3
Netherlands	(H)	58.3	0.3	56.9	0.4	62.6	0.2	63.1	0.3
Israel	(I)	57.4	0.3	56.0	0.3	62.5	0.2	63.4	0.3
New Zealand	(K)	57.6	0.3	56.4	0.3	61.6	0.2	62.9	0.3
Poland (pure)	(B)	57.8	0.3	57.0	0.3	62.2	0.2	63.1	0.3
Total		57.7	0.1	56.7	0.1	62.4	0.1	63.2	0.1
Significant differences*		E-A,D,C,G,I,				K-G,F,E,H,C,I,D,A			
		H-A,D							
		F-A							

* Differences between strains significant at: $P \leq 0.05$ — single underlining
$P \leq 0.01$ — double underlining

the lower proportion of Polish blood, overall, in this generation (since pure Polish were above average). The two strains most below average (NZ and CND) in one generation were also most below in the other, and Swedish carcasses were, on the evidence of both generations, most above average, followed closely by Polish bulls (Fig. 2.26).

The average *fat content* of the carcasses of F_1 bulls was 18.2 percent (Table 2.32) (17.0 at 450 kg and 19.4 at 550 kg). In general, in the comparison of strains (Fig. 2.27) those which had the most fat also had least lean (with the New Zealand strain the fattest and USA and Israeli strains least fat) but there were interesting exceptions (e.g. Polish, British and Dutch carcasses were above average in both fat and lean content though by small amounts). Because the percentage deviations in the case of fat relate to a lower average than in the case of lean proportion, the strain differences look more dramatic in the case of fat but, as seen in Table 2.32 only the extreme values differed significantly from each other in the F_1 generation (and not when slaughter weights were considered separately) and none in the R_1.

In the backcrosses the mean fat content in the carcass was 17.7 percent.

The average percentage of *bone* in F_1 bulls was 17.2 percent (Table 2.32) (17.8 and 16.3 respectively at 450 and 550 kg). A highly significant lower bone content was found in the carcasses of Polish bulls as well as of British and Dutch Friesian F_1 males, but the differences were most marked at the 450 kg slaughter weight. On average, the highest bone content was found in the carcasses of F_1 progeny of USA, Canadian, German, Danish and Israeli strains (Fig. 2.28).

The carcasses of R_1 bulls had a higher bone content (18.2 percent) (a result to which the lower proportion of Polish blood in this generation may have contributed), but only slightly so at equivalent slaughter weights. The effect of strain was not significant. Again, however, the Holstein group of strains had more bone in the carcass, and Polish, British, Dutch and Swedish less.

Ratios. The *lean:fat* ratio was 3.66 in the carcasses of F_1 males and 3.72 in those of the backcrosses (Table 2.33). Differences among strains were not significant in either generation but, not surprisingly, the strains which had above average lean content and below average fat such as the USA Holstein crosses and Swedish Friesian crosses had above average ratios, conversely New Zealand strain crosses were most below average in the ratio of lean to fat (Fig. 2.29). This aspect was not markedly affected by slaughter weight although the F_1 animals slaughtered at 450 kg had, as expected, a higher proportion of lean:fat than at 550 kg.

The *lean:bone* ratio similarly reflects the relative positions of the strains in the two components. Fig. 2.30 shows that carcasses of Polish and British Friesian crosses in particular were above average (more lean in relation to bone)

Fig. 2.28 Proportion of bone in the carcasses of males in the intensive–feeding trial showing each strain as a percentage deviation from the least squares mean
(approx. s.e. F_1 1.2 R_1 1.2)

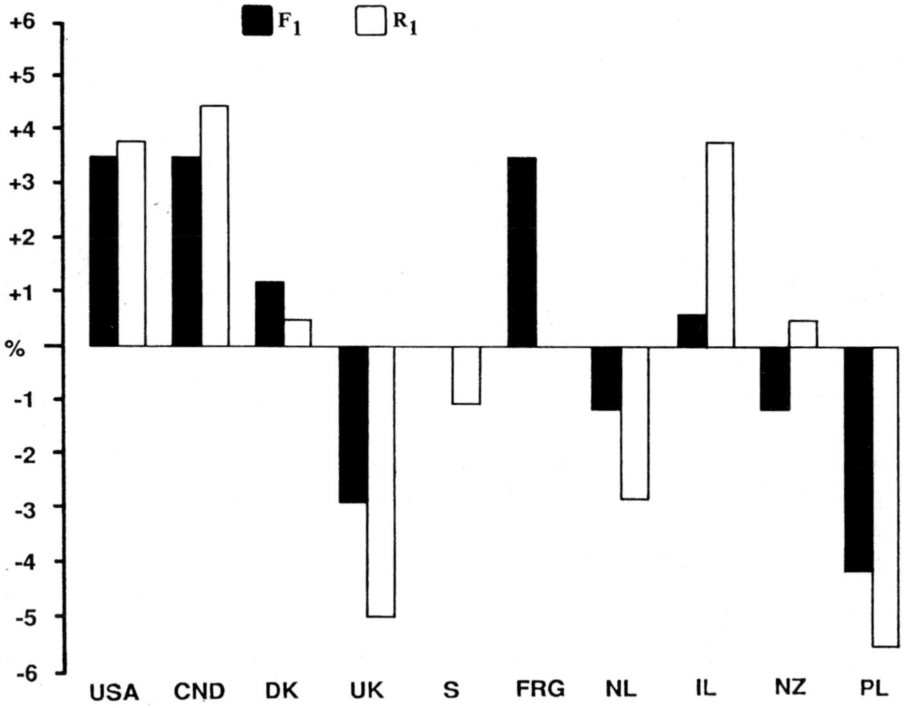

and Canadian and USA Holstein crosses below. In both the F_1 generation (mean ratio 3.77) and the R_1 generation (mean ratio 3.54) particularly the Polish and British strains but also the Dutch and Swedish (the four strains with, on average, the best ratios) differed significantly from those lower down in the scale and particularly from the Holstein strains as shown in Table 2.33. The proportion of lean to bone increased from 450 kg to 550 kg slaughter weight (except, surprisingly, for the Dutch strain) but it did not affect the strain comparison in general.

The area of the *longissimus dorsi* muscle was greater in the F_1 generation (least squares mean 74.8 cm^2), particularly because of the inclusion of the higher slaughter weight (70.0 cm^2 for 450 kg and < 79.8 cm^2 for 550 kg), than it was in the R_1 generation (mean 68.1 cm^2) which was slaughtered at 450 kg only.

Table 2.32 **Least squares means (±s.e.) of carcass tissue composition traits of males of ten different strains at different slaughter weights* (intensive trial)** (from Reklewski *et al.*, 1985)

Paternal strain origin		Lean (%) 450 + 550 kg		450 kg		Fat (%) 450 + 550 kg		450 kg		Bones (%) 450 + 550 kg		450 kg	
		F_1	s.e.	R_1	s.e.	F_1	s.e.	R_1	s.e.	F_1	s.e.	R_1	s.e.
USA	(A)	65.1	0.5	64.0	0.5	17.1	0.5	17.1	0.5	17.8	0.3	18.9	0.3
Canada	(C)	63.9	0.5	62.8	0.7	18.3	0.5	18.2	0.8	17.8	0.2	19.0	0.2
Denmark	(D)	64.0	0.5	64.3	0.5	18.6	0.5	17.4	0.5	17.4	0.2	18.3	0.4
United Kingdom	(E)	64.9	0.6	64.3	0.6	18.4	0.5	18.4	0.6	16.7	0.2	17.3	0.2
Sweden	(F)	65.3	0.5	64.9	0.9	17.5	0.5	17.1	0.9	17.2	0.2	18.0	0.2
Fed. Rep. of Germany	(G)	64.6	0.5	63.7	0.7	17.8	0.5	18.1	0.9	17.8	0.2	18.2	0.2
Netherlands	(H)	64.7	0.5	64.3	0.6	18.3	0.5	18.0	0.6	17.0	0.3	17.7	0.1
Israel	(I)	64.7	0.5	64.8	0.4	18.0	0.5	16.2	0.6	17.3	0.2	18.9	0.2
New Zealand	(K)	63.3	0.5	62.9	0.5	19.7	0.5	18.9	0.6	17.0	0.2	18.3	0.2
Poland (pure)	(B)	65.1	0.5	64.9	0.7	18.4	0.5	17.8	0.6	16.5	0.2	17.2	0.2
Total		64.6	0.2	64.1	0.2	18.2	0.1	17.7	0.2	17.2	0.1	18.2	0.1
Significant differences*						K-A,F,G,I				B-C,A,G,D			
										E-C,A,G			
						D-A				G-K,H			

* Differences between strains significant at: P ≤ 0.05 — single underlining
P ≤ 0.01 — double underlining

Table 2.33 **Least squares means of the lean: fat and lean: bone ratios and those of subcutaneous fat content and *Musculus longissimus dorsi* (MLD) area of males of ten different strains at different slaughter weights* (intensive trial)** (from Reklewski *et al.*, 1985)

Paternal strain		Lean:fat				Lean:bone				MLD area behind last rib (cm²)			
		450 + 550 kg		450 kg		450 + 550 kg		450 kg		450 + 550 kg		450 kg	
		F_1	s.e.	R_1	s.e.	F_1	s.e.	R_1	s.e.	F_1	s.e.	R_1	.e.
USA	(A)	3.89	0.13	3.79	0.19	3.68	0.05	3.40	0.07	72.5	1.8	67.0	1.9
Canada	(C)	3.58	0.13	3.56	0.19	3.60	0.05	3.32	0.07	71.9	1.7	66.5	2.0
Denmark	(D)	3.51	0.12	3.75	0.18	3.71	0.05	3.54	0.06	74.0	1.7	66.8	2.0
United Kingdom	(E)	3.69	0.13	3.56	0.18	3.91	0.05	3.73	0.07	76.2	1.8	68.5	2.0
Sweden	(F)	3.82	0.13	4.00	0.19	3.82	0.05	3.62	0.06	74.9	1.7	68.1	2.0
Fed. Rep. of Germany	(G)	3.78	0.13	3.67	0.18	3.68	0.05	3.51	0.06	76.1	1.7	67.8	1.4
Netherlands	(H)	3.62	0.13	3.68	0.19	3.88	0.05	3.63	0.07	78.2	1.8	72.8	1.3
Israel	(I)	3.71	0.13	4.07	0.19	3.73	0.05	3.43	0.07	76.5	1.8	66.2	1.3
New Zealand	(K)	3.35	0.19	3.38	0.19	3.74	0.05	3.45	0.07	73.4	1.7	65.7	2.2
Poland (pure)	(B)	3.67	0.12	3.74	0.19	3.94	0.05	3.77	0.06	74.6	1.7	71.7	1.2
Total		3.66	0.04	3.72	0.06	3.77	0.02	3.54	0.02	74.8	0.5	68.1	0.6
Significant differences*						B-C,A,G,D,I,K				B-C,A,I,K,G,D			
						E-C,A,G,D,I,K				E-C,A,I,K,G			
						H-C,A,G				H-C,A			
						F-C				C-F,D			

* Differences between strains significant at: P ≤ 0.05 — single underlining
P ≤ 0.01 — double underlining

Table 2.34 **Least squares means of daily lean gains and net energy consumption/kg lean gain of males of ten different strains at different slaughter weights* (intensive trial)** (from Reklewski *et al.*, 1985)

Paternal strain origin		Mean daily lean gain (g)				Net energy intake MJ per kg lean gain			
		450 + 550 kg		450 kg		450 + 550 kg		450 kg	
		F_1	s.e.	R_1	s.e.	F_1	s.e.	R_1	s.e.
USA	(A)	418	0.8	390	0.6	70.0	1.5	69.2	1.1
Canada	(C)	405	0.7	367	0.7	73.5	1.5	74.1	1.3
Denmark	(D)	383	0.7	385	1.0	77.6	1.4	70.3	2.0
United Kinrgdom	(E)	403	0.7	378	0.9	73.5	1.5	70.8	2.1
Sweden	(F)	410	0.7	395	1.2	74.0	1.5	69.2	2.4
Fed. Rep. of Germany	(G)	398	0.7	391	0.9	72.7	1.5	70.3	2.0
Netherlands	(H)	405	0.8	384	1.1	74.5	1.5	73.0	1.7
Israel	(I)	414	0.8	386	0.6	72.1	1.5	69.2	1.3
New Zealand	(K)	375	0.7	354	0.9	80.9	1.4	77.4	2.3
Poland (pure)	(B)	409	0.7	379	0.8	72.8	1.4	70.3	1.9
Total		402	0.2	381	0.3	74.2	0.4	71.3	0.5
Significant differences*		K-A,I,F,B,C,H,E				K-A,I,G,B,E,C,F,H			
		D-A,I,F,B				D-A			

*Differences between strains significant at: $P \leq 0.05$ — single underlining
$P \leq 0.01$ — double underlining

Again, however, the lesser contribution of the Polish strain to the average for the R_1 generation may have contributed to the F_1 vs. R_1 difference since the pure Polish animals showed the highest ratio. Fig. 2.31 shows the results for strains and although the deviations as percentages of the mean appear to be marked they were not statistically different from each other and F_1 and R_1 results for strains did not all support each other.

Daily lean gain can be considered as an index of merit for meat production by combining carcass weight with proportion of lean and the time from birth to slaughter. As shown in Fig. 2.32, the New Zealand Friesians in both F_1 (at both slaughter weights) and backcross generations were clearly below average and this was significant in relation to all the strains which were above average in the F_1 (Table 2.34) but not at the separate slaughter weights. The Danish Frie-

Fig. 2.29 Lean:fat ratio in the carcasses of males in the intensive-feeding trial showing each strain as a percentage deviation from the least squares mean
(approx. s.e. $F_1$3.8 $R_1$5.1)

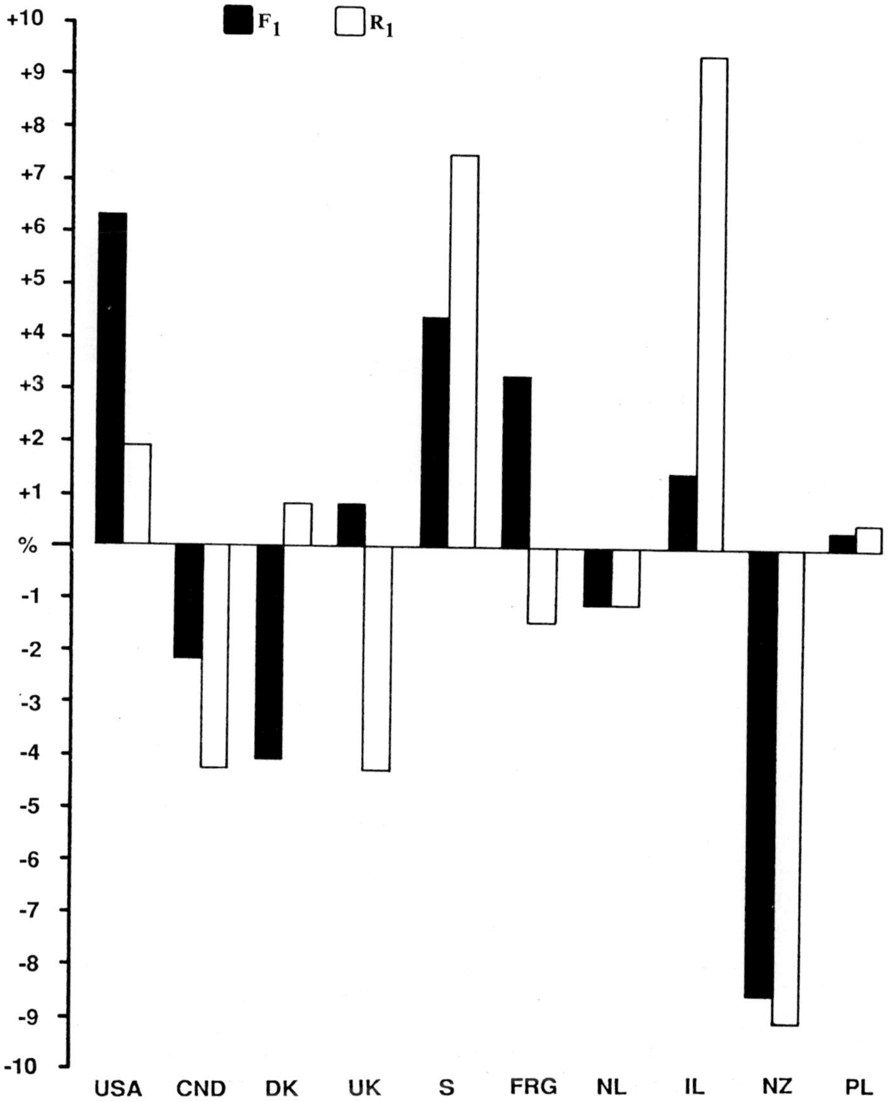

Fig. 2.30 Lean:bone ratio in the carcasses of males in the intensive–feeding trial showing each strain as a percentage deviation from the least squares mean
(approx. s.e. F_1 1.3 R_1 2.0)

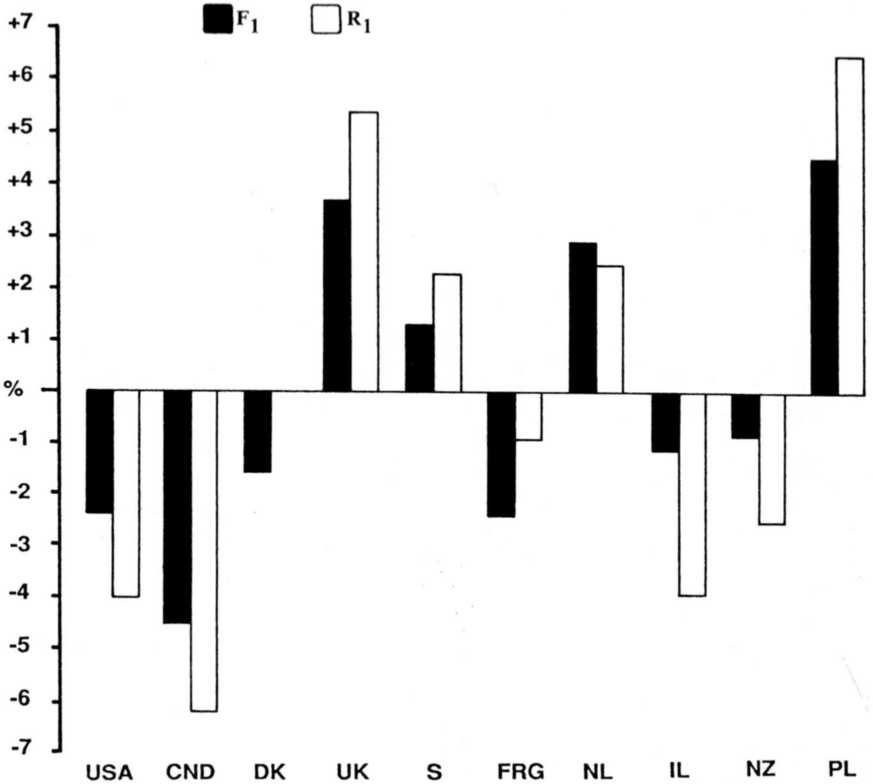

sians in the F_1 were also significantly poorer in daily lean gain than some of the other strains. Unlike for some of the other traits, there was no obvious division between the Holstein and the Friesian strains — some within each type being above and others below average (Fig. 2.32).

The *net energy intake* per kg lean gain (in MJ) differed among strains more or less inversely with daily gain of lean in both generations (and at both slaughter weights of F_1) (Fig. 2.33) and was higher at 550 kg slaughter weight than at 450 kg. Thus the offspring from the New Zealand strain had the highest intakes and the difference from most other strains was statistically significant in the F_1 generation combined (mean 74.2 MJ) (Table 2.34) and separately at he higher

Fig. 2.31 Area of L. dorsi muscle (behind last rib) in the carcasses of males in the intensive–feeding trial showing each strain as a percentage deviation from the least squares mean (approx. s.e. $F_1$0.2 $R_1$0.3)

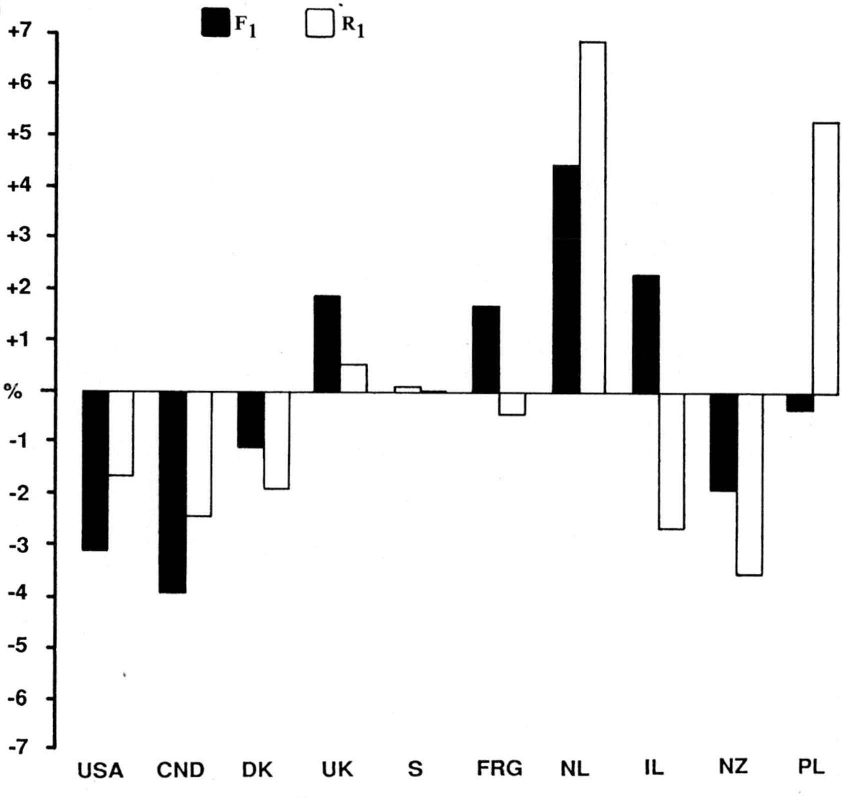

efficiency of the 450 kg slaughter weight. In general, the progeny of USA Holsteins had the lowest intake. Variation among strains was not significant in the R_1 generation (mean 71.3 MJ) and within strains the two generations were not all consistent with each other particularly for the Danish and Canadian strains (Fig. 2.33).

Heterosis estimates. The only indications that strain combinations differed in respect of heterosis estimates was for mean daily gain where crosses with Holstein strains showed a positive deviation from additive effects with the Polish strain ($P < 0.05$), while other crosses showed a negative heterosis. However, the average

Fig. 2.32 Mean daily gain in weight of lean from birth to slaughter of males in the intensive–feeding trial showing each strain as a percentage deviation from the least squares mean (approx. s.e. $F_1$0.2 $R_1$0.3)

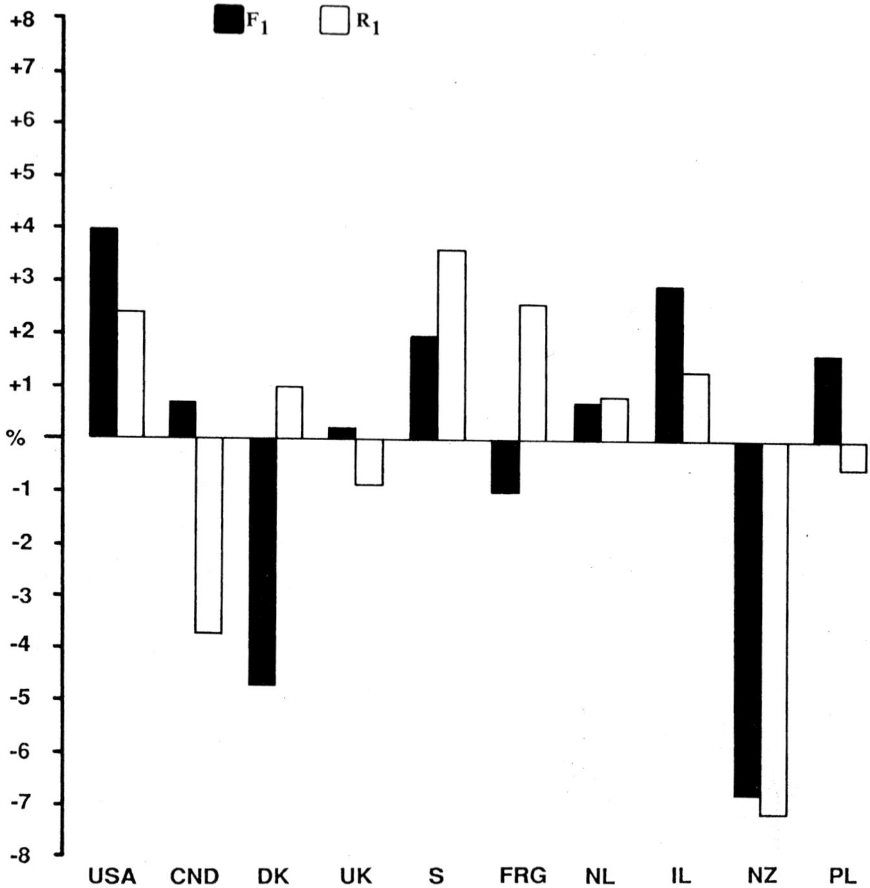

value was trivial at little over 1 percent of the Polish mean. Other beef production and carcass traits were little different from this, and individual heterosis estimates for different strains were not significant. The heterosis estimate for lean to bone ratio, though only 1.6 percent of the mean of the trait, showed significant positive effects for most of the crosses of European strains of Friesian with the Polish. Net energy intake per kg lean gain gave negative estimates of heterosis for the New Zealand and Danish strains (more energy per kg gain) of around 10 percent of the mean of the trait (and statistically significant $P < 0.05$) but an overall average value is difficult to interpret in view of a very wide variation in estimates among strains.

Fig. 2.33 Net energy intake (MJ) per kg of gain in lean from birth in the carcasses of males in the intensive–feeding trial showing each strain as a percentage deviation from the least squares mean (approx. s.e. F_1 2.0 R_1 3.0)

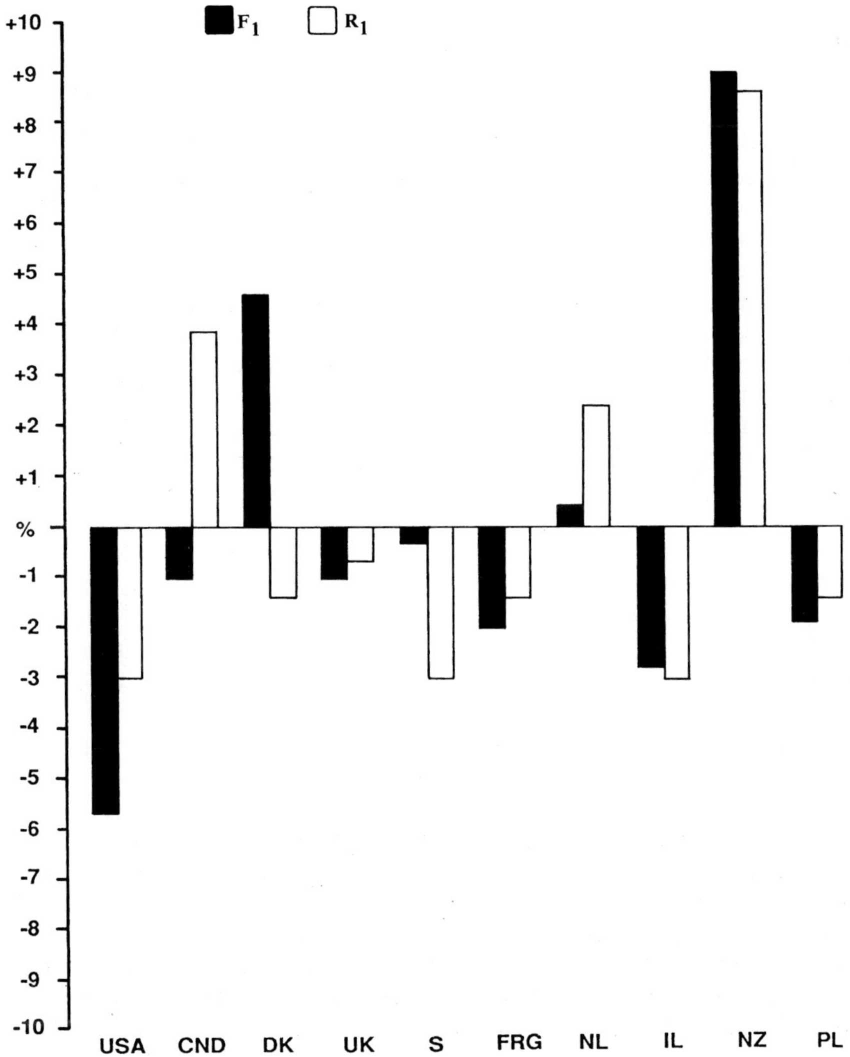

2.5 DISCUSSION — GROWTH IN BOTH FIELD AND INTENSIVE-FEEDING TRIALS AND PERFORMANCE IN BEEF PRODUCTION TRAITS

Growth rate in young cattle is clearly of economic importance not only for beef production but also in rearing breeding females, although the correct balance will depend on the relative values of land, labour and capital which is put into the enterprise. Faster growth rate in females will lower the age of calving and in meat production will speed the turn-round of stock and affect carcass quality. The achievement of faster growth and the development of intensive-feeding practices has been encouraged by the availability of relatively cheap concentrates (Gravert, 1976). The effect of increased feeding is seen clearly by comparing about 700 g/day gain for bulls up to 12 months of age in the field trial, which can be regarded as semi-intensive, with the much faster rates ($<$ 1 000 g/day) for males intensively fed from 110 days old until slaughter. The use of concentrates in the intensive-feeding beef trial was double that in the field trial.

The results also provide interesting evidence on variation in growth among strains. The Holstein strains (USA, Canadian and Israeli) were generally the fastest and Friesian strains grew more slowly. Oldenbroek and Van Eldik (1980) and Geay and Robelin (1979) have shown a higher intake of dry matter for Holstein cattle, possibly on account of a higher innate appetite. Higher intakes in themselves, however, have little importance unless related to gains. As shown in the present experiment, net energy intakes (feed intake per unit of gain) were lower in the faster-growing Holstein strains; in other words, they were more efficient. It is of interest in this context that the faster growth of Holsteins in relation to the other strains was more marked in the intensive-feeding trial than under the less intensive feeding regime of the field trial. This difference in the relative advantages of the different strains may well be associated with the selection objectives and conditions under which the strains were developed in their countries of origin. Thus, by way of illustration from the results — though the comparison is not strictly over the same age period — F_1 males of the USA strain grew less than 1 percent faster than those of the New Zealand strain in the field trial, but 11 percent faster under intensive feeding. The different magnitude of strain differences in body size under the two systems of feeding may also explain why some reports, though not all, suggest that Holstein strains have a considerably greater mature size than European Friesians (10 percent or more — Turton, 1981). At five years old a sample of the intensively-fed animals of these strains also showed the USA strain crosses to weigh approximately 11 percent more than New Zealand strain crosses — which would correspond to a 20 percent or greater difference among the pure strains.

Within the constraints which had to be imposed on even a large-scale experiment, the numbers contributing to the backcross generation (R_1) were only a quarter (approximately) of those contributing to the first crosses (F_1) in the field trial and about half the numbers of the F_1 in the intensive feeding trial. In consequence, much of the apparent variation among strains in the R_1 generation has been found to be statistically non-significant even where that among F_1 strains was significant when results from the two slaughter weights were pooled. However, where the strain rankings in the two generations support each other, there is additional confidence in the reality of the differences and in their potential usefulness in transmitting their breeding value (additive genetic) to crosses.

Growth itself is not, however, the whole story. Size of alimentary tract, gut fill and other factors may also vary and, as shown in the intensive trial, the Holstein strains, in particular, were below average in dressing percentage and some of the European Friesian strains, particularly the British, Dutch and Swedish, were above. Thus the gain in carcass weight and even more especially the gain in lean mass provides a more important comparison of strains. Strain variation in net daily gain was, as expected from the above, less than for daily gain of the whole animal. Since there were differences among strains in the proportion of fat and also in the proportion of bone, strain differences in mean daily lean gain were no longer strictly related to growth rate of the live animal alone; although for a combination of reasons the USA strain ranked first on overall average (in the F_1 generation first at both slaughter weights but behind Sweden and Germany in the R_1 generation) and the New Zealand strain — also the fattest — ranked consistently at the other end.

The ranking of the strains for energy consumption per unit of gain, or lean gain, follows reasonably in parallel that for gain itself. Thus in the absence of direct measures of efficiency, rates of gain provide a useful indicator of merit in efficiency particularly when, as in the case of grazing ruminants, feed intake is only loosely controlled. Caution, however, is required in interpreting the present results in "efficiency" terms because slaughter was at fixed weights. The strains of greater mature size (e.g. the Holsteins) would therefore be at an earlier stage of maturity at that slaughter point and in consequence relatively leaner and faster growing than strains with smaller mature size. In general, feed intake per unit gain declines with increases in total intake and daily gain, but it is the selection of animals which are the exceptions to this generality which would most directly lead to genetic improvement in feed efficiency, especially within breeds.

The present results support those of Weniger *et al.* (1968), Nielsen cited by Duplan (1974), Ernst *et al.* (1973), Huth (1981), Flanagan (1982) and Raue (1983) all of whom found that crossbreds sired by Holstein strains had a greater growth potential than those sired by European Friesians. Slower growth of young bulls of the New Zealand strain was also reported by Flanagan (1982). Oldenbroek (1982) and Brabander *et al.* (1982), however, found Holsteins to do less well in fattening trials both by lower growth and higher feed intake than Dutch Friesians. For Danish performance testing stations Bech Andersen (1982) reported large variation among individuals within strains — thus some Holstein bulls were outstandingly good but others considerably inferior to bulls of the Danish type. Southgate (1982) found that Holstein-Friesians had similar weight gains but higher feed intake relative to British Friesians.

Differences in shape (conformation) are more difficult to demonstrate than those in weight. Among the strains the heaviest were generally also the tallest (height at withers) and had greater chest girth, but it appears also that there were differences in shape since the relative differences in height among the strains are not the same as those for girth. Variation in height at withers after first calving, as shown by Ernst (1968), Witt *et al.* (1971), Jasiorowski *et al.* (1979), Turton (1981) and Gravert (1971), is positively correlated with the onset of lactation yield.

There was little significant variation in components of carcass merit such as the proportion of valuable cuts and area of *l. dorsi* muscle. However, proportions of fat, lean and bone and some of the ratios of these to each other did vary among strains — but significantly so only for the F_1 generation. The absence of significant variation in the backcross generation may be due to the smaller number of animals (half that of F_1) or because strain variation could be less at the lower mean slaughter weight of the backcrosses. (Comparison of the two generations within slaughter weight has not yet been feasible but would also be based on smaller numbers.) Preston and Willis (1974) have shown that fat content, in particular, depends on genotype, level of feeding, slaughter weight and sex of animal.

In the present intensive-feeding beef trial, the USA and Israeli Holsteins and Swedish Friesians had the lowest carcass fat content and, perhaps in consequence, the first two of these the lowest energy costs of producing lean. Conversely New Zealand Friesians were at the opposite end among the strains for both traits. As indicated earlier this could be associated with a lower degree of maturity of the Holsteins relative to the New Zealand and other strains at the fixed slaughter weight.

However, all three Holstein strains and German Friesians (which have a larger percentage of Holstein blood than other European strains) had an above-average proportion of bone in the carcass. This unfavourable point has been previously noted by Duplan (1974), Cook and Newton (1979), Oldenbroek (1982), Flanagan (1982) and Southgate (1982). New Zealand and British Friesians had the lowest percentages of bone and in consequence the best lean to bone ratios.

The highest lean content as such was found in the Swedish and Polish strains of Friesian noted for their dual-purpose use, but the latter strain suffered from above-average fat, thus not providing a particularly favourable lean to fat ratio.

The results show that lean meat can be successfully produced from dairy calves from high producing herds. The more single-purpose (Holstein) dairy strains were thought to score less well on carcass quality points as evidenced by smaller *l. dorsi* muscle area (although strain differences were not statistically significant).

However, total economic efficency depends on a combination of quantitative (daily lean gain and feed efficiency) and qualitative traits in relation to the available feeding system, consumer preferences and beef prices (Colleau and Lienard, 1982). Crossbred bulls fattened for slaughter and sired by Holstein strains from the USA, Canada and Israel produced quantities of meat no less than those from European dual-purpose strains of Friesian. This is a particularly important conclusion for countries where quantity takes precedence over quality. Cunningham and Moioli (1982), however, reported that in investigations under European conditions the profitability of beef production from European Friesians was between 5 and 10 percent greater than that based on Holsteins. This difference in monetary value of the two types of carcass is greater than might have been deduced from the differences in carcass composition observed in the present experiment. However, such differences will be influenced by the methods of carcass classification and the system of pricing which may not discriminate so efficiently against fat (as is well known in some countries) or unduly favour factors of shape or quality which are not readily quantified.

The present experiment was designed to provide estimates of heterosis by comparisons of first cross with backcross generations. In the field trial, backcrosses were made to both the imported and the Polish bulls (for the intensive feeding trial the results on backcrosses to the Polish strain are available, so far, only for a restricted number of traits and not in respect of those considered here). Under the circumstances of the field trial, the average of all F_1 strains would be expected to have the same additive genetic component as the average of the R_1 strains (with a reservation about the respective samples of bulls used as sires). This average F_1-R_1 difference would normally be interpretable in terms of heterosis — and the observed reduction in R_1 performance relative to F_1 for many traits goes in the direction expected from recombinant loss. However, such interpretation may not be valid here since the two generations are not fully contemporary and a temporary, reduced availability of feed depressed R_1 performance in the field trial. The reduction due to feed supply can be inferred from the decline, observed for many traits, in the performance of pure Polish animals contemporary with F_1 and R_1 generations. Moreover, in the field trial, only a proportion of the sires used were common to both generations; though it has been argued by Jasiorowski *et al.* (in press) that the effect of this partial confounding is minimized by the similar breeding values of the two original groups of bulls involved. However, the procedure adopted here, equally valid for the field and intensive-feeding trials, was to compare strain crosses with the pure Polish strain within generation and apply the appropriate model for additive and non-additive genetic contributions. This approach is less dependent on contemporaneity between the generations, but caution is still necessary in the interpretation.

With these reservations, the present results suggest that, in general, heterosis has been quite unimportant in its contribution to growth traits in field conditions and to growth and carcass traits under intensive feeding conditions and, importantly, based on strain differences within a breed and not as commonly, on breed differences. An absence of heterosis in strain comparisons was noted by Oldenbroek (1982), but Przybylska (1985) found a significant effect of heterosis in other strain comparisons. Dickerson (1962) and Donald *et al.* (1977) reported small but significant effects of heterosis between breeds. However, as noted earlier, specific strains themselves at the more extreme ends of the range of performance in these trials gave a small number of marginally significant heterosis estimates (see also Reklewski *et al.*, 1985). From this and the limited evidence from the literature it can be concluded that crosses among breeds show perhaps twice the heterosis found in crosses among strains (Baumung and Panicke, 1986; Cundiff *et al.*, 1986; and Pedersen and Christensen, 1986).

2.6 RESULTS FROM HEIFERS DURING FIRST LACTATION IN THE FIELD TRIAL AND INTENSIVE-FEEDING TRIAL

2.6.1 Age at first calving

The mean age was 885 days for F_1 and 916 for R_1 heifers in the field trial (Tables 2.35 and 2.36) and somewhat earlier at 870 and 837 days (28.6 and 27.5 months) respectively in the intensive-feeding trial (Tables 2.37 and 2.38) but somewhat later (900 days) in the backcrosses to the Polish strain which are shown separately in Table 2.38.

In both the F_1 and R_1 generation of the field trial, variation among strains shown in Fig. 2.34 was highly significant ($P < 0.01$ as shown in Tables 2.39 and 2.40) with the three Holstein strains and the German strain most clearly below average and other European Friesians, and particularly Polish, calving later. Strain differences in age at first calving of the females in the intensive trial (Fig. 2.35) varied between the F_1 and R_1 generations and did not accord well with those in the field trial, though the differences among strains were not significant (Tables 2.37 and 2.38).

2.6.2 Liveweight and body dimensions of first calvers

In the field trial this was recorded in the first few days after calving; in the intensive trial 10 and 305 days after calving. Average liveweights of animals in the field trial were 493 kg for F_1 and 476 kg for R_1 heifers (Tables 2.35 and 2.36) and heifers in the intensive trial 10 days after calving weighed 474 kg for F_1 and 463 kg for R_1 (Tables 2.37 and 2.38). During lactation in the intensive trial the heifers gained substantially in weight and were 565 kg (F_1) and 543 (R_1) by 305 days after calving. The lower weight of the backcrosses relative to first crosses is consistent in both field and intensive trials.

Fig. 2.34 Age at first calving of females in the field trial showing each strain as a percentage deviation from the overall mean (approx. s.e. $F_1$0.4 $R_1$1.1)

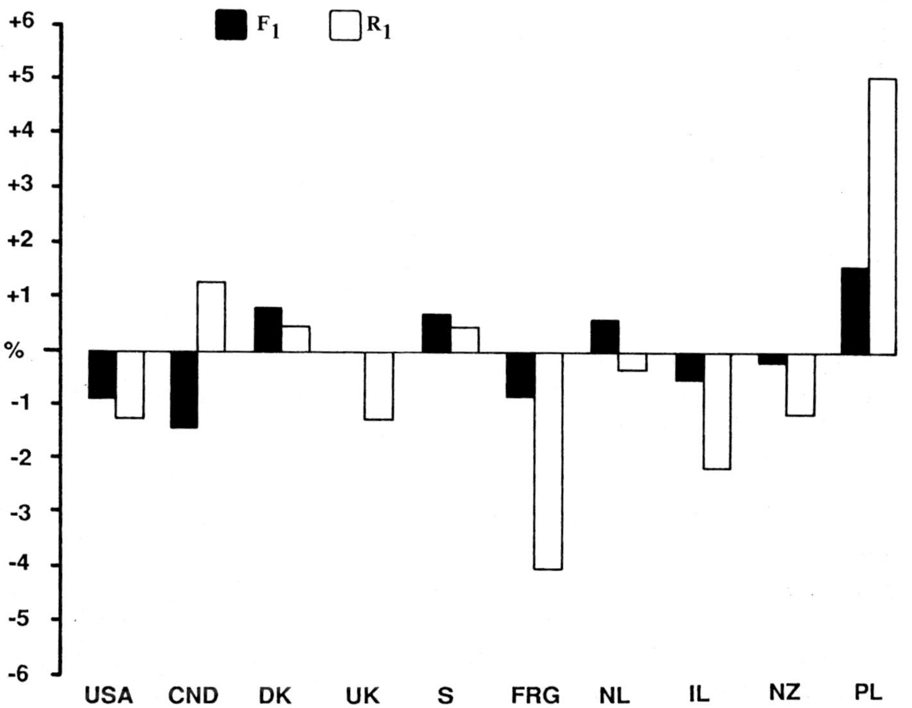

Table 2.35 **Overall mean and s.d. for age at calving, days in milk, body weight and chest girth as well as number for days open of F_1 heifers (field trial)** (from M. Stolzman *et al.*, unpublished)

Paternal strain origin	No. of animals	Age at calving *(days)*	Days in milk	No. of animals	Live-weight *(kg)*	Chest girth *(cm)*	No. of animals	Days open
USA	553	876	333	510	500	184.5	530	123
		97	70		43	7.6		71
CND	619	875	327	556	497	184.3	585	111
		95	65		44	7.9		65
DK	589	884	327	538	491	183.2	568	119
		103	66		44	7.2		74
UK	622	887	326	574	488	182.8	594	117
		98	69		43	7.1		71
S	641	890	322	596	495	183.7	610	112
		104	63		44	7.4		65
FRG	637	873	321	575	493	183.7	610	114
		93	63		43	7.3		67
NL	574	887	318	522	493	183.7	548	111
		101	61		47	7.0		65
IL	634	880	336	582	497	184.2	605	122
		100	71		47	7.7		71
NZ	647	885	333	601	489	183.4	610	125
		97	71		44	7.2		75
Pure Polish	988	902	324	862	493	183.2	925	118
		106	67		45	7.5		72
Total	6 504	885	325	5 916	493	183.6	6 185	117
		100	63		45	7.4		70

Table 2.36 **Overall mean and s.d. for age at calving, days in milk, body weight and chest girth as well as for days open of backcross (R_1) heifers (field trial)** (from M. Stolzman *et al.*, unpublished)

Strain origin	No. of animals	Age at calving *(days)*	Days in milk	No. of animals	Live-weight *(kg)*	Chest girth *(cm)*	No. of animals	Days open
USA	87	898	329	67	493	178.4	84	128
		121	62		49	7.9		70
CND	109	919	325	92	486	178.3	99	119
		135	72		53	7.5		69
DK	89	933	335	71	464	178.7	85	123
		138	62		51	7.2		77
UK	86	925	328	68	473	178.8	76	124
		108	72		42	8.7		74
S	121	909	330	101	483	177.5	112	124
		122	71		47	7.7		68
FRG	105	897	312	85	473	177.5	102	116
		109	59		40	7.5		72
NL	73	924	315	61	464	178.1	69	107
		129	69		49	7.1		61
IL	105	883	333	86	475	178.9	100	126
		111	69		45	6.4		67
NZ	135	915	318	111	475	178.3	125	112
		121	57		48	7.6		62
PL (R_1)	356	932	314	283	472	179.5	333	116
		123	66		42	7.0		69
Pure Polish	198	910	311	144	484	180.0	181	111
		116	65		36	6.6		68
Total	1 464	916	320	1 169	476	178.8	1 366	118
		122	66		45	7.3		67

Table 2.37 **Least squares means (±s.e.) for age of calving and body weight — F_1 heifers (intensive trial)*** (from Jasiorowski *et al.*, 1983)

Paternal origin		Age of calving *(months)*		Liveweight at 10 days after calving *(kg)*		Liveweight at 305 days of lactation *(kg)*	
		LSM	s.e.	LSM	s.e.	LSM	s.e.
USA	(A)	28.6	0.38	486.6	6.3	572.6	6.7
Canada	(C)	28.9	0.38	476.9	6.3	573.8	6.7
Denmark	(D)	28.7	0.37	470.4	6.2	566.7	6.7
United Kingdom	(E)	28.5	0.37	468.2	6.2	556.5	6.6
Sweden	(F)	28.4	0.38	480.4	6.3	573.2	6.9
Fed. Rep. of Germany	(G)	28.9	0.38	473.6	6.4	566.0	6.9
Netherlands	(H)	29.3	0.38	460.7	6.3	553.6	6.9
Israel	(I)	27.7	0.38	489.6	6.3	573.9	7.0
New Zealand	(K)	28.8	0.38	469.9	6.4	558.4	7.1
Poland (pure)	(B)	28.4	0.39	468.5	6.4	556.7	7.0
Total		28.6	0.12	474.5	2.0	565.1	2.2
Significant differences*						H to I, A, F I to D, K, B, E A to B, E	

* Differences between strains significant at: $P \leq 0.05$ — single underlining
$P \leq 0.01$ — double underlining

Table 2.38 **Least squares means (± s.e.) for age of calving, body weights and height at withers — R_1 heifers (intensive trial)** (from Jasiorowski *et al.*, in press)

Paternal strain origin		**Age of calving** *(months)*		**Weight at 10 days after calving** *(kg)*		**Weight at 305 days of lactation** *(kg)*		**Height at withers 200 days after calving** *(cm)*	
		mean	**s.e.**	**mean**	**s.e.**	**mean**	**s.e.**	**mean**	**s.e.**
USA	(A)	28.3	0.67	466	7	544	9	128	0.7
Canada	(C)	26.7	0.65	467	7	548	9	128	0.7
Denmark	(D)	27.9	0.74	465	7	542	11	126	0.8
United Kingdom	(E)	28.2	0.74	457	7	535	10	125	0.8
Sweden	(F)	27.4	0.69	464	7	551	10	126	0.8
Fed. Rep. of Germany	(G)	26.4	0.75	477	8	557	11	126	0.8
Netherlands	(H)	27.5	0.75	453	7	534	11	125	0.8
Israel	(I)	28.3	0.71	469	7	544	10	129	0.8
New Zealand	(K)	27.3	0.69	453	7	535	10	125	0.8
Poland (pure)	(B)	27.4	0.75	462	8	542	11	124	0.8
Total[1]		27.5	0.33	463	3	543	5	126	0.4
Poland R_1		29.6	0.36	457	5	542	5	124	0.3

Significant differences * I to B,H,K,E,D,F,G

C,A to B,H,K,E,D,F

*Differences between strains significant at: $P \leq 0.05$ — single underlining
$P \leq 0.01$ — double underlining

[1] The totals and analysis of differences between strains apply to the main body of the Table which excludes backcrosses to the Polish strain. These backcross results (75% Polish blood, but also including a group of pure Polish animals, and based on a total of 108 animals) are shown, for comparison, at the foot of the Table (H. Jasiorowski and R. Grabowski, unpublished). Within this backcross to the Polish strain there was no significant strain variation except for height at withers ($P < 0.05$).

Variation among strains is shown in Figs. 2.36 to 2.38 and indicates, as did the weights at earlier ages (cf Figs. 2.7 and 2.8), that the three Holstein strains were above average along with — to a lesser extent — Swedish and German strains, with other European strains and the New Zealand Friesians below. Weight differences among strains (expressed as percentages of the mean) were not, however, greater in the intensive trial than in the field trial, although this might have been expected from the greater strain variation in daily gain recorded in the intensive trial (see earlier). Variation among strains was statistically significant for both generations in the field trial (Tables 2.39 and 2.40) but only some of the strains differed significantly from each other in the F_1 generation in the intensive trial 10 days after calving (Table 2.37).

Height at withers and chest girth were both measured 10 days after calving in the intensive trial. The Holstein strains had clearly the larger body frames (Figs. 2.39 and 2.40) but, as before, small differences in conformation may be deduced from these results.

Fig. 2.35 Age at first calving of females in the intensive–feeding trial showing each strain as a percentage deviation from the least squares mean (approx. s.e. $F_1$1.3 $R_1$2.7)

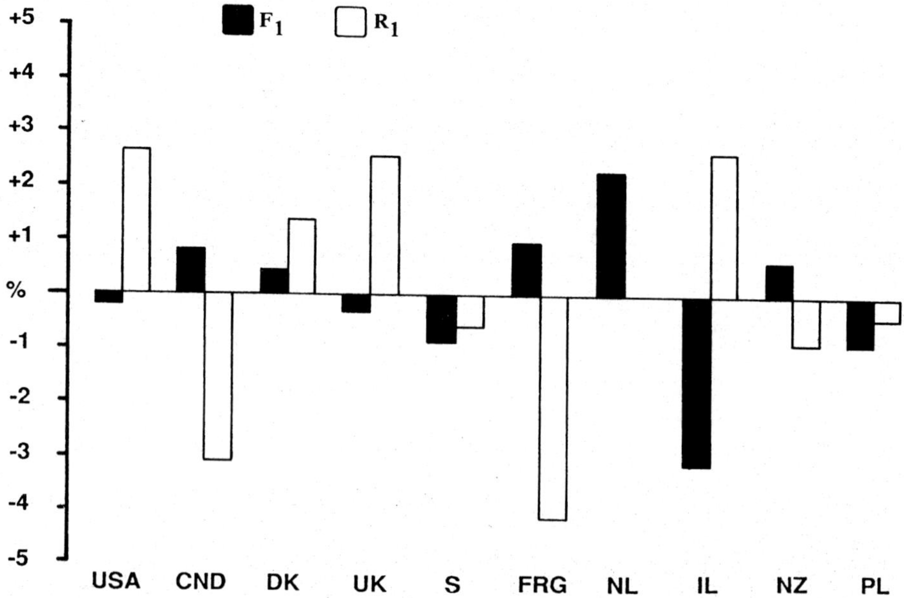

Fig. 2.36 Liveweight at first calving of females in the field trial showing each strain as a percentage deviation from the overall mean (approx. s.e. $F_1$0.3 $R_1$0.09)

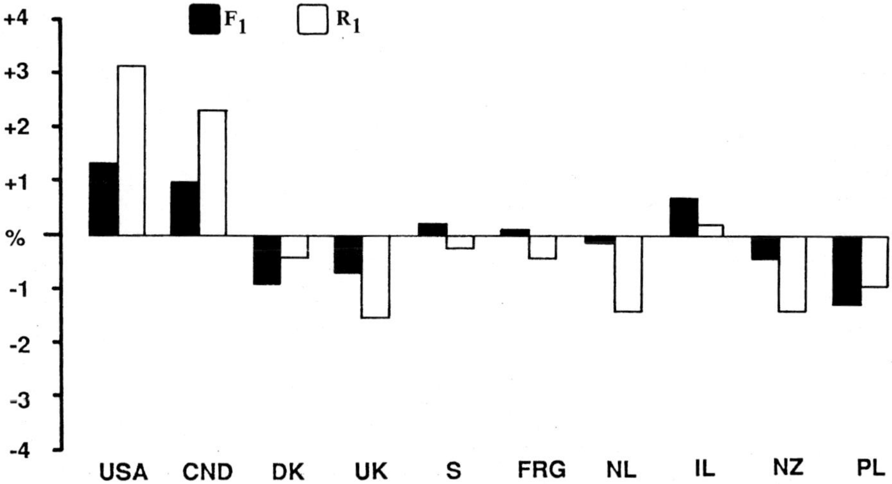

Fig. 2.37 Liveweight 10 days after first calving of females in the intensive–feeding trial showing each strain as a percentage deviation from the least squares mean (approx. s.e. $F_1$1.3 $R_1$1.6)

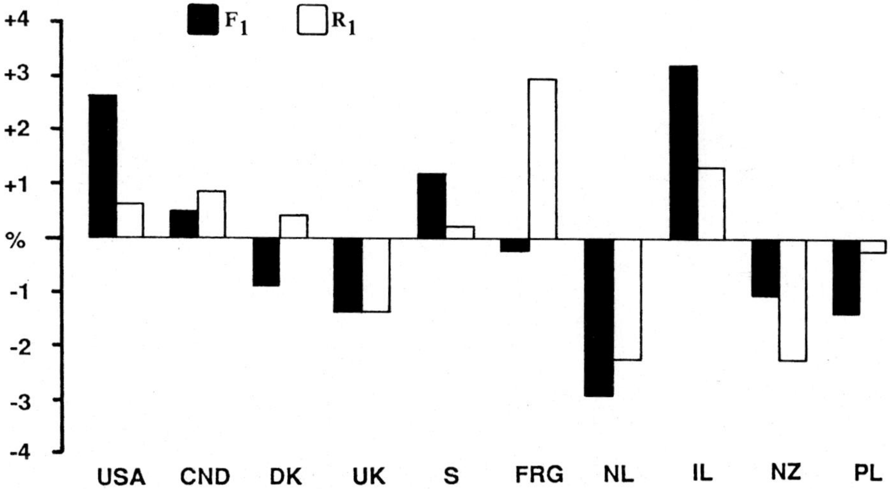

Fig. 2.38 Liveweight 305 days after first calving of females in the intensive–feeding trial showing each strain as a percentage deviation from the least squares mean
(approx. s.e. F_1 1.2 R_1 1.9)

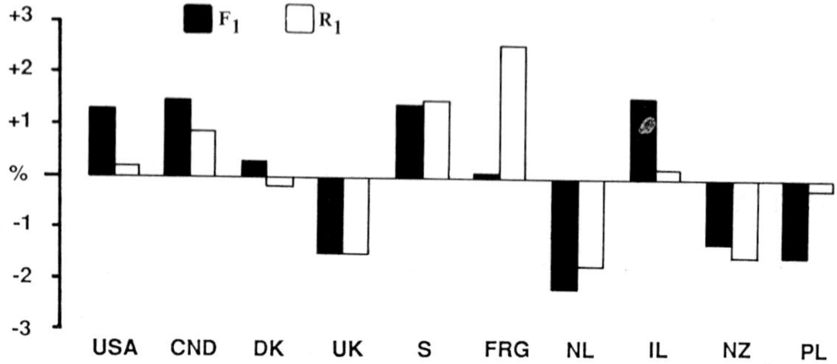

Fig. 2.39 Height at withers 10 days after first calving of females in the intensive–feeding trial showing each strain as a percentage deviation from the least squares mean
(approx. s.e. F_1 0.4 R_1 0.6)

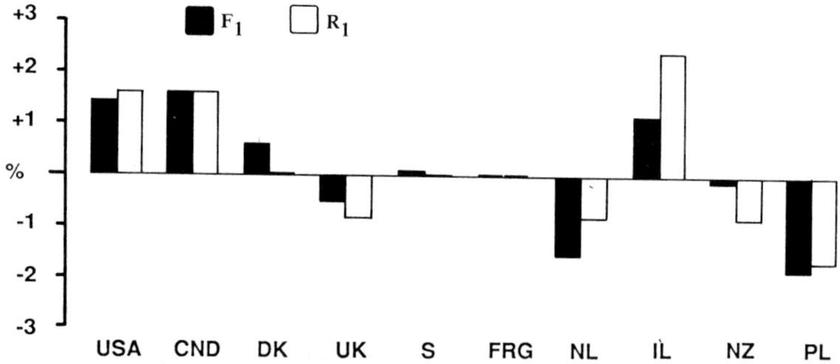

Fig. 2.40 Chest girth at first calving of females in the field trial showing each strain as a percentage deviation from the overall mean
(approx. s.e. F_1 0.1 R_1 0.4)

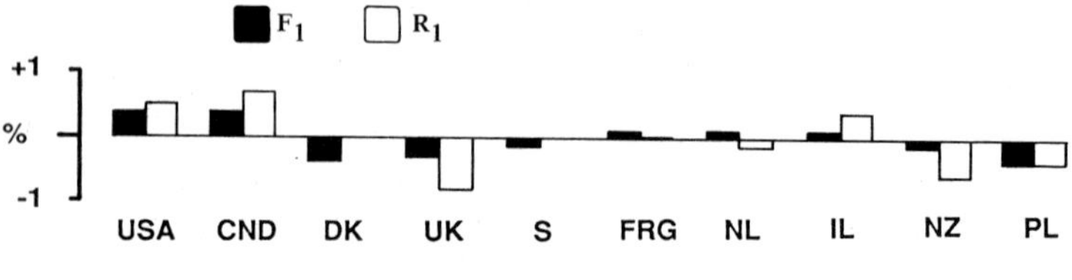

Differences among strains in height at withers were statistically significant (Tables 2.38 and 2.41). In the field trial, chest girth alone could be measured. Variation among strains was very small in percentage terms (Fig. 2.40) but was nonetheless significant (P   0.01) in the F_1 generation. The mean girth for F_1 first-calf heifers was 183.6 cm and 178.8 cm for R_1 animals (Tables 2.35 and 2.36).

2.6.3 Milk yield

The average 305-day yield for first lactation heifers in the field trial was 3 468 kg for the first crosses (Table 2.42) and about 500 kg lower at 2 949 kg for the backcrosses (Table 2.43). Corresponding means in the intensive trial were clearly higher, as might be expected from the difference in feeding, at 4 970 kg for F_1 and 4 726 kg for R_1 heifers but the mean for the backcrosses to the Polish strain was also lower at 3 968 kg and inclusion of this figure would make the average performance of R_1 females in the field and intensive trials less dissimilar (Tables 2.44 and 2.45). As seen from Figs. 2.41 and 2.42, under both sets of feeding conditions the 3 Holstein strains gave between 5 and 11 percent more milk than the average and the Polish strain was most clearly below. It is noteworthy that the New Zealand strain was above average in milk yield under field but at the average level under intensive conditions.

Crosses of the USA and Canadian strains provided estimates of *heterosis* of 8 percent (s.e. 5.8 percent) of the mean yield of Polish females in the field trial. For the intensive trial all estimates of heterosis were lower than their standard errors but averaged 4.6 percent for milk yield.

2.6.4 Butterfat yield and fat content of milk

The mean performance was 139 kg butterfat for F_1 and 115 kg for R_1 heifers in the field trial (Tables 2.42 and 2.43) and 196 and 188 respectively in the intensive trial (Tables 2.44 and 2.45) — proportionately a very similar increase over the field trial as for milk yield alone because the fat content was not greatly influenced by the feeding regimes (4.01 percent in F_1 and 3.90 percent in R_1 of field trial, and 3.97 percent and 3.99 percent respectively in the intensive trial (cf Tables 2.42 with 2.43 and 2.46 with 2.47)). Again, the backcrosses to the Polish strain in the intensive trial (results appended to Tables 2.45 and 2.47) show a lower performance and if included in the average would slightly reduce the difference between the averages quoted for the two trials.

In terms of variation among strains the picture mirrors closely that seen for milk yield although the New Zealand strain is seen to be more clearly above

Table 2.39 **Body size and breeding traits of first cross (F_1) heifers in the field trial — the effects'[1] of paternal strains as deviations (±s.e.) from the mean** (from M. Stolzman *et al.*, unpublished)

Paternal strain origin	Liveweight after calving *(kg)***		Chest girth after calving *(cm)***		Age at first calving *(days)***		Days open*	
	Effect	s.e.	Effect	s.e.	Effect	s.e.	Effect	s.e.
USA	6.6	1.59	0.82	0.26	-8	3.5	8.8	3.03
Canada	5.0	1.56	0.86	0.26	-12	3.4	-3.3	2.98
Denmark	-4.4	1.57	-0.66	0.26	7	3.5	-0.2	3.00
United Kingdom	-3.5	1.52	-0.56	0.25	0	3.3	-0.2	2.90
Sweden	1.0	1.47	-0.11	0.24	6	3.3	-4.7	2.88
Fed. Rep. of Germany	0.7	1.52	0.16	0.25	-7	3.3	-4.9	2.88
Netherlands	-0.8	1.56	0.18	0.26	5	3.4	-4.4	2.99
Israel	3.3	1.54	0.21	0.25	-4	3.4	4.5	2.96
New Zealand	-1.8	1.39	-0.17	0.23	-1	3.2	6.4	2.76
100% Polish	-6.0	1.40	-0.72	0.23	14	3.1	-2.0	2.67

Significance of variation among strains: * ($P < 0.05$)
** ($P < 0.01$)

Approximate least significant differences[2]

$P < 0.01$	5.90	0.97	13	11.3
$P < 0.05$	4.56	0.75	10	8.8

[1] [2] See footnotes Table 2.20

Table 2.40 **Body size and breeding traits of backcross (R_1) heifers in the field trial — the effect of paternal strain as deviations (±s.e.) from the mean** (from M. Stolzman *et al.*, unpublished)

Paternal strain origin	Liveweight after calving *(kg)***		Chest girth after calving *(cm)*		Age at first calving *(days)***		Days open*	
	Effect	**s.e.**	**Effect**	**s.e.**	**Effect**	**s.e.**	**Effect**	**s.e.**
USA	14.7	4.34	0.98	0.77	-11	10.5	11.6	8.84
Canada	11.0	4.10	1.33	0.72	12	10.3	13.5	8.67
Denmark	-2.1	4.56	-0.10	0.80	5	10.9	-3.1	9.20
United Kingdom	-7.1	4.81	-1.46	0.85	-11	11.4	-1.0	9.77
Sweden	-0.9	3.93	-0.08	0.69	5	9.7	5.7	8.28
Fed. Rep. of Germany	-1.9	4.01	-0.01	0.71	-37	9.9	-5.6	8.24
Netherlands	-6.4	4.56	-0.13	0.80	-3	11.7	-15.5	9.70
Israel	0.8	3.98	0.65	0.70	-19	10.2	13.9	8.49
New Zealand	-6.3	3.48	-1.11	0.61	-10	9.6	-5.5	7.68
Poland (R_1)	-4.3	3.01	-0.64	0.53	47	7.5	-2.4	6.38
100% Polish	2.5	3.62	0.57	0.64	24	8.7	-11.7	7.47

Significance of variation among strains: ** ($P < 0.01$)

Approximate least significant differences[2]

$P < 0.01$	14.4	2.65	37	31.4
$P < 0.05$	11.2	2.05	29	24.3

2 See footnotes Table 2.20

Fig. 2.41 Milk yield of females in first lactation in the field trial showing each strain as a percentage deviation from the overall mean (approx. s.e. $F_1$0.8 $R_1$2.4)

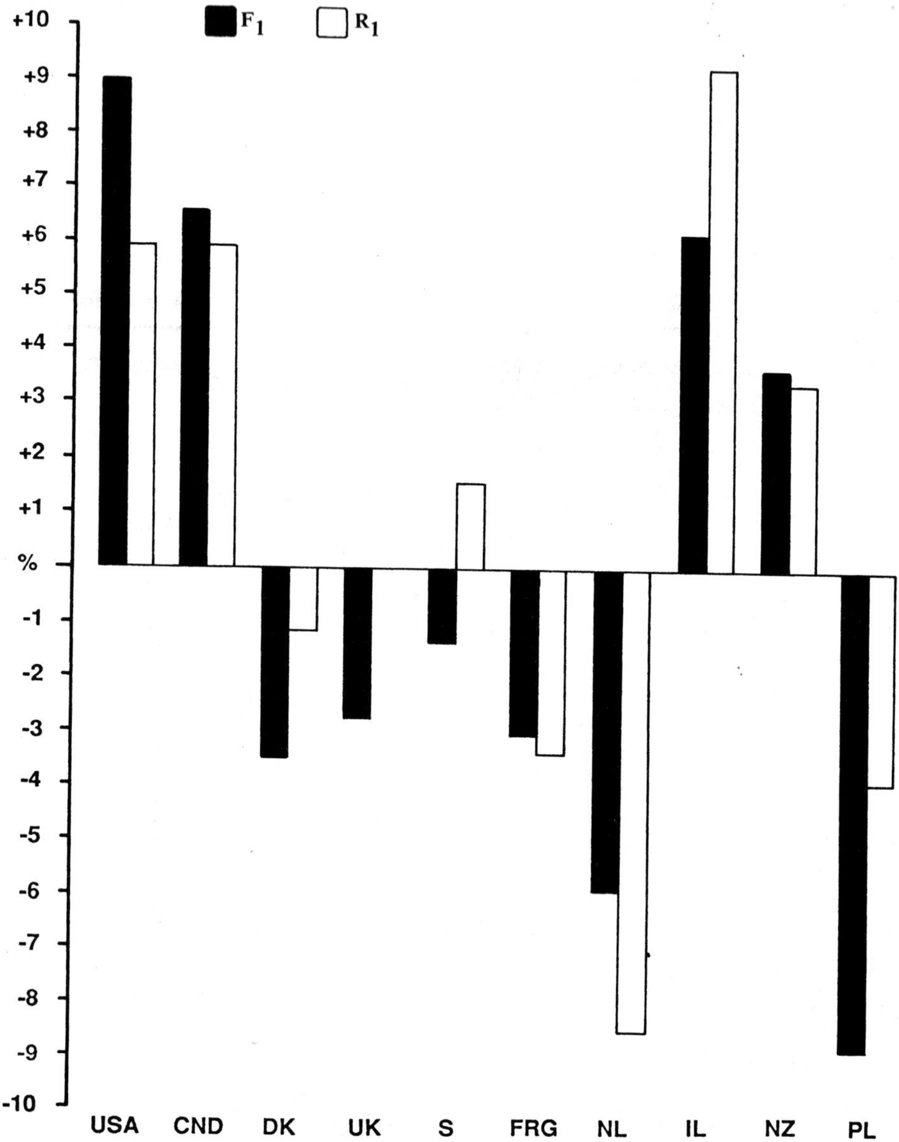

Fig. 2.42 Milk yield of females in first lactation in the intensive-feeding trial showing each strain as a percentage deviation from the least squares mean
(approx. s.e. $F_1$2.9 $R_1$3.8)

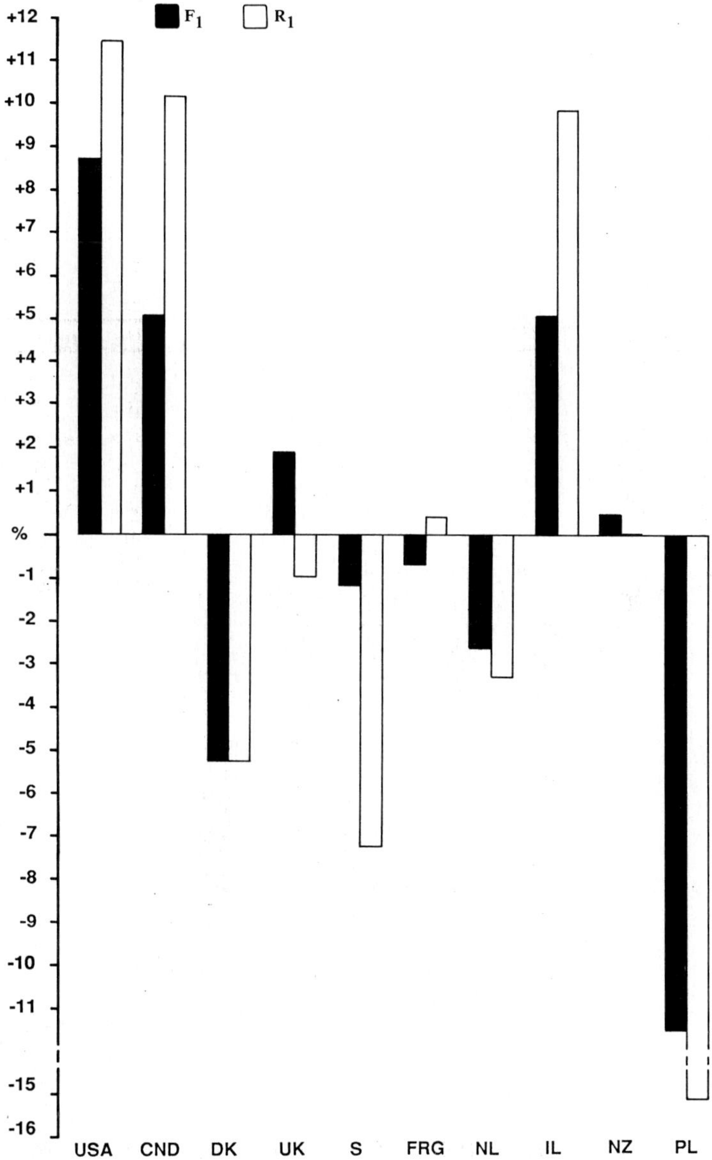

A selection of F_1 cows with sires of the different paternal strains from dams of the Polish Black and White breed

Sire from Canada
Average milk yield
5 525 kg, 4.0% fat.

Photo
Alojzy Zawadski

Sire from Denmark.
Average milk yield
5 110 kg, 4.0% fat.

Sire from the Federal Republic of Germany. Average milk yield 4 218 kg, 4.16 fat

Sire from Israel. Average mmilk yield 5 487 kg, 3.89 fat

Sire from Netherlands. Average milk yield 4 180 kg, 4.43% fat

Sire from New Zealand. Average milk yield 8 364 kg, 4.30% fat

Sire from Poland (pure Polish strain). Average milk yield 5 663 kg, 4.23% fat

Sire from Sweden. Average milk yield 8 364 kg, 3.99% fat

Sire from the United Kingdom. Average milk yield 4 869 kg, 3.91% fat

Sire from the United States of America. Average milk yield 5 816 kg, 4.07% fat

Table 2.41 **Least squares means ($\pm$s.e.) for body measurements — F_1 heifers (200 days after calving)* (intensive trial)** (from Jasiorowski *et al.*, 1983)

Paternal origin		Height at withers (cm)		Chest girth (cm)	
		LSM	s.e.	LSM	s.e.
USA	(A)	129.5	0.52	195.7	1.09
Canada	(C)	129.8	0.52	195.1	1.09
Denmark	(D)	127.0	0.51	193.8	1.07
United Kingdom	(E)	127.1	0.52	192.9	1.08
Sweden	(F)	127.8	0.52	194.6	1.09
Fed. Rep. of Germany	(G)	127.8	0.53	195.1	1.10
Netherlands	(H)	125.8	0.53	192.4	1.10
Israel	(I)	129.3	0.53	195.5	1.09
New Zealand	(K)	127.6	0.53	193.3	1.10
Poland (pure)	(B)	125.5	0.53	193.9	1.11
Total		127.7	0.17	194.2	0.35

Significant differences*

B to C,A,I,F,G,K,E,D

C to F,G,K,E,D

H to C,A,I,F,G,K

I to G,K,E,D

A to F,G,K,E,D

*Differences between strains significant at: $P \leq 0.05$ — single underlining
$P \leq 0.01$ — double underlining

Table 2.42 **Overall mean and s.d. for 305-day lactation yield of F_1 heifers (field trial)** (from M. Stolzman *et al.*, unpublished)

Paternal strain origin	305-day lactation						
	No. of animals	Milk yield *(kg)*	Fat yield *(kg)*	Fat content *(%)*	No. of animals	Protein yield *(kg)*	Protein content *(%)*
USA	553	3 783	148.9	3.93	544	121.5	3.19
		1 065	44.1	0.29		34.7	0.17
CND	619	3 695	147.1	3.98	616	119.0	3.21
		968	40.3	0.30		32.4	0.19
DK	589	3 371	135.7	4.02	584	109.5	3.24
		896	38.5	0.29		29.6	0.18
UK	622	3 378	135.3	4.01	611	110.2	3.25
		936	38.5	0.29		32.6	0.19
S	641	3 438	137.7	3.99	628	111.8	3.24
		899	38.8	0.28		30.4	0.18
FRG	637	3 332	132.8	3.99	621	108.0	3.23
		916	37.5	0.29		29.6	0.17
NL	574	3 256	131.4	4.03	569	106.3	3.25
		873	36.5	0.29		29.5	0.18
IL	634	3 713	147.5	3.97	625	119.3	3.21
		977	40.3	0.32		32.5	0.19
NZ	647	3 576	146.8	4.11	637	116.4	3.25
		939	39.7	0.34		31.4	0.19
Pure Polish	988	3 265	131.5	4.02	970	107.1	3.27
		860	36.5	0.27		28.7	0.17
Total	6 504	3 468	139.0	4.01	6 405	112.6	3.24
		948	39.6	0.30		31.4	0.18

Table 2.43 **Overall mean and s.d. for 305-day lactation yield of backcross (R_1) heifers (field trial)** (from M. Stolzman *et al.*, unpublished)

Paternal strain origin	305-day lactation						
	No. of animals	Milk yield (kg)	Fat yield (kg)	Fat content (%)	No. of animals	Protein yield (kg)	Protein content (%)
USA	87	3 229	123.9	3.82	36	116.9	3.19
		1 018	41.1	0.31		35.7	0.18
CND	109	3 373	128.9	3.85	53	126.3	3.21
		1 028	38.3	0.32		31.0	0.13
DK	89	3 092	120.8	3.88	41	117.4	3.24
		1 026	43.8	0.34		40.6	0.14
UK	86	2 966	116.2	3.92	28	108.6	3.24
		1 050	42.5	0.33		39.4	0.17
S	121	3 161	122.5	3.88	70	104.0	3.20
		922	36.6	0.33		34.3	0.15
FRG	105	2 803	109.9	3.90	50	99.2	3.22
		954	40.4	0.35		32.6	0.15
NL	73	2 692	107.7	3.99	33	95.5	3.23
		766	32.5	0.34		26.0	0.16
IL	105	3 253	123.9	3.80	48	116.8	3.17
		967	39.8	0.31		31.9	0.16
NZ	135	3 103	123.3	3.96	70	107.7	3.26
		928	39.1	0.35		29.6	0.18
PL (R_1)	356	2 684	105.5	3.92	186	97.3	3.27
		953	39.7	0.36		35.7	0.18
Pure Polish	198	2 775	109.7	3.93	121	98.2	3.28
		845	36.0	0.32		26.6	0.18
Total	1 464	2 949	115.3	3.90	736	105.0	3.24
		978	39.9	0.34		34.2	0.17

Table 2.44 **Least squares means (±s.e.) for milk production — F_1 heifers* (intensive trial)** (from Jasiorowski *et al*, 1983)

Paternal origin		Milk (kg)		Butterfat (kg)		Protein (kg)		Total solids (kg)	
		LSM	s.e.	LSM	s.e.	LSM	s.e.	LSM	s.e.
USA	(A)	5 402	144	197.2	5.7	182.7	4.8	649.6	17.2
Canada	(C)	5 225	145	204.8	5.7	178.5	4.8	645.9	17.3
Denmark	(D)	4 706	141	187.9	5.6	168.8	4.7	592.3	16.8
United Kingdom	(E)	5 065	143	200.6	5.7	177.8	4.8	635.0	17.1
Sweden	(F)	4 909	142	194.3	5.6	175.2	4.7	619.2	17.0
Fed. Rep. of Germany	(G)	4 933	146	192.5	5.8	172.9	4.9	616.8	17.5
Netherlands	(H)	4 839	144	199.2	5.8	174.6	4.8	617.6	17.2
Israel	(I)	5 222	144	198.0	5.7	181.8	4.8	645.9	17.3
New Zealand	(K)	4 996	145	206.0	5.7	180.1	4.8	633.6	17.3
Poland (pure)	(B)	4 397	145	177.1	5.7	157.2	4.8	555.0	17.3
Total		4 970	—	195.8	2.2	174.9	1.8	621.1	6.6

Significant differences*

— milk yield: B to A,C,I,E,K,G,F,H; D to A,C,I; A to K,G,FH

— total solids yield: B to A,C,I,E,K,F,H,G; D to A,C,I; A to H

— butterfat yield: B to K,C,E,H,I,A,F,G; D to K,C

— protein yeld: B to A,I,K,C,E,F,H,G; D to A,I

*Differences between strains significant at: $P < 0.05$ — single underlining
$P < 0.01$ — double underlining

Table 2.45 **Least squares means (± s.e.) for milk production R_1 heifers (intensive trial)** (from Jasiorowski *et al.*, in press)

		Milk (*kg*)		Fat (*kg*)		Protein (*kg*)		Total solids (*kg*)	
		mean	s.e.	mean	s.e.	mean	s.e.	mean	s.e.
USA	(A)	5 268	143	195.5	5.8	173.9	4.6	628.8	17.2
Canada	(C)	5 209	142	202.2	5.8	173.9	4.5	628.9	17.1
Denmark	(D)	4 477	150	181.3	6.1	157.1	4.8	566.1	18.1
United Kingdom	(E)	4 678	150	187.3	6.1	159.5	4.8	577.1	18.1
Sweden	(F)	4 379	147	173.5	6.0	152.5	4.7	542.9	17.7
Fed. Rep. of Germany	(G)	4 746	152	180.3	6.2	160.7	4.9	567.4	18.4
Netherlands	(H)	4 570	154	185.2	6.3	154.3	4.9	563.8	18.5
Israel	(I)	5 192	151	197.8	6.2	175.6	4.8	632.6	18.2
New Zealand	(K)	4 725	143	203.6	5.9	168.0	4.6	603.2	17.3
Poland (pure)	(B)	4 012	147	169.7	6.0	142.8	4.7	507.0	17.7
Total[1]		4 726	50	187.6	2.0	161.8	1.6	581.8	6.0
Poland R_1		3 968	83	155.0	3.1	134.4	2.6	511.9	13.2

Significant differences*

— milk
A,C to B,F,D,H,E,K,G
I to B,F,D,H,E,K,G
B to G,K,E,H,D

— fat
K to B,F,G,D,H,E
C to B,F,G,D,H
I to B,F,G,D
A to B,F
E to B

— protein
I to B,F,H,D,E,G
C,A to B,F,H,D,E,G
K to B,F,H
B to G,E,D

— total solids
I to B,F,H,D,G,E
C,A to B,F,H,D,G,E
K to B,F
B to E,G,D,H

*Differences between strains significant at: P ≤ 0.05 — single underlining
P ≤ 0.01 — double underlining

[1] See footnote to Table 2.38

Table 2.46 **Least squares means for milk composition — F_1 heifers (intensive trial)** (from Jasiorowski *et al.*, 1983)

Paternal origin		Butterfat *(%)*		Protein *(%)*		Total solids *(%)*	
		LSM	s.e.	LSM	s.e.	LSM	s.e.
USA	(A)	3.69	0.05	3.40	0.03	12.11	0.08
Canada	(C)	3.94	0.05	3.47	0.03	12.42	0.08
Denmark	(D)	3.99	0.05	3.59	0.03	12.59	0.08
United Kingdom	(E)	3.98	0.05	3.53	0.03	12.54	0.08
Sweden	(F)	3.97	0.05	3.57	0.03	12.63	0.08
Fed.Rep. of Germany	(G)	3.93	0.05	3.55	0.03	12.52	0.08
Netherlands	(H)	4.14	0.05	3.62	0.03	12.78	0.08
Israel	(I)	3.83	0.05	3.49	0.03	12.39	0.08
New Zealand	(K)	4.14	0.05	3.59	0.03	12.71	0.08
Poland (pure)	(B)	4.06	0.05	3.59	0.03	12.65	0.08
Total		3.97	—	3.54	—	12.53	—

Significant differences[1]

— butterfat %
A to H,K,B,D,E,F,C,G,I
I to H,K,B,D,E,F
H to D,E,F,C,G
K to D,E,F,C,G

— protein
A to H,K,D,B,F,G,E,I
G to H,K,D,B,F
I to H,K,D,B
E to H

— total solids %
A to H,K,B,F,D,E,G,C,I
I to H,K,B,F
C to H,K,B
H to E,G

[1] Differences between strains significant at: $P \leq 0.05$ — single underlining
$P \leq 0.01$ — double underlining

Table 2.47 **Least squares means for milk composition R_1 heifers (intensive trial)** (from Jasiorowski *et al.*, in press)

Paternal origin		**Fat** *(%)*		**Protein** *(%)*		**Total solids** *(%)*	
		Mean	**s.e.**	**Mean**	**s.e.**	**Mean**	**s.e.**
USA	(A)	3.74	0.07	3.31	0.03	11.92	0.11
Canada	(C)	3.89	0.07	3.34	0.03	12.13	0.11
Denmark	(D)	4.06	0.07	3.50	0.04	12.56	0.12
United Kingdom	(E)	4.00	0.07	3.42	0.04	12.32	0.12
Sweden	(F)	3.97	0.07	3.49	0.04	12.43	0.12
Fed. Rep. of Germany	(G)	3.81	0.07	3.39	0.04	11.98	0.12
Netherlands	(H)	4.06	0.07	3.38	0.04	12.36	0.12
Israel	(I)	3.80	0.07	3.38	0.04	12.11	0.12
New Zealand	(K)	4.31	0.07	3.56	0.03	12.80	0.11
Poland (pure)	(B)	4.24	0.07	3.56	0.04	12.64	0.12
Total[1]		3.99	0.02	3.43	0.01	12.33	0.04
Poland R_1		3.92	0.04	3.41	0.02	13.24	0.09

Significant differences*

— fat
- K to A,I,G,C,F,E,D,H
- B to A,I,G,C,F,E
- H,D to A,I,G
- E to A,I
- F to A

— protein
- B,K to A,C,I,H,G,E
- D to A,C,I,H,G
- F to A,C,I,H,G
- E to A

— total solids
- K to A,G,I,C,E,H,F
- B to A,G,I,C,E
- D to A,G,I,C
- F to A,G,I
- H to A,G
- E to A,G

* Differences between strains significant at: $P \leq 0.05$ — single underlining
$P \leq 0.01$ — double underlining

[1] See footnote to Table 2.38

average, approximately equalling the Holsteins in the field trial and exceeding them in the intensive trial (Figs. 2.43 and 2.44).

As for milk yield, strain crosses of USA and Canadian type gave estimates of *heterosis* of about 9 percent (related to Polish fat yield) (s.e. 6 percent) (compared with an estimate of general heterosis of 6 percent in the intensive trial) but no estimates in field or intensive trial approached significance.

Butterfat content of the milk (percent) (Figs. 2.45 and 2.46) was clearly highest in the New Zealand strain although both the Polish and Dutch strains made up with fat content for some of their lower yield in milk quantity. Fat content of the USA Holsteins was lowest on average and thus their merit relative to other strains, so clear on milk yield alone, was less for butterfat yield in the field trial and small in the intensive trial. Variation among strains was highly significant for butterfat yield and fat percentage in both the field trial and the intensive trial and in both generations (except for fat percentage of backcross heifers in the field trial) (Tables 2.48 and 2.49; Tables 2.44 to 2.47).

2.6.5 Protein yield and protein content of the milk

Protein yield in the field trial was 113 kg for F_1 heifers and 105 kg for R_1 heifers (Tables 2.42 and 2.43) with a common protein content in milk of 3.24 percent in both generations. As expected from the higher milk yield in the intensive trial, protein yield was 175 kg for F_1 heifers and 162 kg for R_1 not including the backcrosses to the Polish strain (134 kg) (Tables 2.44 and 2.45). However the protein content was also somewhat higher — 3.54 percent in F_1 and 3.43 percent in R_1 (3.41 in Polish R_1) (Tables 2.46 and 2.47) and thus contributes to the higher protein yield in the intensive feeding trial compared to the field trial.

Protein yield (Figs. 2.47 and 2.48) provides a similar picture to butterfat yield since Holstein strains in both field and intensive trials tended to be below average in protein content of milk and the New Zealand and some of the European strains were above average (Figs. 2.49 and 2.50). Strain variation was statistically significant (Tables 2.48 and 2.49) except for backcrosses in the field trial where a rather smaller number of animals were included for protein determinations than for other traits (see Tables 2.42 and 2.43). In view of these strain differences in content, New Zealand Friesians join the Holstein strains in above-average production of protein per lactation (Figs. 2.47 and 2.48) with field and intensive trials showing a similar picture.

2.6.6 Total solids

The total solids content was estimated for the animals in the intensive trial only (Tables 2.44 to 2.47). The mean production was 621 kg (12.53 percent) for

Fig. 2.43 Butterfat yield of females in first lactation in the field trial showing each strain as a percentage deviation from the overall mean (approx. s.e. $F_1$0.8 $R_1$2.6)

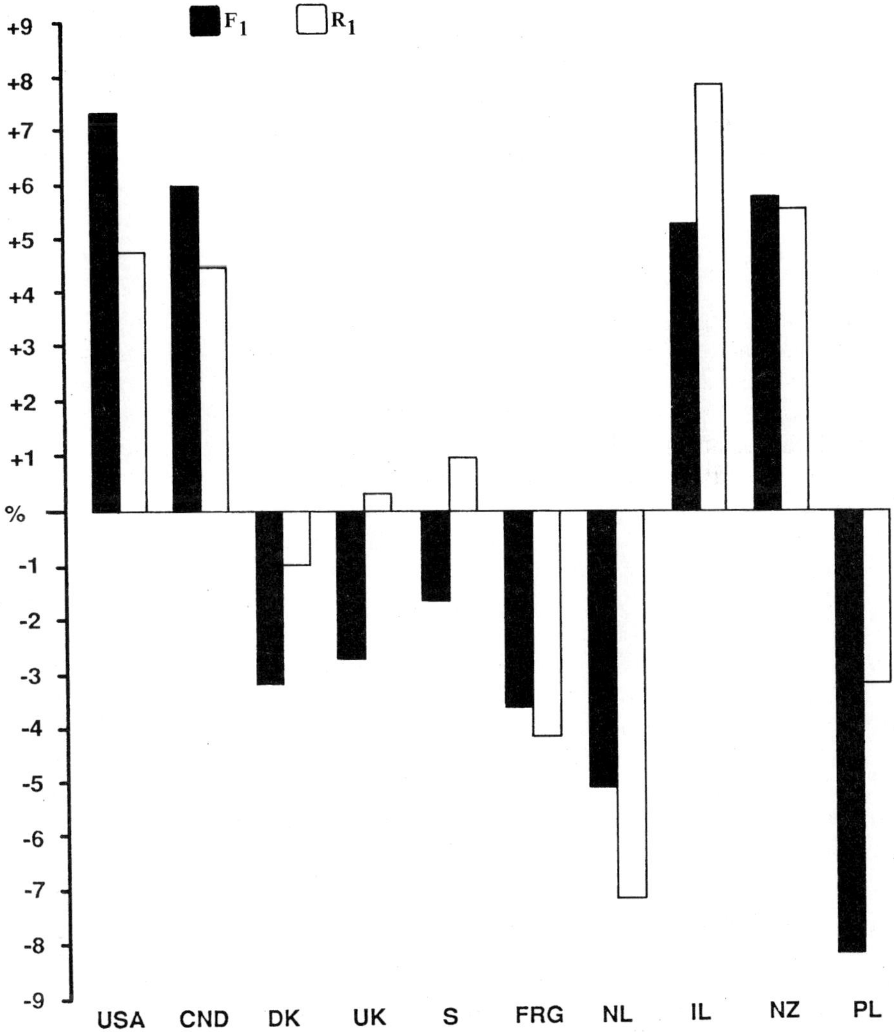

Fig. 2.44 Butterfat yield of females in first lactation in the intensive–feeding trial showing each strain as a percentage deviation from the least squares mean
(approx. s.e. $F_1$2.9 $R_1$3.3)

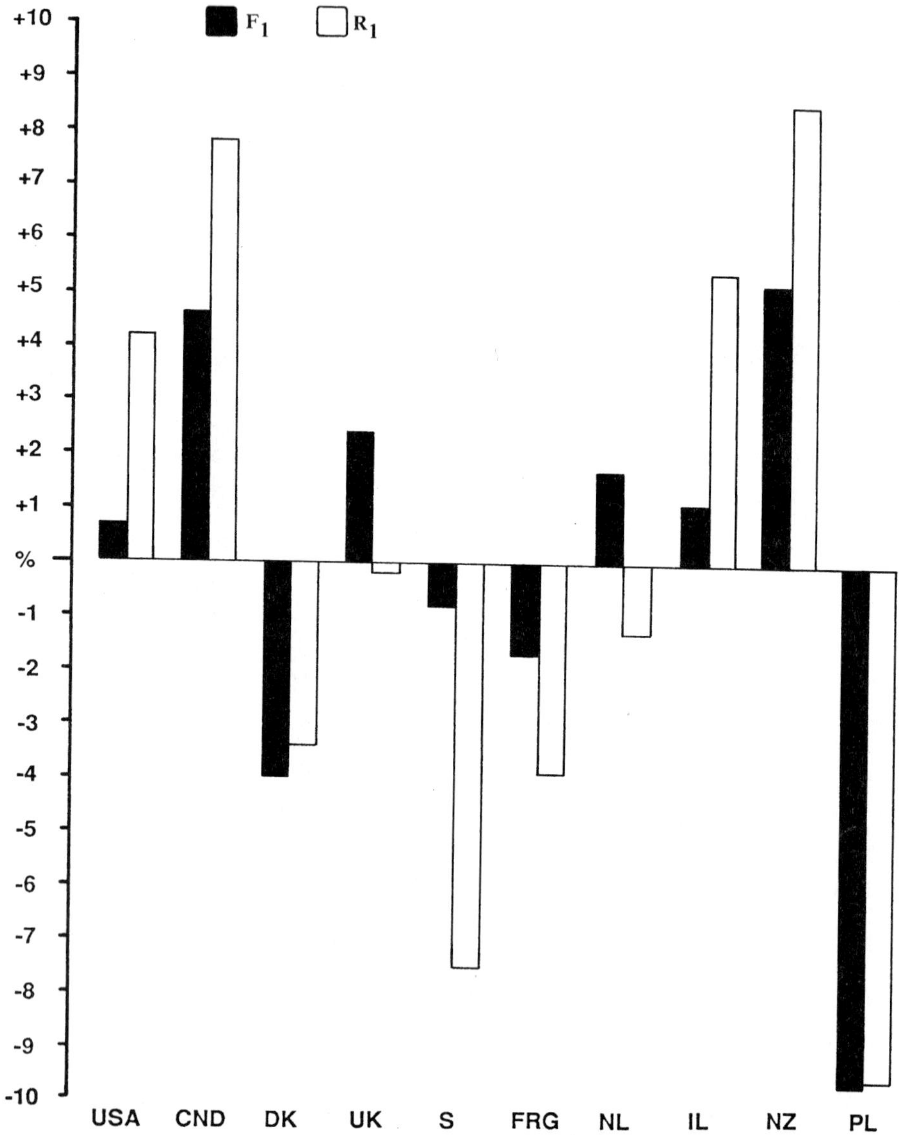

Fig. 2.45 Fat content (%) of milk of females in first lactation in the field trial showing each strain as a percentage deviation from the overall mean (approx. s.e. $F_1$2.7 $R_1$9.5)

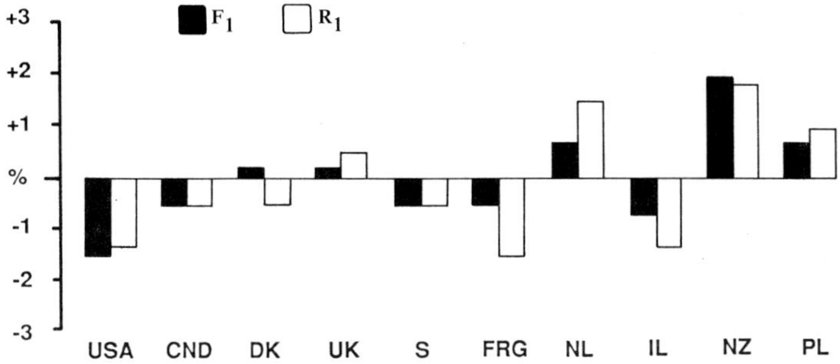

Fig. 2.46 Fat content (%) of milk of females in the intensive–feeding trial showing each strain as a percentage deviation from the least squares mean (approx. s.e. $F_1$1.3 $R_1$1.8)

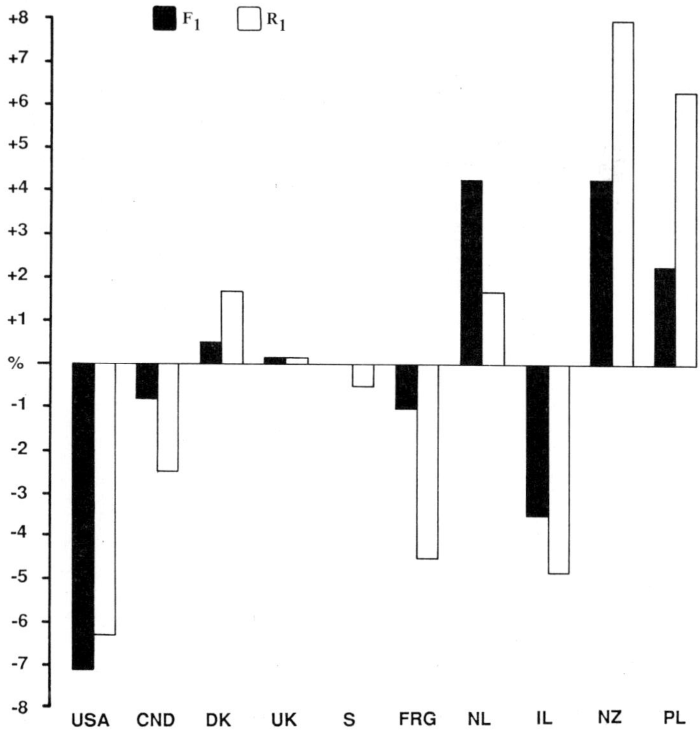

Fig. 2.47 Protein yield of females in first lactation in the field trial showing each strain as a percentage deviation from the overall mean (approx. s.e. $F_1$0.8 $R_1$3.6)

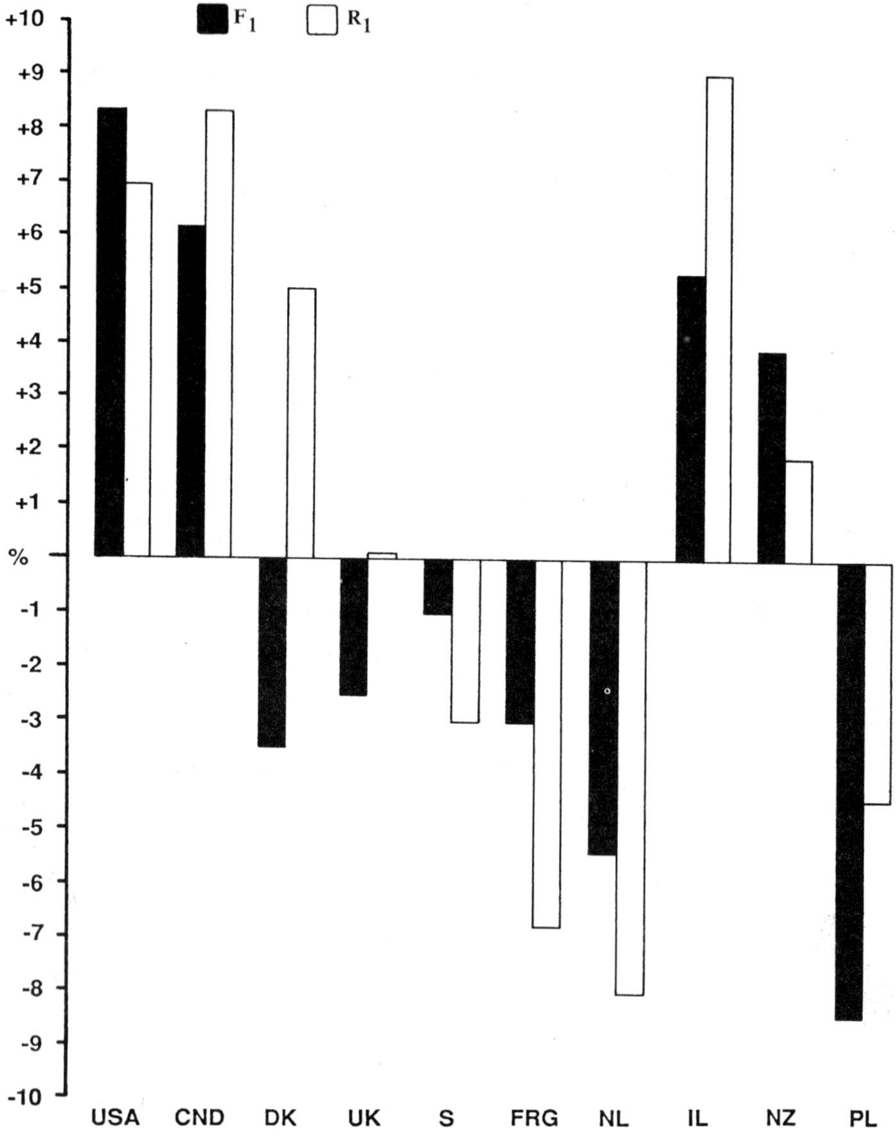

Fig. 2.48 Protein yield of females in first lactation in the intensive feeding trial showing each strain as a percentage deviation from the least squares mean (approx. s.e. $F_1$2.8 $R_1$3.0)

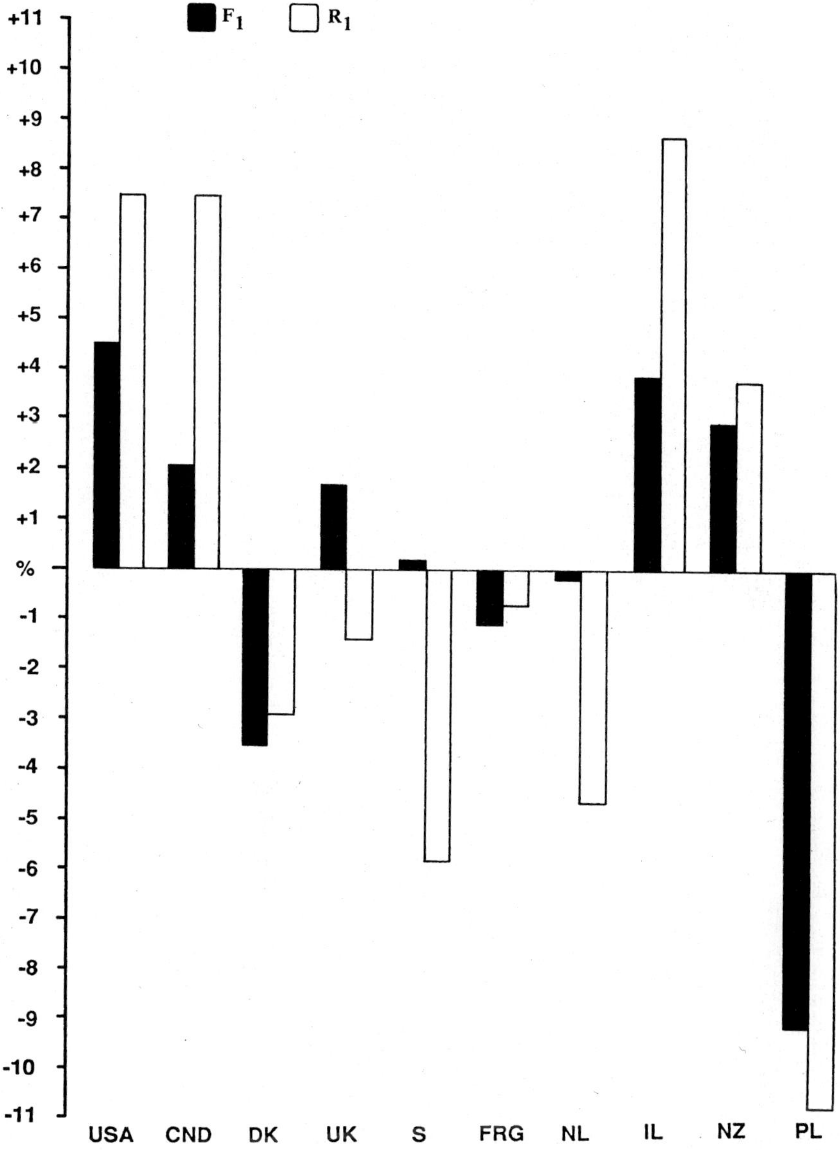

Fig. 2.49 Protein content (%) of milk of females in first lactation in the field trial showing each strain as a percentage deviation from the overall mean (approx. s.e. $F_1$0.2 $R_1$0.9)

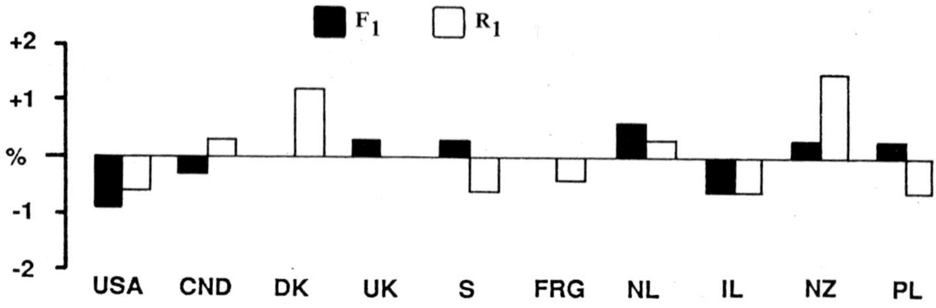

Fig. 2.50 Protein content (%) of milk of females in first lactation in the intensive–feeding trial showing each strain as a percentage deviation from the least squares mean (approx. s.e. $F_1$0.9 $R_1$1.1)

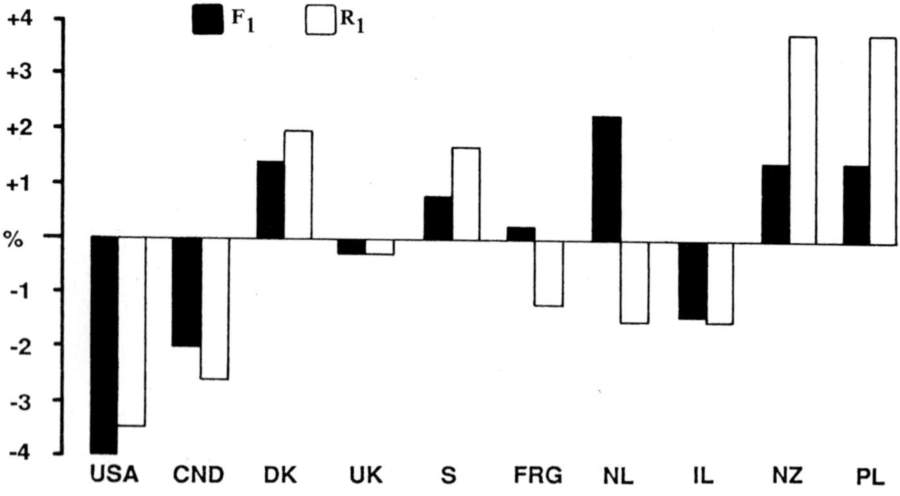

Fig. 2.51 Total solids yield of females in first lactation in the intensive–feeding trial showing each strain as a percentage deviation from the least squares mean
(approx. s.e. F_1 2.8 R_1 3.0)

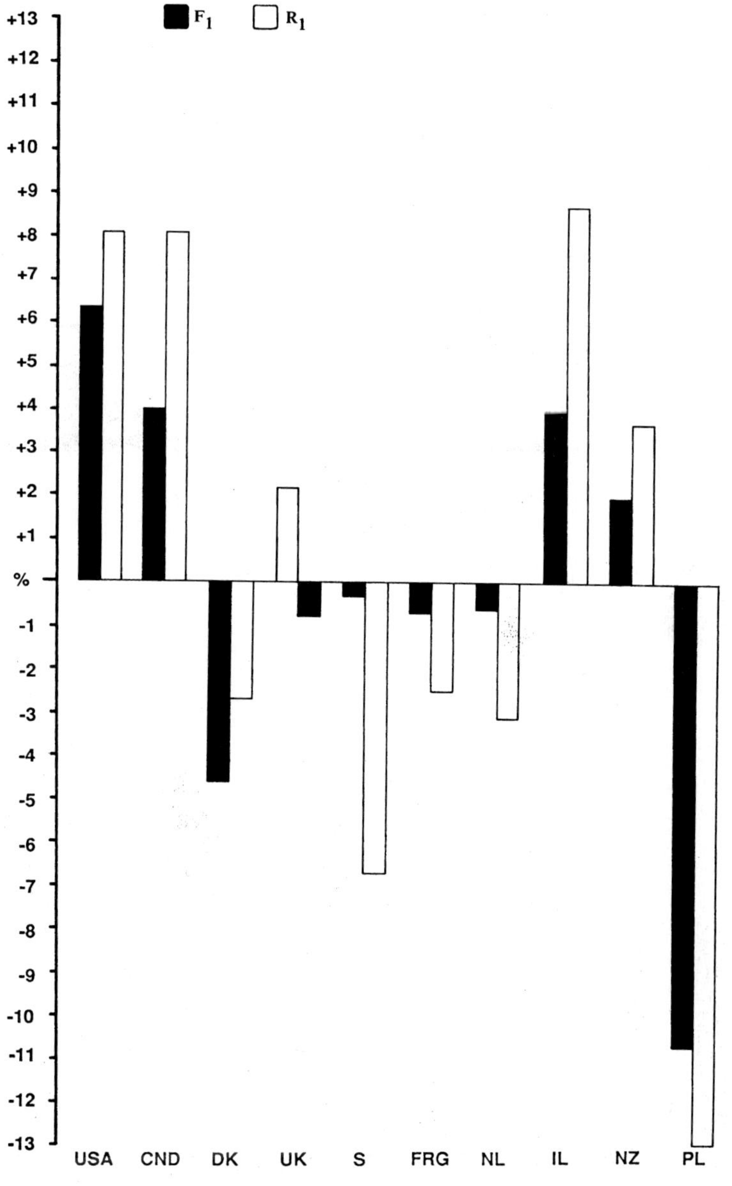

Table 2.48 **Milk production (305-day) and composition and days in milk of first cross (F_1) heifers in the field trial — the effects' of paternal strains as deviations (±s.e.) from the mean** (from M. Stolzman *et al.*, unpublished)

Paternal strain origin	Milk *(kg)***		Fat *(kg)***		Protein *(kg)***		Fat %**		Protein %**		Days in milk**	
	Effect	s.e.	Effect	s.e.	Effect	s.e.	Effect	s.e.	Effect	s.e.	Effect	s.e.
USA	313	29.5	10.3	1.21	9.4	0.95	-0.06	0.011	-0.03	0.006	9.1	2.77
Canada	229	28.9	8.3	1.17	6.9	0.92	-0.02	0.011	-0.01	0.006	4.2	2.69
Denmark	-122	29.3	-4.5	1.20	-3.9	0.94	0.01	0.011	0.00	0.006	-3.4	2.74
United Kingdom	-97	28.2	-3.8	1.16	-2.8	0.91	0.01	0.011	0.01	0.006	-2.1	2.65
Sweden	-47	27.6	-2.3	1.13	-1.1	0.90	-0.02	0.010	0.01	0.006	-4.7	2.59
Fed. Rep. of Germany	-106	27.9	-5.0	1.14	-3.4	0.91	-0.02	0.011	0.00	0.006	-6.8	2.62
Netherlands	-204	28.9	-7.1	1.18	-6.1	0.93	0.03	0.011	0.02	0.006	-6.5	2.71
Israel	216	28.7	7.4	1.17	6.0	0.93	-0.03	0.011	-0.02	0.006	9.6	2.69
New Zealand	127	26.3	8.0	1.08	4.4	0.88	0.08	0.010	0.01	0.006	6.4	2.47
100% Polish	-309	25.7	-11.4	1.06	-9.5	0.83	0.03	0.010	0.01	0.005	-5.6	2.42

Significance of variation among strains: ** ($P < 0.01$).

Approximate least significant differences[2]

	Milk	Fat	Protein	Fat %	Protein %	Days in milk
$P < 0.01$	110	4.5	3.5	0.04	0.02	10.3
$P < 0.05$	85	3.5	2.7	0.03	0.02	8.0

[1] [2] See footnotes Table 2.20

Table 2.49 **Milk production (305-day) and composition and days in milk of backcross (R_1) heifers in the field trial — the effects' of paternal strains as deviations ($\pm$s.e.) from the mean** (from M. Stolzman *et al.*, unpublished)

Paternal strain origin	Milk *(kg)***		Fat *(kg)***		Protein *(kg)***		Fat *(%)*		Protein *(%)*		Days in milk	
	Effect	s.e.	Effect	s.e.	Effect	s.e.	Effect	s.e.	Effect	s.e.	Effect	s.e.
USA	174	76.8	5.5	3.14	7.2	4.70	-0.05	0.039	-0.02	0.034	9.6	7.97
Canada	174	74.4	5.3	3.04	8.7	4.02	-0.02	0.038	0.01	0.029	12.0	7.72
Denmark	-35	79.1	-1.2	3.24	5.3	4.56	-0.02	0.040	0.04	0.033	1.5	8.21
United Kingdom	-1	82.5	0.4	3.37	0.1	5.09	0.02	0.042	0.00	0.037	0.6	8.56
Sweden	47	70.5	1.2	2.88	-3.2	3.64	-0.02	0.036	-0.02	0.027	3.4	7.31
Fed. Rep. of Germany	-101	72.0	-4.9	2.94	7.1	4.09	-0.06	0.036	-0.03	0.030	-10.0	7.46
Netherlands	-254	84.8	-8.3	3.46	8.4	4.67	0.06	0.043	0.01	0.034	-7.9	8.79
Israel	274	73.9	9.1	3.02	9.6	4.01	-0.05	0.037	-0.02	0.029	17.2	7.66
New Zealand	104	65.7	6.4	2.69	2.0	3.76	0.07	0.033	0.05	0.027	-6.3	6.81
Poland (R_1)	-116	55.4	-3.7	2.26	-4.6	3.01	0.04	0.028	-0.02	0.022	-8.1	5.74
100% Polish	-266	63.4	-9.8	2.59	-9.5	3.21	0.02	0.032	0.00	0.023	-12.0	6.58

Significance of variation among strains: ** ($P < 0.01$)

Approximate least significant differences[2]

	Milk	Fat	Protein
$P < 0.01$	260	11.1	14.7
$P < 0.05$	201	8.5	11.4

[1] [2] See footnotes Table 2.20

the first lactation of F_1 heifers and 582 kg (12.33 percent) for that of the backcrosses, to the "imported" strains (25 percent Polish blood) and 512 kg for those to the Polish strain (75 percent Polish blood).

Differences between some strains were highly significant ($P < 0.01$) both for content and total yield and, as seen in Figures 2.51 and 2.52, mirror the results from fat and protein, since the remaining principal solids component of milk (lactose) follows total milk yield almost exactly.

2.6.7 Days in milk and "days open"

The two traits are related since the interval from calving (onset of lactation) to the next conception (days open) influences the number of days in milk. Results are available only for the field trial. Drying-off policy was to allow cows to continue milking until yield had fallen below 1 kg per milking and to allow cows a dry period before their next calving of at least 6 weeks and usually 2 months. Within these constraints on lactation length, the average number of days in milk was 325 for F_1 first lactation and 320 days for R_1 with 117 days and 118 days respectively "open" as seen in Tables 2.35 and 2.36.

Strains differed significantly in the F_1 generation in both traits. Figure 2.53 shows that the Holstein strains and the New Zealand (F_1) had slightly longer lactation lengths and also, in general, longer intervals between pregnancies (Fig. 2.54). The European strains of Friesians had pregnancy intervals below or close to average.

2.6.8 Udder measurements and speed of milking

A number of observations were taken in the field trial on teat placement and distance of the udder from the ground. The results are shown in Tables 2.50 and 2.51 for the F_1 and R_1 generations. In the F_1 generation the variation among strains was highly significant for all the measurements on the udder, but in the R_1 generation for the distance from the floor only. Distance between teats from side to side was greatest in progeny of New Zealand bulls. Front to rear spread was lowest in the Canadian-sired progeny. Females sired by USA bulls had udders further off the ground. The difference between horizontal teat placement areas (TPAD) before and after milking was greatest for the USA and Canadian Holstein strains. The measurement is regarded as an indicator of udder efficiency since it is related to the proportion of glandular tissue which will "shrink" after milking-thus the larger the difference the better.

Fig. 2.52 Total solids content (%) of milk of females in first lactation showing each strain as a percentage deviation from the least squares mean (approx. s.e. $F_1$0.7 $R_1$1.0)

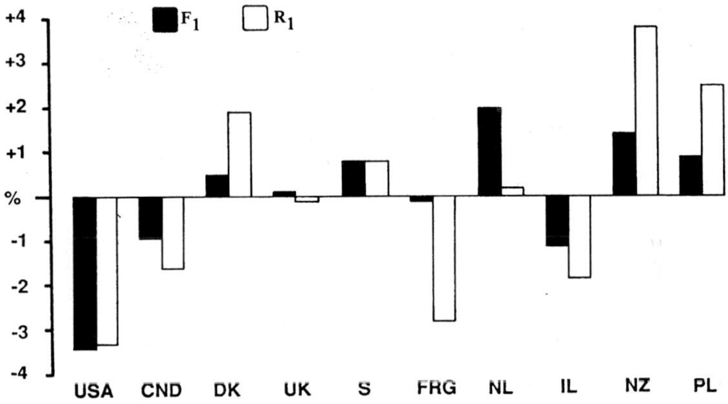

Fig. 2.53 Days in milk of females in first lactation in the field trial showing each strain as a percentage deviation from the overall mean (approx. s.e. $F_1$0.8 $R_1$2.4)

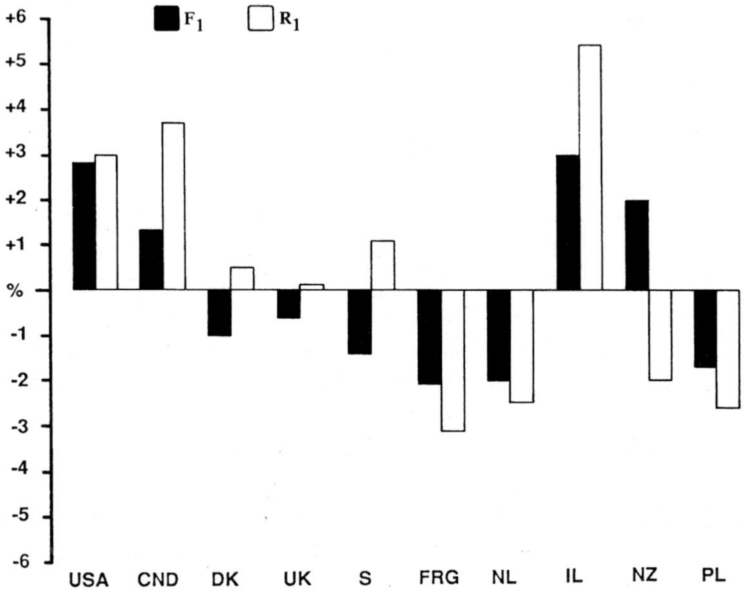

Fig. 2.54 Interval from first calving to next conception of females in the field trial showing each strain as a percentage deviation from the overall mean (approx. s.e. $F_1$2.5 $R_1$7.2)

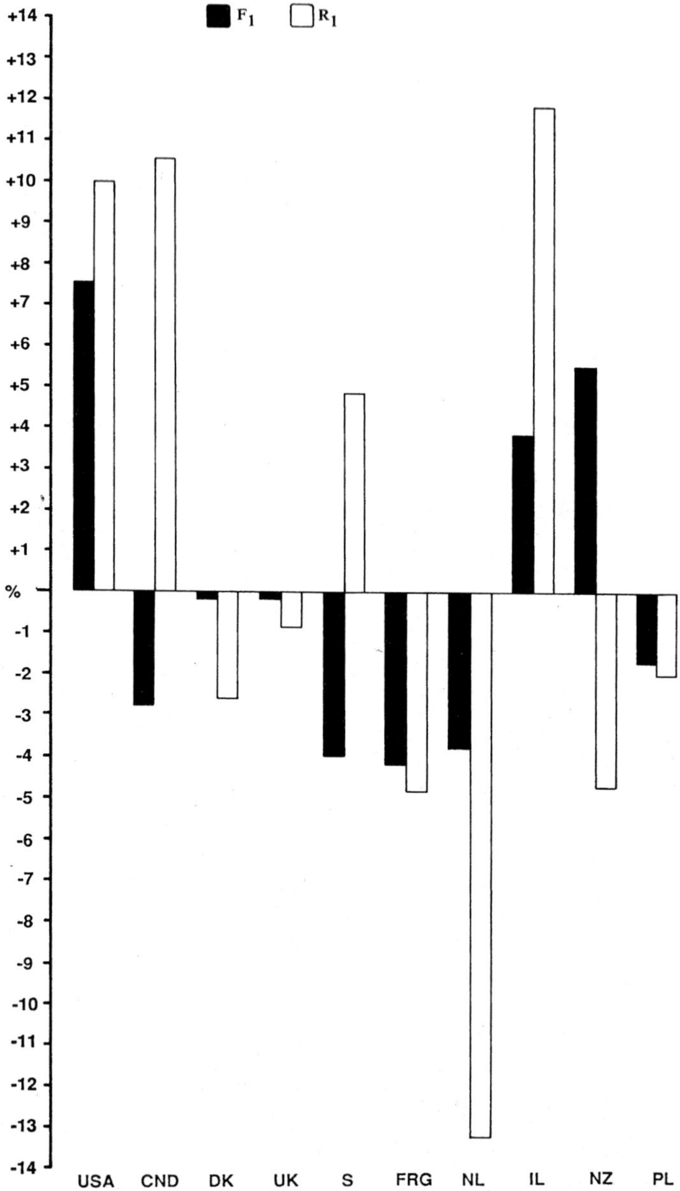

Table 2.50 **Least squares means (± s.e.) for some udder measurements and for speed of milking — F_1 heifers (field trial)** (from M. Stolzman *et al.*, unpublished)

		Spread of teats		Distance from the floor			
Paternal strain origin	No. of animals	Side** *(cm)*	Front** *(cm)*	Front** *(cm)*	Rear** *(cm)*	TPAD* *(cm²)*	Average milk flow per minute *(kg)*
USA	133	9.07	15.94	61.64	60.40	51.26	1.88
		0.15	0.25	0.30	0.30	2.74	0.03
Canada	125	9.50	13.13	61.00	59.48	53.94	1.82
		0.15	0.25	0.30	0.31	2.76	0.03
Denmark	166	8.85	16.94	60.11	58.35	45.05	1.82
		0.14	0.22	0.27	0.27	2.61	0.02
United Kingdom	140	8.89	16.56	59.39	57.77	49.30	1.84
		0.15	0.24	0.29	0.29	2.71	0.03
Sweden	127	8.99	17.26	60.39	58.07	46.99	1.84
		0.15	0.25	0.30	0.30	2.73	0.03
Federal Republic of Germany	149	9.15	16.99	60.28	58.65	46.35	1.85
		0.15	0.23	0.29	0.29	2.69	0.03
Netherlands	137	8.97	16.48	59.84	58.38	49.22	1.90
		0.15	0.25	0.30	0.30	2.73	0.03
Israel	143	9.12	17.22	60.16	58.97	48.56	1.86
		0.14	0.23	0.28	0.28	2.67	0.03
New Zealand	146	9.39	16.47	59.42	57.46	51.26	1.85
		0.14	0.23	0.28	0.29	2.68	0.03
Pure Polish	231	8.59	17.21	58.74	57.03	47.68	1.83
		0.12	0.20	0.24	0.24	2.51	0.02
Total	1 497	9.05	16.72	60.10	58.46	48.96	1.85
		0.05	0.09	0.11	0.11	2.13	0.01

Effect of strain significant at: * ($P < 0.05$)
** ($P < 0.01$)

Average milk flow per minute did not differ significantly among strains in the field trial. For the intensive trial, milk flow was also recorded and for F_1 heifers showed some significant differences between strains-with the Israeli and British strains milking fastest (but ranking in the backcrosses was not the same) (Table 2.52). The proportion of milk obtained from the front quarter was approximately 44 percent in heifers of both generations with no significant variation among strains.

Table 2.51 **Least squares means ($\pm$ s.e.) for some udder measurements and speed of milking — R_1 heifers (field trial)** (from M. Stolzman *et al.*, unpublished)

Paternal strain origin	No. of animals	Spread of teats		Distance from the floor			Average milk flow per minute
		Side *(cm)*	Front *(cm)*	Front* *(cm)*	Rear* *(cm)*	TPAD *(cm²)*	*(kg)*
USA	11	8.57	14.15	61.24	60.97	50.82	1.97
		0.41	0.70	0.78	0.82	3.34	0.16
CND	21	8.56	15.92	58.84	58.01	45.53	1.71
		0.33	0.56	0.62	0.65	2.66	0.12
DK	20	8.73	15.75	58.50	57.87	46.82	1.93
		0.32	0.54	0.60	0.64	2.59	0.12
UK	16	8.48	14.81	60.24	59.37	47.50	1.88
		0.35	0.60	0.67	0.70	2.89	0.13
S	26	8.31	14.90	59.59	58.31	42.69	1.91
		0.28	0.48	0.54	0.56	2.30	0.11
FRG	25	8.52	16.25	59.66	58.64	44.70	1.82
		0.29	0.50	0.55	0.58	2.36	0.11
NL	13	8.27	15.65	60.20	59.54	45.68	1.91
		0.39	0.68	0.75	0.79	3.23	0.15
IL	32	8.59	15.20	59.57	58.87	46.03	2.00
		0.28	0.47	0.52	0.55	2.27	0.11
NZ	38	8.80	15.18	60.29	59.30	43.93	1.99
		0.26	0.44	0.49	0.52	2.11	0.10
Pure	35	8.15	15.74	58.79	57.79	45.31	1.73
Polish		0.26	0.46	0.50	0.53	2.19	0.10
Total	237	8.50	15.35	59.69	58.87	45.90	1.89
		0.16	0.27	0.30	0.32	1.30	0.06

* Effect of strain significant at $P < 0.05$

Table 2.52 **Least squares means (±s.e.) for milking ability of F_1 and R_1 heifers* (intensive trial)** (from Jasiorowski *et al.*, 1983; Jasiorowski *et al*, in press)

Paternal strain origin		F_1				R_1			
		Milk in front quarters as percentage of total		Mean milk flow min^{-1}		Milk in front quarters as percentage of total		Mean milk flow min^{-1}	
		LSM	s.e.	LSM	s.e.	LSM	s.e.	LSM	s.e.
USA	(A)	45.2	1.07	1.81	0.09	44.1	1.8	2.56	0.17
Canada	(C)	44.9	1.10	1.80	0.10	44.6	1.8	2.27	0.17
Denmark	(D)	46.6	1.06	1.67	0.9	44.1	1.8	2.70	0.17
United Kingdom	(E)	44.6	1.08	2.02	0.10	43.9	2.0	2.17	0.19
Sweden	(F)	43.1	1.09	1.71	0.10	44.2	1.8	2.48	0.17
Fed. Rep. of Germany	(G)	46.0	1.09	1.80	0.10	41.8	1.7	2.08	0.16
Netherlands	(H)	43.7	1.09	1.82	0.10	43.9	1.6	2.31	0.16
Israel	(I)	45.6	1.07	2.02	0.09	44.4	1.7	2.19	0.16
New Zealand	(K)	43.0	1.11	1.85	0.10	40.8	1.8	2.40	0.17
Poland (pure)	(B)	43.2	1.08	1.57	0.10	45.9	1.8	2.78	0.17
Mean		44.6	0.43	1.81	0.04	43.8	0.6	2.39	0.06
Significant differences*				I to F,D,B					
				E to F,D,B					
				K to B					
				H to B					

* Differences between strains significant at: $P \leq 0.05$ — single underlining
$P \leq 0.01$ — double underlining

2.7 OBSERVATIONS ON SECOND AND THIRD LACTATIONS

All F_1 heifers which completed their first lactation in the intensive-feeding trial at the Experimental Stations Brwinow and Jastrzebiec were brought to Brwinow. During the second and third lactations, the animals were fed according to the system employed on this farm which involved pasture grazing and green forage in the summer, maize silage and hay in the winter. In addition, each cow consumed between 1.2 and 1.5 tons of concentrates annually. This feeding was less intensive than that used during the first lactation.

Preliminary analysis showed no interaction of strain × lactation; it was decided therefore to combine data from the second and third lactations. Statistical analysis took the form of least squares with factors for the effects of strain, lactation, year and season of calving.

Table 2.53 presents the least squares means with standard errors for milk yield and fat and protein content in the milk of cows in the combined second and third lactations. The total number of lactations was 417.

The highest milk yields were from cows sired by Israeli, USA and Canadian bulls, and the lowest from cows by Dutch, German and Polish bulls, a difference between these two groups of almost 1 000 kg of milk. As seen earlier from first lactation results, the fat and protein content of the milk was lowest in the high milk producing Holstein strains (and vice versa for the New Zealand and European strains). In consequence the ranking of breeds on yield of fat and protein differed from that on milk alone. On fat and protein yield the New Zealand and Swedish Friesian strains equalled or exceeded the Holsteins in the later lactations (the New Zealand strain having also exceeded all three Holstein strains in combined fat and protein production in first lactation when intensively fed). Ranking of strains for dairy traits was similar to that in the first lactation.

Table 2.54 shows the least squares means with standard errors for the weight of cows at 5 years of age and their body measurements. The heaviest were those sired by German bulls and the lightest by New Zealand bulls. The difference was 100 kg. The tallest and longest were from the three Holstein strains (USA, Canadian, Israeli) and the least tall the New Zealand crosses; the shortest were Danish F_1 cows.

2.8 LACTATION CURVES IN THE TEN BLACK AND WHITE STRAINS

The shape of the lactation curve is important because it shows whether milk production is, for example, highly peaked and production therefore very different in different parts of the lactation or, by contrast, flatter and the production more consistent over a longer period of time. To the extent that the shape of the curve is a characteristic of cows, or breeds of cow, knowledge of the shape would assist in herd management and planning and changes of curve might become a breeding objective. In earlier studies the heritability of the shape of the lactation curve has been found to be relatively low (Shanks *et al.*, 1981; Schneeberger, 1978) but some differences have been found between breeds (Wood, 1980a; Madalena *et al.*, 1979), though Abubakar and Buvanendran (1981) found no differences between 1/2 and 3/4 bred Friesians in Nigeria and Gaskins and Anderson (1980) found differences among three crosses with Angus cattle only in the linear but not in the quadratic coefficients derived to describe the lactation curves.

Wood (1969) noted that a reliable algebraic model of the lactation curve could be useful 1) for allowing a more precise measurement of response to (feeding) treatment changes than an analysis of unadjusted records; 2) for estimating total lactation production from incomplete records and 3) for planning in the dairy herd by a reliable forecast of month by month output.

The opportunity was taken in the present experiment to derive lactation curves for 387 first lactations of F_1 heifers in the intensive-feeding trial from weekly records. The shapes for both milk yield and milk composition were compared for the ten strains of Friesian cattle.

For the description of the shape of lactation curves, the algebraic model of Wood (1967) was used:

where:
$$Y_{(n)} = an^b e^{cn},$$

$Y_{(n)}$ is the daily milk yield, fat or protein content in the nth week of lactation;

- *a* is the constant defining the overall production level;
- *b* is the constant parameter of pre-peak curvature;
- *c* is the constant parameter of post-peak curvature.

In addition to the constants for each individual lactation, the goodness of fit (R^2), maximum milk yield and minimum fat or protein content were calculated as well as the week of lactation in which the maximum or minimum was achieved. The models of Wood (1976), and Shanks *et al.* (1981) were used.

Twenty-seven abnormal lactation curves for milk yield and 46 such curves for fat and protein content were excluded from further statistical analysis. This is a lower proportion of atypical lactation curves than given by other authors and described in some detail by Shanks *et al.* (1981).

Analysis of variance for constants and other indices of the lactation curves included the effects of strain, herd, month and year of calving. The results of the analysis of variance are presented in Tables 2.55 to 2.57.

The analyses showed that the effect of strain was significant for all three constants (*a*,*b*,*c*) for milk yield and for maximum daily milk yield, but not for fat content or protein content (except minimum protein content). The differences between strains in respect of maximum daily yield and of the constant *a* are related to differences in milk production between the strains. The difference in the constants *b* and *c* are related to differences in the shape of the lactation curve. Table 2.58 presents the least squares means ($\pm$s.e.) for each of the ten strains for milk yield and shows that the greatest difference was found between the Swedish and Danish strains. The curves for these two strains are shown in Fig. 2.55 and 2.56 and demonstrate graphically a flatter, more desirable, shape for the Swedish crossbred cows. The differences are particularly apparent in the post-peak part of the curve

Table 2.53 **Least squares means and standard errors of milk production characteristics for the second and third lactations combined (for a period of 305 days)** (from Jasiorowski *et al.*, 1982a)

Paternal strain origin	n[1]	Milk yield* mean	Milk yield* s.e.	Fat Yield* mean	Fat Yield* s.e.	Fat Percentage** mean	Fat Percentage** s.e.	n[1]	Protein Yield* mean	Protein Yield* s.e.	Protein Percentage** mean	Protein Percentage** s.e.
USA	49	5 267	184	188	7	3.61	0.05	39	171	6	3.30	0.04
Canada	42	5 214	194	202	8	3.89	0.06	39	169	6	3.33	0.04
Denmark	34	4 675	216	186	8	3.99	0.06	33	162	7	3.46	0.04
UK	38	4 689	202	189	8	4.05	0.06	34	159	7	3.47	0.04
Sweden	40	5 037	199	197	8	3.91	0.06	35	175	7	3.49	0.04
Fed. Rep. of Germany	45	4 419	193	171	7	3.92	0.06	36	162	7	3.47	0.04
Netherlands	41	4 450	204	179	8	4.06	0.06	35	156	7	3.52	0.04
Israel	50	5 296	182	204	7	3.88	0.05	43	180	6	3.39	0.03
New Zealand	33	4 762	217	197	8	4.15	0.06	30	165	7	3.48	0.04
Poland (pure)	45	4 334	192	176	7	4.07	0.06	36	145	7	3.46	0.04
Total		4 814	87	189	3	3.95	0.03		164	3	3.44	0.02

[1] n is the number of second and third lactations

* significant influence of strain (P 0.05)

** highly significant influence of strain (P 0.01)

Table 2.54 **Least squares means and s.e. for liveweight and body measurements of F1 cows at 5 years of age** (from Jasiorowski *et al*, 1982a)

Paternal strain origin	No. of cows	Liveweight* (*kg*)		Height at withers * (*cm*)		Body length* (*cm*)		Hip width (*cm*)		Chest girth (*cm*)	
		mean	s.e.	mean	s.e.	mean	s.e.	mean	s.e.	mean	s.e.
USA	19	600	16.6	130	0.9	154	1.2	59	0.7	200	1.9
Canada	20	579	16.2	130	0.9	153	1.2	58	0.7	199	1.9
Denmark	29	598	15.9	128	1.0	148	1.3	58	0.8	199	2.1
United Kingdom	19	594	16.6	127	1.0	149	1.3	58	0.8	201	2.0
Sweden	17	619	17.3	127	1.0	151	1.3	59	0.8	201	2.0
Fed.Rep. of Germany	20	633	16.2	128	0.9	150	1.2	59	0.7	203	1.9
Netherlands	17	563	17.3	126	0.9	150	1.2	58	0.7	196	1.9
Israel	19	587	16.6	129	1.0	152	1.3	60	0.8	201	2.0
New Zealand	14	538	18.8	127	1.2	151	1.5	56	0.9	195	2.3
Poland (pure)	17	600	17.3	128	1.1	152	1.4	59	0.9	200	2.2

*Significant influence of strain ($P < 0.05$)

Table 2.55 **Analysis of variance of lactation curve parameters for milk yield** (from Brzozowski *et al*, in press)

Effect	DF	Constants a MS	Constants b MS	Constants c MSx10	R^2 MS	Week of lactation wit peak yield MS	Maximum yield MS
Strain	9	59.3**	0.040*	0.004**	0.032	8.4	47.8**
Herd	1	209.1**	0.000	0.000	1.048**	10.6	352.5**
Month of calving	11	55.4**	0.015	0.004**	0.155**	53.7**	54.5**
Year of calving	2	153.1**	0.301	0.022**	0.131**	94.4**	24.1
Error	336	12.0	0.001	0.001	0.023	8.6	11.2

* $P < 0.05$
** $P < 0.01$

Table 2.56 **Analysis of variance of lactation curve parameters for fat content** (from Brzozowski *et al.*, in press)

Effect	DF	Constants a MS	Constants b MSx10	Constants c MSx100	R^2 MS	Week of lactation with the owest fat content MS	Minimum content MS
Strain	9	0.517	0.120	0.0060	0.015	75.4	0.685
Herd	1	0.086	0.184	0.0107	0.283**	51.7	1.992**
Month of calving	11	2.042**	0.406**	0.0279**	0.180**	187.7**	0.303*
Year of calving	2	0.579	0.128	0.0104	0.006	91.4	0.744**
Error	317	0.438	0.092	0.043	0.030	69.6	0.165

* $P < 0.05$
** $P < 0.01$

Table 2.57 **Analysis of variance of lactation curve parameters for protein content** (from Brzozowski *et al.*, in press)

Effect	DF	Constants				Week of lactation with the lowest protein content	Minimum content
		a	b	c	R^2		
		MS	MSx10	MSx100	MS	MS	MS
Strain	9	0.129	0.024	0.0016	0.028	75.4	0.185**
Herd	1	0.029	0.117*	0.0143**	2.547**	185.9	0.090
Month of calving	11	0.346**	0.177**	0.0109**	0.400**	296.7**	0.123**
Year of calving	2	0.899**	0.046	0.0002	0.713**	36.9	0.172*
Error	295	0.082	0.022	0.0015	0.029	85.7	0.042

* $P < 0.05$
** $P < 0.01$

indicating greater persistency of production for the Swedish than the Danish crossbreds. Tables 2.59 and 2.60 show the corresponding results for fat and protein content.

The month of calving significantly affected all constants and indices (with the exception of *b* for yield) both in respect of milk yield and milk composition. This is in agreement with the findings of Wood (1969, 1970, 1980a, 1980b) whose results were obtained under similar climatic conditions, but it is perhaps a matter of some surprise since in the present case a uniform feeding system was applied. Wood (1970) also found significant differences among herds in all three constants but there were only few significant differences between the two herds of the present study probably because both herds, at experimental stations, had the same feeding and management system applied.

Year of calving significantly affected the constant *a* for milk yield and protein content and constant *c* for yield (Tables 2.55 and 2.57).

As shown in Tables 2.58 to 2.60 the value for goodness of fit (R^2) is lower than that presented by Wood (1970, 1976) for Friesian cows or by Abubakar and Buvanendran (1981) for Friesian-Bunaji crosses. It is similar, however, to the value calculated by Madalena *et al.* (1979) for crosses of Holstein-Friesian with Gir. Flat curves without a distinct peak are characterized by a poorer goodness of fit. In general this is typical of heifer lactations (Kellogg *et al.*, 1977; Shanks *et al.*, 1981; Abubakar and Buvanendran, 1981) and provides an explanation for the relatively low values of R^2 found here.

Table 2.58 **Least squares means (±s.e.) for lactation curve parameters for milk yield for ten strains of Friesians** (from Brzozowski *et al.*, in press)

Strain	No. of lactations	Constants						R^2		Week of lactation with the peak yield		Maximum yield	
		a		b		c							
		LSM	s.e.	LSM	s.e.	LSM x10	s.e. x10	LSM	s.e.	LSM	s.e.	LSM	s.e.
mean	360	16.10	0.31	0.220	0.010	-0.278	0.010	0.716	0.013	7.80	0.26	20.37	0.30
USA	36	16.20	0.63	0.259	0.021	-0.310	0.022	0.740	0.028	8.28	0.54	21.69	0.62
CND	35	18.44	0.65	0.216	0.021	-0.259	0.022	0.738	0.029	6.89	0.55	21.85	0.63
DK	36	14.86	0.63	0.283	0.021	-0.344	0.021	0.728	0.028	8.52	0.54	20.43	0.61
UK	36	16.50	0.64	0.203	0.021	-0.261	0.021	0.736	0.028	7.31	0.54	20.47	0.62
S	38	16.13	0.62	0.185	0.019	-0.239	0.021	0.668	0.028	8.10	0.53	19.67	0.61
FRG	29	16.02	0.68	0.210	0.023	-0.261	0.023	0.685	0.031	8.16	0.58	20.24	0.66
NL	35	15.28	0.65	0.218	0.021	-0.265	0.022	0.724	0.029	7.77	0.55	19.45	0.63
IL	38	16.87	0.62	0.224	0.019	-0.296	0.021	0.756	0.027	7.57	0.52	21.40	0.60
NZ	36	16.94	0.64	0.186	0.021	-0.251	0.022	0.676	0.029	7.49	0.55	20.48	0.63
PL (pure)	41	13.71	0.61	0.242	0.019	-0.298	0.029	0.708	0.027	7.91	0.52	18.04	0.52

Table 2.59 **Least squares means (±s.e.) for lactation curve parameters in fat content of milk for ten strains of Friesians** (from Brzozowski *et al.*, in press)

Strain	No. of lactations	Constants						R^2		Week of lactation with the peak yield		Maximum yield	
		a		b		c							
		LSM	s.e.	LSM	s.e.	LSM x10	s.e. x100	LSM	s.e.	LSM	s.e.	LSM	s.e.
mean	34.1	4.239	0.059	-0.1151	0.009	0.1323	0.059	0.465	0.015	10.10	975	3.64	0.036
USA	35	4.190	0.123	-0.1439	0.018	0.1440	0.122	0.457	0.032	9.85	1.55	3.38	0.075
CND	35	4.047	0.124	-0.0845	0.018	0.1020	0.123	0.428	0.032	11.79	1.56	3.63	0.076
DK	39	4.291	0.118	-0.1206	0.017	0.1432	0.117	0.463	0.030	9.72	1.49	3.68	0.073
UK	35	4.350	0.123	-0.1313	0.018	0.1440	0.123	0.495	0.032	9.37	1.56	3.56	0.076
S	31	4.208	0.128	-0.0983	0.019	0.1189	0.128	0.452	0.034	12.62	1.62	3.75	0.079
FRG	27	4.240	0.135	-0.1251	0.019	0.1370	0.134	0.481	0.036	9.51	170	3.56	0.082
NL	30	4.507	0.132	-0.1226	0.019	0.1356	0.131	0.451	0.036	12.03	1.67	3.87	0.080
IL	34	4.127	0.123	-0.1254	0.018	0.1392	0.123	0.500	0.032	9.51	1.56	3.49	0.076
NZ	34	4.284	0.126	-0.1070	0.019	0.1326	0.125	0.483	0.034	8.49	1.59	3.77	0.077
PL (pure)	41	4.149	0.118	-0.0923	0.017	0.1263	0.117	0.460	0.030	8.15	1.48	3.75	0.073

Table 2.60
Least squares means (±s.e.) for lactation curve parameters in protein content of milk yield for ten strains of Friesians (from Brzozowski *et al.*, in press)

Strain	No. of lactations	Constants						R_2		Week of lactation with the peak yield		Maximum yield	
		a		b		c							
		LSM	s.e.	LSM	s.e.	LSM x10	s.e. x100	LSM	s.e.	LSM	s.e.	LSM	s.e.
mean	34.1	3.471	0.028	-0.0512	0.0047	0.07445	0.038	0.504	0.0168	9.85	0.91	3.309	0.0202
USA	35	3.424	0.056	-0.0664	0.0094	0.0849	0.077	0.512	0.034	0.48	1.84	3.20	0.040
CND	35	3.399	0.056	-0.0526	0.0094	0.0707	0.076	0.487	0.034	9.40	1.82	3.21	0.040
DK	39	3.503	0.055	-0.0492	0.0090	0.0776	0.074	0.488	0.033	9.48	1.75	3.38	0.039
UK	35	3.397	0.056	-0.0394	0.0093	0.0697	0.075	0.0547	0.034	9.18	0.81	3.30	0.040
S	31	3.523	0.063	-0.0448	0.0100	0.0633	0.081	0.441	0.036	3.19	1.95	3.38	0.043
FRG	27	3.408	0.058	-0.0479	0.0096	0.0752	0.079	0.539	0.034	0.55	1.88	3.29	0.041
NL	30	3.549	0.059	-0.0536	0.0098	0.0734	0.080	0.494	0.035	1.37	1.90	3.37	0.042
IL	34	3.449	0.056	-0.0618	0.0093	0.0829	0.074	0.533	0.033	8.08	1.79	3.22	0.040
NZ	34	3.488	0.061	-0.0396	0.0101	0.0654	0.081	0.491	0.036	8.84	1.96	3.37	0.043
PL (pure)	4.1	3.567	0.051	-0.0565	0.0085	0.0813	0.069	0.510	0.031	7.88	1.67	3.38	0.037

Fig. 2.55 Lactation curve – first lactation milk yield, Swedish strain (from Brzozowski *et al.*, 1987)

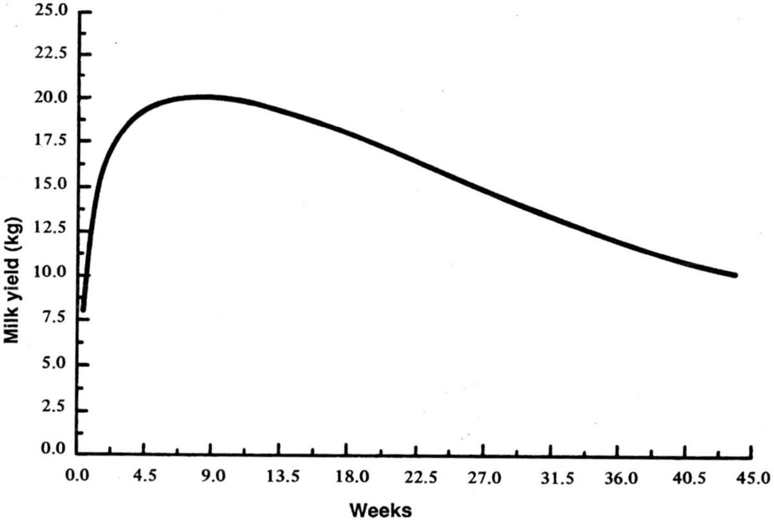

Fig. 2.56 Lactation curve – first lactation milk yield, Danish strain (from Brzozowski *et al*, 1987)

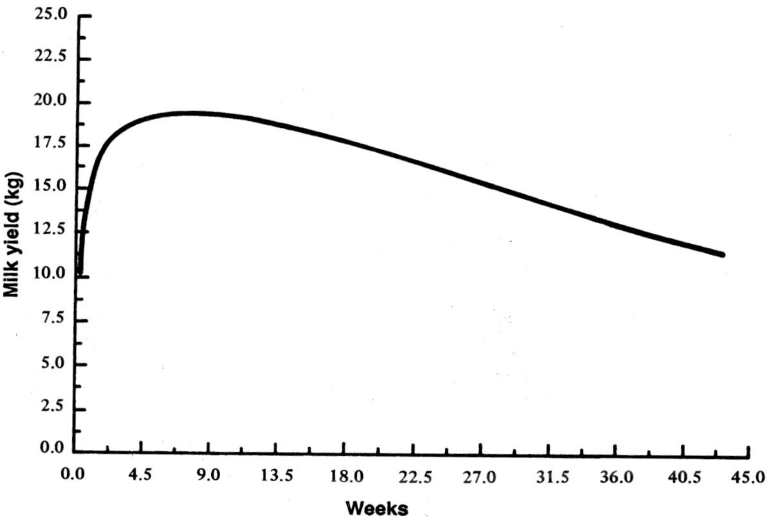

Peak yield was observed in approximately the eighth week of lactation (Table 2.58), earlier than by Shanks *et al.* (1981) but later than by Wood (1976). Minimum fat and protein content were recorded on average in week 10 (Tables 2.59 and 2.60) in agreement with Wood (1976). The differences among strains when maximum yield or minimum fat or protein content were observed were small and statistically not significant.

It must be stressed that the study of lactation curves undertaken here applies to heifers in the intensive-feeding trial. It is not known whether these strain differences in curves might have been similar, greater or smaller in the field part of the strain comparisons where the cattle were much more subject to seasonal variation in, for example, feeding and management.

2.9 HEALTH OF COWS FROM THE INTENSIVE-FEEDING TRIAL

F_1 and R_1 cows of the paternal strains of Friesian from the intensive-feeding trials at the experimental farm at Brwinow also had records taken on disease incidence for later lactations. This was made possible because the herd was under close veterinary supervision. The total average number of cow/years in a 5-year period included in this study was 1 194 (May 1980 to May 1985).

In Table 2.61 the frequency of diseases per 100 cows/year was calculated from the total number of cow/days in the herd divided by 365. The analysis of variance included factors for year and season of calving, lactation sequence and strain. The strain effect on the incidence of diseases was not significant. Reproductive problems were the most numerous, followed by leg and hoof disorders and by mastitis.

2.10 OBSERVATION OF MASTITIS IN THE INTENSIVE-FEEDING TRIAL

Observations were made on the incidence of mastitis and on the number of somatic cells in milk, as well as on the organisms causing infection of 415 F_1 cows in their first and second lactation at Brwinow farm. These were all cows which had taken part in the intensive-feeding trial.

The animals were kept in sheds without bedding. A 4 percent solution of iodoform was used for washing udders before milking and a 33 percent solution to disinfect teats after milking.

Milk samples from the whole udder were taken monthly in the same way as for milk recording. Bacteriological tests on milk from individual quarters and clinical examination of udders were done twice in each lactation at different stages. The number of somatic cells in milk samples was determined by means of a Fos-

Table 2.61 **Frequency of diseases (per 100 cows/year) for cows from the intensive trial at Brwinow** (from W. Empel, unpublished)

	Paternal strain origin										
	USA	CND	DK	UK	S	FRG	NL	IL	NZ	PL[1]	ALL STRAINS
Average number of cows totalled over a 5 year period	245	112	73	84	82	104	86	86	87	225	1 194
Diseases of the reproductive tract	92.8	85.4	99.3	93.5	100.7	100.6	92.3	111.5	106.5	92.5	96.1
Udder (mastitis)	27.8	27.6	29.0	30.8	36.4	31.6	18.7	20.9	25.9	26.7	27.5
Legs and hooves	42.1	44.5	34.5	56.8	44.9	44.1	40.9	30.2	48.6	41.8	42.8
Digestive tract	12.3	3.6	13.8	2.4	12.1	6.7	5.8	11.6	5.2	4.0	7.7
Other diseases	5.7	5.3	5.5	8.3	7.3	8.6	15.2	10.5	7.2	5.3	7.3

[1] Purebred

somatic apparatus (product of Foss Electric in Denmark). It was assumed that a pathological condition was reached when the number of somatic cells exceeded 500 000 per 1 ml of milk. Bacteriological and cytological tests and clinical examination of udders were combined to provide a description of the various forms of mastitis using the IDF classification (IDF, 1967).

Table 2.62 shows the mean number of somatic cells in the first and second lactations of the ten strains, ranked by the mean number found in the first lactation. The number of animals per strain was very similar.

All strains showed a higher mean number of cells in the second lactation than in the first, with the exception of the Canadian strain (which might have had a disproportionately high rate of infection in the first lactation, or low in the second). Both Wilton *et al.* (1972) and Miller *et al.* (1976) have shown an increasing incidence and acuteness of mastitis with advancing lactations as mirrored in the present results.

Table 2.62 **The mean number of somatic cells in udder milk of cows of ten Friesian strains ($x\ 10^3/ml$) ranked according to counts in first lactation in the intensive trial** (from Jasiorowski *et al.*, 1982b)

	Paternal strain origin										
	PL[1]	FRG	NL	USA	DK	UK	S	NZ	IL	CND	mean
First lactation	155	165	181	184	199	229	229	280	305	340	227
Second lactation	341	334	360	357	343	390	446	435	422	310	370

[1] Purebred

An attempt was made to characterize the incidence of the several forms of mastitis by combining the various observations made. The results are shown in Table 2.63. Strains are ranked according to the proportion of healthy (no mastitis) cows showing crossbred cows from Danish bulls with the highest proportion and those from British bulls with the lowest. The incidence of clinical mastitis was in fact quite low (2.2 percent) and the single largest category was for non-specific subclinical mastitis (9.1 percent) showing the general difficulty which still exists in determining clearly the pathenogenesis of mastitis.

The major pathogens, streptococci and staphylococci, isolated from infected quarters were in a minority but as seen in Table 2.64, across the strains of cattle, the higher the proportion of quarters infected by the major pathogens the lower, by and large, the proportion infected by other pathogens.

The difficulty in interpreting the apparent strain variation suggested by some of the present results relates to whether they indicate real strain differences in the susceptibility to mastitis or not. Since strain variation in actual clinical mastitis was very small, not surprisingly in view of the low overall average incidence, it would have to be assumed that all the other criteria represent aspects of susceptibility to mastitis and that strains do not differ in the likelihood that, say, a "latent" infection would develop into a clinical case. There is no evidence to assume such a difference in the present case. In general, genetic resistance to mastitis infection is still a matter of controversy. However, a number of studies suggest that genetic variation exists (e.g. Ryniewicz (1972), in breed comparisons; Stolbaek-Pedersen (1975), in Danish breed comparisons; Skolasinski and Tyszka (1973), from variation among sire progeny groups, and, more recently Solbu (1984) on the basis of sire variation in Norway). Thus the present results might be used to support the hypothesis of genetic variation in susceptibility (or resistance) to mastitis.

2.11 A COMPARISON OF THE GROWTH, WEAR, CONFORMATION AND CHEMICAL COMPOSITION OF THE HOOVES OF THE TEN STRAINS OF FRIESIANS IN THE INTENSIVE-FEEDING TRIAL

The impact of sound feet on the performance of the dairy cow is not always adequately considered and there is little or no information on the relative merits

Table 2.63 **The various forms of mastitis found among infected cows shown as percentages of the total number of cows of each paternal strain, ranked by proportion of healthy cows in the intensive trial** (from Jasiorowski *et al.*, 1982b)

	Paternal strain origin										
	DK	**FRG**	**NL**	**PL**[1]	**USA**	**CND**	**IL**	**S**	**NZ**	**UK**	**Mean**
Healthy (%)	88.0	87.5	79.8	79.4	75.0	75.0	72.9	68.8	67.1	64.8	76.7
Latent infection[2]	2.2	1.6	5.9	8.8	10.2	5.5	7.3	7.8	5.2	10.2	6.5
Non-specific mastitis	4.3	6.7	7.1	4.4	4.6	12.5	10.4	12.5	21.0	14.8	9.1
Subclinical mastitis[3]	1.1	0.8	3.6	2.9	1.8	2.8	4.2	10.9	2.6	4.5	3.2
Clinical mastitis[3]	3.3	1.6	2.4	2.2	3.7	0	4.2	0	1.3	2.3	2.2
Blind quarters	1.1	1.6	1.2	2.2	4.6	4.2	1.0	0	2.6	3.4	2.2
Total mastitic quarters	12.0	12.3	20.2	20.5	24.9	25.0	27.1	31.2	32.7	35.2	23.3

[1] Purebred
[2] [3] Only the bacteria pathogenic for the udder (major pathogens) were taken into account (Griffin *et al.*, 1977)

Table 2.64 **Occurrence of bacteria in samples of quarter milk in ten Friesian cattle strains, ranked according to major pathogens (as a percentage of infected quarters) in the intensive trial** (from Jasiorowski *et al.*, 1982b)

	Paternal strain origin										
	NZ	FRG	DK	NL	IL	USA	S	PL[1]	CND	UK	Mean
Major pathogens[2]	1.3	1.6	3.3	3.6	4.1	5.5	6.2	6.6	6.9	10.2	4.8
Others[3]	31.6	25.6	29.3	30.9	37.5	30.5	34.3	33.8	29.2	26.1	30.7
Total	32.9	26.7	32.6	34.5	41.7	36.1	40.6	40.4	36.1	36.3	35.7

[1] Purebred
[2] Streptococci, staphylococci
[3] *Corynebacterium* bovis, *Micrococcus* sp., enterococci, mixed flora.

in this respect of different strains of Friesian. The opportunity was taken, therefore, to examine the F_1 heifers included in the intensive-feeding dairy trial for hoof conformation, growth, wear and chemical composition (Drodz, 1981). The results of these observations are presented below.

2.11.1 Conformation of hooves

The measurements taken for hoof conformation characteristics are shown in Figure 2.57 and overall means for each component trait in Table 2.65.

Measurements were made on 1) wall to sole angle; 2) wall height; 3) length of front slope; 4) heel height; 5) the ratio of (3) to (4); 6) wall area; 7) sole length; 8) sole width; 9) sole area; and 10) body weight pressure on the sole area. The measurements are shown in Figure 2.58. The overall means for each trait are shown in Table 2.65.

Wall sole angle

At the beginning of the experiment, after the hooves were trimmed, the wall-sole angles of foreleg and hindleg hoof were similar, and no significant differences were observed between inside and outside hooves of the same leg.

Fig. 2.57 Measurements on hooves taken at the beginning of the experiment and 12 months later (from Drozdz, 1981)

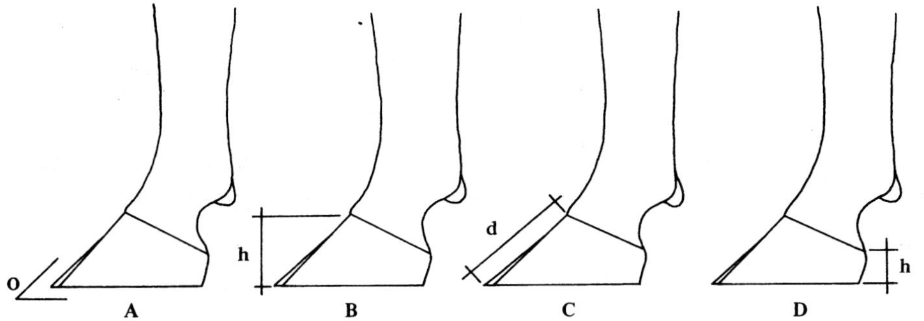

A - Wall sole angle (0)

B - Hoof wall height (h)

C - Hoof front slope length (d)

D - Heel height (h)

E - Hoof sole area

F - Sole length (a), sole width (b)

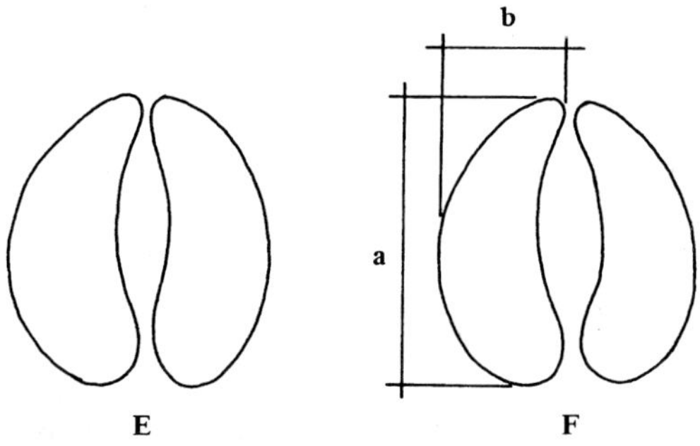

Fig. 2.58 Monthly measurements of hoof horn growth and wear (from Drozdz, 1981)

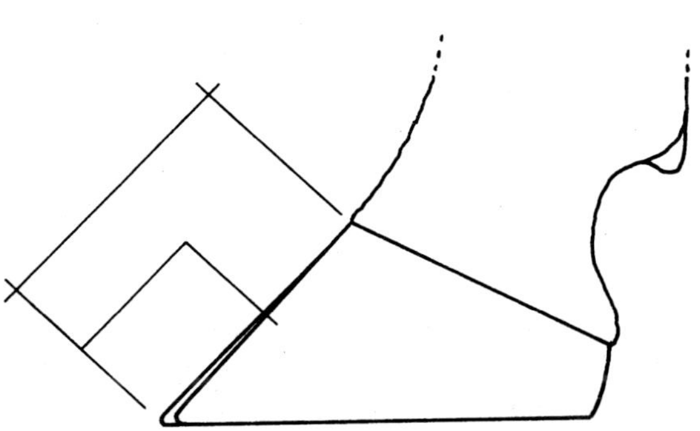

Table 2.65 **Overall means for hoof measurements** (from Drozdz, 1981)

		Foreleg				Hindleg			
		inside		outside		inside		outside	
		beginning of experiment	end of experiment	beginning of experiment	end of experiment	beginning of experiment	end of experiment	beginning of experiment	end of experiment
Wall-sole angle	degrees	47.4	42.7	48.3	44.8	47.7	45.1	48.7	44.9
Hoof front slope length	mm	73.8	78.0	77.3	86.5	73.6	75.9	80 .2	80.6
Heel height	mm	34.9	33.9	38.7	37.1	24.9	30.0	24.5	29.3
Hoof wall length to heel height ratio		2.13	2.36	2.02	3.04	3.04	2.59	3.39	2.86
Hoof wall area	cm^2	72.3	83.1	67.4	74.4	61.1	66.9	66.1	74.2
Hoof sole area	cm^2	37.8	46.8	37.0	43.8	32.2	36.8	37.0	43.0
Sole length	mm	104.8	116.8	99.4	107.1	100.7	100.9	108.8	111.3
Sole width	mm	47.0	51.8	48.4	50.3	40.8	42.4	47.4	47.8

After 12 months, without further trimming, the wall-sole angle had significantly decreased both inside and outside. Only the German crossbreds differed in angle from the Polish control group with a much higher inside angle after the 12 months.

Hoof wall height

Relative to the Polish control group, the New Zealand strain had a significantly lower hoof wall height at the beginning of the experiment for both foreleg and hindleg while for hindlegs alone, Israeli crosses had a lower height at the beginning and Dutch crosses throughout. There were no other strain differences. The overall average for the front legs was 53.3 mm and 54.7 mm for the hindlegs.

Hoof front slope length.

For the foreleg, at the end of the experiment, Israeli and New Zealand crossbreds had significantly longer front slopes than the Polish group. No significant differences were observed in the hindleg slope length

Heel height

At the start there were no significant differences among groups but after 12 months, New Zealand crossbreds had a significantly lower heel height on the outside hoof of the foreleg than the Polish group.

Hoof wall length to heel height ratio

Only New Zealand crossbreds differed appreciably from the other groups for the foreleg. No significant differences were observed for the hindleg. Over the 12 months of the experiment the ratio showed a tendency to increase in all groups of cows. For the foreleg of Swedish, Dutch and Israeli crossbreds the increase was significant.

Hoof wall area

For the foreleg at the beginning of the experiment and for the hindleg at both beginning and end, no significant differences were found in hoof wall area of different groups. For the foreleg, German crossbreds had a significantly greater area of the inside hoof wall at the end of the 12 months.

Sole length

The foreleg of USA crossbreds had the longest sole at the beginning of the experiment. No significant differences were observed between the Polish group and the other breed groups for this trait for the hindleg, although USA crosses had the highest values.

Sole width

At the beginning of the experiment, USA crossbreds had significantly wider soles of the foreleg inside hoof compared to the other crossbreds and to the Polish control group. Canadian crossbreds had a significantly lower sole width than the Polish group at the beginning of the experiment. For the hindleg no significant differences between groups were observed in sole width, except for the Swedish group where the outer hoof was narrower.

Hoof sole area

USA crossbreds had the greatest hoof sole area of foreleg and hindleg at the beginning and one of the greatest areas at the end of the experiment. For the combined areas of the hooves of each animal, there were some strain differences with USA Friesian crosses having the largest and Swedish and Dutch Friesian crosses the smallest areas (Table 2.66).

Body weight pressure per 1 cm^2 of hoof sole area

The overall means for pressure were 2.05 kg/cm^2 at the beginning and 1.74 kg at the end of the experiment. In Canadian and Swedish crossbreds the pressure of body weight per 1 cm^2 of hoof sole area was significantly higher than in the Polish control group and Swedish crossbred cows also differed significantly from USA, Israeli and New Zealand crossbreds (Table 2.66). USA crossbreds with the largest hoof area had the lowest mean body weight pressure per 1 cm^2 of hoof area.

2.11.2 Growth and wear of horn on the hooves

Growth and wear were estimated from regular monthly measurements taken for one year, from the beginning of the experiment, with a slide caliper along the front edge of the hoof wall 1) from the base to the coronet (Measure I) and 2) from the base to the mark (Measure II) (see Fig. 2.58).

The difference between Measures I and II was the distance from the mark to the coronet. Horn growth was estimated by differences in that distance in consecutive months. Rate of wear was estimated from monthly differences in Measure II.

Table 2.67 shows average monthly growth of horn on the hoof in the ten groups of crossbreds.

Table 2.66 **Combined areas of hoof soles per animal (cm) and body weight pressure per cm2 of total sole area averaged for each of the ten strains at the beginning of the experiment** (from Drozdz, 1981)

Paternal strain origin	Combined areas (cm^2) of hoof soles	Body weight pressure per unit area of sole (kg/cm^2)
USA	343.0 c	1.76 a
Canada	278.3 b	2.23 cd
Denmark	282.4 b	2.12 bcd
United Kingdom	279.7 b	2.11 bcd
Sweden	252.4 a	2.31 d
Fed. Rep. of Germany	287.4 b	2.06 bc
Netherlands	273.9 ab	2.14 bcd
Israel	300.4 b	2.00 abc
New Zealand	288.7 b	2.01 abc
Poland	290.6 b	1.96 ab

Means followed by different letters differ significantly ($P < 0.05$)

Table 2.67 **Average monthly growth of horn (mm) on the hoof over a period of one year in different groups of crossbreds in the intensive trial** (from Drozdz, 1981)

Groups	**Foreleg hooves**				**Hindleg hooves**			
	inside		**outside**		**inside**		**outside**	
USA	3.96	cd	3.90	cd	4.00	ab	4.02	a
Canada	3.50	a	3.48	a	3.82	ab	3.88	a
Denmark	4.09	d	4.06	d	4.07	b	4.08	a
UK	3.95	cd	3.91	cd	4.06	b	4.06	a
Sweden	3.63	ab	3.60	abc	3.86	ab	3.86	a
Federal Republic of Germany	3.90	bcd	3.81	bcd	3.91	ab	3.92	a
Netherlands	3.62	ab	3.56	ab	3.83	ab	3.85	a
Israel	3.69	abc	3.67	abc	3.76	a	3.83	a
New Zealand	3.78	abc	3.76	ad	4.07	b	4.06	a
Poland (control group)	3.90	bcd	3.87	bcd	4.00	ab	4.02	a

Average horn growth in different strains marked with different letters differ significantly ($P < 0.05$)

The average growth rate of horn of the foreleg hoof was similar in all groups except in the Canadian crossbreds, where it was significantly lower than in the control Polish group. The highest growth (4.06-4.09 mm per month) was observed in the Danish group. There were almost no significant differences in horn growth of inner and none of the outer hoof of the hindlegs between the groups. Table 2.68 shows average monthly horn wear on the hoof in different groups of crossbreds. For the foreleg, hoof wear was greater on the inside than the outside of the hoof whereas for the hindleg this was, in general, the other way round arising in all probability from the differences in sole size and loading between front and rear.

Table 2.68 **Average monthly wear of horn (mm) on the hoof for one year in different groups of crossbreds in the intensive trial** (from Drozdz, 1981)

Groups	Foreleg hooves				Hindleg hooves			
	inside		outside		inside		outside	
USA	3.69	b	3.37	c	3.48	ab	3.71	ab
Canada	3.03	a	2.53	a	2.90	ab	3.24	ab
Denmark	3.69	b	3.15	bc	3.43	ab	3.66	ab
UK	3.61	b	3.20	bc	3.25	ab	3.58	ab
Sweden	3.35	ab	2.80	ab	3.53	b	3.24	ab
Federal Republic of Germany	3.27	ab	2.69	ab	3.32	ab	3.43	ab
Netherlands	2.87	a	2.56	a	3.11	ab	3.36	ab
Israel	2.81	a	2.45	a	2.72	a	3.02	a
New Zealand	2.99	a	2.44	a	3.41	ab	3.46	ab
Poland (control group)	3.73	b	2.94	abc	3.49	ab	3.79	b

Average horn wear in different strains marked with different letters differs significantly ($P < 0.05$)

Only a few differences between breed groups were statistically significant but crossbreds from Israeli bulls had the lowest rate of wear.

In general, growth of horn exceeded the rate of wear on the hooves and horn length, if left intact, would therefore increase with age under these conditions.

2.11.3 Mineral composition

Concentrations of magnesium (Mg), calcium (Ca), zinc (Zn), iron (Fe), copper (Cu), and phosphorus (P) in the wall and sole of the hooves were estimated.

Concentrations of all but Fe were lower (generally) in the sole than the wall of the hoof. There were relatively few strain differences in concentration, but relationships with wear and general soundness of feet might emerge with further study. Details of the results are given in Table 2.69.

2.12 DISCUSSION — LACTATION PERFORMANCE AND RELATED MATTERS

During the first half of the twentieth century Friesian cattle were brought to and developed in various countries differing in their systems of cattle management,

Table 2.69 **Concentration of Mg, Ca, Zn, Fe, Cu and P (in PPm) in the horn of hoof wall (w) and sole (s)** (from Drozdz, 1981)

Paternal strain origin	Mineral elements											
	Mg		Ca		Zn		Fe		Cu		P	
	w	s	w	s	w	w	s	w	s	w	w	s
USA	194	94	974	967	131	61	28	29	5.2	2.8	443	215
Canada	183	95	1 054	722	142	81	22	37	4.8	2.4	370	183
Denmark	235	130	1 134	936	126	71	26	59	4.5	2.3	399	145
United Kingdom	243	136	1 214	1 326	129	66	24	62	4.9	2.4	405	169
Sweden	217	160	1 098	1 325	142	71	21	47	4.3	2.4	495	180
Federal Republic of Germany	175	100	962	893	137	77	17	49	4.2	3.5	447	186
Netherlands	200	87	1 029	785	128	68	17	78	4.1	2.9	439	170
Israel	223	148	1 139	1 016	138	65	21	59	4.1	3.0	530	180
New Zealand	237	129	1 181	913	135	55	21	57	4.0	3.1	546	181
Poland	186	99	1 136	828	123	64	33	35	4.7	3.1	440	256

in their perceptions of the requirements from cattle, in their marketing structures and outlets for milk and also in their systems of pricing milk. This in turn is likely to have affected breeding aims and selection policies and it is not surprising if differences have arisen among strains of Friesians in aspects of performance, body size and conformation. During the 1960s in particular, a further change took place in Europe with extensive use of Holstein-Friesian cattle from the USA and Canada for crossbreeding with European strains. Relatively high and secure prices for milk coupled with a relative decline in concentrate feed costs ensured a long period of more or less guaranteed profitability for milk.

As other costs related to milk production increased, however, this, together with the potential profitability of production, led to increases in herd size and to a move to improve yields by better feeding and by an emphasis on the genetic potential for higher milk yield in the choice of bulls at the expense of beef traits. Thus Holstein-Friesians were widely used in Italy and Switzerland and also in the Federal Republic of Germany, France and Denmark. Introduction of Holstein-Friesians to the Netherlands started somewhat later and in the United Kingdom and Sweden has been practised to a lesser extent so far.

The procedure for introducing Holstein-Friesian blood has generally been through the repeated importation of semen particularly from the USA and Canada and further dissemination through AI stations by the use of bulls with proportions of Holstein-Friesian blood.

This increasing international exchange of semen led to interest in comparing the genetic value of particular sub-populations of Friesian cattle in the areas to which they had been brought but in which they had not been developed. This in turn led FAO to sponsor the much more comprehensive strain comparison of the Polish trials, reported here, than the comparison previously possible. The earlier comparisons, however, clearly showed that Friesian strains differed. Thus, Zarnecki and Hinks (1976) showed that Danish Friesians at testing stations, mostly in the 1960s, yielded more milk and also had a higher production of fat and protein than crossbred cows from several European strains of bulls mated to Danish females. Table 2.70 shows a summary of their results.

Subsequent investigations in Denmark showed that crossbreds sired by Holstein-Friesian bulls exceeded the yield of Danish Friesians by 639 kg milk and 33 kg fat (Nielson, 1971, quoted by Ernst *et al.*, 1973). In Germany, Langholz and Jongeling (1977) analysed field data and reported the value of using Holstein-Friesians for total milk output.

Politiek *et al.* (1982) compared Dutch Friesians with their crosses from Holstein-Friesian or British Friesian bulls (5 bulls per strain). Results for two lactations are summarized in Table 2.71. Both Holstein-Friesian and UK Friesian crosses yielded more milk than pure Dutch Friesians but had lower fat and protein percentages, with the Holstein-Friesian crosses having the highest yield by a con-

Table 2.70 **Least squares means for milk production traits of the progeny of bulls of various strains of Friesian mated to Danish Friesian cows** (from Zarnecki and Hinks, 1976)

	Paternal strain origin					
Trait	**DK**	**NL(FRS)**	**NL(NRS)**	**S**	**FRG**	**Mixed**
Milk yield (kg)	5 200	4 904	5 117	5 094	4 474	4 936
Fat yield (kg)	227	215	225	218	195	214
Fat (%)	4.37	4.40	4.39	4.29	4.35	4.33
Protein (%)	3.67	3.69	3.67	3.74	3.74	3.67
Protein yield (kg)	191	180	187	189	167	181
Liveweight of cows at the beginning of testing (kg)	546	532	555	550	536	528
at the end of testing (kg)	581	571	578	570	583	574

Table 2.71 **Least squares means of milk production traits of Dutch Friesians and their crosses with Holstein Friesian (HF) and British Friesian (BF)** (from Politiek *et al.*, 1982)

	Lactation							
	I				II			
	n	**Milk** *(kg)*	**Fat** *(%)*	**Protein** *(%)*	**n**	**Milk** *(kg)*	**Fat** *(%)*	**Protein** *(%)*
Average	218	4 931	4.14	3.34	180	5 912	4.14	3.36
Dutch Friesians	68	-360	+0.13	+0.02	57	-523	+0.12	+0.03
F_1 crossbreds NL x HF	70	+462	-0.14	-0.03	54	+582	-0.12	-0.05
F_1 crossbreds NL x BF	80	-102	+0.01	+0.01	69	-58	+0.00	+0.03

siderable margin, but the poorest composition. Politiek *et al.* (1982) estimated the genetic difference of the Holstein-Friesian bulls used to have been +1 600 kg milk relative to the Dutch bulls but the latter to have been genetically superior for fat content (+0.5 percent) and protein content (+0.1 percent). The negative correlation between milk yield and both fat and protein content of milk, which this suggests across strains, mirrors the negative genetic correlation between yield and concentration reported for example within the Holstein-Friesian breed in the USA by Van Vleck (1978) and Hargrove *et al.* (1981).

Gaunt (1973) and Van Vleck (1978) have shown that increases in total fat or protein yield are achieved most rapidly by genetic selection on milk yield alone. However, this may change if pricing of milk were to alter the balance directly in favour of compositional criteria, either by penalty for low concentrations of milk constituents or bonuses for high. Moreover, particularly when milk is used largely for processing — the production of butter and cheese — transport and other costs may favour high fat or protein content at the expense of milk volume. High milk volume is associated with a high production of lactose for which there is now a lower market demand. Any move to reduce the energy costs required to produce surplus lactose may therefore become a further factor favouring milk with higher concentrations of protein and fat.

It is clear from the experiments and field trials with ten strains of Friesian cattle conducted in Poland that there are real differences in milk yield and composition among the strains. This is seen particularly in the comparison of the three Holstein-Friesian strains (USA, Canadian and Israeli) with the New Zealand Friesian. The Holstein-Friesian group were clearly superior in total milk yield and the New Zealand strain in composition such that on fat yield and protein yield the New Zealand strain ranked at or close to the top. Milk in New Zealand is, of course, mostly produced for processing and thus the selection of the New Zealand Friesian, like the Jersey which it is displacing has stressed high fat and protein content. It is interesing, however, that the crosses with the New Zealand strain did particularly well in the intensive trials, although in New Zealand the cows are mainly geared to pasture utilization.

Table 2.72 shows the ranking of the strains in both the intensive feeding and the field trial for both generations — the first crosses (F_1) and the backcrosses (R_1). In terms of the ranking of the strains the results from the two systems of feeding and management show a substantial degree of agreement and this suggests that interactions of strain with level of feeding are unlikely to be of major importance. The strain ranking on first lactation results is also supported in general by the preliminary results from later lactations.

From the study of lactation curves for milk yield in the intensive-feeding trial it can be concluded that though there were small but statistically significant differences among strains, these differences must be set alongside other differeces

Table 2.72 **The ranking of strains in the Polish trials under intensive feeding and in the field trial for the combined yields of fat and protein — showing deviations (kg) from the mean for each class**

	Intensive feeding trial					Field trial	
	Generations						
	F_1		R_1		F_1		R_1
NZ	+ 15	CND	+ 27	USA	+ 20	IL	+ 19
CND	+ 13	IL	+ 24	CND	+ 15	CND	+ 14
USA	+ 9	NZ	+ 22	IL	+ 13	USA	+ 13
IL	+ 9	USA	+ 20	NZ	+ 12	NZ	+ 8
UK	+ 8	UK	- 3	S	- 3	DK	+ 4
NL	+ 3	NL	-10	UK	- 7	UK	+ 1
FRG	- 5	FRG	- 8	DK	- 8	S	- 2
S	- 1	DK	-11	FRG	- 8	FRG	-12
DK	-14	S	-23	NL	-13	NL	-17
PL	-36	PL	-37	PL	-21	PL	-19
Mean	371[1]		349[1]		252[2]		220[2]

[1] Least squares mean
[2] Overall mean

among the Friesian strains in deciding what is desirable and what is not. It would seem safe to assume that in any consideration of choice of strain for crossbreeding in, for example, developing countries, the relatively small differences in the shapes of lactation curves among the strains are likely to be of lesser importance than some of the other differences in performance. However, in situations where yield, as such, is of lesser importance, the choice of shape of lactation curve may assume significance as an aid to herd management.

Decisions about strain substitution or strain crossing will require consideration of factors associated with body size and meat quality characteristics, as well as other factors contributing to herd life and the costs of production such as aspects of health survival, or the shape and wear of hooves, examined here. Differences among the crosses representing different strains have not been clearly demonstrated in the latter characteristics perhaps because the numbers required to do so may be greater than those recorded. However, as shown earlier, there are real differences in body size and some components of beef production from different strains. The largest and smallest strain crosses differed by at least 11 percent in liveweight or weight gain in the intensive trial (though much less in the field trial) suggesting about twice that difference attributable to the pure strains. Whether this affects the maintenance costs of production will depend in practice on the system and type of feeding used, although in theory the smaller strains might expect to have a slight advantage. It is interesting in this context that the combined production of fat and protein of the different strains when expressed per 100 kg liveweight hardly affects the ranking of strains in the present trials and leaves the relative differences between the top and bottom ranking strains virtually the same as when it is judged on yield alone.

It is apparent from the results of the field trial that backcrosses produced less milk, fat and protein than did the first crosses but, as indicated in relation to liveweight and growth, this may be attributable largely to unintended changes in feeding. The reduction in yield of the pure Polish groups contemporary with the R_1, relative to the F_1 (490 kg milk) provides a crude measure of the effect, subject only to sampling differences between the two groups. A corresponding change for pure Polish females in the intensive trial, though smaller in magnitude (387 kg milk) is more difficult to explain in terms of changes in feeding, but as stated before, the R_1 generation was not contemporary with the F_1 although both were thought to have been fed and managed similarly.

As for growth and carcass traits, heterosis estimates were derived from deviations of crosse from pure Polish, within generations, and comparing F_1 with R_1 deviations. Only for milk yield and butterfat yield were there two estimates (USA and Canadian strains) (averaging around 8 percent) in the field trial approaching significance, but these are worth pointing out perhaps mainly because they represent crosses with Holstein-Friesian strains which may be considered as the most distant genetically from the other strains in these comparisons. On this hypothesis, however, it is surprising that the third of the Holstein-Friesian strains investigated, the Israeli strain, provided no hint of an effect of heterosis. Over all strains, the effect of heterosis was also not significant in either trial or for any of the dairy traits although estimates are of the order also found in the literature. Thus, for the intensive-feeding trial, general heterosis effects were about 6 percent for content and 1.4 percent for protein content. For the field trial, as noted earlier, crosses from USA and Canadian bulls gave the only specific heterosis estimates approaching significance, namely 8 percent for milk yield and 9 percent for fat yield.

It is important to bear in mind that the absence of heterosis in these strain comparisons should be seen for what it is. As judged from the strain differences and the estimation of heterosis genetic differentiation within the Friesian breed appears to have been primarily in relation to genes acting additively when expressed under conditions of good management and feeding. This in no way undermines the thesis that heterosis may be of considerable importance when crosses of very diverse breeds are involved and conditions, as in many developing countries, are harsh — as emerges in later reviews in this book.

Considerable care was taken in the initial planning of the Polish trials and in the importation of semen from the countries of origin to ensure that the sample of bulls of each strain used in Poland was large enough (approximately 40 per strain) to be representative of the then current genetic status of the young bulls being tested in their respective countries. In the event it seems, as shown earlier (Table 2.4), that the bulls used in the project were a good random sample characterizing the AI population in the participating countries. Since the process of genetic improvement in the various countries had continued over the past decade or more — that is since the time that the bulls in the present trials were the latest available — and since there has been a continuing influx of Holstein blood into the European Friesian strains, it is possible that the strains if re-sampled now would not have precisely the same performance difference. It would indeed be surprising, however, if the general principles established here and by earlier studies were in any way altered: namely that real differences exist among the Friesian strains of the world and that what is best for one set of circumstances may not be best for another.

References

ABUBAKAR, B.Y. and BUVANENDRAN, V. Lactation curves of Friesian-Bunaji crosses in Nigeria.
1981 *Livest. Prod. Sci.*, 8: 11-19.

ARORA, D.N. and SHARMA, J.S. Factors affecting some of the economic traits in Holstein-Friesian
1983 cattle. *Indian Vet. J.*, 60: 820-823.

BAUMUNG, A. and PANICKE, L. *Ergebnisse der Heterosisschätzung beim Milchrind.* 37th Annual
1986 Meeting of the European Association for Animal Production, Budapest. 1-4 September.

BECH ANDERSEN, B. Comparison of Danish Friesian, Friesian and Red Danish for beef production.
1982 *In* More O'Ferrall, G.J., ed. *Beef production from different dairy breeds and dairy beef crosses.* Martinus Nijhoff, The Hague.

BRABANDER, D.L., BOUQUE, Ch. V., CASTEELS, M., AERTS, J.V. and BUYSSE, F.X. Ability of Hols-
1982 tein Friesian, Dutch Black and White and Belgian White-Red bull calves for veal and beef production. *In* More O'Ferrall, G.J., ed. *Beef production from different dairy breeds and dairy beef crosses.* Martinus Nijhoff, The Hague.

BRZOZOWSKI, P., JASIOROWSKI, H. and REKLEWSI, Z. Lactation curves of ten strains of Friesian
1987 cattle. *Annals of Warsaw Agricultural University — SGGW-AR. Anim. Sci.* (In press).

COLLEAU, J.J. and LIENARD, G. The economic comparison between the Holsteins and Norman breeds.
1982 *In* More O'Ferrall, G.J., ed. *Beef production from different dairy breeds and dairy beef crosses.* Martinus Nijhoff, The Hague.

CUNDIFF, L.V., GREGORY, K.E. and KOCH, R.M. *Genetic variation, heterosis and complementarity*
1986 *between different breeds and types of cattle.* 37th Annual Meeting of the European Association for Animal Production, Budapest. 1-4 September.

CUNNINGHAM, E.P. and MOIOLI, B. Economic and genetic optimisation of dual-purpose bull testing
1982 and selection. *In* More O'Ferrall, G.J., ed. *Beef production from different dairy breeds and dairy beef crosses.* Martinus Nijhoff, The Hague.

COOK, K.N. and NEWTON, J.M. A comparison of Canadian Holstein and British Friesian steers for
1979 the production of beef from an 18-month grass/cereal system. *Anim. Prod.*, 28: 41-47.

DICKERSON, G.E. Implications of genetic-environment interaction in animal breeding. *Anim. Prod.*,
1962 4: 47-63.

DONALD, H.P., GIBSON, D. and RUSSELL, W.S. Estimations of heterosis in crossbred dairy cattle.
1977 *Anim. Prod.*, 25: 193-208.

DROZDZ, A.A. *Hoof box traits of 10 groups of crossbred Friesian cattle.* Annual Meeting of the Euro-
1981 pean Association for Animal Production, Munich.

DUPLAN, J.M. La race bovine Pie-noire en 1973. *Elevage et Insémination,* No. 139.
1974

ERNST, E. Vergleichende Untersuchungen der Leistungen von Deutschen Schwarzbunten, Holstein-
1986 Friesians und deren Kreuzungen (I). *Tierzüchter,* 20: 798-800.

ERNST, E., LANGLET, J.F. and MARTIN, H. Vergleichende Untersuchungen an Deutschen Schwarz-
1973 bunten, Holstein-Friesians sowie den Kreuzungen aus beiden Populationen. *Schriftenreihe der Agrarwiss. Fakultät, Kiel.* Heft 50.

FLANAGAN, J. Comparison of beef performance of different Friesian strains from progeny test results
1982 in Ireland. *In* More O'Ferrall, G.J., ed. *Beef production from different dairy breeds and dairy beef crosses.* Martinus Nijhoff, The Hague.

GAUNT, S.N. Genetic and environmental changes possible in milk composition. *J. Dairy Sci.*, 56:
1973 270-278.

GASKINS, C.T. and ANDERSON, D.C. Comparison of lactation curves in Angus-Hereford, Jersey-Angus
1980 and Simmenthal-Angus cows. *J. Anim. Sci.*, 50: 828-832.

GEAY, Y. and ROBELIN, J. Variation of meat production capacity in cattle due to genotype and level
1979 of feeding: genotype-nutrition interaction. *Livest. Prod. Sci.*, 6: 263-276.

GRAVERT, H.A. *Der Einfluss der Körpergrösse auf die Effizienz der Milcherzeugung bei*
1971 *Rinderkreuzungen.* Kurzbeitrag zum X Internationalen Tierzuchtkongress, Paris-Versailles.

GRAVERT, H.A. *The early calving of heifers and its impact on beef production.* 27th Annual Meeting
1976 of the European Association for Animal Production, Munich.

GRIFFIN, T.K., DODD, F.H., NEAVE, F.K., WESTERGARTH, D.R. and KINGWILL, R.G. A method
1977 of diagnosing intramammary infection in cows for large experiments. *J. Dairy Sci.*, 44: 25-45.

HARGROVE, G.L., MBAH, D.A. and ROSENBERGER, J.L. Genetic and environmental influences on
1981 milk and milk component production. *J. Dairy Sci.*, 64: 1593-1597.

HUTH, F.W. Auf die Waage Enttäuschen Schwarzbunte nicht. *Deutsche Schwarzbunte,* 4: 8-10.
1981

INTERNATIONAL DAIRY FEDERATION. *Definition of mastitis.* Document No. 34, Brussels.
1967

JASIOROWSKI, H., GRABOWSKI, R. and GRODZKI, H. *Origin and genetic value of the Black and White*
1979 *Lowland cattle in Poland.* International Symposium on Utilization of the Holstein-Friesian Breed for the Improvement of Cattle in Socialist Countries. Warsaw.

JASIOROWSKI, H., GRABOWSKI, R., REKLEWSKI, Z., STOLZMAN, M. and DLUGOLECKI, L. *Inter-*
1982a *national testing of various strains of Friesian cattle in Poland. II. Differences in milk production in successive lactations.* 33rd Annual Meeting of the European Association for Animal Production, Leningrad.

JASIOROWSKI, H.A., SCIUBISZ, A. and KOZANECKI, M. *Differences in the incidences of mastitis in*
1982b *various Friesian cattle strains.* 33rd Annual Meeting of the European Association for Animal Production, Leningrad.

JASIOROWSKI, H., REKLEWSKI, Z., and STOLZMAN, M. Testing of different strains of Friesian cattle
1983 in Poland. I. Milk performance of F_1 paternal Friesian strain crosses under intensive feeding conditions. *Livest. Prod. Sci.*, 10: 109-122.

JASIOROWSKI, H.A., REKLEWSKI, Z., STOLZMAN, M. GRABOWSKI, R. and LUKASZEWICZ, M.
1987 Testing of different strains of Friesian cattle in Poland. III. Milk peformance of the strain backcrosses under intensive feeding conditions. *Livest. Prod. Sci.* (In press).

KELLOGG, D.W., URQUHART, N.S. and ORTEGA, A.J. Estimating Holstein lactation curves with a
1977 gamma curve. *J. Dairy Sci.*, 60: 1308-1315.

KRAMER, C.Y. Extension of multiple range tests to group correlated adjusted means. *Biometric,* 13:
1957 13-18.

LANGHOLZ, H.J. and JONGELING, C. Stand und Entwicklungstendenzen der Rinderzucht Nord-
1977 deutschlands. *Züchtungskunde*, 49: 3-20.

MADALENA, F.E., MARTINEZ, M.L. and FREITAS, A.F. Lactation curves of Holstein-Friesian × Gir
1979 cows. *Anim. Prod.*, 29: 101-107.

MILLER, R.H., OWEN, J.R. and MOORE, E.D. Incidence of clinical mastitis in a herd of Jersey cattle.
1976 *J. Dairy Sci.*, 59: 113-119.

OLDENBROEK, J.K. Meat production of Holstein-Friesians in comparison to Dutch Friesians and Dutch
1982 Red and White. *In* More O'Ferrall, G.J., ed. *Beef production from different dairy breeds and dairy beef crosses.* Martinus Nijhoff, The Hague.

OLDENBROEK, J.K. and VAN ELDIK, P. Differences in feed intake between Holstein-Friesian, Dutch
1980 Red and White and Dutch Red cattle. *Livest. Prod. Sci.*, 7: 13-23.

PEDERSEN, J. and CHRISTENSEN, L.G. *Heterosis for milk production traits by crossing Red Danish,*
1986 *Finnish Ayrshire and Holstein-Friesian cattle.* 37th Annual Meeting of the European Association for Animal Production, Budapest. 1-4 September.

POLITIEK, R.D., VOS, H. and KORZER, S. Comparison of Friesian cattle from different origins. *Z.*
1982 *Tierz. Züchtungsbiol.*, 99: 272-285.

PRZYBYLSKA, H. *Wolyw systemu wychowu na wzrost i pózniejska uzytkowozc mleczna jalówek o róz-*
1985 *nym udziale genotypu rasy holsztynsko-fryzyjskiej.* Praca doktorska, IGiHZ PAN.

PRESTON, T.R. and WILLIS, M.B. *Intensive beef production* (2nd edition). Pergamon Press, Oxford,
1974 New York, Toronto, Sydney.

RAUE, F. Welche Rasse für die Rindermast. *Agrar Ubersich,* 4: 50-54.
1983

REKLEWSKI, Z., JASIOROWSKI, H., STOLZMAN, M., LUKASZEWICZ, M. and DE LAURANS, A.
1985 Testing of different strains of Friesian cattle in Poland. II. Beef performance of male crossbreds of different Friesian cattle strains under intensive feeding conditions. *Livest. Prod. Sci.*, 12: 117-129.

RYNIEWICZ, Z. *Badania nad dziedzicznym i srodowiskowym uwarunkowaniem podatnosci krów na stany*
1972 *zapalne gruczolu mlecznego.* Praca doktorska, Biuletyn IGiHZ PAN.

SHANKS, R.D., BERGER, P.J., FREEMAN, A.E. and DICKINSON, F.N. Genetic aspects of lactation
1981 curves. *J. Dairy Sci.*, 64: 1852-1860.

SCHNEEBERGER, M. *Der Verlauf der Laktationskurve und die Schatzung der Milchleistung beim*
1978 *schweizerischen Braunvieh.* Dis no. 6168. Swiss Fed. Inst. Technol., Zurich.

SKOLASINSKI, W. and TYSZKA, Z.U. Mozliwosc zwalczania mastitis przy pomocy metod hodowlanych.
1973 *Przeglad Hodowlany,* 41: 3-4.

SOUTHGATE, J.R. A note on the comparison of Holsteins and Friesians for growth, feed efficiency
1982 and carcass traits. *In* More O'Ferrall, G.J., ed. *Beef production from different dairy breeds and dairy beef crosses.* Martinus Nijhoff, The Hague.

SOLBU, H. Disease recording in Norwegian dairy cattle. II. Heritability estimates and progeny testing
1984 for mastitis, ketosis and "all diseases". *Z. Tierz. Züchtbiol.*, 101: 51-58.

STOLBAEK-PEDERSEN, P. *Prevention of mastitis in cows.* Proceedings of Seminar on Mastitis Control,
1975 Reading University, UK. 7-11 April. p. 249-258.

STOLZMAN, M., JASIOROWSKI, H., REKLEWSKI, Z., ZARNECKI, A. and KALINOWSKA, G. Com-
1987 parison of ten Friesian strains in Poland under field conditions. I. Strain comparison for growth rate. *Livest. Prod. Sci.* (in press).

TURTON, J.D. Crossbreeding of dairy cattle — a selective review. *ABA,* 49: 293-300.
1981

VAN VLECK, L.D. Breeding for increased protein content — milk. *J. Dairy Sci.*, 61: 815-824.
1978

WENIGER, J.H., STEINHAUF, D. and PAHL, G.H.M. *Muskeltopographie der Schlachtkörper.* BLV
1963 Verlagsgesellschaft, Munich. pp 1-144.

WENIGER, J., RIECHERS, R., DIETERT, W., ZEDDIES, J. and ENGELKE, F. Beobachtungen und
1963 Betrachtungen über den Einsatz von Holstein-Friesian Bullen in der Zucht des Deutschen Schwarzbunten Rindes. II. *Tierzüchter,* 1: 5-6.

WILTON, J.W., VAN VLECK, L.D., EVERET, R.W., GUTHRIE, R.S. and ROBERTS, S.J. Genetic and
1972 environmental aspects of udder infections. *J. Dairy Sci.*, 55: 183-193.

WITT, M., ADREAE, U., HUTH, F.W., KALLWEIT, E., PFLEIDERER, U.E., RAPPEN, W., SCHUTZ-
1971 BAR, W., WERHAHN, E., RÖSELER, W. and SELHAUSEN, D. Einfluss der Kreuzung von Vatertieren der Holstein-Friesian Rasse mit Deutschen Schwarzbunten Kühen auf

Körperentwicklung, Milchleistung und Muskelbildungsvermögen. *Schr. Reihe Max-Planck Inst. Tierz. Tiernähr.*, 56: 1-90.

WOOD, P.D.P. 1967 Algebraic model of the lactation curve in cattle. *Nature,* 216: 164-165.

WOOD, P.D.P. 1969 Factors affecting the shape of the lactation curve in cattle. *Anim. Prod.*, 11: 307-316.

WOOD, P.D.P. 1970 A note on the repeatability of parameters of the lactation curve in cattle. *Anim. Prod.*, 12: 535-538.

WOOD, P.D.P. 1976 Algebraic models of the lactation curves for milk, fat and protein production with estimates of seasonal variation. *Anim. Prod.*, 22: 35: 40.

WOOD, P.D.P. 1980a Breed variations in the shape of the lactation curve of cattle and their implications for efficiency. *Anim. Prod.*, 31: 133-141.

WOOD, P.D.P. 1980b A note on the lactation curves of some high-yielding British Friesian cows. *Anim. Prod.*, 30: 299-302.

ZARNECKI, A. and HINKS, C.J.M. 1976 Strain differences in milking performance, food consumption and body size in Danish test stations. *Livest. Prod. Sci.*, 3: 165-172.

3. History, distribution and performance of Friesian cattle in the developed world

The ten countries participating in the strain comparisons in Poland will be dealt with in some detail (sections 3.1 to 3.10). The information has been largely supplied and in some cases compiled by the breed association or breeding organization in each of the countries concerned. The development of Friesian cattle in other developed countries (section 3.11) has been recorded more briefly but is also generally based on information supplied by associations in the countries themselves.

3.1 CANADA

3.1.1 History

Importations of pedigree Holstein-Friesian cattle into Canada began in 1881 (Ayrshires had been imported in 1821, Shorthorns in 1825, Jerseys in 1868 and Guernseys in 1873). According to USA archives a bull and a cow were sold in 1881 from the USA to Archibald Wright in Winnipeg. In the same or next year Michael Cook and Son, Autsville, Ontario, bought one bull and nine cows from the USA.

In 1883, the Journal "The Farmer's Advocate" stated that a company, Cook and Lord, had started to import Holstein-Friesian cattle and sell them to American and Canadian farmers. On 18 September 1884 the first annual meeting of the Dominion Holstein Breeders' Association took place. Michael Cook was the first President. In the first few years after the Association was created, herdbook registrations of Holsteins in Canada were included in the USA herdbook. In 1892 the name of the Association was changed to The Holstein-Friesian Association of Canada, and, in 1984, to the Holstein Association of Canada. The first volume of the Holstein-Friesian Herdbook was published in 1892. The last importation of cattle from the Netherlands was in 1905.

The Holstein-Friesian Association of Canada is a non-profit membership corporation with, at present, over 15 000 breeders of Holstein-Friesian cattle. It operates under national charter and is the only organization authorized to register Holstein cattle in Canada. Of the registered cattle 95 percent are Holsteins.

3.1.2 Milk recording

There are two milk recording programmes:

— Canadian Record of Performance for Dairy Cattle (ROP) operated by the Canadian Department of Agriculture. Milk weight from each cow in a herd is determined by an Inspector at intervals of approximately one month. Butterfat testing is done by the milk testing laboratories. Holstein bulls can be registered if the dam is officially milk recorded.

— Supervised Dairy Herd Improvement (DHI) sponsored by the Provincial governments and producer groups offers two types of DHI services: Supervised (once a month a DHI supervisor visits a farm to weigh the milk of each cow at all milking times) and Owner-Sampler (the owner weighs the milk and a DHI Supervisor collects the samples and sends them to the milk testing laboratories). The Holstein-Friesian Association recognizes Supervised records only. Owner-Sampler records cannot be used in Sire Proving.

In 1984, 8 891 herds, with a total of 440 965 dairy cows, were included in the official milk recording programmes (ROP and DHI). The yield for all dairy cows tested in the federal programme averaged 6 842 kg milk and 257 kg butterfat per lactation, while in 1973 the figures were 5 700 and 214 respectively. Hence, the yield increase over the last eleven years was 1 142 kg milk and 43 kg butterfat.

Table 3.1 shows the average milk yields for individual breeds in use in Canada.

Official milk records of cows are expressed as a production index called Breed Class Average (BCA). All records completed in 1951-54 were divided into classes by age and month of calving. Base production level (100 BCA) is 4 915 kg milk and 182 kg fat which was the average of 49-month-old cows which calved in April in the above-mentioned years. Since that time the yield of every cow of that age calving in April is compared to this production level and expressed as a percentage. An older cow, e.g. 70 months old, calving in April, needs to produce 5 187 kg milk and 192 kg fat to achieve 100 percent BCA. The national BCA average for records completed in 1984 was 140 for milk and 140 for fat. Bull ratings are also expressed in BCA.

3.1.3 Artificial insemination

In 1984, ten insemination centres were operating which kept dairy bulls and produced semen. In dairy herds 1 098 498 inseminations were performed, all by professional inseminators. In addition 417 720 doses of semen were distributed for owner use. A total of nearly 1.5 million doses of dairy bull semen was used. Substantial quantities of Holstein-Friesian bull semen were exported –1.4 million doses in 1984 and 989 000 in 1985 to 49 countries. The principal buyers were the USA, USSR, Switzerland, the Netherlands, UK, Federal Republic of Germany and Italy. In 1984, Canada imported over 104 000 doses of Holstein-Friesian bull semen from the USA and 3 000 from New Zealand for a Canadian/New Zealand comparison trial.

In 1984, insemination centres bought 223 young Holstein-Friesian bulls. It is recommended that one bull should be bought for 1 500 first inseminations. The goal is to have 20 percent of inseminations performed with the semen of young bulls. In the Holstein breed the semen of one tested bull should be used on 400 to 500 cows in production tested and classified herds. Minimum production and type information accepted as reliable proof is equivalent to 20 daughters in 5 or more herds. Proofs must have repeatability of 55 percent or more before they are reported.

Table 3.1 **Milk yields of cows tested in 1984 in Canada**

		305 days' production		BCA[1]			
Breed	**Days in milk**	**Milk** *(kg)*	**Fat** *(kg)*	**Milk**	**Fat**	**No. of records**	**Percentage of each breed**
Ayrshire	332	5 631	228	157	155	13 965	7.8
Brown Swiss	343	5 584	231	140	145	604	0.3
Canadienne	325	4 130	178	142	135	968	0.5
Guernsey	340	5 106	245	151	145	2 796	1.6
Holstein	337	6 842	257	143	146	152 157	85.0
Jersey	328	4 484	227	145	137	8 353	4.7
Shorthorn	295	4 431	169	153	146	199	0.1

[1] Relative to Breed Class Average — see text

3.1.4 Breeding programme

The breeding goal for Holstein cattle is the maximum profitable use of dairy cows. This means that first calving should take place before the age of 27 months and cows are expected to calve every 12 to 13 months: milk yield should be high and its composition good. The animals should have a long life expectancy, be free of disease and easy to keep and feed.

For the most important production traits (milk yield and body conformation) an annual genetic progress of approximately 1.0 percent has been estimated. Only the traits with the highest impact on cattle profitability are approved as primary traits. For the Canadian Holstein these are: milk yield per lactation and lifetime yield and body conformation (compared to the true type identified in the final classification).

Two parameters were considered to be secondary traits in the selection programme; the percentage of protein and butterfat. These do not require improvement as much as the primary traits and are antagonistic to the milk yield.

Fig. 3.1 Selection scheme for Holstein AI bulls in Canada (from Chesnais, 1984)

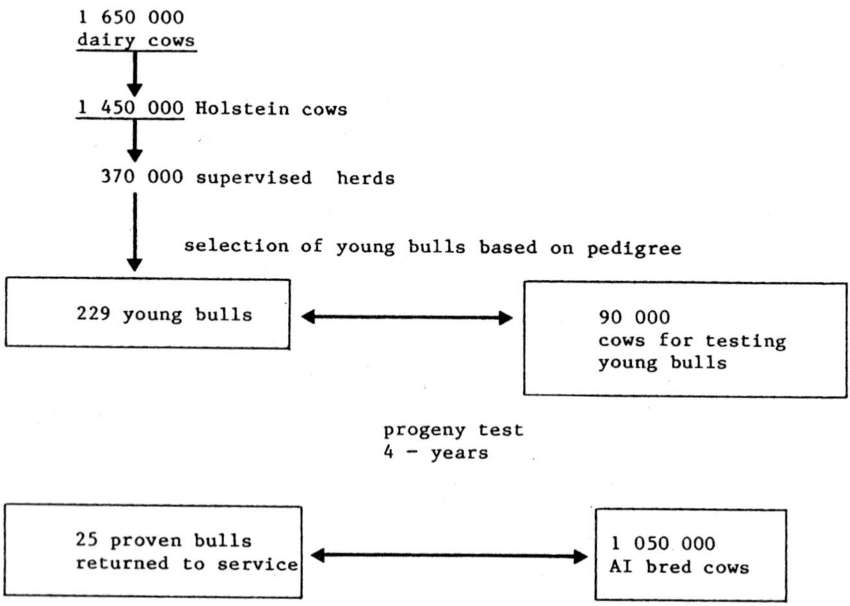

Traits such as milking speed, easy calving, fertility, body size, temperament, resistance to diseases, hereditary defects and feed use should also be monitored in a population.

The basic elements of the Holstein bull selection programme are presented in Figure 3.1 published by Chesnais (1984).

The effectiveness of the Canadian breeding programme was judged by Schaeffer (1983) as shown in Table 3.2.

For nearly two decades genetic progress in milk yield has been obvious. Over the last few years the number of young tested bulls has almost tripled. Despite this substantial increase it has been possible to improve simultaneously the average genetic value of the tested bulls. As indicated in Table 3.3 the value of indices has increased highly.

Table 3.2 **Annual rate of phenotypic and genetic change for milk yield in the Holstein breed in Canada from 1960 to 1979 for cows of all ages** (from Schaeffer, 1983)

Period	**Phenotypic change** *(kg)*	**Genetic change due to sires** *(kg)*	**Total genetic change** *(kg)*
1960—69	79.5	9.2	12.6
1969—79	74.2	20.0	25.9

Table 3.3 **Number of young Holstein bulls entering Canadian AI studs from 1973 to 1984 and evaluation of their average pedigree index for milk yield (1980 base)** (from Chesnais, 1984)

Year of entry	**Number of bulls**	**Pedigree index**	**Average proof of sires**	**Average proof of dams**
1973	82	-4.22	-3.42	-2.56
1975	99	-3.89	-3.96	-2.35
1977	85	-1.95	0.25	0.11
1979	115	-1.09	1.90	1.47
1981	175	0.19	4.15	1.95
1983	229	2.56	7.85	3.79
1984	285	4.61	6.41	3.55

3.1.5 Type classification

The Holstein-Friesian Association started the programme for the evaluation of conformation or type in 1925. Since that time the animals' conformation was the important factor in cattle improvement. In 1982, the Holstein Association, as the first in the world, adopted the linear method in type classification. This refers to the description of physical traits by scoring them from one extreme to the other on a scale from 1 to 9. More than 100 000 animals are classified each year, the period between classification being nine months. Each classifier evaluates 8 000 to 10 000 animals per year. Sires are ranked and published for all characteristics. Only those above average are considered for extensive use as proven sires.

3.1.6 National Identification Programme (NIP)

Since 1972, the Joint Dairy Breeds Committee has provided the identification system for grade dairy females which aims to identify grade cattle for selection within a herd and for sire proving. Each identified cow receives a NIP number with the following code letters:

A	— Registered sire and dam same breed as sire
B,C,D,E,	— Two, three, four, five or more generations of registered sires on record, respectively
U	— Parentage of animal uncertain or unidentified
X	— Sire and dam not same breed.

Holstein cows identified in the NIP program as ''D'' may qualify for designation as a ''Qualified Dam'' if requirements according to type classification and production level are fulfilled. Daughters of Qualified Dams can be recognized as purebred.

In 1972-83 a total of 863 245 grades (of which 96 percent were Holsteins) were identified. Bulls from females at various stages of the grading-up process are not accepted for herdbook registration.

3.2 DENMARK

3.2.1 History

Indigenous Black and White cattle have long been present in the Jutland Peninsula region, and have been known as Jutland cattle or Black and White Jutland cattle.

Detailed data concerning their origin are not available (Korkman, 1961). It is thought by some that the people of Jutland and Friesland have common ancestry. The cattle they have bred probably also originated from one population (Duplan, 1974). Formerly, Jutland cattle were known as animals of compact body conformation, not very tall, with a strong skeletal structure and considerable body weight (Korkman, 1961). The black and white or dark grey and white colour has prevailed: however, as in Friesland, animals of different colours could be found and of these red and white occurred most frequently.

At the beginning of the nineteenth century the economic situation was favourable for beef production. Meat performance ranked first because fattened animals were sold for export. At that time Black and White cattle were bred throughout the whole Jutland Peninsula territory. Later on, market demands for beef export diminished and a decreased interest in this breed of cattle was noted. In 1845, at Randers, the breeding goal for the Black and White breed was discussed, but the question whether it should be selected for development of milk traits or bred as dual-purpose cattle remained unsolved. As a result of this lack of agreement on objectives, the Black and White Jutland breed was presented, at cattle shows, in two types, i.e. dairy and dual-purpose animals. However, this situation changed and since the end of the nineteenth century all the animals have been treated as one population of dairy cattle. At that time the population of Black and White cattle was considerably reduced.

In the middle of the nineteenth century the breed was displaced by Shorthorns from the South, whereas in the East the Red Danish breed dominated. Since 1860 Black and White cattle have been imported from Holland. The imported animals were maintained as a separate population until 1949. There were conside-

Table 3.4 **Dairy cows of different breeds under performance testing and their average milk yield in 1984 in Denmark** (from Nat. Committee on Danish Cattle Husbandry, Summary of Annual Report, 1984)

Breed	Percentage of cows under milk recording	Average yield 1983-84		
		Milk *(kg)*	Fat *(kg)*	Fat *(%)*
Red Danish	14.4	5 918	244	4.13
Danish Black and White	59.0	6 274	251	4.00
Jersey	16.5	4 454	274	6.14
Red and White	0.9	5 661	219	3.87
Crossbreds	9.2	5 878	251	4.26

rable differences between strains: the Jutland Black and White breed was favoured for its better milk composition, mainly the fat content. The average annual milk yield in 1948-49 in this breed was 3 489 kg of milk with 3.93 percent of fat, while that of the Dutch strain was 4 268 kg and 3.66 percent respectively. Further, the Dutch strain was supposed to have better external traits. In 1949, Breeders' Associations of both strains were amalgamated and since then the name of Black and White Danish cattle (SDM) has been accepted.

In later years bulls from the Netherlands, Germany and Sweden were imported. In the 1950s sires were evaluated by the testing station method. Average milk yield for the first lactation of daughters sired by Danish bulls and tested under station conditions up to 1955 amounted to 3 790 kg of milk with 4.34 percent of fat, while for those sired by imported bulls it was 4 589 kg of milk with 4.28 percent of fat (Hansen, 1957 quoted by Korkman, 1961).

3.2.2 Milk recording

In 1984 the cattle population in Denmark reached 2.75 million head including 951 000 dairy cows and 59 000 cows of beef breeds. The Black and White Danish Dairy Breed (SDM) is the most numerous constituting approximately 60 percent of the total population, whereas the Red Danish breed and the Jersey constitute about 15 percent each. About 701 100 cows were milk recorded in 1984. Their yield averaged 5 881 kg of milk with 4.31 percent of fat and 3.44 percent of protein. Milk recording is carried out widely comprising almost 74 percent of the cows. The average herd size is not large and a dairy herd of 27 cows is quite common, whereas in herds included in milk recording programmes, the average number of cows is 38 (Summary of Annual Report, 1984. NCDCH).

Average milk yields in dairy herds of different breeds are presented in Table 3.4.

In recent years the Black and White Danish cattle population has been markedly penetrated by Holstein-Friesians. The breeding goal is to select a modern type of cow with a large productive potential and good workability traits.

3.2.3 Breeding programme

Approximately 350 breeding bulls are evaluated in Beef Performance Testing Stations annually. The criteria used include daily weight gain (T-value) measurements of *Musculus longissimus dorsi* (MLD) area (U-value), and general breeding ability (I-index). A breed average equal to 100 is assumed.

Approximately 30 percent of all inseminations have used young bull semen. In particular provinces, use of the tested bull semen varies from 20 to 40 percent. About 80 percent of inseminations with the tested bull semen occurs in herds included in milk recording. Within herds, semen from tested bulls is used at random. The bulls' evaluation programme assumes that in testing of young bulls their semen will be used in the proportion of 3 000 inseminations of cows per bull. In 1984, 239 young bulls went into dairy progeny testing.

Comparison of the breeding value of tested bulls is carried out on an S-index basis. Assumptions underlying the use of this new index were prepared in 1983. The S-index is an aggregate breeding value which in given circumstances is the best possible estimate of an animal's total merit, that is the true breeding value for the traits which should preferably be improved in the breed in question. The S-index is a standarized breeding value with an average of 100 and standard deviation of 5.

In 1984 the initial evaluation of 114 Black and White bulls was performed on the S-index basis. The contemporary S-index can be received by bulls having the final R-values and at least 15 first calvers classified according to their body conformation. The R-value expresses the breeding value for a bull for butterfat yield. It is based on the 305-day first lactations. Since January 1984 the contemporary R-values are published for bulls with at least 60 effective daughters. The final S-index is received by bulls with final R-values and full classification for a batch of daughters having second lactation records.

Table 3.5 shows the traits included in the S-index with the economic/biological weight per relative breeding value unit for each trait. The weighting factors of the different traits are determined by their economic importance, the heritability and genetic variation of the traits and, when available, the genetic correlations among the traits. The total value of one S-index is 145 Dkr (Danish kroner) for the SDM breed.

If the progeny tested bulls are selected on the basis of their S-indices, all traits included in the S-index will be improved. This is shown in Table 3.6 from which it appears that the average breeding value of the selected bulls (above 22 percent selection) is above 100 for all traits included in the S-index. The butterfat yield contributes most to the breeding value of a bull.

In the SDM breed there is a relatively high positive correlation between butterfat yield and the traits included in the classification of daughters' groups. Such traits as U-value, F-value, calving index, milkability and temperament make only slight contributions, whereas feet and legs as well as udder and teat qualities make a more substantial contribution to the total superiority expressed in Danish kroner (Dkr).

Table 3.7 presents the economic return of selecting progeny tested bulls on the basis of individual traits included in the S-index. The estimates are based on the results of approximately 22 percent of the best bulls for each trait.

Direct selection for butterfat yield may lead to a lower total economic return than the selection for a comprehensive S-index. In the SDM breed all the traits included in this index are positively correlated with the general index value. A correlation between the butterfat yield and body conformation is high, between milkability and temperament moderate, whereas the correlations between the remaining traits are low. The selection of progeny tested bulls on the basis of S-index is the selection criterion which results in the highest breeding improvement. Increased butterfat yield will be almost the same as that with selection based on the R-value alone.

The average S-index value of progeny tested bulls in 1984 was 108.7. Breeding progress achieved in the Danish population of Black and White cattle in recent years has been large. Within the decade 1973-83 the milk yield in the recorded cow population increased by 940 kg of milk. The average annual increase in respect of fat yield was 3.9 kg.

Annually, approximately 20 bulls are progeny tested for beef traits. A bull is tested on the basis of records of 12 sons tested in the Beef Performance Testing Station. The average daily gains of progeny of 17 bulls, evaluated in 1984, were 1 250 g, with a dressing percentage of 52.7 percent and an MLD area of 49.3 cm^2.

Thus, the improvement programme includes a number of beef performance traits. Denmark belongs to those few countries where not only the fattening ability (growth rate and feed conversion) but also the carcass quality is being improved, since the MLD area is included in the S-index.

Table 3.5 **Traits included in the S-index and their economic/biological weight in Denmark**

Trait	Weight per relative value unit	Relative weight
Butterfat yield (R-value)	0.40	1.00
Daily gain (T-value)	0.18	0.45
MLD area (U-value)	0.18	0.45
Fertility (F-value)	0.23	0.58
Calving index	0.20	0.50
Feet and legs	0.17	0.43
Udder shape	0.22	0.55
Teat size and shape	0.15	0.38
Teat position and placement	0.23	0.58
Milking ability	0.07	0.18
Temperament	0.09	0.23

Table 3.6 **Average breeding values for Danish Friesian (SDM) bulls with S-indices of at least 106 in Denmark.**

Trait[1]	**Average breeding value**	**Superiority of the bull** ***(Dkr)***
Butterfat (R-value)	112.7	736
Daily gain (T-value)	101.6	42
MLD area (U-value)	100.4	10
Fertility (F-value)	100.7	23
Calving index	100.5	15
Feet and legs	103.7	91
Udder shape	104.8	153
Teat size and shape	104.0	87
Teat position and placement	104.4	147
Milking ability	102.5	25
Temperament	102.0	26
S-index	109	1 355

[1] For weights attached to each trait included in the S-index see Table 3.5

Table 3.7 **Effect of selection for single traits in Denmark**

Trait	**Direct effect** ***(Dkr)***	**Total effect** ***(Dkr)***	**Proportion of S-index** *(%)*
Butterfat yield	911	798	63
Daily gain	117	15	1
MLD area	120	-30	-2
Fertility	132	276	22
Calving index	116	203	16
Feet and legs	160	653	52
Udder shape	214	769	61
Teat size and shape	155	653	52
Teat position and placement	231	769	61
Milking ability	62	435	34
Temperament	77	406	32
S-index	—	1 267	100

3.2.4 Herdbook registration

In Denmark herdbook registration includes:

1. EDP-registration of all milk recorded cows. Information is included on identification number and birth date, sire and dam, insemination date, calving date, sire and sex of a calf, type and conformation score.
2. All bulls have a herdbook number which is registered in a "Bull Herdbook".
3. On the basis of registered data, a "Cow Herdbook" is kept for those cows that meet certain minimum requirements as to yield, type, conformation scores and ancestry.

"The Bull Herdbook" 1984 included 1 400 bulls for the last 5 years, whereas 729 cows were registered in the last issue of the "Cow Herdbook".

3.2.5 Organization of artificial insemination

The Federation of Danish AI Associations includes 24 local Associations. Basic facts concerning the Federation's activities are given in Table 3.8. In 1984, AI Stations used 69 Black and White bulls with the dairy progeny test and 46 bulls with additional evaluation for conformation traits. In 1984, a total of 14 855 doses of bull semen were exported to 16 different countries.

3.3 GERMANY

3.3.1 History

In the first description of East Friesland cattle given by Weber (1811, cited: Kräusslich, 1981) the population was characterized as follows: animals black-pied or spotted, large milk producers but also fully useful for fattening and because of this used to improve the German breeds. The developing industry stimulated interest in tallow and beef production and Shorthorn bulls were imported to improve beefing abilities. The first Shorthorn bull was brought to Eiderstadt, North Friesland in 1843. The results of crossing with Shorthorns were considered positive although there was a drop in milk production. For this reason crossing of the Shorthorn with the Black and White Friesian was rather limited. Planned breeding aiming at an improvement of the Friesian began in 1858. The first herdbook associations were founded in Ostfriesland (1878), Schleswig-Holstein (1879) and Oldenburg (1880). Soon, many other regional breeding associations were created. Consolidation of the

German Friesian stock began after 1891 with the closing of the German-Dutch border dictated by veterinary regulations aimed at eliminating the entry of diseases. Big subpopulations, differing in production type, have existed within the German Friesian population since the 1920s. The dual-purpose type with stressed dairy characters has predominated but, in the region of Oldenburg, animals with clearly accentuated beef performance traits were maintained until the 1960s.

Payment for milk according to butterfat content was introduced in Germany in the 1930s, creating a stimulus for fat content. In the period 1949-58, fat content increased by 0.43 percent.

Crossbreeding with Holstein-Friesians marked a new era for the German Friesians. Crossbreeding started in the 1960s when Munich University introduced Holstein-Friesians into its Experimental Farm in Oberschleissheim.

At present, Black and White Friesian cattle are maintained in all States (Länder) of the Federal Republic of Germany and breeding work is conducted by twelve regional breeding associations. The Friesian population numbers 2 million cows and is the biggest amongst the breeds in the Federal Republic of Germany accounting for up to 40 percent of the cattle population. The largest number of Black and White are found in Lower Saxony where the other breeds are almost totally absent, thereafter, the states of Schleswig-Holstein, North Rhineland-Westphalia and Hessen have the most numerous Black and White stock, 80 percent of all the Black and White cattle being located in these four states. Herd size and average milk production in these various regions is shown in Table 3.9.

Table 3.8 **Activities of the Federation of Danish AI Associations**

	1984
Inseminated females	1 100 881
Herds	41 701
Bulls, alive on 1 Oct. 1984	1 810
Bulls with I-values[1]	789
Bulls with R-values[1]	102
30-60 days' non-return %	72.5
Membership fee, Dkr per female	108.99

[1] For definitions see text

3.3.2 Development of milk production — nature and economic conditions and herd structure

Over 50 percent of farms which maintained cattle in the Federal Republic of Germany in 1960 gave up cattle breeding during the following 25 years, particularly in the good soil regions. This change has led to an intensification of dairy production on remaining farms. Farms with inefficient production have had to abandon dairy cattle breeding. The greatest increase in milk production and in the concentration of dairy cattle breeding has been observed in regions with favourable conditions for dairy cattle.

According to a study by Politiek and Bakker (1982) bulls supply about 52 percent of the total beef marketed in the Federal Republic of Germany as the main category of slaughter stock. Culled cows make up 29 percent, slaughter heifers 12 percent, calves 5 percent and steers 2 percent.

The improvement of beefing abilities has parallelled the improvement of dairy production although the former has been much slower. Average production figures are shown in Table 3.10.

As noted by Gravert (1983), both the breeding goal and the strategy for exploiting cattle are influenced by economics. The price ratio of milk to grain as well as the price relations between nutrients contained in concentrates and roughages have a decisive influence. The cheaper price of concentrates relative to that of milk and roughages, recently seen in the EEC as well as the Federal Republic of Germany, has made the feeding of larger amounts of concentrates more profitable even when the increase in milk production has been less than 2 kg per 1 kg concentrates. The concentrates to milk price ratio remained, in the 1970s, at a level which had previously been observed only in the USA. An equivalent of 1.2 kg milk paid for 1 kg concentrates in 1972-77 whilst in 1976-77, 1 kg concentrates was equivalent to 0.8 kg milk. Despite the considerable demand for concentrates on world markets, but due to good crops, the prices of soybean meal and other basic components of concentrates for dairy cows have not increased. As a result the price ratio of concentrates to milk in the EEC countries is presently 1:0.7 which is similar to USA conditions. These price relations, of course, encourage feeding a higher proportion of concentrates in the dairy cow diet which further stimulates an increase in milk production.

Another economic factor which affects the efficiency of milk production is the cost of labour. According to Gravert (1983) the cost of one hour's labour to attend dairy cattle was equivalent to 13.2 kg milk in 1972-73 and had risen in 1980-82 to 19.3 kg milk. With limited farm areas, farmers are having to face a decision on what to place the greater emphasis — beef or milk — with production based on their own home produced feed. Until now, the market situation favoured dairy production. During the period 1965-82 beef prices increased by 37 percent and milk prices by 55 percent. Thus price relations had shifted toward the profitability of milk production.

3.3.3 The Friesian population — herdbook breeding

Black and White cattle, with more than 2 million head, are the predominant breed among the 11 dairy breeds bred and selected in the Federal Republic of

Table 3.9 **Herd size and milk yield in different regions of the Federal Republic of Germany**

	Number of dairy cows per herd			Average milk yield per cow *(kg)*		
	1969	1979	1984	1969	1979	1984
Schleswig-Holstein	14.8	26.9	35.3	4 261	4 652	4 780
Lower Saxony	7.9	15.6	21.6	4 268	5 105	5 503
Northrhine-Westphalia	7.9	13.4	17.8	4 072	4 696	4 977
Hessen	5.2	8.6	10.9	3 844	4 464	4 557
Rheinland-Pfalz	4.8	9.5	13.1	3 504	3 977	4 172
Baden-Württemberg	5.5	9.0	11.3	3 299	3 907	4 120
Bavaria	7.7	10.8	13.2	3 504	4 073	4 258
Fed. Rep. of Germany	7.7	9.8	15.3	3 379	4 396	4 607

(Source: P. Grothe, 1985, personal communication)

Table 3.10 **Beef production in the Federal Republic of Germany**

	Average production of beef and veal per cow *(kg)*		Average carcass weight of fattened bulls *(kg)*		
	1960	1980	1960	1970	1980
Northern region	147	276	243	267	317
Region of the central uplands	174	222	287	302	314
Southern region	145	247	286	321	343
Average in the country	145	247	268	294	328

Germany. Black and White represent (P. Grothe, 1984, personal communication):

Percentage	**Category**
43.4	Herdbook registered animals
32.9	Artificially inseminated animals
46.4	Milk recorded cattle
67.0	Active breeding population

The average herd size is 24.8 cows with significant regional differences. Breeders of the Black and White cattle are organized into 12 regional breeders' associations. Their main task is to assist their members in the breeding work, organizing auction sales and keeping the herdbook. The National Cattle Breeding Programme, designed and regularly updated by Government agencies, is implemented by these breeders' associations together with AI centres. The AI centres were founded mainly by farmers. To a large extent they are cooperatives, associations or societies; some are owned by dairy enterprises or have different legal status. Table 3.11 presents the organizational state of the Black and White Friesian breeders' associations in the country (P. Grothe, 1984, personal communication).

Since starting to crossbreed with Holstein-Friesians, the breeding goal has been the creation of a highly productive cow which would combine high milk yield (7 000 kg) with high fat (4.0 percent) and protein (3.5 percent). Reproductive characters should not deteriorate. Type classification should favour animals with a good bone frame, good movement, and with a functional udder. The German Friesians, over many years, as a dual-purpose breed with emphasis on dairy performance, should be well adapted to production under various conditions regardless of herd size and microclimate of buildings.

Protein production has been included in bull evaluation and selection since 1977. The development of dairy production within the recorded herdbook population of the Black and White Friesian cows is shown in Table 3.12.

Maintenance of the high fat content alongside a considerable increase in milk production has been a real success of German breeding. It also testifies to the proper choice of Holstein-Friesian bulls for the crossbreeding programme. The phenotypic and genetic gains estimated for the largest Friesian region of Lower Saxony are presented in Table 3.13.

In relation to 1972, the average annual genetic gain was 1.73 kg fat and 31 kg milk. This improvement of almost 1 percent per year is regarded as good to have been obtained in practice, regardless of the optimistic assumptions made at the time of formulating the breeding programme.

Holstein-Friesian bulls have gained importance in German breeding during recent years. This situation is well illustrated by information on the Friesian population from Lower Saxony in Table 3.14.

Table 3.11 **Breeding Associations for Black and White cattle in the Federal Republic of Germany**

Association	Number of members	Number of animals
Schleswig-Holstein	1 374	1 173
Hannover	7 234	212 204
Oldenburg	1 766	57 382
Ostfriesland	4 214	120 839
Osnabrück	2 433	41 899
Emsland	2 475	60 023
Südoldenburg	642	19 936
Westfalen	3 359	65 567
Rheinland	2 105	58 074
Hessen	1 844	36 270
Baden-Württemberg	1 266	15 747
Bavaria	578	17 616
Total 1984	29 289	765 430
(total 1972)	42 384	450 804

(Source: P. Grothe, 1985, personal communication)

Table 3.12 **Milk performance of cows in the German Friesian herdbook**

Year	No. of animals	Milk *(kg)*	Fat *(kg)*	Fat *(%)*	Protein *(%)*
1962	425 458	4617	182	3.94	—
1967	405 332	4793	189	3.94	—
1972	372 472	5161	206	3.99	—
1977	412 975	5507	218	3.95	—
1982	552 757	5969	239	4.01	3.38
1985	622 001	6174	250	4.05	3.37

(Source: P. Grothe, 1985, personal communication)

Table 3.13 **Phenotypic trend in the performance of heifers of the Black and White breed and the mean breeding value of their sires in the years 1977 to 1985 in Lower Saxony, Federal Republic of Germany - in relation to 1977** (from *Jahres Statistik*, 1985)

	1977 (actual performance)	1979	1980	1981	1982	1983	1984	1985
Phenotypic trend								
Milk (kg)	(4 745)	+ 221	+ 295	+ 341	+ 451	+ 523	+ 468	+ 613
Fat (%)	(3.97)	+ 0.06	+ 0.02	+ 0.03	+ 0.04	+ 0.02	+ 0.05	+ 0.10
Fat (kg)	(188)	+ 12	+ 13	+ 16	+ 20	+ 22	+ 22	+ 30
Protein (%)	(3.43)	-0.06	-0.05	-0.10	-0.12	-0.12	-0.12	-0.12
Protein (kg)	(163)	+ 5	+ 8	+ 6	+ 9	+ 11	+ 10	+ 15
Sire breeding value								
Milk (kg)		+ 101	+ 165	+ 206	+ 258	+ 259	+ 283	+ 320
Fat (%)		+ 0.02	+ 0.02	+ 0.04	+ 0.04	+ 0.07	+ 0.10	+ 0.11
Fat (kg)		+ 5.9	+ 9.2	+ 12.1	+ 14.7	+ 16.3	+ 19.0	+ 21.3
Protein (%)		+ 0.02	+ 0.0	-0.02	-0.02	-0.02	-0.03	-0.03
Protein (kg)		+ 2.0	+ 3.9	+ 5.6	+ 7.6	+ 8.3	+ 9.3	+ 10.4

Table 3.14 **Proportion of heifers by year and sire group, Federal Republic of Germany** (from *Jahres Statistik*, 1984)

Sire group	1972 (%)	1976 (%)	1978 (%)	1980 (%)	1982 (%)	1984 (%)
German Black and White (G)	38.2	24.9	18.1	7.7	3.8	1.6
Dutch Black and White (D)	8.8	16.1	18.9	11.1	6.7	3.3
Holstein-Friesian (HF)	1.3	16.9	27.3	46.3	64.7	75.1
Crossbred B & W × Crossbred HF	2.8	8.3	10.7	7.4	3.3	1.1
G × HF crossbreds	2.3	13.8	20.3	20.6	16.1	10.7
G × D × HF	0.0	1.1	1.9	4.7	3.7	5.3
Unidentified	46.8	18.9	2.8	2.2	1.7	2.9

Table 3.15 **Causes of culling of cows in Lower Saxony, Federal Republic of Germany** (from *Jahres Statistik*, 1982)

	Milk yield class				
Reason of culling	To 3 999 kg	4 000-4 990 kg	5 000-5 990 kg	Above 6 000 kg	Total
Sale for breeding	20.4	9.7	7.0	6.9	8.2
Age	5.7	9.5	8.9	8.5	8.7
Low productivity	4.2	11.2	13.0	12.1	11.8
Infertility	11.1	30.4	31.7	31.7	30.4
Infectious diseases	0.1	0.4	0.4	0.5	0.4
Udder diseases	3.2	8.3	9.8	10.9	9.6
Milking ability	0.5	2.4	3.1	3.5	2.9
Metabolic diseases	0.6	1.6	2.1	2.3	2.0
Feet problems	1.0	3.7	4.9	5.6	4.7
Other	53.2	23.1	19.1	18.0	21.3

Whilst in 1972 only 3.6 percent of heifers were sired by Holstein-Friesian bulls or Holstein-Friesian crossbreds, twelve years later 85.8 percent represented this genotype which shows the changes to which the German Black and White cattle have been exposed.

According to Gravert (1983), besides a considerable increase in milk yield, some less desirable characteristics have appeared. During the last two decades the annual rate of cow culling has increased from 25 to 30 percent. In the last 10 years the proportion of cows above 6 years of age decreased from 38 to 28 percent. Data from Lower Saxony demonstrate these tendencies more clearly in Table 3.15.

The reasons for culling show that in herds of high production level the number of animals culled for infertility, udder diseases, and feet problems is higher. Deterioration of reproduction in herds of high production has also been noted by the artificial insemination stations. For example, as shown in Table 3.16, in more than 83 percent of herds with yields below 4 400 kg per cow up to 1.7 services were required per conception, whereas in a similar proportion of herds with yields above 6 800 kg per cow, up to 1.9 services were required per conception.

Thus, Gravert (1983), when formulating the future breeding goals for the German Friesian population, stressed the fact that any further increase of production should be tightly linked with other economically important traits. A high yielding cow should also be capable of being used over a greater number of lactations without causing health and reproductive problems.

Table 3.16 **Insemination index by herd production level, Federal Republic of Germany** (from *Jahres Statistik,* 1985)

	Milk yield class					
Insemination index[1]	**To 4 400 kg**	**4 401-5 200 kg**	**5 201-6 000 kg**	**6 001-6 800 kg**	**Above 6800 kg**	**Total**
≥ 1.3	43.1	33.4	24.2	18.5	13.3	23.8
1.4	11.7	15.2	16.0	14.3	12.3	14.5
1.5	13.0	13.9	16.4	16.6	16.1	15.8
1.6	9.7	11.4	13.7	14.9	14.8	13.5
1.7	5.6	8.3	9.6	10.9	12.2	9.8
1.8	5.6	6.0	7.0	8.2	10.3	7.6
1.9	3.4	4.1	4.3	5.7	6.7	5.0
2.0	3.3	3.3	3.1	4.0	5.0	3.7
≥ 2.1	4.6	4.4	5.7	6.9	9.3	6.3

[1] Number of inseminations per conception

3.3.4 International importance of German Friesians — sale of animals and semen

Appreciation of the Friesian breeders in the Federal Republic of Germany is evidenced by a growing interest in importation of animals from the country. It suggests the efficiency of the breeding programme in matching type and performance characteristics to actual needs and production capacities. Trade in breeding animals is described in Table 3.17.

In 1984 German Friesians were exported to 31 countries. Average price for the Black and White breeding stock was increasing from year to year until the EEC milk quota regulations were introduced. In 1985 average prices for animals sold through auction sales were as follows:

Animal	**Price (DM)**
Bulls	2816
Cows	2464
Pregnant heifers	2501
Young animals	1318
Calves	761

An important quantity of semen (169 221) doses was also exported from the Federal Republic of Germany in 1984.

The relatively early decision to cross the local Black and White stock with Holstein-Friesians from North America is doubtlessly the main reason for the present wide interest in the German Friesians. Additional assets of the German strain are high fat content and relatively good beef performance. These factors, together with lower prices of animals and lower transport costs as compared to the USA and Canada, lead many countries to purchase animals from the Federal Republic of Germany.

Table 3.17 **Development of Friesian cattle sales in the Federal Republic of Germany in the years 1962-84**

Sales	1962	1967	1972	1977	1982	1985
Auction	37 293	41 565	28 427	33 486	43 759	38 889
Off farm	1 222	4 829	18 733	23 029	31 974	75 629
Export	1 202	6 512	12 965	14 098	29 089	67 161

(Source: P. Grothe, 1985, personal communication)

3.3.5 Organization of artificial insemination and bull selection

Forty-nine artificial insemination centres exist in the country most of which have been established by the farmers in the form of cooperatives. In 1984, 5 870 719 first inseminations were done which covered 93.1 percent of cows and heifers. The efficiency of artificial insemination is measured by the non-return rate between the 60th and the 90th day from service which was 67 percent in 1984. In 1984 the numbers of cows and heifers inseminated by breed were as follows: Fleckvieh — 2 403 000, Black and White Friesians — 2 018 000, Red and Whites — 873 000, and Brown Swiss — 390 000.

The high usage of AI and the widespread use of milk recording (50.6 percent of all cows) facilitate bull progeny testing. Breeders' associations and AI centres cooperate in effecting breeding programmes.

Out of 717 000 herdbook cows 21 500 potential bull dams are selected. They are mated to 50 bull sires (of which about 30 are from abroad). One thousand bull calves are performance tested and selected with an intensity of 1:2. Some 500 bulls are purchased for AI and progeny tested (in 1984, 534 test bulls were used). With semen from each young bull 600-800 cows (first calvers) are inseminated. After progeny testing 100 bulls with positive test results are used as cow sires. Out of these bulls about 20 are classed as elite bulls, serving as bull sires.

The characteristics taken into consideration in the breeding programme are: milk, fat and protein, body conformation (with special emphasis on udder and legs), ease of milking, fertility, calving ease and, for males, also beef traits.

The use of embryo transfer is increasing. In 1984, 5 619 embryos were obtained from 611 donor cows.

3.4 ISRAEL

3.4.1 History

In the area of Palestine the local people farmed a local primitive cattle breed, the Baladi, producing annually about 500 kg of milk. At first the local cattle breed was improved by the use of bulls brought from the environs of Damascus (Levi, 1977). The crossbred produced about 1 800 kg of milk. At the beginning of the twentieth century German Templars imported Black and White bulls from Germany for crossing with Baladi and Damascus cows.

The process of creating the Israeli Holsteins can be divided into four stages: 1) research period, 2) upgrading period using Black and White Dutch bulls,

3) Holstein-Friesian period, 4) creation of the Israeli Holstein strain. These stages are marked by quite different milk yields. Some of the production data are shown in Table 3.18.

From 1921 Black and White Friesian cattle were imported from the Netherlands. The initial period was marked by problems of adaptation. The acclimatization of European cattle encountered difficulties. Ticks transmitted many diseases which were difficult to control. European systems of feeding and maintenance of dairy cows did not suit Israeli conditions. In 1926 the Israel Cattle Breeders' Association (ICBA) was founded and this organization solved existing problems to a large extent. It worked out breeding programmes and outlined the principles on which to create Israel's own strain of dairy cattle. This process of improvement was based mainly on upgrading through the importation of very good bulls. In the years from 1936 to 1938 there were some sizeable importations of cows from the Netherlands, Switzerland, Romania and Germany. In 1937 ICBA took the initiative in promoting AI.

In 1945 the evaluation of bulls using the daughter-dam comparison method was begun. During this time cooperative insemination stations were organized. In 1946, 11 Holstein-Friesian bulls and 60 cows were imported from Canada. The bulls were distributed to 11 regional AI stations and used heavily. From 1950 to 1960 there was a rapidly increasing demand for milk and milk products. At that time a total of 18 000 animals were imported, the majority of which were Holstein-Friesians from the USA.

Table 3.18 **Increase in dairy producion of milk recorded cows in Israel** (from Levi, 1977)

Year	Number		Yield			Stage of breeding work
	Herds[1]	Cows	Milk *(kg)*	Fat *(%)*	Fat *(kg)*	
1924-25	12	248	1 725	—	—	Indigenous stage
1934-35	20	1 222	3 916	3.70	145	Upgrading by Dutch bulls
1944-45	77	15 937	4 356	3.62	158	
1954-55	184	14 273	4 632	3.56	165	Initial use of Holstein bulls
1964-65	204	26 489	5 847	3.25	190	Transition to Holsteins
1970-71	219	38 376	6 563	3.24	213	Israeli Holstein period
1974-75	231	57 995	7 191	3.18	229	

[1] A village of smallholder herds is counted as one herd

Regional AI centres gradually merged into two cooperatives of dairy producers: ON in the north and Hasherut in the south. Payment for AI service is on a cow/year basis with choice of semen of the best bulls at no additional price. Inseminations are made by full-time inseminators except in the southern Negev desert.

In 1953 the first bulls were tested by the contemporary comparison method. In 1954 the bull laying-off system was introduced. Since that time almost all breeding of cows and heifers is by AI and about 75 percent of all inseminations by proven bulls.

In 1962 there was a final importation of bulls from the USA for AI.

3.4.2 Background to production

Israel is situated in a subtropical climate with temperatures exceeding 30°C during the summer period and with rain only in the winter months, starting at the end of November until March. On non-irrigated soil, vegetation and feed production is only possible from December to April. The country has a hilly terrain with mostly rocky soil and almost half of its area has desert features. In such climatic conditions most animal feed comes from irrigated fields which is the reason why forage is so expensive. In this particular situation the basic feed for milking cows is agricultural industry wastes, concentrates and a small supplement of hay and silage (Dori *et al.*, 1977).

The average feed ration for a cow with a high milk yield is 19 kg of dry matter and containing 16 percent crude protein. The ration is composed of 10 kg of concentrates, 2 kg of whole cotton seed, 2 kg dry matter of orange peel, 4 kg dry matter of maize, wheat or green fodder silage and 2 kg of peanut, lucerne or vetch hay. The proportion of forage in rations is 35 percent. The content of raw fibre in the diet is low, approximately 15 percent. This feeding system adversely affects milk fat content but does not affect protein content.

3.4.3 Milk recording

Milk recording up to 1975 covered about 50 percent of dairy cows in Israel. Two types of herds exist: large cooperative industrial farms (Kibbutz) which own from 150 to 550 cows and small herds with 15 to 50 cows in family farms (Moschav-Herden) (Levi, 1977).

In 1984, Israel had 707 herds of dairy cattle numbering in total 183 500 registered females of the Israeli Holstein breed, including 187 large industrial farms of the Kibbutz type. The remaining herds belonged to individual farms associated in cooperative farming organizations (Moschav). In addition there were 62 beef herds totalling 30 230 cattle.

According to Bar-Anan (1971) in the years 1957-67 annual genetic improvement in the heifer population was 65 kg milk and 1.2 kg fat; thus, a bull which had a breeding value of +100 kg of 3.5 percent fat-corrected milk in one year was no better than average 4 years later.

In the years 1965-70 the genetic gains in kg fat exceeded the gain in milk. The average annual genetic gain over the 5-year period was: 27 kg milk, 4 kg fat, or 70 kg 3.5 percent fat-corrected milk. During the years 1976-85 the annual phenotypic and genetic gains in heifer lactations were 168 and 91 kg for milk and 6.5 and 2.1 kg for fat respectively. This rate of improvement is unequalled elsewhere. The improvements have been maintained over a long period of time in conditions of very high productivity and a constantly increasing number of recorded cows. In 1984 approximately 60 percent of the cows were milk recorded.

Table 3.19 shows the number of recorded cows and average production over the years.

3.4.4 Breeding programme

At the end of the 1960s insemination with frozen semen began and computers were used for selection. Accuracy of estimates in testing bulls was increased by enlarging the scope of information and incorporating milk recording results of females in second lactation (Bar-Anan, 1971). Prospective bull dams are first selected on the basis of being 2 standard deviations above herd average.

Table. 3.19 **Milk yield of recorded cows** (according to Bar-Anan, 1971, and Summary of 1984 *Milk and Butter Fat Recording Israel* CBA 1984)

Year	Cows	Milk *(kg)*	Fat *(kg)*	Fat *(%)*
1949	7 833	4 044	141.2	3.65
1959	16 917	5 347	187.0	3.49
1969	34 132	6 271	204.4	3.26
1974-75	54 239	7 191	229.0	3.18
1983	61 440	8 672	285.0	3.29
1984	66 000	8 734	287.0	3.29

About one third of the preliminary selected cows, after the evaluation of type and conformation, qualify as bull dams and account for 1 percent of the population. About 30 percent of inseminated bull dams are inseminated with imported semen. The remaining cows are mated to bulls from the national testing programme (Levi, 1977).

Daughter progeny tests are carried out three times per year by BLUP procedure for the first three lactations. The following traits are evaluated: kg milk, kg fat, kg ECM (economical fat-corrected milk), percent fat, linear type traits, percent conceptions of daughters, percent incidence of mastitis and of percent culling for mastitis, low yield or infertility. Calving performance tests for direct and maternal effects on difficult calvings and prenatal calf mortality are computed for first calvings. Progeny test for percent protein and incidence of health problems are under consideration and at the research stage.

All bull calves from the dairy herds are used for fattening. Beef progeny tests by BLUP procedure are carried out on live and carcass weights of yearling bull calves at slaughterhouses.

Fifty young bulls are brought annually into AI. To avoid inbreeding, one third of the young bulls are by semen imported from selected sires from outside Israel. There is no importation of bulls or cows, and no importation of semen beyond that for the breeding of the bull dams. On average, five proven bulls are returned to service after progeny testing. Twenty-five percent of the inseminations in milk recorded herds are by young bulls and the remainder by proven bulls. Heifers calve at two years of age, thus tests based on part-time lactations become available when the bulls are 4.5 years old. At the age of five years, most of the bulls are already slaughtered and the very best ones returned to service. The main culling of bulls is during first and second lactations of daughters. The final selection ratio of bulls is 1:12.

Young bulls are, as a rule, mated with heifers after the first calving in order to randomize progeny testing. One thousand inseminations are carried out in milk recorded herds in order to produce 100 first lactation daughters per bull. The positively proven bulls are at their return to service mated with heifers for testing their direct effect on heifer calvings. Tests for difficult calving and calf mortality from first parity heifers are not used as a criterion for sire culling, but are used for nominating proven bulls for inseminations of maiden heifers (Bar-Anan, 1984).

Older cows are eligible for registration in the Elite Cow Register; the registration requirement is a life-time yield of above 100 000 kg of milk.

Ninety percent of the animals in Israel are enrolled in a cooperative animal insurance company, the ''Hachaklair'', which provides a comprehensive veterinary service. Cows are pregnancy checked and in many herds all veterinary diagnoses and treatments are recorded on computers in the cow barns for follow-up inspection. The data are available for progeny testing. All cow data on milk recording, inseminations, pregnancy checks and non-government veterinary care are integrated in one computerized data bank which belongs to the Israel Cattle Breeders' Association.

3.5 NETHERLANDS

3.5.1 History

Black and White Friesian cattle had their origins in an area of rich pasture grounds situated along the North Sea coast between West Friesland and the Jutland Peninsula.

As early as the fourteenth century, historical sources mention cattle from this area, stressing their qualities (Krüger, 1961). In the Middle Ages there were many local breeds adapted to local environmental conditions. The above-mentioned area specialized in cattle breeding because Friesland had good conditions for feed production on grasslands but not for grain farming on a large scale. The Netherlands has good climatic conditions for cattle breeding. Median yearly temperature is +10°C and the rainfall is 722 mm. Soil is mainly sandy, but not very rich. In the provinces of Friesland, North Holland, Seeland and Groningen there are rich soils on the coast and polders composed mainly of sea moss. In the Middle Ages the main profit from cattle breeding came from the sale of draught oxen. It was not until the sixteenth century that horses began to play an increased role as a labour force. At that time the demand for butter increased which was due to the rapid development of cities. It caused milk production to gain greater economic significance. In the seventeenth century pastoral farming was the predominant activity in Friesland. The population of breeding cattle in this area was already known as a local breed as indicated by the numerous sales of animals exported to different European countries.

Serious work on the improvement of productivity of the Friesian breed was undertaken in the Netherlands in the eighteenth century. The East Friesian breed population was isolated in the nineteenth century when the Dutch/German borders were closed.

It is believed that the original native type of the Black and White Friesian breed has its roots in the Netherlands, Germany and Denmark. The first Dutch Black and White Friesian cattle herdbook was created in 1874. Milk recording was begun at the turn of the twentieth century. Initially local recording services were set up. The Central Milk Recording Service began its work in 1943. The Netherlands was the first country in the world to introduce the recording of milk protein content in 1958.

3.5.2 Description of production

The total milk production from 2.5 million dairy cows amounts to 13.2 billion kg milk per year.

The milk pricing system is based on payment for butterfat and protein yield (both of great importance in selection) and a negative price for water content (volume) of milk. Transport and manufacturing costs are mainly dependent on volume. The consumption of liquid milk is 1.7 billion kg. Milk products produced (billion kg) are: cheese 0.5, condensed milk 0.6, milk powder 0.4, butter 0.2.

Cattle breeding in the Netherlands has its specific character. The pattern observed in other countries does not always correspond to Dutch breeding. In the decade 1973-83 the number of milking cows increased by 20 percent and reached 2.5 million. The position of the Black and White breed is, however, different: whereas in most countries Black and White cattle are on the increase, in the Netherlands they have decreased by about 6 percent. The Red and White (MRIJ) breed population has grown faster and this is related to the better beef performance of this breed which gave a more favourable profit index over recent times.

Practically all Friesian and the major part of the MRIJ bull calves are fattened for white meat (lean meat). Bull beef production (about 500 kg liveweight) is based on dual-purpose MRIJ calves. In the eastern and southern provinces with sandy soils and smaller farms the MRIJ cattle is predominant. In these regions the intensity of land use increased more than in the western part.

During the last decade there has been a strong tendency to concentrate and specialize dairy cattle stock. The majority of herds had their cow maintenance systems changed by the introduction of loose housing. This has affected the labour output and has stimulated herd growth to an average size of about 42 animals. The number of farms decreased from 92 000 in 1975 to 61 000 in 1983.

Intensive farming on grasslands and rotational grazing are the basic grazing patterns. Many herds are rounded up at night into barns for complementary feeding. Only a small number of herds is kept indoors all the year round.

There has been an increase in concentrate feeding. The process of intensifying production on the farms was accompanied by changes in the organization of the breeding service and supervisory institutions (Thomas and van Burg, 1981).

Small local milk recording centres and insemination stations were merged. In 1951 the Federation of AI Associations was established. The Federation of Milk Recording Associations has been functioning since 1960. The National Computer Centre was founded in 1973.

3.5.3 Milk recording, herdbooks and organization of breeding work

Since the end of the second world war the number of milk recorded cows and their yield have increased rapidly (Table 3.20). In the first 25 years annual improvement was 40 kg of milk and 0.02 fat percent. In the next 13 years it was 76 kg of milk and 0.013 fat percent. The number of milk recorded cows increased eightfold. Milk recording now covers about 75 percent of the dairy cow population.

The average yield over all lactations for the Black and White breed in the year 1984 was 5 717 kg of milk with 4.2 percent fat and 3.34 percent protein. For the Red and White breed the average yield was 5 457 kg of milk with 4.01 percent fat and 3.39 percent protein. Cows of the Groningen breed yielded 5 238 kg of milk with 4.03 percent fat and 3.41 percent protein content.

Politiek (1978) gives an interesting account of performance of two subpopulations of Friesians. Milk performance of Dutch Friesians (DF) from the Friesland and North Holland provinces was evaluated. The average yield of the group of animals tested in 1972-74 at the Experimental Farm Flevoland was 4 440 kg of milk with a fat content of 4.25 percent and protein content 3.42 percent in the first lactation. Heifers from North Holland yielded 530 kg more milk than their contemporaries born in Friesland. The estimated genetic difference in milk yield was 12 percent in favour of North Holland.

In 1984, 818 892 cows were registered in the herdbooks. Black and White registered cows had a milk yield of 5 822 kg with a 4.22 percent fat content and 3.35 percent protein content. In comparison with non-registered animals, breeding cattle yielded 329 kg more milk with a more desirable fat content. In 1984, a total of 356 732 calves of the Black and White breed were registered in herdbooks.

Eleven provincial organizations (including the Breeding Association NRS and FRS) created a Syndicate (Royal Society Nederlands Rundvee Syndicaat) which started to work on 1 January 1984. The Syndicate offers central com-

Table 3.20 **Number of milk recorded cows and their milk yield in the Netherlands** (from Jaarverslag, 1984)

Year	No. of recorded cows	Milk *(kg)*	Fat *(kg)*	Fat *(%)*	Protein *(%)*
1945-46	218 186	3 628	127.7	3.52	—
1969-70	823 901	4 639	183.7	3.96	3.33
1983-84	1 794 998	5 626	232.9	4.14	3.36

puterized registration of all breeding data, provides the identification of animals and oversees milk recording and AI centres. Provincial milk recording associations are members of the Syndicate. The Syndicate employs type classifiers and has a Sire Advisory Programme.

The National Committee for testing milking ability has been in operation since 1962. In 1984, 77 groups of half-sibs of 77 Black and White bulls were evaluated on the basis of progeny. The overall average maximum milk flow per minute was 2.7 kg and considered optimal. The proportion of milk in the forequarters (Udder Index) was 44.4 percent. In addition the milking ability was estimated in 601 cows, mostly future bull dams.

3.5.4 Breeding goal

The objective is to breed high yielding cows of milk type with special emphasis on fat and protein yield. The purpose is (i) to obtain, for the entire Black and White population, an average yield of 6 000 kg milk with a 4.25 percent fat and 3.75 percent protein content (Stapel, 1980) and (ii) to ensure that a further increase in milk yield should not lead to a deterioration of meat performance.

Since 1980, the cow and bull indices for yield have been based on a payment system (fat and protein yield minus manufacturing costs, and including feeding costs for fat, protein and carrier (lactose)). The so-called ''Netto-Guilder-Index'' (Dommerholt, 1980) is used. The new selection criterion is intended to stimulate more accurate economic selection and gives a greater interest in the increase of protein and fat concentration in milk. To this end a criterion applied upon buying milk, the so-called negative basic price for the amount of milk delivered, was introduced. Producers delivering milk with a high protein and fat content are given bonuses. Consequently the net economic value of percentage content of basic milk components increased. The economic value for fat and protein yield is dependent on cheese and butter prices which are approximately in the proportion of 1 to 1. The breeding value of a bull (^{1}B) was established on the basis of the following index:

$$^{1}B = 0.316\ ^{BV}\text{milk} + 260\ ^{BV}\text{fat\%} + 500\ ^{BV}\text{protein\%}$$

BV is breeding value for each trait considered in the index. The application of the new index resulted in full agreement between the breeding objective and the criteria of breeding value estimation. The ranking of bulls evaluated on the basis of the new index may not be the same as on the previously applied criteria of the total of protein and fat yield in kg. Similarly, this will influence selection effectiveness. This is illustrated in Table 3.21.

Coefficients of genetic correlations indicate that the net index shows considerably closer positive correlations with the fat and protein content. The selection carried out on the basis of the new index is expected to cause a faster improvement in milk composition while progress in milk yield might be slowed down. It should be pointed out that since 1984 the Dutch farmers have reduced the milk quota by 8.65 percent per herd. In this situation the cow population is now decreasing by about 10 percent. The value of an increase in milk yield is then decreased to some extent in favour of fat and protein content and of beef traits.

3.5.5 Bull evaluation programme

The Netherlands has four stations for young bull performance testing, with 437 places. The growth rate is tested over 8 months, from day 120 to day 365 of life. Results of yearling bulls in the course of the test are shown in Table 3.22 (Jaarverslag, 1984). They show the slightly better performance of the Red and White breed.

About 370 young bulls of the Black and White breed undergo annual tests on progeny. About 20 percent of all inseminations are done with the test bull semen. About 1 000 doses of semen are used on average for a test bull. Breeding evaluation of a bull is carried out on the basis of milk performance of 100 daughters. Preliminary indices of test bulls are estimated on the basis of breeding value of the maternal sire and grandsire. Average preliminary indices for tested bulls of the Black and White breed from 1984 are given in Table 3.23 (Jaarverslag, 1984), and the cow index in Table 3.24.

Table 3.21 **Genetic correlation between the breeding value for each feature and the net index or the fat and protein yield in kg in the Netherlands** (from Dommerholt, 1980)

	Milk yield	Fat (%)	Protein (%)	Fat yield (kg)	Protein yield (kg)
Net index	0.64	0.24	0.52	0.81	0.89
Fat and protein yield (kg)	0.91	0.01	0.07	0.97	0.96

Table 3.22 **Performance results of selected AI yearling bulls under test in the Netherlands** (from Jaarverslag, 1984)

Breed	**DF**		**MRIJ**	
Selected for AI	yes	no	yes	no
Number of bulls	105	63	117	100
Weight at 365 days (kg)	401	379	423	411
Rump height at 365 days (cm)	130	128	128	126
Growth/day (g)	1 111	1 064	1 174	1 135

Table 3.23 **Average pedigree index of Black and White test bulls in the Netherlands (1984)**

Test Bulls	**Milk** *(kg)*	**Fat** *(%)*	**Protein** *(%)*	**Net Guilder value** **1984**	**1983**	**Dutch Friesian**	**Holstein-Friesian**
400	+ 877	+ 0.09	-0.02	292	247	20%	80%

Table 3.24 **Average cow index for the Black and White breed for 1982-83 in the Netherlands**

	No. of records	**Milk** *(kg)*	**Fat** *(%)*	**Protein** *(%)*	**Net Guilder value**	**Increase 1984-83 net Guilder value**
Cow index	529 023	+ 167	+ 0.03	+ 0.01	+ 65	+ 9
Average breeding value of sires	529 029	+ 237	+ 0.07	+ 0.00	+ 95	+ 15

The selection intensity in the breeding programme is as follows:

- young AI bulls, selection on breeding value for beef production — 1:2
- selection of cow sires after progeny testing on breeding value for milk and type — 1:10
- selection of bull sires on breeding value for milk and type — 1:20
- selection of bull dams on breeding value for milk and type — 1:50

3.5.6 Insemination

In 1984 there were 2 177 265 first inseminations (about 78 percent of the population). Artificial insemination is even more widely applied in the milk recorded population where it amounts to over 81 percent (Jaarverslag, 1984). The effectiveness of artificial insemination is very high with a non-return rate of 72 percent. The semen of Black and White bulls was used for 62 percent of inseminations and the Red and White for 33 percent in 1984. In Dutch insemination stations in the same year, 599 Black and White bulls were used and 254 Red and White, 9 Groningen, 10 Jersey, and 18 of remaining breeds. The Netherlands started to use Holstein-Friesians much later than other European countries, but now semen from Holstein-Friesian bulls is used intensively. A considerable amount of semen continues to be imported. The purchase in 1984 of Holstein-Friesian semen from the USA and Canada amounted to 566 000 doses and from the Federal Republic of Germany to

Table 3.25 **Inseminations by breed of bull (%) in 1984 in the Netherlands**

Dutch Friesian (DF)	10.8
Holstein-Friesian (HF) — USA and Canada	40.5
Crossbreds (HF × DF)	10.5
	61.8
Dutch Red and White (DRW)	29.9
Red HF	2.0
Crossbreds (HF × DRW)	0.8
	32.7
Piemontese (Italian beef breed)	2.5
Other beef breeds	0.9
	3.4
Other breeds	2.1

Photographs of the ideal cows as selected by the appropriate breed association

Canada. Huntsdale Lester Delight. Photo Canadian Livestoc Photography Inc.

Denmark. Photo Jørgen Bak Rasmussen

Federal Republic of Germany. Evchen 4424381

Israel

The Netherlands. Bemma 20. Photo Veeteelt

New Zealand

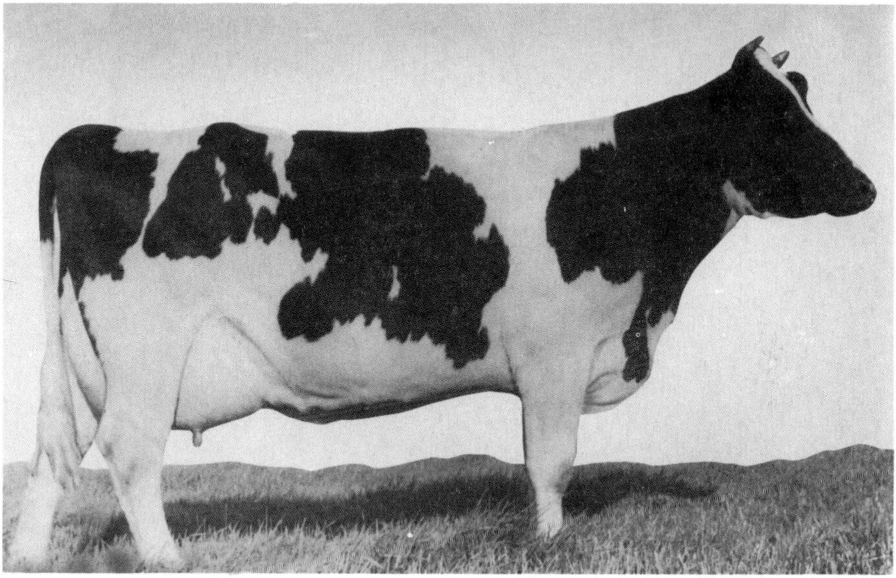

Poland. Photo Alojzy Zawadzki

Sweden

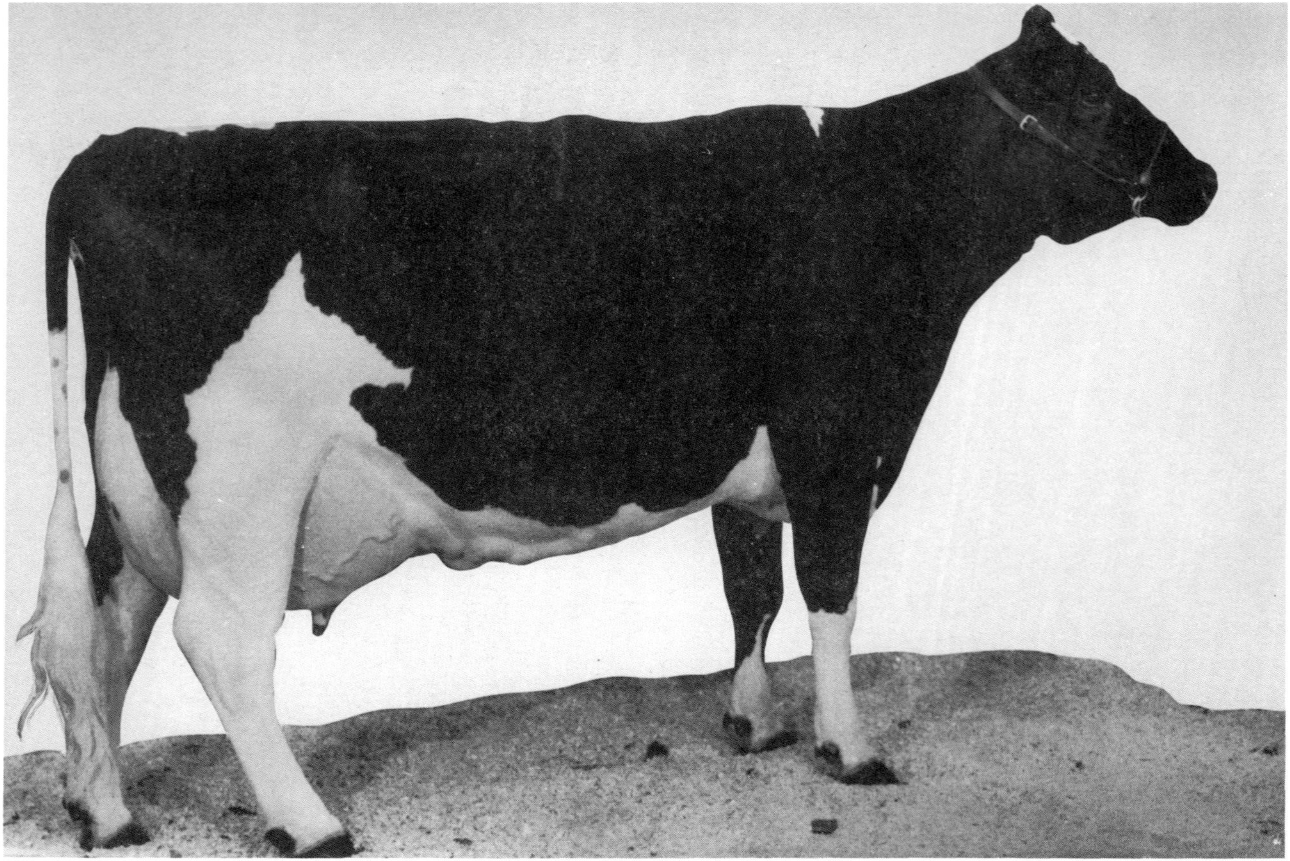

United Kingdom. Holmland Ruby 42. Supreme Champion 75th Anniversary Show of the British Friesian Cattle Society. Photo Douglas Low.

United States of America. Zehrview Arlinda Polly. Photo Agri Graphics Ltd.

63 600 doses. The total number of first inseminations by Holstein-Friesian bull semen was 881 990. The proportions of inseminations by breed of bull are shown in Table 3.25.

3.6 NEW ZEALAND

3.6.1 History

The first Friesians to enter New Zealand were imported from the Netherlands in 1884. Subsequent importations came from a number of countries. However, by far the most important influence on the development of the breed in New Zealand came from the cattle imported from Canada and the USA in the first decade of this century and from later importations of bulls from these countries. Subsequent selection from this base has developed the present New Zealand type Friesian. The New Zealand Friesian Association (now renamed The New Zealand Holstein-Friesian Association) was formed in 1910, but primarily because of the dominance of the Jersey breed in manufacturing milk herds, expansion of the breed was slow.

3.6.2 Production conditions

The territory of New Zealand, including offshore islands, comprises 27 million hectares, the total distance from north to south is 1 600 kilometres while its widest point is 450 kilometres from east to west. The mean annual temperature varies according to location from 10 to 16°C. Average annual rainfall is 1 200 mm, varying over different parts of the country from 600 mm to 1 500 mm. The temperate, maritime climate is very favourable for pasture growth and consequently for livestock production based on pasture. Annual pasture yields can reach as high as 14 000 - 16 000 kg dry matter.

New Zealand has about 3.3 million dairy cattle (of which about 2.3 million are cows in milk), 4.6 million beef cattle (about 1.5 million breeding cows) and nearly 70 million sheep.

The human population is 3.25 million and therefore only a small proportion of annual livestock production is required for the domestic market. The livestock industries have been developed largely to produce for export. The dairy industry is a good example of this. About 90 percent of milk produced is used for manufacturing and about 80 percent of manufactured dairy products are exported.

Dairy farmers produce milk either for the liquid milk market (about 9 percent of dairy farmers) or for the manufacturing milk market. Those producing milk for manufacturing operate on a seasonal basis, calving all their cows in the late winter-early spring, and milking no cows for about 2 months in the winter.

Manufacturing milk is paid for on the basis of the quantity of fat, or fat plus protein, and the price is determined by the price received for these products on the world market. Consequently the price producers receive is much lower than in other dairying countries. In 1985 it was the equivalent of about 10 US cents per kg milk (compared to Canada 41 cents, UK 24 cents, Netherlands 27 cents).

Because of the low milk price, New Zealand dairy farmers need a low-cost system of farming. This is achieved by utilizing to the full the natural climatic advantages with low capital, low labour input systems of dairying. Cows are grazed on pasture supplemented when required by pasture hay and silage; concentrates are not fed. Average herd size is about 150 cows and the farm is usually operated with family labour.

All heifers are mated to calve at about 24 months of age and the annual replacement rate is about 18 percent. Some 500 000 surplus calves are reared each year for meat purposes (mainly 'bull beef') with other surplus calves being slaughtered at one week of age.

The comparatively small number of farmers who produce milk for liquid consumption contract to supply milk throughout the year and consequently they calve their cows on a year-round basis. To compensate them for the higher costs they incur, they receive a somewhat higher price for their milk than do those producing manufacturing milk and, although the major feed supply is still pasture, concentrates may be fed, particularly during periods when grass is in short supply.

Traditionally the Friesian breed was confined almost entirely to the herds producing milk for liquid consumption with the Jersey breed dominating in manufacturing milk herds. However, since 1960 there has been a steady swing to the Friesian in manufacturing milk herds and in 1985 about 65 percent of all artificial insemination services were made from Friesian bulls. As a result, whereas "liquid milk" herds have continued to be dominantly Friesian, "manufacturing milk" herds now contain a mixture of Friesian, Jersey and Friesian × Jersey animals. The swing to the Friesian has also resulted in a rapid increase in the number of registered pedigree (herdbook) Friesian cows and whereas 30 years ago the majority of these cows would be in "liquid milk" herds the majority are now in "manufacturing milk" herds.

Herd improvement work in New Zealand, including the operation of herd (milk) recording services, is the responsibility of the Livestock Improvement Council of the New Zealand Dairy Board which sets the policies carried out by the Board's Livestock Improvement Division. The members of the Council represent the

Council's six operating Regions, the parent body, the New Zealand Dairy Board and the Ministry of Agriculture and Fisheries. The major activity of the Council is the operation of the Herd Improvement Plan which is concerned with four major areas of activity: data collection; research; services such as herd recording, progeny testing and artificial insemination; and field advisory services.

3.6.3 Milk recording

There are 14 600 dairy farms in New Zealand producing milk for manufacturing and 1 300 dairy farms producing milk for direct consumption.

Milk recording commenced in 1908 but the present cooperative milk recording service did not start until 1922. In 1984-85 about 60 percent of dairy cows were production-recorded. There are four systems of recording differing in the frequency of recording: monthly, six-weekly, alternate month, and production ranking. Each is available under either a "standard" (recording officer on farm) or "self sample" basis. In principle, breed associations approve the use of standard, monthly, six-weekly and alternate-month records for registered pedigree cows, but there are some differences between associations.

The production figures in Tables 3.26 and 3.27 illustrate a number of aspects of the New Zealand dairy industry:

— the comparatively low yields resulting from almost complete reliance on pasture as a feedstuff.
— the influence of the Jersey breed, as illustrated again by the relatively low average milk yield but high fat content.
— the comparatively short average lactation length due to the seasonal calving pattern and consequently the common drying-off date for all or most of the cows in a herd regardless of their calving date.

Table 3.26 **Production of recorded cows in New Zealand**

Year	Number of cows *('000)*	Milk *(kg)*	Fat *(kg)*	Fat *(%)*	Average lactation length *(days)*
1955-56	474	2 745	139	5.06	257
1975-76	775	3 205	152	4.76	251
1981-82	922	3 326	159	4.78	250
1984-85	1 300	3 470	166	4.78	247

3.6.4 Breeding programme

The goal of dairy cattle breeding in New Zealand is to achieve easily managed cows which produce large quantities of milk, fat and protein under conditions of pasture feeding. Stichbury (1982) groups the traits which might be considered in selection into three categories: production traits; utility or management traits; aesthetic traits.

1. Production traits

Because of the predominant requirement for manufacturing milk, most emphasis in selection has been on fat yield. This has resulted in milk and protein yields also being increased but with the increase in the economic importance of protein more emphasis is now being placed on selection for protein yield. However, this will not result in marked changes in the cattle selected as research has shown that in both the Friesian and Jersey breeds there is a very high correlation between milk, fat and protein yields.

2. Utility or management traits

These are the non production traits which are of economic importance to dairy farmers because of their effect on the ease of management of the cow. They would include aspects such as udder shape and teat placement and traits such as temperament, ease of milking and resistance to mastitis.

3. Aesthetic traits

These include traits, such as colour, which often form part of a breed standard but which are of little or no relevance to commercial dairying. It is recognized that to achieve economic progress it is important to minimize the emphasis given to aesthetic traits in favour of production.

Table 3.27 **Production of New Zealand Herdbook Friesians**

Season	Number of cows	Milk *(kg)*	Fat *(kg)*	Fat *(%)*	Average lactation length *(days)*
1955-56	7 735	4 061	152	3.74	271
1971-72	21 942	4 468	180	4.03	280
1981-82	41 099	4 553	189	4.14	271
1984-85	48 488	4 729	198	4.18	270

Selection of both cows and bulls is based on production and breeding indices for cows, and breeding indices for bulls. At birth, breeding indices are calculated on the basis of ancestry (average of the breeding indices of sire and dam). These ancestry indices are subsequently modified in the case of cows by their own production indices and in the case of bulls by the performance of their daughters.

Relative to the base year of 1960 (index 100) the current average breeding index (fat) of progeny-tested Friesian bulls is 118.

3.6.5 Artificial insemination

New Zealand has four artificial breeding services. The major service is operated by the Livestock Improvement Council but there are also three privately operated services. About 90 percent of dairy farmers use one or more of these services with about 70 percent using the Livestock Improvement Council's service for all or the majority of their inseminations. In 1984 some 1 375 000 cows (60 percent of the total dairy cow population) were inseminated under the Council's service with over 90 percent of all inseminations being from proven bulls. The average breeding index of the proven bulls used was 136.

The testing and selection of bulls follow similar principles to those used in other major dairying countries. The starting point is the selection of high breeding index cows (registered and non-registered) which are also satisfactory in other important traits. These cows are mated under a contract scheme to selected sires.

Each year some 150-160 bulls, mainly Friesian and Jersey, are obtained in this way. They are then progeny tested for production and other important traits through a sire proving scheme, and when proven, the best 5 percent of them are selected for extensive use. This intensity of selection is made possible by the very large number of inseminations achieved per bull through the use of fresh semen with high dilution rates in the very concentrated mating season applicable in most herds. This allows over 15 000 inseminations to be achieved from some sires during a three-month mating period. Within-herd contemporary comparisons of the production of daughters of different bulls indicate that the production improvements expected from the bulls' breeding indices are being achieved.

3.6.6 The New Zealand Holstein-Friesian Association

The New Zealand Holstein-Friesian Association is involved in:

- Maintaining the herdbook
- Breed improvement

— The operation of systems for comparing the conformation of cows and assessing the conformation being transmitted by bulls
— Breed publicity.

To be eligible for entry in the herdbook, animals must be the progeny of sires and dams registered in the herdbook. Both Black and White and Red and White animals are eligible for registration but Red and White animals carry the suffix "Red". Females described as "off colour" (e.g. black markings extending right down to the hoof) are eligible for entry in the Herdbook where they carry the suffix "OC". Male "off colour" animals are not eligible for registration.

The Association also operates a supplementary register under which female cattle can, through a grading-up procedure, eventually have their female descendants entered in the Herdbook, where they carry the suffix SR. At present there is no provision for male animals to qualify for entry in the Herdbook through such a grading-up scheme.

3.7 POLAND

3.7.1 History

The Polish Black and White breed was created by crossing and grading-up local breeds with the Friesian breed imported from western Europe, primarily from the Netherlands (Jasiorowski *et al.*, 1979). First mention of an importation of cows from Friesland into Poland was made by Mikolaj Firlej and dates back to 1570. A little later, in 1630, Friesian cows were purchased by Lech Sapieha. With time, more and more Friesian cattle were bought, and people came to call the better herds "Dutch" cattle. Quite substantial numbers of Black and White cattle were also brought to western Poland and Silesia from Germany, primarily Oldenburg. Before the the first world war the imported population of Friesian cattle accounted for approximately 5 000 cows, about 95 percent of which could be traced back to the Netherlands. Imported pedigree cattle, together with local stock which had a proportion of Friesian genotype, provided the bulls which were widely used in the general cattle population.

The first Milk Recording Circles were organized in central Poland in the years 1905-06. In a short time, milk recording, together with efficient methods of dairy cow feeding, led to a significant increase in yield. As early as 1910, there were a few herds which had a yield of over 4 000 kg of milk per cow. At that time the fat content in milk was not recorded. After Poland regained its independence in 1918, supervision of breeding was organized. The law on animal breeding, passed in 1934, created uniform rules for entering into herdbooks and for milk recording. During that same year, there were four breeders' associations in Poland which organized the owners of cattle (Konopinski, 1949).

Over the years from 1925 to 1938 there were regular importations of breeding animals from the Netherlands, Germany and also on a small scale from Sweden. Milk yield increased rapidly. In the year 1936-37 the average yield from over 53 000 tested cows was 3 497 kg milk with a fat content of 3.35 percent — a substantial improvement over average levels in the 1920s.

During the second world war Poland suffered one of the greatest material losses among European nations: 67 percent of the cattle population was destroyed, and almost all pedigree animals were killed. As a part of war compensation Poland received 112 000 cows, of which 75 000 were received from the Red Army, 20 000 came from UNRRA gifts, and 16 500 were sent from Sweden. Most of the gifts were handed to private farmers. The cattle received from the Red Army came from Pomerania and the UNRRA gifts included cattle primarily of the Holstein-Friesian breed (Jasiorowski *et al.*, 1979).

Artificial insemination was organized in 1954. In the same year the Union of Pedigree Animal Breeding (UPAB) was founded. State farms which specialized in breeding became members of this organization. The task of UPAB was the improvement of animals in their own herds and the production of sires for mass breeding needs. At the time, it was commonly believed that the best method for improvement of Black and White cattle was crossbreeding with the Dutch strain. For that reason, the importation of large numbers of Dutch bulls was resumed in 1956 and by 1967 over 200 bulls had been imported from the Netherlands for breeding herds. The Dutch bulls improved the milk yield and fat content.

Heifers were also imported from the Netherlands and a small number from the Federal Republic of Germany. The body conformation of the Polish Black and White cattle became more and more like that of the Dutch dual-purpose cattle. The cows were short (the average height at withers was 128 cm), deep (chest girth was 195 cm) and well-muscled. Adult animals attained an average weight of 600 kg. In the late 1950s the average milk yield was low. Approximately 560 000 cows were milk recorded, and the yield was about 3 000 kg milk with 3.61 percent fat content. Cows registered in herdbooks gave approximately 500 kg more milk. The low milk yield was associated with the fact that cattle breeding was not profitable enough. In order to determine whether this yield level was caused by low genetic merit or by environmental factors (primarily the level of feeding), samples of the cattle population were tested in testing stations which functioned on similar lines to Danish bull testing stations. It was found that the production potential of Polish cattle was much higher. Heifers from state farms had lactations of approximately 3 600 kg milk with a fat content of 3.69 percent, and heifers from private herds 3 300 kg milk with an average 3.5 percent fat content (Jasiorowski *et al.*, 1979).

3.7.2 Production conditions

The climatic and soil conditions and the extent of grazing differ from region to region in Poland. The most favourable conditions for cattle breeding are near the Baltic coast where there are good soils, mild climate, an average annual temperature of approximately +8°C, and a total annual rainfall of approximately 550 mm. In central Poland the soils are predominantly sandy and the climate is more continental. Average annual temperatures are +7.8°C, and total annual rainfall 530 mm. During the summer, periods of drought often occur and impair the quality of pastures. In the north-east region annual rainfall is almost 100 mm higher.

About 80 percent of arable land belongs to private farmers. The average size of a private farm is 5 ha. In the eastern and southern parts of the country the size of farms is small. Larger farms are located in the western and northern parts of Poland. The remaining arable land belongs to state and cooperative farms.

The cattle population in Poland was approximately 11 million in 1984 out of which 5.4 million were dairy cows (82 percent Black and White). Cows on state farms accounted for 12.5 percent of the total. The total milk production in 1984 was 15.8 million litres, 70 percent of which was purchased by cooperative dairies. Fattening of cattle is usually performed in an extensive system since price structures do not favour using grain for fattening. Young beef cattle are usually slaughtered after reaching a weight of 420 kg. Weight gain during fattening does not normally exceed 0.6 kg per day.

The number of cattle slaughtered totalled 1.2 million tonnes (liveweight) of which 45 percent came from culled cows. The limited use of concentrates and the low profitability of fattening cause a large proportion of calves to be slaughtered at the age of two weeks. The average annual production of carcass weight of beef and veal was about 57 kg averaged over the total number of cattle of all types, ages and both sexes.

In 1983 the price relation of 1 kg liveweight of slaughter animals to the average price of 1 kg milk was 5:1. On the other hand, the ratio of the price of milk to the price of fodder barley was 0.92:1.

Low profitability of cattle breeding did not encourage the setting-up of dairy farms. The average number of cows in private farms included in the milk recording scheme was about seven. At the present time there is a slight tendency to concentrate production in larger herds.

3.7.3 Milk recording

Table 3.28 shows the results of milk production for cows in the official milk recording system from 1950-84 based on monthly visits by official recorders (method A).

Milk recording covered approximately 14 percent of the total cattle population. Practically all state herds (4 339) with relatively large numbers of cows each were recorded. However, in the private sector milk recording is practised on a limited scale with only about 2 percent of the cattle population evaluated. In the last few years there has been a substantial decrease in the number of cows milk recorded in the private sector. The actual breed composition of the milk recorded animals in 1984 was as follows:

Breed	Percentage
Black and White	94.2
Red and White	5.4
Simmental	0.2
Polish Red	0.1
Others	0.1

Production improvement in the milk recorded cattle population during the last 35 years was approximately 18 kg milk per year. More pronounced gains were made in the fat content of milk, which increased in this period by 0.60 percent — an annual increase of 0.017 percent. In those 35 years, production improvement in fat yield was 1.26 kg annually.

Improvement indices for cattle registered in the herdbooks were even lower, as shown in Table 3.29, but the number of registered cows increased over that period 30-fold.

3.7.4 Breeding programme

The objective of breeding has been to produce dual-purpose cattle. Selection is aimed at reaching a milk yield of 4 500 kg milk with a fat content of 4.1 percent from the productive cattle population. The animals are expected to be of average size, their height at withers should be 132 cm and body weight 550-600 kg. The animals regarded as most desirable are characterized by rapid growth, good feed utilization and a well-muscled body.

A bull testing and selection programme was introduced in 1971. Initially the programme was supervised by the Animal Husbandry Institute. In 1975, following its establishment, the Central Animal Breeding Office supervised the programme. The Central Animal Breeding Office also coordinated the activities connected with milk recording, artificial insemination and maintained the herdbooks.

The number of dams of the Black and White breed used to breed bulls is estimated to be 8 900 head, 65 percent of which are in state herds and 35 on private farms. The average yield of herds from which the bull dams come was 4 459 kg of milk (Szelag and Nahlik, 1985). The average superiority of dams of bulls over their contemporaries in the same herds was 19 percent in fat yield. In 1983 the average yield of bull dams was 5 891 kg milk, 250 kg fat, and 4.25 percent fat when the highest yielding lactations from each bull dam were averaged.

The growth rate of male calves from the bull dams is evaluated in performance testing stations. The calves which qualified for further evaluation are tested from day 121 to day 360 of their lives. In Poland there are 25 bull rearing units with a total capacity of about 2 800 bulls. The rearing centres evaluate

Table 3.28 **Average yield from milk recorded cows in the years 1950-85 in Poland** (from CSHZ, 1986)

Year	All recorded cows			Cows owned by private farmers only		
	Number of cows	Milk yield *(kg)*	Fat *(%)*	Number of cows	Milk yield *(kg)*	Fat *(%)*
1950	62 151	3 023	3.39	5 473	3 045	3.73
1975	759 427	3 409	3.87	171 828	3 963	3.91
1985	754 923	3 668	3.98	114 094	4 350	4.02

Table 3.29 **Average yield of cows registered in herdbooks in the years 1950-85 in Poland** (from CSHZ, 1986)

Year	Number of cows	Milk yield *(kg)*	Fat content *(%)*
1950	11 469	3 535	3.42
1975	321 391	3 669	3.91
1985	317 189	3 865	3.99

the weight gains compared with the rotational average for contemporaries, as well as body conformation, size, and musculature. The average liveweight of the bulls after completing the test is approximately 430 kg and the daily weight gain during the testing period is 1 250 g. The best performing young bulls are placed on AI stations for progeny testing and the rest used either for natural service or are slaughtered. Table 3.30 gives more information on bulls which have qualified for AI centres after completing the test. As can be seen, the growth rate of bulls of the Holstein-Friesian breed was greater than that of the other types.

About 2 400 bulls of the Black and White breed were performance tested in 1983. From these, approximately one quarter are selected for AI centres. Progeny testing of about 500 young bulls is performed in state herds. On an average 350 artificial inseminations are made from each young bull being tested for the first time. All in all, approximately 200 000 cows are used for bull progeny testing. While waiting for the test results, about 30 000 semen doses are stored from each bull. In 1984, a total of 581 bulls were tested, 361 of them for the first time. The breeding value estimation of a bull for milk characteristics is performed on the basis of the superiority of daughters over their contemporaries using the contemporary comparison method. The basis for selecting sires is a fat yield index and also their rating on the fat content of the milk. The final selection of sires is made by a committee. The information about the best young bulls evaluated in a given year is expanded to include information about the type, size, exterior, and quality of udders of a group of 20 daughters each. Each year about 20 bulls of the Black and White breed are finally selected from the national programme for use at AI stations as the sires of future bulls. In addition, a list of bulls from which semen is to be imported is drawn up. In the last few years imported semen was used for artificial insemination of

Table 3.30 **Characteristics of bulls which qualified for AI centres following the performance test in Poland**

Genotype	**Weight at 360 days** *(kg)*	**Daily weight gain** *(g)*	**Height at withers at 360 days** *(cm)*
Black and White	429	1 236	118.6
F_1 Holstein-Friesian	440	1 276	122.0
Holstein-Friesian	461	1 336	127.4

about 16 percent of bull dams. Most of the semen was imported from the USA, Canada, Sweden and the Federal Republic of Germany. Small quantities of semen were also purchased in Denmark, UK and New Zealand. About half of the dams were mated to bulls of the Holstein-Friesian breed.

3.7.5 Organization of artificial insemination

In Poland there are 51 AI centres with approximately 2 100 bulls in total. A hundred of these bulls are of beef breeds used for commercial crossing to about 10 percent of cows in dairy herds. The breeds included are: Charolais, Simmental and Limousin. The Black and White breed is represented by 1 800 bulls, including 100 Holstein-Friesian and 150 with different degrees of Holstein-Friesian blood. Artificial insemination is used for about 70 percent of the cow population, including 3.2 million cows of the Black and White breed. In the mid-1970s artificial insemination was on a much larger scale, and used for practically 90 percent of cows. In 1984, semen from progeny tested bulls was used for artificial insemination of 66 percent of the cows (in 1975 this had been only 5 percent).

The average calving interval for the adult cattle population is 392 days. Annually about 25 percent of heifers are introduced into a herd, and the average age at first calving is 940 days. In herds which belong to private farmers the replacement rate is lower at about 18 percent (WOWUB, 1985).

3.7.6 Pedigree breeding

Registration in herdbooks is used primarily to record the parentage of animals. In 1983, 379 255 animals were registered in herdbooks, consisting of almost half of the cows from the adult population. Most of the pedigree animals — that is 65 percent — were the property of private farmers. Specialized state farms dealing with animal breeding had 54 000 cows and, in 1984, the yield from cows on these farms was 4 500 kg milk with 4.0 percent fat content.

3.8 SWEDEN

3.8.1 History

In the middle of the nineteenth century the Swedish cattle population was dominated by Red and White cattle improved by the use of Shorthorns and

Ayrshires. At that time there were already some Black and White herds. In 1877 a farmers' association in Malmö decided to import Black and White animals from Friesland and assigned funds for that purpose. Three hundred animals of the Friesian breed were purchased within four years. There was no great interest in beef production toward the end of the nineteenth century and the improvement of Black and White cattle was focused on dairy performance. Many Black and White cattle were imported from Friesland in about 1890 and at the very beginning of the twentieth century (1904-07), many animals were imported from the Netherlands. Some importation of Friesian bulls took place in the 1920s, the animals originating from Germany and the Netherlands. The Swedish Breed Society for Friesian Cattle was founded in 1913.

The first cattle herdbooks were established in 1892. At first they were kept by the Farmers' Associations as county herdbooks. Since 1913 the national cattle herdbooks have been published yearly. A milk recording programme was begun in the region of Malmö in 1898. Artificial insemination was started in the southern province of Skaane in 1943 and in the province of Halland in 1944.

The Swedish Black and White Friesians originate mainly from the Netherlands, although Germany and lately American Holsteins have also contributed to the formation of this strain. Because of strict veterinary and health regulations the importation of animals and semen to Sweden is restricted. The first semen doses from the USA and Canada were imported in the early 1960s. During 1975-85 about 50 000 doses were imported. The influence is considerable at the moment, because 25-30 percent of the young bulls in AI are out of USA sires.

Small herds prevail in Sweden, the average herd size being 19 cows in 1985. Milk recorded herds are somewhat bigger, 25 cows on average. Red and White cows (SRB) make up the majority of the national dairy population (56 percent). The share of the Black and White breed was 35 percent in 1985. During the last decade the Black and White Swedish Friesian population has increased by about 9 percent (Philipsson, 1984).

3.8.2 Milk recording

During the period 1900-55 the average milk yield increased by 20 kg per year. As from the mid-1950s, modern breeding methods have been introduced. An important role in the acceleration of breeding improvement must be attributed to progeny testing of bulls and artificial insemination. Consequently, the improvement in production has increased, reaching approximately 60 kg of milk per year. Genetic progress has been estimated to be about 1 percent per year. In 1984 the average yield of milk recorded cows was 6 326 kg with 4.20 percent fat and 3.4 percent protein. Milk recording comprises 67.9 percent of the dairy cow population. The average milk yield is, at present, the highest in Europe.

Statistics from milk recording herds of Black and White cows are summarized in Table 3.31.

3.8.3 Registration of Friesian animals

A Standard Register and, especially for bull dams, an Elite Herdbook are kept by the Breeding Association. About 4 000-5 000 milk recorded cows are registered in the Standard Register and about 1 000-1 200 cows in the Elite Herdbook annually. Bull dams are selected from the entire milk recorded population. About 500 bulls are registered in the herdbooks annually but only some 20 percent of them enter artificial insemination for progeny testing.

A very high proportion (94 percent) of recorded cows is artificially inseminated thus facilitating breeding work to a great extent. Essential for the proper conduct of breeding programmes is a close integration of activities between milk recording, artificial insemination, and herdbook organizations. Such an approach makes it possible to include secondary traits (reproduction, type, and health) into the genetic improvement programme.

3.8.4 Breeding programme

The Swedish Friesian strain of cattle represents a dual-purpose type. Although 90 percent of beef comes from the dairy herd, the most important product is milk as

Table 3.31 **Milk recording results for Swedish Black and White cows in Sweden** (from *Ars Statistik Meddelande* No. 133)

	1960	1970	1980	1985
Number of recorded cows	64 363	74 665	126 267	153 418[1]
Percentage of Black and White cows in the milk recorded herd	22.7	21.7	30.8	34.2[1]
Milk yield (kg)	5 019	5 571	6 175	6 529
Fat yield (kg)	199	221	246	266
Fat content (%)	3.96	3.97	3.98	4.08

[1] In 1984

shown by the milk to beef price ratio of 1:9. Relative to EEC countries this ratio is low, indicating relatively higher prices for milk or low prices for beef in Sweden (Philipsson, 1984). An equivalent ratio in EEC countries is usually close to 1:7.

All the bulls going forward for progeny testing are performance tested before they are allowed into test matings (170-180 bulls yearly). Growth rate and feed conversion are measured from day 60 to day 360 of life. The average liveweight at 360 days is 430 kg for Black and White bulls. Only the bulls which have gained above the station average are normally allowed into test matings.

About 40 percent of the active population is mated to the test bulls. The semen of one test bull provides 1 600-1 700 inseminations. During the waiting period about 35 000 semen doses per bull are collected and after collection the bull is slaughtered in most cases. Laying off bulls as an alternative to automatic storing of frozen semen is now being discussed. The bull's fertility of his female progeny as heifers at first calving is recorded.

The combined bull breeding value index in 1984 is shown in Table 3.32 (evaluation based on 100-150 daughters for production and reproduction, 30-40 daughters for other traits). The bull breeding value estimation is very comprehensive but with the stress on milking ability traits.

Some other additional traits are included in the selection programme which would aim to improve the longevity of the population and decrease the frequency of some diseases. In view is a reduction in the incidence of mastitis, and an improvement in general resistance to diseases of dairy cows (Philipsson, 1984). This has become possible because, since 1984, all veterinary treatments of dairy cows in the country have been recorded and centrally processed.

3.9 UNITED KINGDOM

3.9.1 History

Black and White dairy cattle were imported into the United Kingdom from the Netherlands and Germany from 1850 onwards to provide milk and dairy products for the population of the newly established towns created by the industrial revolution. In addition to a number of importations from the Netherlands in 1914, 1936 and 1950 (a total of 144 bulls and 70 females) there was an importation of 40 bulls and 63 females from South Africa in 1922. The dairy qualities of the Dutch cattle were readily appreciated and in 1909 a breed society was formed which later became the British Friesian Cattle Society of Great Britain and Ireland (BFCS). The aim of this Society has remained unchanged since it was formed "to promote the Friesian breed". In furthering this aim the Society has established an open herdbook which registers all Friesian and Holstein-Friesian pedigree animals, from any part of the world. This gives members access to the full spectrum of genetic material to breed the cows most suited to their farm environment. The Society does not differentiate between Friesians and Holstein-Friesians and the result is there are many types of "Black and White" cows which vary from the extremely angular North American type to the blockier and beefy type.

In 1946, to repair the ravages of the second world war, 54 Canadian Holstein-Friesian bulls and 166 females were imported and at this time also a second herdbook was established. This is a closed herdbook registering only Holsteins.

Table 3.32 **Characteristics in bull evaluation and breeding value in 1984 in Sweden**

Characteristic	Breeding value	Economic value
Milk index (fat + protein)	112	+12
Daily gain (Own perf.)	102	+ 1.0
Daughters' fertility	110	+ 3.0
Stillbirth as sire	—	- 2.0
Stillbirth as mat. grandsire	M	0
Ease of milking	97	- 0.75
Udder conformation	112	+ 3.0
Udder — floor distance	107	+ 1.75
Teat length	96	- 0.60
Teat placement	109	+ 2.25
Extra teats	103	0
Leg conformation	101	+ 0.10
Temperament	100	0
Bull index		+20

Table 3.33 **Number of dairy and beef cows in UK (thousands)** [from *June Census, Dairy Facts and Figures* (UK), 1985]

	England and Wales	Scotland	Northern Ireland	United Kingdom
Dairy cows	2 573	272	288	3 133
Beef cows	691	438	202	1 331
Total cows	3 264	710	490	4 464

Now known as the British Holstein Society (BHS) it remains a relatively small organization. Animals may be registered with both BFCS and BHS and when this occurs they are known as "dual registered" animals. Approximately one third of the Friesian/Holstein population is registered with the breed societies.

The Friesian/Holstein breed dominates the UK dairy industry. Ninety percent of the nation's milk comes from the Black and White herds and 60 percent of the beef. The beefing aspect remains an important part of the dairy farmer's income. It is thus the ability of the breed to meet the demands of modern dairying systems combined with good secondary beefing qualities which has led to its pre-eminence in the UK.

3.9.2 The dairy cow population

In 1985 the total number of cows in the UK was 4.4 million comprising 3.1 million dairy cows and 1.3 million beef cows (Table 3.33).

Commercial crossing is widely used in dairy herds with 62 percent of home-produced beef originating from that source. The percentage of beef inseminations done in the UK fluctuates between 30 and 40 percent. Table 3.34 shows the number of first inseminations to dairy and beef bulls in the period April 1984 to March 1985.

The British Friesians are the most numerous dairy breed and constitute about 90 percent of the total dairy population. Other milking breeds are Dairy Shorthorn, Ayrshire, Jersey and Guernsey. There are also small numbers of Red Poll and Dexter (Table 3.35).

Table 3.34 **Number of first inseminations in UK, 1984-85 (thousands)** [from *Dairy Facts and Figures* (UK), 1985]

	England and Wales		Scotland		Northern Ireland	
	(No.)	*(%)*	*(No.)*	*(%)*	*(No.)*	*(%)*
Dairy	918	57.1	64	60.4	56	27.7
Beef	686	42.8	42	39.6	137	68.2
Dual purpose	1	0.1	—	—	8	4.1
Total (all breeds)	1 605	100	106	100	201	100

The dairy herd in the UK at June 1985 (3.13 million) was 4.5 percent down on the previous year and 6.0 percent below the peak level of June 1983. The national dairy herd is now at a lower level than at any time in the last 25 years. This can be considered as a reaction to the volume-based quota introduced in April 1984. Prior to 1985 the dairy population had remained fairly static (3.24 million in 1975 and 3.28 million in 1984) but during this period the average herd size increased from 43 to 64 cows.

As more than 90 percent of British Friesians are located in England and Wales, the reported characteristics of this strain are mainly based on data from the Milk Marketing Board of England and Wales (MMB).

Milk recording schemes embrace 42 percent of dairy cows and 31 percent of all producers (Table 3.36). In England and Wales, the average size of herds in the National Milk Records (NMR) scheme in 1984 was 90 cows.

Table 3.35 **Dairy cow breed distribution in UK (%)** [from *Dairy Facts and Figures* (UK), 1985]

Breed	England and Wales (1984)	Scotland (1984)	Northern Ireland (1982)
Ayrshire	2.1	22.2	1.3
Friesian	90.5	54.7	96.8
Holstein	2.9	1.6	NK[1]
Guernsey	1.9		NK[1]
		0.3	
Jersey	1.6		NK[1]
Dairy Shorthorn	0.4	NK[1]	1.5
Friesian/Ayrshire	—	16.0	—
Others	0.6	5.2	0.4

[1] NK = not known

Table 3.36 **Number of cows and herds in official milk recording schemes 1984 in UK** [from *Dairy Facts and Figures* (UK), 1985]

	Number of herds	Percent of producers	Number of cows	Percent of total dairy cows
England & Wales	13 751	35	1 242 716	46
Scotland	1 305	40	132 221	45
Northern Ireland	211	3	16 247	5
United Kingdom	15 267	31	1 391 184	42

Table 3.37 shows the average production of cows, by breed, in the NMR scheme. In the last decade the milk yield of British Friesian milk-recorded cows (in England and Wales) has increased by 849 kg, fat plus protein yield by 65 kg, fat content by 0.12 percent and protein content by 0.01 percent.

3.9.3 The breeding programme

The objective of the dairy breeding programme is to breed cows which are:

— efficient producers of milk fat and milk protein. The main selection objective in England and Wales prior to the introduction of quotas was yield of fat and protein. Whilst this remains the prime objective, the effect of the volume-based quota has been to place increased emphasis on compositional quality.
— of high value as cull cows and capable of producing calves of good beef potential. As increased profitability cannot be sought through increased volume of milk, it is important to consider the additional income which can be obtained from cull cows and surplus calves.
— of sound dairy conformation, particularly for mammary and locomotive traits.

There are a number of progeny testing schemes in the UK differing only in the emphasis given to particular characters and in the scale of their operation. The scheme run by the MMB for its Friesian/Holstein bulls is the largest and is described to illustrate the approach used (G. Swanson, 1984, personal communication) (see Figure 3.2).

Table 3.37 **Lactation averages by breed (England and Wales)** (from *B & P Report* 1984-5)

Breed	Milk *(kg)*	Fat & protein *(kg)*	Fat *(%)*	Protein *(%)*	Milk *(kg)*	Fat & protein *(kg)*	Fat *(%)*	Protein *(%)*
Ayrshire	4 297	308	3.85	3.31	5 043	365	3.90	3.33
Friesian	4 744	327	3.68	3.21	5 593	392	3.80	3.22
Holstein	5 452	*	3.69	*	6 252	432	3.75	3.16
Dairy Shorthorn	4 145	285	3.57	3.30	5 019	350	3.62	3.27
Guernsey	3 596	288	4.48	3.52	4 022	329	4.63	3.55
Jersey	3 367	292	4.90	3.76	3 849	345	5.18	3.79

* Insufficient numbers

Fig. 3.2 AI breeding programme in UK

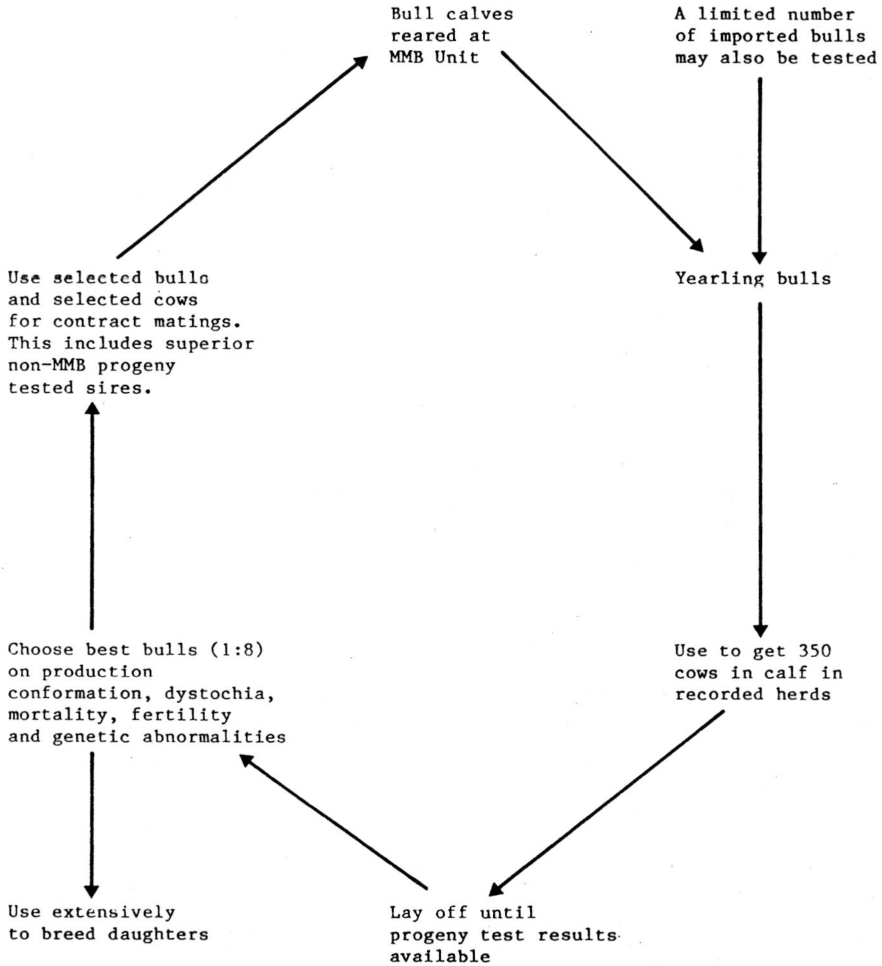

(G. Swanson, 1985, personal communication)

Approximately 100 pedigree Friesian/Holstein bulls are progeny tested each year in cooperating milk-recorded herds. The owners of these herds have contracted with the MMB to offer a specified number of cows for mating to young bulls each year and to rear and milk record for the first lactation at least 50 percent of the resulting heifers. In 1984-85 a total of 34 918 animals in 2 410 herds were on contract. The Ministry of Agriculture limits the number of doses of semen from a progeny test bull which can be distributed to 750. A bull's test insemination are

Table 3.38 **Pedigree estimates of young bulls entering progeny testing scheme in UK** (from *B & P* Report 1984-85)

Year of entry into test	SIRES[1] Average ICC of fat plus protein *(kg)*	DAMS[2] Average ICC fat plus protein *(kg)*	DAMS[2] Average CGI	TEST BULL Estimated ICC fat plus protein *(kg)*
1978	+ 18.4	+ 12.4	624	+ 15.5
1979	+ 20.0	+ 15.4	654	+ 17.6
1980	+ 22.2	+ 16.8	668	+ 19.5
1981	+ 29.0	+ 18.4	685	+ 23.2
1982	+ 27.7	+ 19.0	690	+ 23.1
1983	+ 31.8	+ 24.3	743	+ 27.9
1984	+ 31.4	+ 24.7	747	+ 28.2

[1] Does not include American Holstein sires
[2] Does not include Holstein dams

Table 3.39 **Average merit of MMB proven Friesian/Holstein stud in UK** (from *B & P Report* 1984-85)

	Average Improved Contemporary Comparison				
	Fat		Protein		Milk
Year	*(kg)*	*(%)*	*(kg)*	*(%)*	*(kg)*
1972-73	+ 7.7		+ 4.8		+ 179
1975-76	+ 9.5		+ 6.3		+ 246
1978-79	+ 10.8	+ 0.02	+ 7.6	-0.01	+ 257
1981-82	+ 15.4	+ 0.08	+ 9.0	-0.01	+ 291
1982-83	+ 16.3	+ 0.08	+ 9.6	-0.02	+ 319
1983-84	+ 16.2	+ 0.07	+ 9.8	-0.03	+ 337
1984-85	+ 15.5	+ 0.07	+ 9.8	-0.01	+ 319

carried out over a period of six months on average and the aim is to obtain 250 progeny per bull in 40-50 herds. At the end of the test, each bull has between 60 and 70 daughters with completed lactations, giving an average weighting of 46 effective daughters (repeatability of 80 percent). After carrying out the initial test inseminations, the bulls are laid off and not used again until the progeny test results are available.

All calvings of the mates of the bulls are monitored and information on abnormalities, calving difficulty and mortality collected. Data on the production of daughters for milk, fat, protein yield, fat and protein percentage are obtained from the NMR and conformation assessments for 16 main traits plus beef shape are carried out by MMB classifiers in conjunction with the breed panel members. Information is also collected from owners on temperament and ease of milking. The fertility of active bulls is monitored by the 49-day non-return rate.

Minimum progeny test standards in terms of accuracy and the minimum number of herds across which daughters must be spread are set by the Ministry of Agriculture. Approximately 1 in 8 bulls are chosen to return to the stud as proven bulls. The current active stud of bulls numbers 45 with the average number of doses of semen distributed per bull per year being approximately 20 000 (range 5 000 to 80 000).

The average merit (ICC) of MMB proven Friesian/Holstein bulls in 1984-85, weighted by the number of inseminations done by each bull, relative to 1974, was:

+319 kg milk	
+ 15.5 kg fat,	+0.07 percent fat
+ 9.8 kg protein,	−0.01 percent protein

In January 1986 the UK Friesian/Holstein base was moved and ICCs published after this date will be 220 kg milk, 0 kg fat and 6 kg protein lower than before.

The best available progeny tested sires, irrespective of ownership and including imported semen, are used on selected cows to breed sons for progeny testing. In general 8-10 sires are selected each year and the aim is to obtain 13 sons per sire. Bull mothers are selected on the basis of their cow genetic index (CGI) for weight of fat plus protein (Hill and Swanson, 1983). These cows are contracted to the MMB and sons are purchased at about 12 weeks of age to be reared at the MMB Bull Rearing Unit. The minimum CGI accepted in a contract mating cow is 680 (i.e. a cow in the top 10 percent of the breed). Milk composition, particularly protein percent, is also considered, as is the cow's conformation.

In 1984-85 contracts were signed for 169 cows, of which 10 were superovulated for embryo transfer. Of the 169 cows, 164 had CGIs which averaged 779, or 27.9 kg fat plus protein (a single cow with a CGI of 779 would be in the top 5 percent of the breed). The other five were purebred Holsteins. 89.9 percent of contract mating cows were classified Very Good or Excellent for conformation.

The average merit of bull sires entering the Bull Rearing Unit in 1984 was +31.4 kg fat plus protein. For the dams the equivalent figure was +24.7 kg. Thus the estimated merit of young bulls in 1984-85 was 28.2 kg fat plus protein which is 0.3 kg up on 1983-84 (see Table 3.38).

3.9.4 Bull evaluation

Up to the present, progeny test evaluations (known as 'Improved Contemporary Comparisons' — ICCs — in the UK) have been carried out for some 4 000 Friesian/Holstein bulls for widespread use and 30 000 for limited use. Bulls in widespread use are mainly those in AI with daughters in 20 or more herds. Limited-use bulls are those with daughters in few herds.

Because of the volume of data it is necessary to use different methods for calculating ICCs for each group. For widespread-use bulls the technique used is BLUP with the incorporation of maternal grandsire effects. Bulls are grouped according to year of registration or country of origin. The progeny of limited-use bulls are included in the BLUP analysis by treating all such bulls registered in any one year as a single bull, i.e. a year-group mean. For limited-use bulls a contemporary

Table 3.40 **Percentage Holstein blood in young AI bulls (MMB) in UK** (from *B & P Report* 1985)

Intake year	Total bulls	Bulls with Holstein blood	Percentage of total	Percentage Holstein blood in intake
1977-78	134	54	40	8.4
1978-79	141	24	17	7.1
1979-80	151	62	41	17.6
1980-81	127	61	48	29.9
1981-82	121	84	69	29.2
1982-83	115	86	75	43.5
1983-84	114	98	86	50.0

comparison is subsequently calculated after adjusting the records of daughters and contemporaries for the estimated genetic merit of their sires provided by the BLUP evaluation. All results are expressed as transmitting abilities and all bulls are ranked relative to the same fixed base.

Table 3.39 shows the average merit of the MMB proven Friesian/Holstein stud since 1974. These figures are weighted average ICCs taking into account the number of inseminations per bull.

In 1984-85, in England and Wales, the MMB and five other Associated AI Centres effected 1.91 million inseminations of which 57.2 percent were from dairy bulls and 54.1 percent from Friesian/Holstein bulls.

The percentage of Holstein blood in proven Friesian bulls (weighted by number of inseminations) fell from 31.2 percent in 1983-84 to 25.7 percent in 1984-85 as a reaction to quotas. However, the average percentage of Holstein blood in the progeny test intakes over the same period increased from 43.5 percent to 50.0 percent (see Table 3.40). Thus the average Holstein blood in proven bulls is expected to increase over the next few years.

3.9.5 Type evaluation

Type classification is done by the breed societies and also by the AI organizations. Linear assessment was introduced in 1983. In the first two years the MMB classified 28 243 cows (in the first and second lactation) for 16 conformation traits and the beef shape on a linear scale of 1 to 9. Bull breeding values for the traits are estimated using BLUP techniques with bulls allocated to one of four groups on the basis of their percentage of Holstein blood. There is also a fifth group for imported New Zealand Friesians. Those bulls with no Holstein blood constitute the base group and all breeding values are expressed as deviations from this. In 1985, 1 045 bulls were evaluated.

3.10 UNITED STATES OF AMERICA

3.10.1 History

The first importations of Friesian cows to America were made by Dutch immigrants in 1613 and by the West India Company in 1625. The history of the breed also records importations in 1795 and 1810 but all these animals were lost through crossbreeding. The start of the breed is considered to be the purchase of a Dutch cow by Winthrop W. Chenery, Belmont, MA. This cow was brought by a sailing vessel from Holland to Boston in 1852. In the years 1857-61 the same breeder imported 12 more Dutch Friesians including a bull. He was followed by other farmers and in the years 1852 to 1905 some 6 500 Friesians were imported. Since that time there have not been any importations from outside of North America. The Holstein-Friesian Association was created in 1885.

A noticeable increase in the rate of improvement in production has occurred during the last twenty years, by putting into practice population genetic methods in breeding programmes, the wide use of artificial insemination, and modern methods of feeding and housing of dairy cows. As a result Holstein-Friesian cattle have achieved a significant place among dairy cattle populations of the world.

3.10.2 The influence of environmental and economic factors on the improvement of individual yield

Dairy cattle husbandry in the USA has undergone large changes during the last 25 years. The most noteworthy outcome of these is the increase of production per cow. It is impossible to enumerate all the stimuli that have caused these changes but Crowley and Niedermeier (1981) consider the drop in the number of dairy herds with the parallel increase in their size, changes in price of fodder and energy, transportation capabilities, as well as economic and social development to have had the most important effect. These changes are still taking place and will probably influence milk production in the years to come to a greater extent than has been observed up to now. As recently as the 1950s there were 5 million dairy farms but in 1979 this number had decreased to 352 000 (Crowley and Niedermeier, 1981). At the same time the population of dairy cows decreased from 21 million to

Table 3.41 **Milk production and changes in feeding of dairy cows in the USA** (from Waldo and Jorgensen, 1981)

Year	Number of dairy cows	Yearly milk production (tons)	Feed consumption of particular fodder types (million corn units)				
			Concentrates	Hay	Silage	Pasture	Total
1953	23 869	55 169	15 954[1]	13 883	9 584	17 946	57 367
1958	20 132	54 473	18 194	13 853	9 239	16 605	57 891
1963	17 628	57 322	21 836	14 281	10 918	18 509	65 542
1968	13 804	52 724	22 762	13 959	11 995	23 023	71 739
1973	11 282	52 346	24 123	10 604	10 545	14 242	59 514
1978	10 853	52 232	24 586	10 623	10 908	11 299	57 416

[1] Grain units

11 million head. Such a drastic reduction of the dairy herd has, however, not decreased the total milk production. The individual yield has increased from 2 000 kg to nearly 5 200 kg of milk. Table 3.41 shows some of the changes in the size of population and feeding of dairy cows during this period. In 1982 the number of farms with dairy cows was 312 000 of which 165 000 were commercial dairy farms. At present a programme aimed at reduction of the cow population by 10 percent is being realized.

The total intake of nutrients has remained almost unchanged. Yet, the average composition of the diet of a dairy cow has drastically changed and nowadays, expressed in corn units, it is as follows (taking the situation 25 years ago as equal to 100 percent): concentrates — 154 percent, hay — 77 percent, other preserved roughages — 114 percent, pasture — 63 percent. Other preserved roughages included primarily corn silage — in 1954 the production of corn silage was 47.7 million tons; in 1974 — 101 million tons, while production of grass silage in 1954 was 6.4 million tons, and in 1974 — 8.3 million tons. According to Waldo and Jorgensen (1981) the average consumption of fodder per dairy cow, expressed in corn units, over the year 1980, was as follows (figures in brackets give percentages relative to 1954): concentrates — 2 265 (339), hay — 979 (169), other preserved roughages — 1 005 (25), pasture — 1 041 (139), a total of 5 290 (220).

According to Reid (1977), quoted by Waldo and Jorgensen (1981), much of the increase of individual milk yield should be attributed to the improvement of fodder quality and to the change of the diet structure. Reid claims that 40 percent of the milk yield increase was due to a higher consumption of concentrates and another 20 percent was due to better quality roughages which together influence

Table 3.42 **Percentage of feeding units from different components of the dairy cow diet in the USA** (from Waldo and Jorgensen, 1981)

Year	Average milk yield	Corn units (%)				
		Concentrates	Hay	Silage	Pasture	Total
1953	2 309	27.8	24.2	16.7	31.3	2 401
1958	2 706	31.4	23.9	16.0	28.7	2 876
1963	3 252	33.3	21.8	16.7	28.2	3 718
1968	3 819	31.7	19.5	16.7	32.1	5 197
1973	4 640	40.5	17.8	17.7	23.9	5 275
1978	5 089	42.8	18.5	19.0	19.7	5 290

Table 3.43 **Average farm and herd size and the number of dairy farms with annual sales at or over US$ 2 500 in selected States and in the USA** (from Nott *et al.*, 1981)

State	1954	1964	1974
Number of farms			
California	12 452	5 571	3 046
Wisconsin	107 350	71 754	46 951
New York			
USA	397 977	349 244	196 057
Average farm size (ha)			
California	60.3	89.5	113.4
Wisconsin	63.2	78.5	92.7
New York	85.0	105.7	129.6
USA	65.3	94.3	111.7
Average herd size			
California	58	125	245
Wisconsin	19	27	36
New York	25	35	50
USA	20	37	48

Table 3.44 **Efficiency of labour and the cost of 1 hour to attend an animal in the USA** (from Nott *et al.*, 1981)

Year	Time of labour per cow per year *(hours)*	Milk production per hour of labour *(kg)*	Cost per hour of labour *(US$)*
1956	91	51.3	1.46
1965	63	94.3	1.85
1975	51	130.2	6.29
1978	50	136.1	9.01

the digestibility of fodder and enhance its consumption. Table 3.42 shows changes in feed usage over the years.

Concentration and specialization of production has led dairy herds producing less profitably — such as small herds of low individual production — to go out of business. This has further accelerated changes in dairy cattle husbandry and increases in productivity.

These changes have been accompanied by greater regionalization of production to the Great Lakes, the northwest, northeast and California. In many States the size of the dairy herd has decreased but in those regions where favourable conditions for dairy husbandry exist, the number of dairy cows remains stable, as in Wisconsin which is known as America's Dairy Land. In recent years almost 50 percent of milk has been produced in five States: Wisconsin — 17 percent, California — 10.2 percent, New York — 8.6 percent, Minnesota — 7.4 percent, and Pennsylvania — 6.5 percent. There are big differences between these States in concentration and specialization of production. Family farms prevail in the North and in the Central West, while in Central and Southern California big commercial dairy farms predominate. Some regional data are summarized in Table 3.43 (Nott *et al.*, 1981).

It is difficult to compare the situation in California with other regions of the USA. Thanks to its exceptionally favourable climate it is possible to create very large dairy farms relatively inexpensively. It is also possible to cultivate alfalfa on irrigated fields almost the whole year round. Natural conditions also favour good quality hay production. On the big farms of this State the feeding area per cow is less than 0.5 ha. Specialized dairy farms not only buy concentrates but more and more

Table 3.45 **Comparison of milk yield and monetary returns per cow of different cattle breeds in the USA for 1984**

Breed	Herd size	Milk yield *(kg)*	Fat yield *(kg)*	Protein yield *(kg)*	Value of milk minus feed cost *(US $)*
Ayrshire	43	5 673	225	197	941
Brown Swiss	46	5 976	243	213	933
Guernsey	58	5 195	242	189	903
Holstein-Friesian	88	7 230	264	232	1 090
Jersey	73	4 947	240	191	906
Shorthorn	35	5 115	187	172	652

Table 3.46 **Cow population under the National Cooperative Dairy Herd Improvement Programme, USA** (from Norman, 1984)

	Official Programme				Unofficial Programme				
Year	Herds	Average cow number per herd	Percent of population	Total number of cows	Herds	Average cow number per herd	Percent of population	Total number of cows	Percent of population in both programmes
1970	34 308	61.9	17.2	2 122 011	23 281	36.1	6.8	839 343	25.3
1985	39 100	83.4	29.9	3 262 174	24 733	55.3	12.5	1 367 969	42.4

Table 3.47 **Milk production of cows covered by the official and unofficial NCDHIP as compared to other cows in the USA**

	Official		Unofficial		All other cows	
Year	Milk *(kg)*	Fat *(kg)*	Milk *(kg)*	Fat *(kg)*	Milk *(kg)*	Fat *(kg)*
1970	5 897	220	5 630	209	4 013	145
1984	7 071	262	6 576	249	4 837	174

often have to buy roughages — above all hay. Concentration and specialization of production also influenced the use of labour as shown in Table 3.44.

Efficiency has been improved mainly by changes in housing and milking systems on the bigger farms. Considerable improvement has been achieved by introducing free-stall cowsheds with separate milking parlours.

Most of the milk produced in the USA comes from the Holstein-Friesian. This genotype is best adapted to the conditions of intensive production where the economic superiority of this strain is reflected in Table 3.45, comparing the breeds maintained in the USA.

3.10.3 Milk recording

The first milk recording association was started in 1906 in Michigan and by 1929 all the 48 states had cow testing associations (Voelker, 1981). The Dairy Herd Improvement Association (DHIA) was founded in 1927. A national scheme covered 8 states in 1965 and thereafter spread to include 44 states.

Cow numbers covered by the official and unofficial National Cooperative Dairy Herd Improvement Programme (NCDHIP) increased from nearly 3 million cows in 1970 to 4.6 million in 1985 (Table 3.46). The proportion of Holstein-Friesians in the NCDHIP amounts to 92 percent. The size of the recorded population has an important influence on both the ability to carry out the testing programme and on the rate of genetic improvement. Almost 42 percent of cows have been covered by the programme during recent years.

There is a clear, positive relationship between the participation in the NCDHIP and milk production as shown in Table 3.47. Thus the superiority of herds participating in the official programme compared to other cows has increased from 1 884 kg in 1970 to 2 234 kg in 1984. A similar trend is observed in the case of the cows under unofficial NCDHIP. According to Norman (1984) 51 percent of officially tested lactations are potentially available for bull evaluation and genetic improvement. In the Holstein-Friesian breed about 53 percent of cows had an identified sire.

3.10.4 Artificial insemination

At present, about 68 percent of the dairy cow population in the USA is artificially inseminated. The central body coordinating artificial insemination activities across the USA is the National Association of Animal Breeders (NAAB). In 1979 there were 25 artificial insemination organizations, 11 of which produced 92 percent of

Table 3.48 **Quantity of produced and exported semen by bulls of different breeds in 1979 within NAAB, USA — million doses** (from Miller, 1981)

	Semen production	Export	Frozen semen order by breeders
Holstein	11 262	1 594	0.528
Guernsey	0.328	0.017	0.018
Jersey	0.523	0.045	0.055
Brown Swiss	0.226	0.165	0.040
Ayrshire	0.069	0.001	0.015
Milking Shorthorn	0.020	—	0.005
Other dairy breeds	0.039	—	0.021
Beef breeds	1 086	0.264	1.125
Total	13 553	2 086	1.806

Table 3.49 **Average number of young Holstein-Friesian bulls bought and tested in the artificial insemination centres in the USA** (from Norman, 1984)

Year	Number of bulls
1960	113
1965	198
1970	227
1975	518
1980	912
1983	1 082

Table 3.50 **Average predicted differences (relative to 1982) of Holstein-Friesian bulls in January 1986 in the USA** (from Shainline, 1986)

Bulls	Number of bulls	Predicted Differences			
		Milk *(kg)*	Fat *(%)*	Fat *(kg)*	US$
Marketed in AI	415	+ 336	-.01	+ 11	+ 85
Non AI	7 594	-182	+ .01	-6	-45
All bulls	11 179	-110	.00	-4	-28

total semen supply. According to Miller (1981) semen production in the USA reached a total of 17.4 million doses — 86 percent from dairy and 14 percent from beef bulls. In 1979 about 12 percent of the yearly production was exported and 10 percent was the property of breeders. Table 3.48 gives information for different breeds.

Particular AI centres run their own young bull sampling and testing programmes. The average yearly replacement of young Holstein-Friesian bulls in the artificial insemination centres is given in Table 3.49.

The average transmitting ability of the Holstein-Friesian bulls is indicated by the Predicted Differences related to the base year 1982 and shown in Table 3.50.

The data in Table 3.50 show that the transmitting ability of the bulls used in AI is much better than that of those used for natural mating. Preliminary indices, combining parental and sib information, have been used for the selection of young bulls in recent years. Thanks to these indices the quality of young bulls has much improved. The average value of young Holstein-Friesian bulls, ranked according to Pedigree Means, was −131 kg milk and −3 kg fat in 1976 while in 1983 the figures were, respectively, 223 kg and 6 kg.

In the last decade a boom has occurred for Holstein-Friesian breeding material and semen. According to the information published in the National Association of Animal Breeders' Annual Report 1985, 2.78 million dairy bull semen doses, out of which 84 percent had been produced by the Holstein-Friesian bulls, were exported from the USA to about 60 countries in 1984. The value of that export was about US$ 21.5 million, the largest importer being EEC countries.

3.10.5 The Holstein Association

According to the report of the Holstein Association of America (1984) the total number of members was 45 135 and the number of registered animals 492 380. In that year 5 630 registered animals were sold to foreign countries.

The average milk yield per lactation of a registered cow was 5 877 kg at 3.65 percent and 313 kg fat on a 305-day, 2 × milking mature equivalent basis.

Over the years, the popularity of the Holstein-Friesian in the USA has grown, the number of new registrations in the herdbooks testifying to this (Table 3.51). In 1970 over 70 percent of the breeding animals registered in the herdbooks belonged to the Holstein-Friesian breed. Fifteen years later this breed made up 76 percent of registered stock.

The Association developed the idea of the "True Type" model of a Holstein animal. The first "True Type" model was developed in 1922. Type classification was initiated in 1929. In 1967 the Holstein Association introduced the "Descriptive Classification" system. Since that time more than 4 million animals were classified for type on a scale from 50 to 100 points. As a result animals were classified into 6 classes, from poor to excellent. Since 1983 the "Linear Classification" system became effective and replaced the "Descriptive Classification". It is used to assess 15 primary and 14 secondary traits. Is is now a basis for genetic evaluation and to assign a final type score. The current breed average score for a mature Holstein cow in its prime is about 81 points.

For bulls the "Predicted Difference for Type" (PDT) is estimated. It measures a sire's transmitting ability for improving conformation score expressed as a prediction of the average final score difference above or below herdmates. The PDT index ranges from –3.50 to +3.50 among USA Holstein sires. The Holstein Association also provides a "Total Performance Index" (TPI) using a production classification and their own type classification data. It combines milk production, fat test and type in a 3:1:1 ratio. In 1984 the Holstein Association official type inspectors classified 515 301 animals in 10 522 herds.

3.11 FRIESIAN CATTLE IN OTHER DEVELOPED COUNTRIES

Apart from the countries which participated in the international comparison of Friesian strains of cattle in Poland, there are many other countries in which the Friesian is the main dairy breed (Table 3.52). In Western Europe this includes Ireland, Greece, Spain, France, Italy and Luxembourg. In several other countries Friesians are an important component of the dairy cattle population (Table 3.52). In Eastern Europe the German Democratic Republic has been traditionally a "Friesian" country, but the existing population consists of the synthetic dairy breed (Schwarz buntes Milchrind — SMR) with a share of Jersey blood. In the Soviet Union, the Friesian population is rapidly increasing, from

Table 3.51 **Animals registered in the breed association herdbook by year in the USA** (from Norman, 1984)

Year	Ayrshire	Guernsey	Holstein	Jersey	Brown Swiss	Milk Short-horn	Red and White	Total
1970	15 069	43 383	281 574	37 097	16 416	5 410	—	399 349
1985	11 120	25 106	394 506	65 357	13 590	3 311	5 283	518 273

Table 3.52 **Numbers and yield of Friesian cows in some West and East European countries as well as in Japan and Australia**

Country	Year	No. of dairy cows	Proportion of Friesians	Proportion of milk recording	Yield of recorded Friesian cows					Increase of production 1974-1984		
					Milk	Fat		Protein		Milk	Fat	
		('000)	(%)	(%)	(kg)	(kg)	(%)	(kg)	(%)	(kg)	(kg)	%
Ireland	1985	1 528	96	7.5	5 104	180	3.52	167	3.27			
Greece	1985	220	70	4.5	4 571	166	3.64			[7]429	21	0.11
Spain	1982	1 958	65	[4]5	(5 836	193	[5]3.30)					
France	1984	6 761	62	35	5 175	199	3.85	166	3.23	[8]1 047	[8]46	0.11
Italy	1986	3 076	53.3	25.7	5 945	209	3.51	187	3.14	986	26	0.07
Finland	1985	623	19.8	50	5 785	242	4.19	186	3.21	794	33	0.08
Switzerland	1985	830	12	60	6 026	233	3.86	189	3.14	1 043	45	
Portugal	1985	310	10	90	4 529	152	3.37			288	12	
Austria	1984	1 000	1.7	32	5 600	227	4.05	174	3.13	880	38	
German Democratic Republic	1985	2 005	100	95	3 823	153	4.00			350	17	[9]0.09
USSR[1]	1980	69 078	25.2		(5 559	212	[1]3.82)					
Hungary	1985	587		73	[6]5 132	[6]188	[6]3.66					
Czechoslovakia	[2]1985	1 880	39.4	99	3 780	152	4.00					
Yugoslavia	1984	2 974	4.2	6.1	5 175	193	3.72			735	30	0.05
Japan	1984	1 322	99.8	34	6 375	236	3.70			979	47	
Australia	1985	[3]2 800	55	45	4 246	171	4.03	139	3.29	[4]1 000	45	

1 From Ernst and Dmitriev (in press), yield figures for 111 000 cows in leading herds only, in 1984
2 Friesians and crossbreds
2 Total number of dairy cattle
4 Approximate estimate only
5 Herdbook registered cows
6 Pure HF and crossbreds of HF x Simmental
7 Difference in yield 1980:1984
8 1974/85
9 Difference in yield 1977:1984

23 percent in 1981 to 31.8 percent in 1986 (Grothe, 1986). In other countries like Hungary, Czechoslovakia, Romania, Bulgaria or Yugoslavia, Black and White cattle are used mainly for up-grading of local breeds (mainly Simmental) or located on dairy farms around big cities for fresh milk supply (Stolzman, 1980 and 1985). From outside Europe, the Friesian populations in Japan and Australia need special mention — the first being among the highest yielding in the world, and the second making significant progress in numbers and production.

Table 3.53 **Progeny testing of Friesian bulls in some West and East European countries as well as in Japan and Australia in 1984 to 1985**

Country	Proportion of Friesian females bred by AI %	No. of young bulls tested annually	No. of females mated per young bull	Average no. of recorded daughters per bull	Selection intensity of bulls after progeny testing	Waiting bull system
Ireland	60	45	800	60	1:4	lay off
Greece	70					
France	84	588	350-400	90	1:5	lay off
Italy	67	150	1 000		1:5	semen storage
Finland	100	65-70		100		semen storage (30-40 000 doses)
Switzerland	100	30	600	100	1:7	semen storage (35 000 doses)
Portugal	36					
Austria	86					
German Democratic Republic	100	600	350	50	1:6	
Hungary	100	120	250	50	1:5	10 000 doses for storage then lay off
Bulgaria	92					
Czechoslovakia	99	120	1 500	150	1:3	semen storage
Yugoslavia	90	50	350-400		1:4	different systems in different republics
Japan	[1]99.4	100	480	50	1:3.5	semen collection
Australia	46	250	500	60	1:6	lay off

[1] 1983

Additional statistics on the Friesian populations in these countries are given in Tables 3.52 to 3.54. Table 3.52 shows the proportion of Friesians in the dairy cattle population and provides some production data. Table 3.53 presents information on the testing of bulls and Table 3.54 gives figures for the import and export of Friesian semen. The more detailed descriptions which follow, concerning Friesians in various countries, are based on information provided by breeding organizations in these countries. The authors gratefully acknowledge this help.

Table 3.54 **Export and import of semen from Friesian bulls in some West and East European countries as well as in Japan and Australia in 1984**

Country	Semen export: No. of doses	Semen export: where to	Semen import: No. of doses	Semen import: from where
Ireland[1]	9 000	UK	46 300	UK, Canada
Greece			18 000	USA, Canada, New Zealand, France
Spain			15 710	Canada[2]
France	30 000		80 000	USA, Canada, various European countries
Italy	25 000	Spain, Austria Fed. Rep. of Germany, Portugal, France, Switzerland	230 000	Canada, USA
Finland	8 000	Cyprus	1 000	USA
Switzerland	33 000	Far East, Africa South America	45 000	USA, Canada
Portugal			15 350	Canada, USA, Netherlands
Austria			1 972	Canada[2]
USSR			70 512	Canada[2]
Hungary			61 850	USA, Canada, Federal Republic of Germany
Czechoslovakia			6 700	USA, Canada, Federal Republic of Germany, Netherlands
Yugoslavia			3 250	USA
Japan			15 300	USA
Australia			94 350	Canada[2]

[1] 1985

[2] No information about importations from other countries

In *Ireland* the population of dairy cows numbered 1.53 million in 1985 — an increase of more than 20 percent above the numbers in 1974. The vast majority of the dairy cattle are Friesian, and approximately 60 percent are mated by AI (M. Kearney, personal communication). Family farms with 40 to 80 cows predominate. Cows generally calve in spring and are fed seasonal forage. The use of concentrates is very low. The production of an average cow is about 3 000 kg of milk (Koesling, 1984) but the average for recorded cows exceeds 5 000 kg of milk with 3.52 percent fat (Table 3.52).

The size of the milk recorded population is relatively small and this places constraints on bull progeny testing. As shown in Table 3.53, only 45 bulls were entered for progeny testing annually. Selection of Friesian bulls is based on milk yield and body conformation of daughters, and on the beef conformation and growth rate of bulls themselves. In the selection of bull dams, conformation and beef shape are taken into account in addition to the yields of fat and protein.

Beef plays an essential role in the Irish economy so the selection objective is for dual-purpose cattle. The proportion of Holstein genes in the population of young bulls is low and semen from Holstein bulls is made available only upon request (More O'Ferrall, 1984). According to More O'Ferrall, in the Irish economic conditions, only farmers who are producing over 4 500 kg of milk/cow/year could gain substantially from a change to Holsteins, but a major change in the national dairy herd from Friesians to Holsteins would be expected to bring heavy losses in the profitability of beef production.

In Ireland, commercial crossing with beef bulls is widely used. Fifty-four percent of artificially-bred Friesian females are inseminated with semen from beef bulls. Friesian semen import and export takes mainly place between Ireland and the United Kingdom (Table 3.54), but 13 percent of the semen imported in 1985 was purchased from Canada.

The Friesian population in *Greece* is relatively small at about 150 000 cows, although this represents 70 percent of all dairy cows (Table 3.52). Milk yield of recorded cows is in the range of 4 600 kg and the increase in milk production in the years 1980 to 1984 has been substantial (in excess of two percent per year). Importation of semen was at a relatively high level with 18 000 doses in 1985 with 77 percent of that from Canada and the rest from the USA, New Zealand and France.

Spain has a rapidly increasing Friesian population. In 1982 the numbers of Friesian and Friesian crossbred cows were 1.27 million, compared to 334 000 in 1955 (Grothe, 1986). Herds registered in the herdbook (the nucleus) accounted for 0.3 percent of the cows, multiplication and test herds for 12 percent; the rest were commercial herds. In the early 1980s milk recording was practised to only a very small extent. Bulls used for artificial insemination in 1980-81 were mainly of American origin (65 percent) and 35 percent of European origin, but many of them with Holstein-Friesian blood (Alberro, 1982). In 1984 and 1985 Spain imported from Canada 35 000 doses of semen further increasing the move toward a Holstein-Friesian type of cow in Spain. More than 75 percent of the Friesian cow population is served by AI (Ordonez, personal communication). Livestock has also been imported in large numbers mainly from European countries, like the Federal Republic of Germany, Denmark, the United Kingdom, the Netherlands, France and Ireland. In the years 1965 to 1984 a total of 48 000 Friesian animals were imported. The milk yield of the "average" cow increased from 2 173 kg in 1965 to 3 280 kg in 1984. The herdbook registered cows produced 5 836 kg of milk in 1984 but with a fat content of only 3.30 percent.

The neighbouring country, *Portugal,* has a small Friesian population of about 30 000 cows. Nearly all of them are milk recorded. Milk yield of Friesians is in the range of 4 500 kg of milk with a low fat content of 3.37 percent (Table 3.52). The AI industry serving one third of the Friesian females is based on imported semen. In 1985 more than 15 000 doses were imported (Table 3.54) with 84 percent from Canada and the USA.

The Friesian population of *France* amounts to more than 4 million cows, one of the biggest among West European countries. It has strongly grown in number, replacing the Normandy breed which was as numerous as the Friesian some 25 years ago but represented only 13.2 percent of the milk-recorded population in 1984. The average milk yield of registered French Friesians (269 284 cows) was 5 881 kg of milk with 3.84 percent of fat in 1985. The French Friesian Association has 5 930 members.

The number of young bulls entering AI was nearly 600 in 1984 (Table 3.53). Semen from young bulls is used to inseminate females after their first calving. Insemination with semen from young bulls is limited to 15 percent of all AI services in any one herd. Breeding value of bulls for milk traits was formerly estimated by use of the MCC method, but in 1986 the BLUP procedure was introduced. France exports both Friesian cattle and semen from Friesian bulls. In 1985, the number of animals exported was 5 750 and of semen 30 000 doses (Table 3.54). The use of American Holsteins is widespread. A total of 80 000 semen doses were imported in 1985, of which nearly 25 000 were from Canada.

In *Italy,* Friesians make up more than half the dairy cow population of over 3 million (Table 3.52). About 25 percent of all dairy cows are milk recorded but the proportion among Friesian cattle is 37 percent with an average yield of 5 945 kg milk containing 3.51 percent fat and 3.16 percent protein (Table 3.52).

Friesians are kept in all regions of the country with the highest concentration in Lombardy and Emilia (Bettini and Nardone, 1976). In the years from 1974 to 1986 the number of milk recorded cows increased three-fold. Average milk yield of Friesian cattle over the period of 1974-85 increased by 986 kg and fat production by 26 kg.

Artificial insemination is widely used — about two thirds of recorded Friesian cattle are bred in this way. Annually, 150 young bulls enter the breeding programme with matings to 1 000 cows each; they are evaluated for milk and conformation traits using a BLUP procedure and 1 in 5 are subsequently selected. In 1984 some 230 000 doses of semen were imported from Canada and the USA as well as several hundred Friesian cattle from both European countries and North America. Italy also exports semen and cattle.

The Italian Friesian Association has more than 10 000 members with an open herdbook and a registered population of over half a million (in 1983). Italy had already started in 1928, probably the earliest among European countries, to use Holstein-Friesians for dairy cattle improvement and this has continued ever since.

In *Belgium,* the proportion of Friesian is also increasing. It rose from 18 percent of the female population in 1976 (Hanset, 1976), as one of five Belgian dairy or dual-purpose breeds, to 26 percent in 1980 (Stolzman, 1985). Twenty years ago this breed was mainly located in the northern part of the country (in the polders) and in the east of the central belt. Although the Belgian market prefers animals with strongly developed musculature in all farm animal species, the Friesian population has been penetrated by American Holsteins and this process has accelerated in recent years as shown by the importation of semen from Canadian bulls, which amounted to 7 000 doses in 1894 and doubled in 1985 (Canada, 1985).

The history of the Friesian breed in *Finland* is short. In 1970 Friesians made up 0.5 percent of the total cattle population but by 1985 this breed accounted for nearly 20 percent of the dairy cow population (Table 3.52). The Friesians compete successfully with the Ayrshire breed (Ettala, 1984). Genetic improvement is based on a total merit index of bulls, estimated by the BLUP procedure. Practically all females are artificially inseminated and one Friesian bull is introduced for progeny testing per 2 250 first AI services. Bull dams are selected from cows with an index of at least 115. The required minimum for the average milk yield of a bull dam is 7 500 kg with 3.8 percent fat and 3.2

percent protein content. The use of Holstein-Friesians for improvement of the Finnish Friesians is only modest.

Friesians are not, of course, native to *Switzerland* but they have increased in number nearly tenfold in the last 25 years, but the breed is still in a minority (Table 3.52). Swiss Friesians are of a dairy type and characterized by an outstanding milk production (over 6 000 kg of milk). The extent of milk recording is high and, as in Finland, all females are bred artificially. In 1984, the Friesian Association had 2 639 members; 58 884 animals were registered in the herdbook. Selection methods involve sire proof evaluation for milk traits (estimated by the BLUP method) reproduction, growth (own performance) and daughters' conformation. Swiss Friesians have a high proportion of Holstein-Friesian blood. Importation of Holstein-Friesian semen was started in 1966 and has continued since (Table 3.54). In 1984, 21 percent of first services (AI) were with imported semen. Switzerland exports both Friesian cattle and semen. In 1984, breeding stock was exported to Italy and semen to Afghanistan, Thailand, Bolivia, Sudan, Zambia, Tanzania and Somalia.

Some 20 years ago Friesian cattle were introduced to the *Austrian* province of Carinthia. It is still only a small population, but with a high milk production and fat content (Table 3.52). The animals are of dairy type, improved by imported semen mainly from Holstein-Friesian bulls.

In *Eastern Europe* the origin of Friesians can be traced back to the seventeenth or eighteenth centuries in the German Democratic Republic, Poland and the European part of the Soviet Union. In Bulgaria, Czechoslovakia, Hungary, Romania and Yugoslavia the first Friesians were imported within the last three decades.

In the *German Democratic Republic* the actual cattle population consists of the dual-purpose Black and White Milk Cattle (SMR). Its genotype includes 50 percent of Holstein-Friesian, 12 to 25 percent of Jersey and the rest of native Friesian blood. The crossbreeding process has been completed and selection is practised within the new breed. The breeding goal is high milk production potential, with high content of fat and protein, good growth rate and satisfactory beef quality. Average yield of herdbook cows (345 000 head), in 1985, was 4 371 kg milk and 183 kg fat (fat content 4.16 percent) all higher than for recorded cows in general (Table 3.52). Approximately 120 000 cows are maintained in breeding centres and their averagage yield amounts to 195 kg of butterfat. The bull dams are selected from this population. The average production of 4 500 bull dams was 7 255 kg milk, 337 kg fat and the fat content was 4.65 percent.

There are excellent facilities for bull progeny testing for dairy characteristics with one of the highest selection intensities (Table 3.53). Females are exclusively bred with the use of AI, 55 percent are raised in herds specializing in this and the number of cattle per herd is very high. To increase beef production and improve beef quality, as well as the uniformity of the SMR breed, a new type classification system and breeding value testing for beef traits was introduced. Growth rate of bulls is tested in performance testing stations and additionally on female half-sibs. Since 1984, all AI bulls have been progeny tested for fattening ability and beef quality. Twenty-one percent of the first services by AI are made with semen from beef bulls (mainly Simmental) and the progeny of both sexes fattened and slaughtered.

The German Democratic Republic is an exporter of livestock; in 1984 the number of SMR heifers exported was 7 100 (Baum, 1986).

The number of Friesian cattle in the *Soviet Union* has been growing steadily. Within its large population of cattle (128 million), which consists of numerous breeds, about half were dairy and dual-purpose types and of these the proportion of Friesians was nearly 32 percent in 1985 — 10 percent up on 1979 (Stolzman, 1980). Older, but more precise figures are shown in Table 3.52. The majority of the Friesian population (92 percent) exists in the European part of the country.

Various strains of Black Pied cattle, approved as a breed in 1925, were developed from crossing local cattle with Dutch Black Pied and East Friesian breeds. There are five large populations: Central (Mid-Russian), Baltic, Ural, Siberian and other zones, particularly Ukraine, Byelorussia, Central Asia and parts of Transcaucasia (Ernst and Dmitriev, in press). These populations differ in milk yield and the fat content of milk. Selection has provided the basis for some highly productive herds, the leading ones, according to Ernst and Dmitriev (in press) averaging 5 559 kg milk with 3.82 percent fat (Table 3.52) with even higher yields at individual stations. The Baltic population comprises the Lithuanian Black Pied and the Estonian Black Pied, the latter with origins in 1885, each with their own herdbook. Milk recorded Lithuanian Black Pied herds averaged 4 000 kg milk with individual farms much better, whilst for Estonian Black Pied cattle, herdbook-registered cows averaged 4 402 kg milk with 3.96 percent fat — and individual cows exceeded 400 kg fat in a 305-day lactation (Ernst and Dmitriev, in press).

The Soviet Union imports Friesian cattle and semen from Holstein-Friesian bulls. From Canada alone, in 1984 and 1985, 418 000 doses were imported (Canada, 1985).

In the present AI bull population, 8 percent are Holstein-Friesian and 29 percent are progeny of Holstein-Friesian sires. Approximately 4 million Friesian cows and heifers are mated to Holstein-Friesian bulls. The estimated number of Holstein-Friesian crossbred females amounts to 1 million. In good feeding and management conditions the crossbreds produce about 500 to 600 kg more milk than their dams. It is intended that, up to 1990, Holstein-Friesian bulls will be mated to 6.5 million Friesian cows and to 3 million crossbreds (Bicz, 1986).

In *Hungary,* the large scale of cattle breeding was thought to require a breed more suitable for milk production than the native Simmentals. The crossing experiments with American Holstein-Friesians started in 1966 and gave satisfactory results (Dohy and Szüllo, 1974). In 1970, importation of American Holstein-Friesians and their semen was started (Bartosiewicz and Gere, 1981). At first, only semen from Red and White Holstein-Friesian bulls was used for crossbreeding with Simmentals, but later on Black and Whites were also admitted. In the early 1970s, Hungary decided for specialization in its cattle population — one part to be selected for dairy traits and the other for beef characteristics. The present composition of the milk recorded population is as follows:

8 percent Simmental
6 percent pure-bred Holstein-Friesian (HF)
38 percent crossbreds (Black and White HF × Simmental)
30 percent crossbreds (Red and White HF × Simmental)
6 percent Hungaro-Fries
2 percent others

The genotype of the Hungaro-Fries (about 32 000 cows) consists of 50 percent Holstein-Friesian, 25 percent Jersey and 25 percent Simmental. A further increase of this population is not foreseen. More than 500 000 cows are upgraded by use of Holstein-Friesian. There is a close cooperation with the Holstein Association of America. Experts from the USA make a type classification and cooperate in the sire selection (Grothe, 1985). Hungary has made very rapid progress in the improvement of milk yield by increasing the average per cow from 2 411 kg in 1975 to 4 145 kg in 1983. Large quantities of semen are imported (Table 3.54) mainly from the USA (65 percent in 1984).

In *Bulgaria* upgrading of local cattle breeds (Simmental, Bulgarian Grey and Bulgarian Red) with Black and White Friesians is widely used. It is planned that by the end of this century Friesian cattle will be the main dairy breed making up 80 percent of the total cattle population, which in 1985 amounted to 1.75 million head. Animals and semen from pure Holstein-Friesian bulls, or with a high proportion of this blood, are imported from North America,

Table 3.55 **Milk recording results for 1985 in Czech and Slovak Republics** (from Malovec, 1986; Matous, 1986)

	Czech Republic					Slovak Republic				
Genotype	No. of cows	Milk *(kg)*	Fat *(kg)*	Fat *(%)*	Calving interval	No. of cows	Milk *(kg)*	Fat *(kg)*	Fat *(%)*	Calving interval
Holstein-Friesians						5 653	3 931	151	3.82	408
Friesians	27 581	4 088	164	4.01	390	11 044	3 540	140	3.95	404
Holstein-Friesians × Simmental						77 669	3 694	142	3.86	399
Friesian × Simmental	370 047	3 776	152	4.01	382	5 158	3 245	129	3.94	404
Friesian × Pinzgauer						9 313	2 953	118	3.95	384
Simmental	311 607	3 563	146	4.08	382	103 565	3 174	124	3.93	398
Pinzgauer						34 861	2 702	107	3.96	389

Israel and European countries, mainly from the Federal Republic of Germany (Grothe, 1987).

The pure Holstein-Friesian population in *Czechoslovakia* is small, amounting to only 2.6 percent of the total dairy cattle. Crossbreeding of local breeds (Simmental and Pinzgauer) with Friesians is widely used and Friesian crossbreds account for 36.8 percent (Table 3.52). Table 3.55 shows the milk yield of Holstein-Friesians, their crossbreds with Simmental and Pinzgauer and that of pure Simmental and Pinzgauer in the Czech and Slovak Republics.

In both republics the Holstein-Friesian had the highest yields. Some details on progeny testing of Friesian bulls are shown in Table 3.53. In 1987-88 the BLUP procedure for sire proof evaluation will be introduced. Milk yield, fat content, ease of milking, beef efficiency and type classification are all considered. All Friesian bulls are tested for growth rate in performance testing stations. Selection of bull dams is based on the following milk production requirements:

Item	Butterfat yield	Fat content (%)	Herd index (%)
Lactation with the highest yield	220	4.00	110
Average of all lactations	200	4.00	100

There are also strict requirements for an udder index (38 to 55 percent), speed of milking (80 percent of milk output during the first three minutes of milking), liveweight (550 kg) and height at withers (130 cm). In 1984 a relatively small amount of semen was imported (Table 3.54).

The Friesian population in *Yugoslavia* is small, amounting to 4.2 percent of the total cow population. The milk yield of recorded cows exceeds 5 000 kg (Table 3.52). Friesians were first introduced to this country in the early 1950s and up to 1960 more than 25 000 females were imported from different European countries. In the 1970s, Holstein-Friesians were imported from the USA. The Yugoslav Friesian population has been upgraded by use of semen from Holstein-Friesian bulls, mainly originating from the USA. Some information on bull dairy progeny testing is shown in Table 3.53.

From among countries of the Far East, Japan and Australia need special mention. In *Japan* the Black and Whites are practically the only dairy breed. The milk yield is very high — the "average" cow yields 5 400 kg and average for the milk recorded population is close to 6 400 kg milk, the highest among countries included in Table 3.52. The average number of cattle per farm is 26. Higher numbers are found on the north island of Hokkaido — 46 head per farm. The Friesian Association has nearly 38 000 members representing 46 percent of all dairy farmers and 213 000 cows are registered in the herdbook. In the progeny testing of bulls (Table 3.53) evaluation of milk traits is supplemented by consideration of growth rate, carcass quality and type classification. Bull dams are selected on milk yield, fat index and type score. These traits are also taken into consideration for maternal granddams. Japanese cattle are of Holstein-Friesian type. Breeding stock, as well as frozen embryos and semen, are imported. In 1984, 15 300 semen doses were imported from the USA, and 6 000 doses from Canada in 1985.

The origin of Friesians on the *Australian* continent goes back to 1886 when the first animals were imported from New Zealand (two bulls and six females). It is interesting that one of these bulls had himself been imported to New Zealand two years earlier as one of the first group of Friesians introduced there. The Australian Friesian Association was founded in 1914 and had 2 579 members in 1985. At present, the Friesians make up more than 50 percent of the dairy cattle population. The majority are grade animals, and only 14 percent of dairy farmers have some registered cows which, in total, numbered 38 000 in 1985. The herdbook is open with four appendices to allow grading-up of cattle into fully registered status both for males and females.

Proofs for milk traits for AI bulls (Table 3.53) are calculated by the BLUP method using a full animal (cow and bull) model. Proofs for type traits are also calculated using the BLUP procedure with a sire model. Selection of dams is made by the AI centres which, however, differ in the criteria they apply.

In the last decade the breed has relied heavily on semen and young bulls imported from Canada, New Zealand and other countries (Table 3.54). Importation of semen from the USA was banned until recently, but in future years considerable interest is expected for this type. Australia is exporting both herdbook-registered and grade animals to Taiwan Republic of China, Saudi Arabia, Mauritius and the Republic of Korea.

References

ALBERRO, M. Structure of the Spanish Friesian cattle population and dissemination of genetic improve-
1982 ment through the Spanish dairy cattle industry. *Zootechnia,* 31 (10-12): 355-364.

BAR-ANAN, R. *Breeding the Israeli Friesians.* Hakirya, Tel-Aviv.
1971

BAR-ANAN, R. Continuous progeny testing, use and selection of proven bulls in Israel. *G. Sc. Agric.*
1984 *Soc. Finland,* 55: 489-495.

BARTOSIEWICZ, L. and GERE, T. Studies of Holstein cattle in Hungary. *Holstein Science Report,* pp.
1981 1-4.

BAUM, N. Ergebnisse des Zuchtprogrammes 1981 bis 1985 und die weiteren züchterischen Aufgaben
1986 zur Intensivierung der Rinder-Schweine-und Schafproduktion. *Tierzucht, 3:* 97-100.

BETTINI, T.M. and NARDONE, A. *Status and application of breeding schemes in Italy.* Seminar on
1976 the Optimization of Cattle Breeding Schemes, Dublin, 26-28 November 1975. EUR 5490e pp. 249-260.

BICZ, A.J. *Bydlo czarno-biale w Zwiazku Radzieckim i metody jego doskonalenia.* Referat wygloszony
1986 na Sympozjum Niemiecko-Radzieckim, Leningrad. 10-14 lipiec.

CANADA. *Artificial insemination in Canada.* Annual Report, 1985.
1985

CENTRALNA STACJA HODOWLI ZWIERZAT (CSHZ). *Wyniki oceny wartósci wzytkoweg bydla w 1985r*
1986 (WOWUB). Warsaw.

CHESNAIS, J.P. *Effectiveness of current selection schemes of production traits in Canadian Holstein*
1984 *cattle.* Benchmark 100. August 1984. Toronto.

CROWLEY, J.W. and NIEDERMEIER, R.P. Dairy production 1955 to 2005. *J. Dairy Sci.*, 64(6): 971-974.
1981

DENMARK. *National Committee on Danish Cattle Husbandry.* Summary of Annual Report.
1984

DOHY, J. and SZÜLLO, B. *Crossbreeding for the specialization of cattle types by the exploitation of*
1984 *world breed gene reserves.* First World Congress on Genetics Applied to Livestock Production, Madrid.

DOMMERHOLT, J. Züchterische Möglichkeiten bei Verschiedenen Verfahren der Milchbezahlung. Tier-
1980 züchter, 32: 270-271.

DORI, S., LEW, B. and KAUFMANN, W. Probleme der Fütterung von Milchvieh in Israel. *Tierzüchter,*
1977 11: 486-488.

DUPLAN, J.M. La race bovine Pie-Noire en 1973. *Elevage et Insémination* No. 139.
1974

ERNST, L.K. and DMITRIEV, N.G. 1. Cattle excluding zebus. In: *Animal genetic resources of the USSR.*
1987 FAO Animal Production and Health Paper series, Rome. (In press).

ETTALA, E. Comparison between Ayrshire, Friesian and Finnish cattle dairy cows on home-produced
1984 feeds. *Journal of the Scientific Agricultural Society of Finland*, 35: 509-524.

GERMANY, FEDERAL REPUBLIC OF. Jahres Statistik.
1982, 1984, 1985

GRAVERT, H.O. Zielvorstellungen in der Schwarzbuntzucht. *Bauernblatt* 07.05.1983: 2216-2218.
1983

GROTHE, P. Ungarn hat Rinderzucht längst auf Leistung ausgerichtet. *Deutsche Schwarzbunte,* 9 (3):
1985 14-16.

GROTHE, P. Spanien: Mit der Schwarzbuntzucht geht es steil aufwärts. *Deutsche Schwarzbunte,* 10 (1):
1986 38.

GROTHE, P. Bulgarien: Rinderhaltung in Umbruch. *Deutsche Schwarzbunte,* 11 (1): 38.
1987

HANSET, R. *Cattle selection in Belgium.* Seminar on the Optimization of Cattle Breeding Schemes,
1976 Dublin, 26-28 November 1975. EUR 5490e. pp. 263-271.

HILL, W.G. and SWANSON, G. A selection index for dairy cows. *Anim. Prod.*, 37: 313-319.
1983

ISRAEL. *Summary of Milk and Butter Fat Recording.* Israel Cattle Breeders Association.
1984

JAARVERSLAG. Royal Society Nederlands Rundvee Syndicaat.
1984

JASIOROWSKI, H., GRABOWSKI, R. and GRODZKI, H. *Origin and genetic value of the Black and White*
1979 *Lowland cattle in Poland.* International Symposium on the Utilization of the Holstein-Friesian Breed for the Improvement of Cattle in Socialist Countries, Warsaw.

KOESLING, T. Milcherzeugung in Irland. *Tierzüchter,* 36 (7): 302-304.
1984

KONOPINSKI, T. *Hodowla Bydla,* Wyd. 2, Poznán.
1949

KORKMAN, H. Rinderrassen in Nordeuropa. *In* Hammond, J., Johansson, I. and Haring, F., eds. *Hand-*
1981 *buch der Tierzüchtung,* Vo. III. Verlag Paul Parey, Hamburg and Berlin.

KRÄUSSLICH, H. *Rinderzucht.* Verlag E. Ulmer, Stuttgart.
1981

KRÜGER, L. *Geschichtliche Entwicklung der Rassen in der europäischen Tierzucht.* Handbuch der Tier-
1961 züchtung s. 25-42, Band III. Paul Parey Verlag, Hamburg.

LEVI, V. Aufbau und Entwicklung der Schwarzbunten Milchviehpopulation in Israel. *Tierzüchter,* 11:
1977 470-473.

MALOVEC, E. Vysledki kontroly úzitkovosti v chove hovädzieho dobytka na Slovensku. *Nas chov,* 8:
1986 328-330.

MATOUS, E. Wysledki kontroly úzitkovosti skotu v CSR za kontrolni rok 1985. *Nas chov,* 8: 325-327.
1986

MORE O'FERRALL, G.J. Holstein or Friesian for Irish dairy farmers. *Farm and Food Research,* 62-64,
1984 (April).

MILLER, P.D. Artificial insemination organizations. *J. Dairy Sci.*, 64 (6): 1283-1287.
1981

NORMAN, H.D. *Overview of dairy cattle breeding in the United States.* Lecture presented in Warsaw.
1984 December 1984.

NOTT, S.B., KAUFMAN, D.E. and SPEICHER, J.A. Trends in the management of dairy farms since
1981 1956. *J. Dairy Sci.* 64 (6): 1330-1343.

PHILIPSSON, J. *European breeding plans.* Benchmark 100. August 1984. Toronto.
1984

POLITIEK, R.D. *Comparison of Friesians of different origin. Optimum methods of cattle breeding for*
1978 *increasing meat and dairy production.* International Symposium, Warsaw. p. 141-158.

POLITIEK, R.D. and BAKKER, J.J. Livestock in Europe. Perspectives and prospects. *Livest. Prod. Sci.*,
1982 9 (1/2) 335 p.

SCHAEFFER, L.R. Effectiveness of model for cow evaluation intraherd. *J. Dairy Sci.*, 66: 874-880.
1983

SHAINLINE, W.E. JR. *Animal improvement programs laboratory,* ARS, UDA.
1986

STICHBURY, J.W. *Herd improvement in New Zealand.* Documentation of conference papers. South
1982 Pacific Friesian conference, New Zealand, 9-16 February 1982.

STOLZMAN, M. *Actual position and development of the Friesians in Eastern Europe.* 5th World Frie-
1980 sian Conference, Bremen, 25-30 August. pp 23-40.

STOLZMAN, M. Bydlo fryzyjskie w swiecie. *Prace i materialy zootechniczne,* 35: 9-16.
1985

SWEDEN. *Ars Statistik.* Meddelande No. 133.
1984

SZELAG, B. and NAHLIK, K. *Analiza wyników oceny i sélekcji buhajów w wychowalniach w roku 1984.*
1985 Instytut Zootechniki i Centralna Stacja Hodowli Zwierzat, Kraków.

THOMAS, H. and VAN BURG, P.F.J. Entwicklung in der Grünlandwirtschaft der Niederlande. *Tier-*
1981 *züchter,* 33 (2): 73-75.

UNITED KINGDOM. *Breeding and Production Division Report 1984-85.* MMB No. 35. Milk Marke-
1985 ting Board, Thames Ditton, Surrey.

UNITED STATES OF AMERICA. *Annual Report.* National Association of Animal Breeders.
1985

VOELKER, D.E. Dairy Herd Improvement Associations. *J. Dairy Sci.*, 64 (6): 1296-1277.
1981

WALDO, D.R. and JORGENSEN, W.A. Forages for high animal production. National factors and ef-
1981 fects of conservation. *J. Dairy Sci.*, 6 (6): 1207-1229.

4. A review of investigations on crossbreeding of European strains of Black and White cattle with Holstein-Friesians in developed countries

4.1 INTRODUCTION

Economic conditions in EEC countries during the past 20 years have favoured milk production relative to beef. The cost of concentrates has been sufficiently low to encourage its greater use to boost milk output, and labour costs could be contained by intensification and increased output per cow. This has encouraged farmers to improve the milk performance of their cattle. As a consequence, in several countries, Black and White Friesian cattle already represent a distinctly marked dairy type. Undoubtedly, in transforming the Friesian from a relatively dual-purpose animal into a more single-purpose dairy type, an important role has been played by crossbreeding with the North American Holstein-Friesian strain. Due to the widespread use of deep-frozen semen and artificial insemination, farmers across the whole world have a choice of bull semen with the highest breeding value. This, also, has been one of the main reasons for the rapid expansion of crossbreeding with Holstein-Friesian semen.

According to Turton (1981) over 300 papers were published during the previous decade on crossbreeding of dairy cattle for the improvement of milk production. However, the number directly related to Friesian and Holstein-Friesian cattle is considerably smaller. Moreover, many of the papers provide inadequate information, or have experimental designs directed at very specific questions, so that extrapolation of their results to provide general conclusions is difficult. Effectively, therefore, the number of studies relevant to the present context of crossing European strains of Friesian with North American is rather limited. Most investigations have taken place in the Netherlands, the Federal Republic of Germany and the Democratic Republic of Germany. Although crossbreeding was undertaken for different specific reasons the following general objectives were involved:

— Development of a new breed or a new type of animal (Schönmuth *et al.*, 1980), providing a moderate inflow of new genes into the population to be improved.

— Upgrading as done in Israel, and also, more recently, in EEC countries.
— Exploitation of any potential heterosis through rotational crossbreeding as done to a certain extent in Norway.

4.2 PERFORMANCE OF FIRST GENERATION CROSSBREDS

Holstein-Friesians were imported in the 1960s and 1970s to most West European countries, as well as to Poland, Hungary, Czechoslovakia, Yugoslavia and USSR (Ernst, 1968; Liebiediev, 1975; Kaczmarek *et al.*, 1976; Gere, 1979; Louda *et al.*, 1979; Rusjajev, 1979; Vagi, 1979; Oldenbroek, 1980; Szücs *et al.*, 1982). Investigations in these countries have shown that the genetic potential of Holstein-Friesian cows for milk production and milking ability is higher than that of local strains. When the cows are intensively fed with concentrates they can reach a level of milk production comparable to that achieved in the USA. Although milk produced by Holstein-Friesian cows contains relatively lower concentrations of fat and protein, the low concentrations are compensated for by a high milk yield. Thus, improvement of milk performance of different local strains by crossbreeding goes back to the early 1960s, when in Lower Saxony and Bavaria, Holstein-Friesian bulls were used first on an experimental scale.

By utilizing the genetic differences between the Holstein-Friesian populations of North America and the Friesian population of Western Europe it has been possible to increase milk yield considerably in a relatively short time as well as change some of the exterior traits, especially the udder and legs. According to Witt *et al.* (1971) a single use of Holstein-Friesian bulls for mating with German Friesian cows resulted in an average increase in milk production of the progeny of 840 kg of milk and 28 kg of fat. On the other hand, milk fat content decreased by 0.16 percent as compared to contemporaries from the control group of German Friesians. One of the first observations on the relative milk yield of Holstein-Friesians comes from Romania (Georgescu *et al.*, 1967), where the authors compared milk production of different strains. Friesian strains from the United Kingdom, Poland and Holstein-Friesians from the USA yielded respectively 4 840 kg, 4 079 kg and 5 349 kg milk per lactation. Comprehensive investigations on milk yield of crosses and control groups of German Friesians and Holstein-Friesians from the USA were conducted by Ernst *et al.* (1973). The results indicated a superiority of 740 kg of 4 percent FCM of German Friesian × Holstein-Friesian crosses over German Friesians. Pure Holstein-Friesian cows in their first lactation showed twice that superiority over German Friesian contemporaries, i.e. 1 425 kg of milk with 4 percent of fat. Some of the comparative results are summarized in Table 4.1.

A number of other authors confirmed that the inflow of Holstein- Friesian blood would be expected to increase milk production considerably (Huth, 1972; Grothe 1974; Droese *et al.*, 1981). Thus, Holstein-Friesian bulls were used mainly for mating with cows of other Friesian strains (Witt *et al.*, 1971; Grothe, 1972; Gravert, 1974; Jasiorowski and Stolzman, 1974; Gravert, 1975; Bicz and Borisova, 1976; Ganss, 1976; Lotthammer and Vierling, 1976; Farzalijev, 1977; Koncar *et al.*, 1977; Oldenbroek, 1977; Kaczmarek *et al.*, 1979; Oldenbroek, 1980; Poczynajlo *et al.*, 1980; Zalewski *et al.*, 1984; Zeltikov, 1980; Stolzman *et al.*, 1981; Politiek *et al.*, 1982). Holstein-Friesians, have, however, also been used in crossbreeding with other dairy breeds (Crettenand and Meyer, 1972; Brandt *et al.*, 1974; Averdunk and Gottschalk, 1975; Colleau, 1978; Pandey *et al.*, 1978; Pawlina, 1978; Turton, 1981).

The results of these various investigations indicate that crossbred cows sired by Holstein-Friesian bulls exceeded the milk production of the contemporary maternal strains and breeds by 600 kg of milk. Turton (1981) concluded that crossing of European Black and White cattle with Holstein-Friesians has led to increased milk yield, due to the greater additive genetic merit for milk yield of the latter, but in general the milk produced by crossbred cows has a lower fat and protein content than that of the maternal strains. However, the crossbred cows had noticeably better conformation and udder attachment, fitting them better to machine milking.

While with abundant feeding of concentrates, crosses of European strains of Black and White cattle with Holstein-Friesians showed gains in yield over the local pure strain, it is essential to know whether the crossbreds were capable of utilizing roughages, particularly in countries with insufficient cereals. Such investigations were conducted in Poland. Crossbred cows were fed hay and maize silage *ad libitum.* This diet was supplemented with urea and a mineral mixture (Jankowski, 1985). The milking performance of these Black and White × Holstein-Friesian crosses is presented in Table 4.2.

As seen from Table 4.2, in intensive feeding conditions Black and White × Holstein-Friesian crossbreds achieved a greater superiority in milk production over the contemporaries of the Polish strain than with roughage feeding. (Cows in the concentrate-fed group received about 1 metric ton of concentrates during a lactation period.) The nutritional regime also significantly affected fat and protein contents in milk with crossbreds fed roughages producing milk with lower fat content, contrary to normal expectation, and lower protein content. However, even under the more "extensive" feeding conditions, milk yield of crossbred cows was significantly higher than that of Black and White cows of the Polish strain. Investigations also indicated that the crossbreds were not inferior to Polish Black and White cattle in their utilization of roughages.

Similar results were obtained by Oldenbroek (1984) in comparisons of

Table 4.1 **Milking performance for the first lactation of Black and White × Holstein-Friesian crossbreds as compared with their Black and White contemporaries**

Author		Country	Productivity									
			Milk *(kg)*		Fat *(kg)*		Protein *(kg)*		Fat (%)		Protein (%)	
			F_1	difference	F_1	difference	F_1	difference	F_1	difference	F_1	difference
Bicz, H.I. and Borisova, T.F.	1976	USSR	4 717	+911	177.0	+31.5	—	—	3.75	—0.02	—	—
Farzalijev, S.G.	1977	USSR	4 153	+774	145.3	+23.7	—	—	3.50	—0.10	—	—
Ganss, L.	1976	GDR	3 956	+517	133.2	+14.2	123.9	+7.2	3.37	—0.09	3.17	—0.13
Jasiorowski, H.A. *et al*	1983	Poland	5 402	+1005	197.2	+20.1	182.7	+25.5	3.69	—0.37	3.40	—0.19
Kaczmarek, A. *et al*	1979	Poland	6 059	+711	262.5	+29.8	206.6	+19.9	4.33	—0.02	3.41	—0.08
Kraüsslich, H.	1975	FRG	3 968	+439	156.9	+19.0	—	—	3.94	+0.05	—	—
Lotthammer, K.H. and Vierling, K.	1976	FRG	6 696	+240	281.0	+5.0	—	—	4.17	—0.13	—	—
Oldenbroek, J.K.	1980	Netherlands	5 576	+497	215.5	+11.4	181.8	+17.4	3.90	—0.15	3.28	0.02
Pasierbski, Z. and Romer, J.	1978	Poland	3 684	+618	145.4	+17.4	119.5	+17.5	3.91	—0.27	3.20	—0.10
Poczynajlo, S. *et al*	1980	Poland	4 200	+1126	162.2	+38.0	134.4	+31.3	3.86	—0.17	3.20	—0.16
Politiek, R.D. *et al*	1982	Netherlands	5 393	+822	—	—	—	—	—	—0.27	—	—0.05
Vos, H. *et al*	1975	Netherlands	5 140	+737	188.2	+7.6	164.1	+19.8	3.68	—0.43	3.21	—0.08
Zalewski, W. *et al*	1984	Netherlands	5 205	+1253	204.0	+44.4	—	—	3.92	—0.08	—	—
Zeltikov, A.	1980	USSR	5 383	+802	199	+33.0	192.0	+31.0	3.70	—0.07	3.56	—0.05
Ernst, E.	1968	FRG	4 947	+661	191	+33.0	—	—	3.86	—0.18	—	—

Holstein-Friesians and Dutch Black and White cows with diets of roughages or containing 50 percent of concentrates. On the roughage diet, heifers of the Holstein-Friesian breed produced, on average, 4 649 kg of milk with 4.06 percent fat and 3.12 percent protein in the first lactation, whilst their Dutch Black and White contemporaries produced 3 887 kg of milk with 4.32 percent fat and 3.33 percent protein. Holstein-Friesian heifers had a higher voluntary feed intake and they converted into milk about 51 percent of the dietary energy compared with only 49 percent for heifers of the Dutch strain.

In addition to any improvement of crossbreds from additive genetic effects, their performance level may also be related to an increase in heterozygosity with resulting heterosis when there is a favourable combination of different genotypes (Langholz, 1976; Beckett *et al.* 1979; Kinghorn, 1982). However, not all productive traits are influenced by heterosis to a similar degree. In general, traits with low heritabilities (low additive genetic variation) display higher heterosis than those with a high heritability (Donald and Russell, 1968; Donald *et al.*, 1977). Under conditions not involving "extremes", the effect of heterosis does not usually exceed 10 percent of the mean of the trait of the two breeds or strains contributing to the first cross (Haiger, 1976; German, 1981). Studies by McDowell (1976) with crossbreeding of eight dairy breeds of cattle showed that the heterosis for milk production of 2-breed crosses amounted to 7.0-7.3 percent. Donald *et al.* (1977) estimated values for 305-day milk yield in first and second lactation of 6.4 percent and 3.8 percent respectively. Similar results were reported by Cunningham (1972 and 1974); Rincon *et al.* (1976); and Touchberry (1978). According to McDowell (1976) and Donald *et al.* (1977) little or no heterosis was found for length of lactation, milk composition and milking ability.

Table 4.2 **Milk yield and milk composition of crossbreds (averages for three lactations) on two types of feeding regime in Poland** (after Jankowski, 1985)

	Experimental (roughage)/Feeding		Control (concentrates)	
	Yield	Superiority of crossbreds	Yield	Superiority of crossbreds
Milk (kg)	3 359	+ 275	4 758	+ 628
Fat (kg)	127.6	+ 9.4	191.3	+ 22.4
Protein (kg)	101.3	+ 5.7	157.9	+ 20.5
Fat content (%)	3.80	—0.10	4.04	—0.07
Protein content (%)	3.03	—0.09	3.31	—0.01

The results of numerous investigations showed that for characteristics of growth and development there is a positive and significant effect of heterosis (Dickinson, 1961; Long *et al.* 1976. Also, a slight, favourable influence of heterosis on reproductive traits was found by Donald and Russell (1968); Cundiff *et al.* (1974); Obritzhauser *et al.* (1979). Most of these authors emphasize that the effect of heterosis on production under their conditions is seen only in the F_1 generation.

Several authors, however, have either failed to find an effect of heterosis in relation to growth and development and milk performance traits or found it very low (Touchberry and Bereskin, 1966; Pearson and McDowell, 1968; Colleau, 1978; Troccon, 1978; Boie, 1981). McDowell and McDaniel (1968) have pointed out, however, that heterosis can be profitable from the economic point of view, even when it is statistically non-significant for individual traits.

Przybylska (1985) investigated the influence of different concentrations of net energy (i.e. 18 and 22 percent of the dry matter) on growth rate of first cross and backcross heifers up to 18 months of age. (The lower energy concentration was achieved by the greater use of forages with only a very small supplement of concentrates.) It was found that Holstein-Friesian × Black and White crosses had a higher growth rate than their contemporaries of the Polish strain while the conversion rate of forages was similar in both groups of animals. The higher growth rate of crossbred heifers was attributable to a significant influence of heterosis. However, no interaction of feeding level with genotype was reported.

The expression of heterosis can be affected by environmental conditions in which the crossbreds are maintained. Thus, Long *et al.* (1979) reported that when growth and development traits of F_1 heifers originating from crossing 5 breeds of cattle (including Holstein-Friesians) were involved, the effect of heterosis was higher in intensively fed animals (14 percent), than in those reared on pasture (7 percent). Similar results were described by Langholz (1976) in an F_1 generation of bulls originating from crossing several breeds with German Friesian cattle and fattened by different systems. The intensively fattened crossbred bulls gained weight 8.8 percent faster than their purebred contemporaries of the local breeds, whereas those fed silage exceeded purebreds by 6.6 percent and those on pasture showed an effect of heterosis of only 1.8 percent. This relationship between heterosis and management regime was interpreted by these authors as a genotype-environment interaction. Numerous other reports describe genotype-environment interactions in both meat performance traits (Langholz and Pabst, 1976; Liboriussen *et al.*, 1976; Andersen, 1978; Langholz, 1978) and milk production (Mao and Burnside, 1969; Richardson *et al.*, 1971; Lamb *et al.*, 1977). In this context it is known that animals of different breeds respond in a different way to feeding restrictions. Thus, animals of early maturing breeds, finishing their growth more quickly, were more seriously affected by lengthy underfeeding than those of late maturing breeds. These findings were confirmed by the results of Swedish investigations demonstrating an interaction between genotype and rearing system on the growth rate and feed conversion of purebred SKG heifers and crossbreds between SLB × SKB and SJB × SKB (Dim, 1977), as well as on age of puberty in SKB, SLB and SJB purebred heifers and SKB × SJB crossbreds (Berglund *et al.*, 1980).

Apart from nutrition, which is a major component of environment in considering the interaction between genotype and environment, the magnitude of this interaction depends on many different effects listed by Dickerson (1962). However, most investigations on the effectiveness of crossbreeding with Holstein-Friesian strains have been performed under intensive feeding conditions, frequently simulating the feeding practice in the USA because of a popular view that Holstein-Friesian cattle were developed to fit such conditions. Thus, the possibility that interactions of genotype with level of nutrition might be important has not been adequately tested. However, there are a few trials in which Holstein-Friesian cattle crossbred heifers were reared on a forage diet (Gardner and Garcia, 1966; Przybylska, 1985; Seath *et al.*, 1966; Grieve *et al.*, 1976), but, unless accompanied by comparison with intensive feeding, this again provides no information on interaction.

4.3 RESULTS OF UPGRADING WITH THE HOLSTEIN-FRIESIAN STRAIN

4.3.1 Milking performance

Changes in the dairy cattle population in EEC countries were analysed by Cunningham (1978), who found that the annual progress in total milk production of these countries in the late 1970s amounted to approximately 1 percent, with a relatively constant number of cows and stagnation in the level of consumption of milk and dairy products. At the same time there was an upward trend in beef production, with an average increase per year of 1.5 percent, while consumption went up in proportion to this increase. This progress was attained by an increase in milk yield per cow, an increase in the slaughter weight of fattened animals, and a reduction in the proportion of calves slaughtered.

In EEC countries the Black and White Friesian population constitutes 60 percent of the total cattle population. The remaining dairy breeds have, at best, remained static in number, or as in several countries, more often declined. For example, distinct changes in favour of Friesians have occurred in Ireland where the trend has been to replace Shorthorns and in the United Kingdom where Ayrshires (and, earlier, Shorthorns) were replaced; in Denmark, the change from Red Danish cattle to Friesians occurred at a slower rate, as in France in relation to the Normandy breed. According to Cunningham (1978) European sub-populations of Friesian cattle in the 1970s were still only moderately influenced by the North American Holstein-Friesian strain.

Cunningham (1978) has considered the rate at which the Holstein-Friesian type might replace European strains in the context of different breeding strategies. As shown in Table 4.3, even the use of only 20 percent of Holstein-Friesian semen with backcrossing would allow that type to dominate the population in less than two decades. Some years ago, five out of eight national Friesian populations in EEC countries used Holstein-Friesian semen either to a high or moderately high degree in crossbreeding programmes which, according to Cunningham (1978), would lead to the replacement of over 50 percent of genes in these populations by Holstein-Friesians within a decade. A similar trend can be observed in Simmental and Braunvieh of the Alpine region where the Red and White Holstein-Friesian type is being used.

In recent years the use of Holstein-Friesian bulls has been intensified, particularly notable being the use of this breed for crossbreeding in the Federal Republic of Germany, Italy, Netherlands, Denmark, Switzerland and France. The most important remaining non-Friesian breeds in Europe — Simmental, Red and White, Normandy and Braunvieh — have changed considerably in numbers but their position is not at present endangered.

Huth (1981) investigated the effects on performance of backcrossing using proven Holstein-Friesian bulls with positive tests from the USA and Canada. Some of the results are in shown in Table 4.4. It is noticeable that milk yield went up with successive increases in the proportion of Holstein-Friesian blood. Huth (1981) also found that crossbreeding resulted in the increase of height at withers by approximately 5 cm, body length by approximately 6.5 cm, and body weight on average by 12 kg compared to purebred German Friesian. Genetic investigations by Langholz and Jongeling (1977) on field data confirmed earlier results and showed that crossbreeding with the Holstein-Friesian increased milk yield of crossbreds compared to German Friesians under normal production conditions.

Backcrossing with Holstein-Friesian bulls has become widespread in the Federal Republic of Germany. According to Boie (1981) in 1979 over 88 percent of Holstein-Friesian bulls used for breeding in the Schleswig-Holstein region had some genes of the Holstein-Friesian strain. The same author found, on the basis of results from field trials, that the highest superiority of crossbreds over German Friesians appeared in the first lactation (583 kg milk and 22.2 kg fat) while in older animals the difference was considerably smaller (391 kg

Table 4.3 **The effect of backcrossing using exotic semen on different proportions of the population assuming 25 percent of annual herd replacement; 42 reared heifers per 100 cows; 90 percent of reared crossbred heifers retained in the herd** (after Cunningham, 1978)

Proportion of inseminations by foreign strain (%)	Proportion of replacement heifers sired by foreign strain (%)	Number of years needed for 50% of genes to be replaced by foreign strain
20	30	13.2
25	38	10.5
30	45	8.9
35	53	7.6
40	60	6.7

Table 4.4 **Performance of German Black and White Friesian and of crosses sired by proven Holstein-Friesian bulls (with positive tests) from the USA and Canada** (after Huth, 1981)

		Holstein-Friesian proportion (%)		
	Black and White German Friesian	50	75	87.5
Number of cows	159	59	62	39
Age at first calving (months)	29.8	28.9	29.9	28.6
Milk yield for 305 days (kg)	4 896	5 155	5 890	6 037
Fat content (%)	4.14	3.90	3.89	3.81
Fat yield (kg)	203	201	227	230
Milk yield corrected for 4% fat (FCM — kg)	4 999	5 078	5 793	5 865
Calving interval (days)	384	387	392	400

milk and 15.8 kg fat in second lactation).

Extensive studies on the performance of different generations of Holstein-Friesian crossbreds have also been carried out for many years in the Netherlands. The investigations were conducted on eight experimental farms maintaining a total of 500 cows (Oldenbroek, 1980 — Table 4.5). In crossbreeding programmes only proven Holstein-Friesian bulls with positive tests from the USA and Canada were used. The results indicate that, as before, the increase in the proportion of Holstein-Friesian genes was accompanied by an increase in milk, fat and protein yields, while fat content decreased. The author found no statistically significant heterosis or maternal effect.

Similar results were obtained from a comparison of milk production of F_1 generation crossbreds and backcrosses, conducted in Poland (Przybylska, 1985) as shown in Table 4.6. The superiority of crossbreds over their herdmates was very clear. Backcross heifers with 75 percent Holstein-Friesian genotype produced in their first lactation almost 1 500 kg more milk than the Polish Black and White contemporaries. In this case backcrossing did not adversely affect milk composition.

4.3.2 Evaluation of beef performance in crossbreds

One of the first European evaluations of meat production from crossbreds originating from German Black and White dams sired by Holstein-Friesian bulls was published by Weniger *et al.* (1968) who found that under intensive feeding conditions crossbreds showed a satisfactory growth rate. Qualitative carcass traits were inferior to the European Friesian strain, having a higher bone percentage and lesser musculature of the carcass.

Detailed investigations on meat production characteristics of the progeny of Holstein-Friesian sires were carried out in the Netherlands on intensively fattened bulls. As seen from Table 4.7, no significant differences in growth rate were found between bulls of the Dutch strain and crossbreds with 25 percent or 50 percent of Holstein-Friesian blood. However, the group of bulls with 75 percent of Holstein-Friesian genotype had daily gains and feed conversion slightly poorer than F_1 crossbreds. In relation to carcass quality, estimated by the fleshiness and fat covering scores, there was a trend for quality to decline with an increasing Holstein-Friesian contribution.

Investigations over several years in the Federal Republic of Germany by Huth (1981) also showed that fattening ability of crossbreds was good but carcass quality traits were slightly poorer relative to German Friesian bulls (treated as control animals). Bulls were fattened up to 78 weeks of age according to normal standards of farm fattening in the Federal Republic of Germany, i.e.

Table 4.5 **Evaluation of first lactation (305-day) milk production in Dutch Friesian cows and crossbreds with different proportions of Holstein-Friesian genotype** (after Oldenbroek, 1980)

	Dutch Friesian	Holstein-Friesian proportion *(%)*		
		25	50	75
Number of cows	48	70	58	70
Milk yield (kg)	5 079	5 276	5 576	5 827
Fat content (%)	4.05	3.89	3.90	3.77
Protein content (%)	3.26	3.25	3.28	3.23
Fat yield (kg)	204.1	203.9	215.5	218.7
Protein yield (kg)	164.4	170.4	181.8	187.8

Table 4.6 **Milk production in first lactation of Polish Black and White cattle and of crossbreds with different proportions of Holstein-Friesian blood** (from Przybylska, 1985)

	Polish Black and White	Holstein-Friesian proportion *(%)*	
		50	75
Number of cows	30	30	30
Milk yield (kg)	3 479	4 614	4 958
Fat (%)	3.91	4.12	4.08
Protein (%)	3.26	3.24	3.29
Non-fat solids (%)	7.70	7.51	7.42
Fat yield (kg)	137.2	189.0	202.4
Protein yield (kg)	113.6	151.5	160.2

corn silage *ad libitum* supplemented daily with 2.5 kg of concentrates per animal. Liveweight at the age of 78 weeks averaged 578 kg in crossbreds, 50 kg heavier than for the German Friesian animals. Body measurements showed that at 78 weeks old the crossbreds were 7 percent taller, 6 percent longer and 5 percent deeper than German Friesians but were 1 percent narrower in width. These proportions in body conformation gave the impression that the crossbreds were

Table 4.7 **Aspects of meat production of Dutch Friesians and crossbred bulls with varying proportions of Holstein-Friesian blood** (from Oldenbroek, 1980)

	Dutch Friesians	Holstein-Friesian proportion (%) 25	50	75
Number of bulls	32	33	28	35
Daily gain (g)	1 063	1 063	1 070	1 043
Starch unit intake per 1 kg gain	3.19	3.24	3.19	3.33
Fleshiness (score)	3.6	3.3	3.0	2.5
Fat covering (score)	3.5	3.4	3.4	3.1
Dressing percentage	59.2	58.4	58.2	57.9

Table 4.8 **Aspects of meat production of German Friesian bulls and crosses sired by Holstein-Friesian** (after Huth, 1981)

	Number of animals	Starch unit intake per 1 kg of gain	Average daily gains from 3 to 75 weeks (g)	Carcass weight (kg)	Dressing percentage (%)
German Friesians	78	3.28	911	300	59.9
HF crossbreds	63	3.24	996	322	59.0

flat and poorer muscled in comparison to German Friesians. An analysis of fattening results (Table 4.8) showed that the crossbreds had a higher slaughter weight (by 7 percent) than the German Friesian control animals, in spite of a 1 percent lower dressing percentage, because they had grown more rapidly to the age of 78 weeks. As seen in Table 4.9, these crossbreds also had a higher internal fat content and slightly lower trim fat content in three rib cuts than those of Black and White Friesians.

An economic evaluation of these trials was made by Wiemer (1981) and is shown in Table 4.10. As expected, the profit margin between income and costs improved with increasing herd life mainly on account of decreased annual replacement costs. The absence of a significant difference between the genetic groups in slaughter values might be surprising in view of the poorer carcass quality of Holstein-Friesian crosses relative to German Friesians, but this appears to be compensated by the higher carcass weights of the former. Also, in the case of beef from cull cows in the Federal Republic of Germany, this usually goes for sausage production and, as pointed out by Gravert (1983), little attention is paid to carcass quality. The main criterion determining price is the quantity of lean meat. The crosses had a greater advantage in annual income from milk the longer the herd life. Perhaps surprisingly, replacement costs of the Holstein-Friesian crossbreds were only very little higher than for German Friesians. However, it must be emphasized that financial assessments are particularly sensitive to economic conditions and may not have widespread validity. For example, a greater price difference between slaughter grades could materially affect the relative values of the two genetic groups.

Table 4.9 **Weights of parts of the carcass and fat cover in German Friesian bulls and crosses sired by Holstein-Friesian** (after Huth, 1981)

	Number of animals *(kg)*	Internal organs [1] *(kg)*	Skin	Head and tongue *(kg)*	Legs *(kg)*	Internal fat [2] *(kg)*	Proportion of fat in 3 rib cuts *(%)*
Black and White Friesians	78	14.2	43.2	18.2	8.8	27.2	23.7
HF crossbreds	63	16.2	44.3	19.5	10.8	32.3	22.0

[1] Heart + liver + lungs

[2] Kidney fat + intestinal fat + stomach fat

Table 4.10 **Economic evaluation of German Friesian cows and crosses sired by Holstein-Friesians according to length of herd life of cow** (from Wiemer, 1981)

	Cow life (lactations)					
	1		**3**		**5**	
	German Friesian	**HF cross-breds**	**German Friesian**	**HF cross-breds**	**German Friesian**	**HF cross-breds**
Mean annual income (DM)						
Milk	2 429	2 512	3 092	3 208	3 233	3 371
Slaughter value of the cow	1 363	1 402	496	523	296	306
Carcass price (DM/kg)	5.43	5.35	5.40	5.30	5.37	5.30
Calf value	376	381	376	381	376	381
Total income	4 168	4 295	3 964	4 112	3 905	4 061
Annual variable costs (DM)						
Replacement cost	2 400	2 500	800	833	480	500
Losses	19	20	19	20	19	20
Machinery & buildings depreciation	113	117	117	122	116	121
Annual roughage costs (DM)						
Winter	245	250	257	265	256	263
Summer	65	66	100	103	100	103
Concentrates	487	513	700	737	745	789
Total annual costs (DM)	3 329	3 466	1 993	2 080	1 716	1 796
Annual profit (DM)	839	829	1 971	2 032	2 189	2 265
Contribution of cow slaughter value % to:						
Total production	32.7	32.6	12.5	12.7	7.6	7.6
Costs of heifer purchase	56.8	56.1	62.0	62.8	61.7	61.8

Studies by Raue (1982), over several years, showed that beef performance of bulls with a high proportion of Holstein-Friesian genotype can be satisfactory if the fattening regime is adjusted to the animal's genotype. Thus, Raue suggested that fattening should be to a final weight of 550-600 kg, which can be reached by bulls at an age of 20-24 months. It was important, however, to employ a moderate feeding level during the first year allowing daily weight gains within the range of 1 100-1 200 g. Crossbreds on such a regime did not accumulate excess fat prematurely and had achieved maximum muscle development by slaughter.

Crossbreeding with Holstein-Friesians also had some negative effects. Apart from the slightly poorer carcass quality, already referred to, unfavourable changes in health, fertility and calf viability have been reported. Gravert (1983) summarizing the results of an almost 20-year period of backcrossing to Holstein-Friesians in the Federal Republic of Germany pointed out that the percentage of cows culled annually during the past ten years had increased, on average, from 25 to 30 percent. At the same time the proportion of cows older than 6 years had decreased from 38 to 28 percent of the herd. Freese and Gravert (1983) have suggested that more attention should be paid to calf rearing and udder health. They found that calf losses in dairy herds were increasing and were positively correlated with milk yield and particularly with sire breeding value for milk performance traits. Calf mortality in the Federal Republic of Germany is estimated at 5-10 percent of calves born. The actual value of these losses was about 200 million DM. Similar results were also reported in the Netherlands (Oldenbroek, 1980).

References

ANDERSEN, H.R. Effect of energy level on growth and efficiency. *In: Patterns of growth and development in cattle.* A seminar in the EEC Programme of Coordination of Research on Beef Production. Ghent, 11-12 October 1977. pp 393-412.
1987

AVERDUNK, G. and GOTTSCHALK, A. Möglichkeiten der Kreuzungszucht beim Zweitnutzungsrind: Brown-Swiss × Braunvieh in Bayern — eine Zwischenbilanz. *Züchtungskunde,* 47: 393-403.
1975

BECKETT, R.C., LUDWICK, T.M., RADER, E.R., HINES, H.C. and PEARSON, R. Specific and general combining abilities for production and reproduction among lines of Holstein cattle. *J. Dairy Sci.*, 62: 613-620.
1979

BERGLUND, B., BRÄNNÄNG, E., HENNINGSSON, T., JANSON, L., JOSEFSSON, G., LINDKVIST, G., MAGNUSSON, B., PHILLIPSON, J., and SWENSSON, C. Comprehensive experiments on traits affecting longevity in Swedish dairy cattle breeds. *Swedish J. Agric. Res.*, 10: 159-166.
1980

BICZ, H.I. and BORISOVA, T.F. Opyt skrieszcziwanija cziorno-piostrogo skota s golstino-frizskim. *Ziwotnowodstwo,* 9: 18-21.
1976

BOIE, D. *Kreuzungsparameter bei HF-verwendung.* Vortragstagung der Deutschen Gesellschaft für Züchtungskunde e.v. und der Gesellschaft für Tierzuchtwissenschaft. Giessen, 24-25 September.
1981

BRANDT, G.W., BRAUNON, C.C. and JOHNSTON, W.E. Production of milk and milk constituents by Brown-Swiss, Holsteins and their crossbreds. *J. Dairy Sci.*, 57: 1388-1393.
1974

COLLEAU, J.J. Comparaison de types génétiques bovins laitiers mixtes ou spécialisés. In: *La vache laitière. Aspects génétiques, alimentaires, pathologiques.* Institut National de la Recherche Zootechnique, Versailles.
1978

CRETTENAND, J. and MEYER, U. Vorläufige Ergebnisse aus dem Kreuzungsversuch Simmentaler Fleckvieh und Red Holstein. *Mitt. Schweiz. Fleckviehzuchtverb.,* 5: 25-36.
1972

CUNDIFF, L.V., GREGORY, K.E. and KOCK, R.M. Effects of heterosis on reproduction in Hereford, Angus and Shorthorn cattle. *J. Anim. Sci.*, 38: 711-727.
1984

CUNNINGHAM, E.P. Components of variation in dairy cow production records. *Ir. J. Agric. Res.*, 11: 1-10.
1972

CUNNINGHAM, E.P. *Crossbreeding strategies in cattle populations.* Proc. Working Symposium on Breed Evaluation and Crossing Experiments, Zeist. pp 107-125.
1974

CUNNINGHAM, E.P. Strukturelle Veränderungen der Rinderzucht in der Europäischen Gemeinsschaft. *Tierzüchter,* 30: 138-140.
1978

DICKINSON, A.G. The genetical basis of conformation in cattle as revealed by crossbreeding. *Z. Tierz.*
1961 *Züchtungsbiol.*, 75: 262-275.

DICKERSON, G.E. Implications of genetic-environment interaction in animal breeding. *Anim. Prod.*,
1962 4: 47-63.

DIM, N.J. Growth rate and feed conversion efficiency of purebred and crossbred dairy cattle. *Swedish*
1977 *J. Agric. Res.*, 7: 29-33.

DONALD, H.P., GIBSON, D. and RUSSELL, W.S. Estimations of heterosis in crossbred dairy cattle.
1977 *Anim. Prod.*, 25: 193-208.

DONALD, H.P. and RUSSELL, W.S. Some aspects of fertility in purebred and crossbred dairy cattle.
1968 *Anim. Prod.*, 10: 465-471.

DROESE, N., FIEDLER, H., BREITENSTEIN, H.G. and BAUMUNG, A. Ergebnisse Züchterischer Unter-
1981 suchungen zur Möglichkeiten Beeinflussung der Milcheiweissleistung. *Tierzucht,* 4: 163-165.

ERNST, E. Vergleichende Untersuchungen der Leistungen von Deutschen Schwarzbunten, Holstein-
1968 Friesians und deren Kreuzungen, I. *Tierzüchter,* 20: 798-800.

ERNST, E., LANGLET, J.F. and MARTIN, H. *Vergleichende Untersuchungen an Deutschen Schwarz-*
1973 *bunten, Holstein-Friesians sowie den Kreuzungen aus beiden Populationen.* Schriftenreihe der Agrarwiss. Fakultät, Kiel. Heft 50.

FARZALIJEV, S.G. Effektiwnost skrieszcziwanija cziorno-piostrogo i golstino-frizskogo skota. *Ziwot-*
1977 *nowodstwo,* 1: 37-40.

FREESE, E. and GRAVERT, H.O. Kälberverluste - ein wachsendes Problem. *Züchtungskunde,* 55 (3):
1983 186-202.

GANSS, L. Leistungseigenschaften von Jungkühen verschiedenen Genotypen bei industriemäsigen Pro-
1976 duktionsbedingungen. *Tierzucht,* 7: 300-303.

GARDNER, R.W. and GARCIA, L.V. Accelerated growth and early breeding studies with Holstein heifers.
1966 *J. Dairy Sci.*, 49: 720 (Abst.).

GEORGESCU, G., PETRESCU, A., SIMIONESCU, D. and STANA, N. *The breeding and productivity of*
1967 *Friesian cattle in Romania.* ABA, 36: 28.

GERE, T. *Izmienienije niekotorych koliczestwiennych priznakow Wiengerskogo piostrogo skota pri*
1979 *skrieszcziwanii s golstino-frizskoj porodoj.* Symposium on the Utilization of the Holstein-Friesian Breed for the Improvement of Cattle in Socialist Countries, Warsaw. pp 253-256.

GERMAN, E. Die Einkreuzung mit Red Holstein beim Schweizerischen Simmentaler Fleckvieh. *Schweiz.*
1981 *Landw. Mh.*, 59: 452-462.

GRAVERT, H.O. Der Genetische Unterschied zur US Holstein-Friesian Population, geschätzt über den
1974 Einsatz durchschnittlicher Besamungsbullen aus Minnesota. *Tierzucht,* 26: 94-96.

GRAVERT, H.O. Possibilities for crossbreeding in dual-purpose cattle: Holstein × Black Pied.
1975 *Züchtungskunde*, 47: 404-412.

GRAVERT, H.O. Zielvorstellungen in der Schwarzbuntzucht. *Bauernblatt*, 2216-2218 (7.5.83).
1983

GRIEVE, D.G., STONE, J.B., MACLEOD, G.K. and CURTIS, R.A. All silage forage programs for dairy
1976 cattle. I. Heifer performance from birth to eighteen months of age. *J. Dairy Sci.*, 59: 912-918.

GROTHE, P.O. Holstein-Friesians in der BRD und anderen europäischen Ländern. *Tierzüchter*, 24:
1972 493-494.

GROTHE, P.O. Weitere Ergebnisse und Erfahrungen aus dem Holstein-Friesian Einsatz in der Deutschen
1974 Schwarzbunt-zucht. *Tierzüchter*, 26: 17-18.

HAIGER, A. *Results of a crossing experiment (Braunvieh × Brown Swiss, Braunvieh × Holstein-*
1976 *Friesian)*. 27th Annual Meeting of the European Association for Animal Production, Zürich.

HUTH, F.W. Untersuchungen über die Kreuzung vos Holstein Friesian Vatertieren mit Kühen der Rasse
1972 Deutsche Schwarzbunte. *Tierzüchter*, 24: 400-401.

HUTH, F.W. Auf die Waage Enttäuschen Schwarzbunte nicht. *Deutsche Schwarzbunte*, 4: 8-10.
1981

JANKOWSKI, W. *Wplyw poziomu zywienia na efektywnosc uzytkowania mlecznego krów rasy cb i*
1985 *mieszanców po buhajach holsztynsko-fryzyjskich*. Ossolineum-Wroclaw-Warszawa.

JASIOROWSKI, H. and STOLZMAN, M. Przeprowadzone w Polsce miedzynarodowe badania nad
1974 porównaniem cech uzytkowych róznych odmian bydla czarno-bialego. *Prz. hod.*, 22: 22-25.

JASIOROWSKI, H., REKLEWSKI, Z. and STOLZMAN, M. Testing of different strains of Friesian cattle
1983 in Poland. I. Milk performance of F_1 paternal Friesian strain crosses under intensive feeding conditions. *Livest. Prod. Sci.*, 10: 109-122.

KACZMAREK, A., APOLINARSKI, E., ROSOCHOWICZ, L., DORYNEK, Z., WOLOSZYNSKI, W. and
1976 KLAUDEL, T. Bydlo holsztynsko-fryzyjskie importowane ze Standów Zjednochzonych. II. *Prz. Hod.*, 19: 12-14.

KACZMAREK, A., ROSOCHOWICZ, L., DORYNEK, Z. and WOLOSZYNSKI, W. *Acclimatization and*
1979 *role of Holstein-Friesian cattle imported to Poland.* Symposium on the Utilization of the Holstein-Friesian Breed for the Improvement of Cattle in Socialist Countries, Warsaw. pp 28-65.

KINGHORN, B. Genetic effects in crossbreeding. I. Models of merit. *Z. Tierz. Züchtungsbiol.*, 99: 59-68.
1982

KONCAR, L., VIDOVIC, J. and DOJII, R. The effect of crossbreeding Yugoslav Pied with Holstein-Friesian cattle on the development of female progeny. *Savremena Poljoprivreda,* 25: 5-10.
1977

KRÄUSSLICH, H. Einfluss der Holstein-Friesian Anpaarung auf die Erstlaktationsleistungen bei Verschiedenen Umweltbedingungen. *Tierzüchter,* 13: 513-515.
1975

LAMB, R.C., WALTERS, J.L., ANDERSON, M.J., PLOWMAN, R.D., MICKELSEN, C.H. and MILLER, R.H. Effects of sire and interaction of sire with ration on efficiency of feed utilization by Holsteins. *J. Dairy Sci.*, 60: 1755-1767.
1977

LANGHOLZ, H.J. *Aspects of generalizing results from breed comparison experiments with special regard to the Danish crossbreeding project.* 27th Annual Meeting of the European Association for Animal Production, Zürich.
1976

LANGHOLZ, H.J. and PABST, W. *Stratified experiment on commercial crossbreeding for beef in dairy herds in the Federal Republic of Germany.* Seminar on Crossbreeding Experiments and Strategy of Beef Utilization to Increase Beef Production. Verden, EUR. 5492e, 317-343.
1976

LANGHOLZ, H.J. and JONGELING, C. Stand und Entiwcklungstendenzen der Rinderzucht Norddeutschlands. *Züchtungskunde,* 49: 3-20.
1977

LANGHOLZ, H.J. Effect of breed and interaction with nutrition. In: *Patterns of growth and development in cattle.* Vol. 2. Current topics in veterinary science. Ed. N. de Boer and J. Martinus Nijhoff, The Hague.
1978

LIBORIUSSEN, T., NEIMANN-SORENSEN, A. and ANDERSEN, B.B. *Genotype-environment interaction in beef production.* EEC Seminar on Crossbreeding Experiments and Strategy of Beef Utilization to Increase Beef Production, Verden. pp 427-441.
1976

LIEBIEDIEV, M.M. Mietody ispolzownija golstino-frizskogo skota. *Ziwotnowodstwo,* 12: 15-20.
1975

LONG, C.R., THOMAS, R.C. and PEGUES, S.S. Growth of straightbred and crossbred heifers of five breeds. In: *Beef cattle research in Texas 1974-75.* Texas Agric. Exp. Station, 12-23.
1976

LONG, C.R., STEWART, T.S., CARTWRIGHT, T.C. and BAKER, J.F. Characterization of cattle of a five breed diallel. II. Measures of size, condition and growth in heifers. *J. Anim. Sci.*, 49: 432-437.
1979

LOTTHAMMER, K.H. and VIERLING, K. Fruchtbarkeit und Milchleistung von Deutschen Schwarzbunten Rindern und von Kreuzungen mit Holstein-Friesian. *Tierzüchter,* 28: 11-14.
1976

LOUDA, F., KRATOCHWIL, L. and SMERNA, J. *The results of breeding of the black-pied (Friesian) cattle and its crosses in Czechoslovakia.* Symposium on the Utilization of the Holstein-Friesian Breed for the Improvement of Cattle in Socialist Countries, Warsaw. pp 177-195.
1979

MAO, J.L. and BURNSIDE, E.B. Sire by herd environment interaction for milk production. *J. Dairy Sci.*, 52: 1055-1062.
1969

McDowell, R.D. Factors affecting performance of Holsteins in subtropical regions of Mexico. *J. Dairy*
1976 *Sci.*, 59: 722-729.

McDowell, R.E. and McDaniel, B.T. Interbreed matings in dairy cattle. III. Economic aspects.
1968 *J. Dairy Sci.*, 51: 1649-1658.

Obritzhauser, W., Haiger, A., Steinwender, R., Greimel, H. and Konrad, S. Vergleichsversuch von Braunvieh mit Brown Swiss und Holstein-Friesian Kreuzungen. 3. Kalbinnenaufzucht. *Bodenkultur Wien*, 30 (4): 417-422.
1979

Oldenbroek, J.K. Vergleich nordamerikanischer Holstein-Friesians mit Niederländischen Schwarz- und Rotbunten. *Tierzüchter*, 29: 374-377.
1977

Oldenbroek, J.K. Breed and crossbreeding effects in a crossing experiment between Dutch Friesian and Holstein-Friesian cattle. *Livest. Prod. Sci.*, 7: 235-241.
1980

Oldenbroek, J.K. Holstein-Friesian, Dutch Friesian and Dutch Red and Whites on two complete diets with a different amount of roughage: performance in first lactation. *Livest. Prod. Sci.*, 11: 401-415.
1984

Pasierbski, Z. and Romer, J. Przydatnosc bydla holsztynsko-fryzyjskiego do doskonalenia miejscowego bydla czarno-bialego. *Prz. hod.*, 24: 8-10.
1978

Pawlina, E. *Badania nad wzrostem i mlecznoscia mieszancow (F_1) uzyskanych od krow nczb unasienianych nasieniem buhajow rasy holsztynsko- fryzyjskiej importowanym z USA.* Rozprawa doktorska. IHiTPZ AR. Wroclaw.
1978

Pearson, L. and McDowell, R.E. Crossbreeding of dairy cattle in temperate zones: a review of recent studies. *ABA*, 36: 1-15.
1968

Poczynajlo, S., Wasilewska, B., Kwiatkowski, J. and Czarnecki, E.T. Ocena uzytkowosci mlecznej mieszancow F_1 krajowego bydla rasy ncb z buhajami holsztynsko-fryzyjskimi. *Pr. Mater. zoot.*, 23: 7-16.
1980

Politiek, R.D., Vos, H. and Korzer, S. Comparison of Friesian cattle from different origins. *Z. Tierz. Züchtungsbiol.*, 99: 272-285.
1982

Przybylska, H. *Wplyw systemu wychowu na wzrost i pozniejska uzytkowosc mleczna jalowek o roznym udziale genotypu rasy holsztynsko-fryzyjskiej.* Praca Doktorska, IGiHZ PAN.
1985

Raue, F. HF und Mastengenicht. *Agrar übersicht*, 3: 56-61.
1982

Richardson, D.O., Owen, J.R., Plowman, R.D. and Miles, J.T. Importance of sire × ration interaction in production and feed intake traits of dairy cattle. *J. Dairy Sci.*, 54: 1518-1525.
1971

Rincon, E.J., McDaniel, B.T. and Robinson, O.W. *Additive, heterozygotic and maternal breed effects on milk yields during first lactation.* Proc. 71st Annual Meeting of the American Dairy Science Association, Raleigh.
1976

RUSJAJEV, A.M. *Rost czisliennosti, razwitje i moloczaja produktiwnost importnogo golstino-frizskogo*
1979 *skota w choziajstwach RSFSR.* Symposium on the Utilization of the Holstein-Friesian Breed for the Improvement of Cattle in Socialist Countries, Warsaw. pp 125-235.

SCHÖNMUTH, G., WILKE, A., SEELAND, G., MICHULITZ, H. and SIEBER, W. Ergebnisse der
1980 Milchrindzüchtung unter Kreuzungsbedingungen. *Archiv für Tierzucht,* 23 (4): 227-235.

SEATH, D.M., COOPER, T. and DEATON, O.W. How an all-roughage ration for heifers affects normal
1966 growth and later milk production. *J. Dairy Sci.*, 49: 717 (Abst.).

STOLZMAN, M., JASIOROWSKI, H., REKLEWSKI, Z., ZARNECKI, A. and KALINOWSKA, G. Friesian
1981 cattle in Poland. Preliminary results of testing different strains. *Wld Anim. Rev.*, 38: 9-15.

SZÜCS, E., MONORI, I. and AES, I. *Evaluation of lactation of milking cows by means of discriminance*
1982 *analysis.* 33rd Annual Meeting of the European Association for Animal Production, Leningrad.

TOUCHBERRY, R. *A comparison of the general merits of purebred and crossbred dairy cattle resulting*
1978 *from twenty years and four generations of crossbreeding.* Proceedings of the 19th Annual National Breeding Roundtable, 18: 1-19.

TOUCHBERRY, R. and BERESKIN, B. Crossbreeding dairy cattle. *J. Dairy Sci.*, 49: (I) 287-300; (II)
1966 647-658; (III) 659-667.

TROCCON, J.L. Genotypes pie-noirs du troupeau expérimental de Theix. In: *La vache laitière. Aspects*
1978 *génétiques, alimentaires, pathologiques.* Institut National de la Recherche Zootechnique, Versailles. pp 31-38.

TURTON, J.D. Crossbreeding of dairy cattle — a selective review. *ABA,* 49 (5): 293-300.
1981

VAGI, I. *Rol gostino-frizskoj parody w skotowodstwie Wengrii.* Symposium on the Utilization of the
1979 Holstein-Friesian Breed for the Improvement of Cattle in Socialist Countries, Warsaw. pp 267-277.

VOS, H., POLITIEK, R.D. and ROOY, J. Kruisengsproeven met sperma van Noordamerikanse zwart-
1975 boute stiren in Niederland. 3. De melkproduktie. *Bedrijfsontwikkeling,* 6: 303-308.

WENIGER, J.H., RIECHERS, R., DIETERT, W., ZEDDIES, J. and ENGELKE, F. Beobachtungen und
1968 Betrachtungen über den Einsatz von Holstein-Friesian Bullen in der Zucht des Deutschen Schwarzbunten Rindes. II. *Tierzüchter,* 1: 5-6.

WIEMER, E. *Bestimmungsfaktoren der Schlachterlosen von Milchkühen.* Vortragstagung der DGZ e.v.
1981 und der Gesellschaft für Tierzuchtwissenschaft. Giessen, 24-25 September 1981.

WITT, M., ANDREAE, U., HUTH, F.W., KALLWEIT, E., PFLEIDERER, U.E., RAPPEN, W., SCHUTZ-
1971 BAR, W., WERHAHN, E., RÖSELER, W. and SELHAUSEN, D. Einfluss der Kreuzung von Vatertieren der Holstein-Friesian Rasse mit Deutschen Schwartzbunten

Kühen auf Körperentwicklung, Milchleistung und Muskelbildungsvermögen. Schr. Reihe *Max-Planck-Inst. Tierz. Tiernähr.*, 56 (92): 1-90.

ZALEWSKI, W., KAMIENIECKI, K. and TRAUTMAN, J. Wstepne wyniki badan nad doskonaleniem bydla
1984 czarno-bialego w kierunku mlecznym przy uzyciu holsztyno-fryzow. *Zesz. probl. Post. Nauk rol.*, 267: 25-40.

ZELTIKOV, A. Effektiwnost skrieszcziwanija cziorno-piostrogo skota s golstino-frizskim. *Molocz. Miasn.*
1980 *Skotowod.*, 2: 36-37.

5. A review of crossbreeding, particularly with Black and White Friesian cattle, in developing countries

5.1 ASIA AND THE PACIFIC

In this section we will deal only with the developing areas of this large continent thus excluding the Asian part of the Soviet Union and Japan. Climatically the region is very diversified with different agro-ecological regions.

As can be seen from the map (Figure 5.1) a large part of the Asian and Pacific continent is occupied by arid and semi-arid areas and by tropical and subtropical forests. Temperate and tropical grasslands, while not including the highlands, occupy rather limited territories.

In developing Asia and the Pacific, the total land area is 1 960 million hectares out of which 19.7 percent is occupied by arable land and permanent crops and 22.6 percent by permanent pastures. The human population of this region is almost 2.5 billion which gives a figure of 1.25 persons per each hectare of total land and 3.1 persons per hectare of agricultural land. This is the highest density of human population in the world. The cattle population of this region is 335 million plus 121 million buffaloes. This distribution can be seen from the map (Figure 1.2).

Milk consumption is highly variable between different countries of the Asian and the Pacific region. This phenomenon has a long tradition and dates back to the era of the large human migrations. In general, the people who drink milk are the descendants of Aryans who moved from Mesopotamia to west and north-west Asia and into Europe. Except for the Mongolians, these people do not traditionally drink milk. However, due to increased incomes and urbanization, milk consumption has increased in south-east and east Asia. Japan, Republic of Korea and Taiwan have increased their milk consumption to a relatively high level of 50-70 kg per person per annum. The people of Thailand, Malaysia, Indonesia and the Philippines now consume an average of 10-20 kg of milk per annum but China and the countries of Indo-China still have the lowest consumption of milk per person per annum.

Beef is usually the by-product from draught animals in south-east and east Asia. It has traditionally been provided from old culled oxen and cows for which the prices are lower than for other meat. Beef consumption in the Indian sub-continent is restricted by religious traditions.

Cattle are an important source of draught power in the Asian and the Pacific region and this has to be taken into account in all breeding programmes.

The importation of Friesian and Holstein-Friesian cattle became popular in Asia mainly for milk production. The cattle are kept purebred or used for crossbreeding. Evidence shows that the optimum proportion of Friesian blood is invariably related to the adversity of the existing animal environment. Very diversified systems of cattle husbandry can be recognized in the region. In many areas cattle are kept mainly for draught power and manure, while milk and beef are produced, often for family consumption.

A frequent system of agricultural production in the high rainfall region is labour-intensive, with small-scale crop farming and livestock reared on a subsidiary basis for food and draught purposes. Cattle in this system are fed with crop residues and other cereal by-products and generally, little cultivated fodder or natural or developed grasslands are available. This system predominates in south, east and south-east Asia.

Fig. 5.1 Asia and the Pacific — main agro–vegetation zones

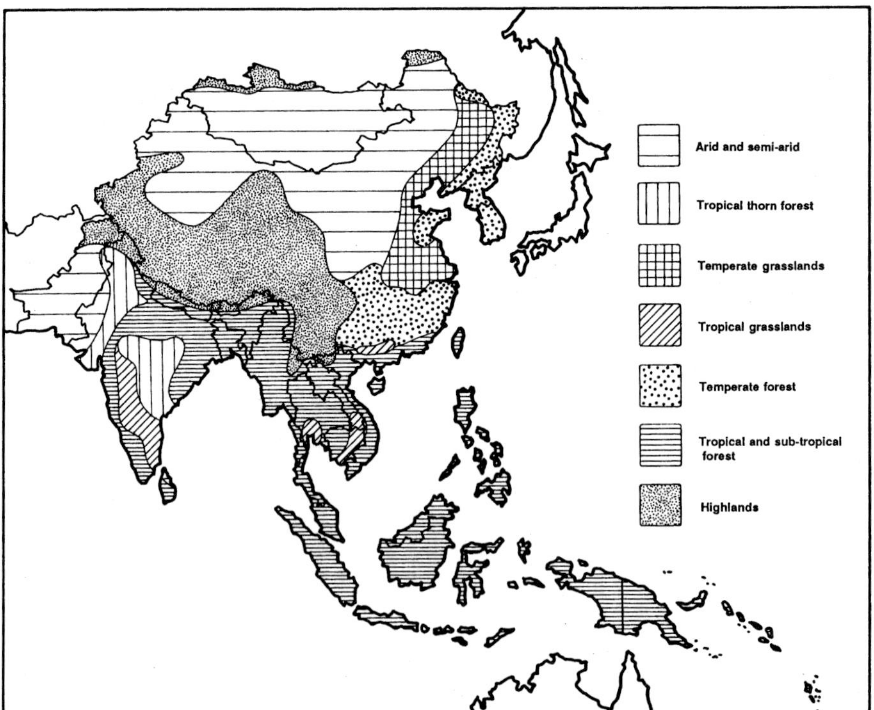

In south-east Asia cattle are found under natural irrigation systems and in south Asia, within artificial irrigation systems. In south Asia, in the semi-humid areas, cattle are kept under the low rainfall cultivation system.

The transhumance system (vertical and horizontal) can be found in west and central Asia and the Himalayas. In this same area, cattle are also kept under the fallow system of agriculture.

In south-east Asia we find cattle in the shifting cultivation system in forest and bush areas.

In Table 5.1, the increase in numbers of cattle, and milk and beef production is shown in the region during the last ten years. For the purpose of better presentation, the region has been divided into the Far East with the following countries: Bangladesh, Bhutan, Brunei Darussalam, Burma, East Timor, Hong Kong, India, Indonesia, Republic of Korea, Laos, Macau, Malaysia, Maldives, Nepal, Pakistan, Philippines, Singapore, Sri Lanka and Thailand; and into the Centrally Planned Economies (CPE) in Asia where the following countries have been included: China, Democratic Kampuchea, Democratic People's Republic of Korea, Mongolia and Viet Nam.

It may be observed that in the Far East the number of cattle and especially dairy cows has increased during the last ten years. The same can be seen for the total beef and milk production. A low share of beef in the total meat production can be

Table 5.1 **Total number of cattle and milk and beef production in the Far East and Asian CPE countries (1975-85)**

	No. of cattle	No. of cows	Total meat	Beef & veal	Total milk	Cows milk
	(million head)		*(million metric tons)*			
FAR EAST						
1975	250.5	28.7	4.1	0.9	36.3	14.4
1980	262.9	33.7	5.2	0.9	43.8	17.5
1985	271.5	38.6	6.2	1.1	55.3	23.9
ASIAN CPEs						
1975	62.5	1.3	10.2	0.3	2.9	1.2
1980	58.3	1.3	15.2	0.3	3.6	1.5
1985	58.5	1.7	22.2	0.5	5.3	2.9

noticed. Less than 50 percent of total milk produced comes from cattle, the rest is produced mainly by buffaloes and partly by sheep and goats. A similar picture can be observed in the Asian CPE countries but the total number of cattle declined during the last ten years by four million.

As shown in Table 5.2, milk production per caput is almost ten times higher in the Far East than in Asian CPE countries, while in meat production per caput the opposite may be observed. However, in both groups of countries the share of beef is very low. Production of beef per head of cattle is low in both sub-regions, but production of milk per cow is much higher in CPE countries. This is possibly due to the fact that in those countries the small number of milk cows are more of the dairy type with exotic blood. In the Far East countries, zebu cows still predominate. It may be seen from Table 5.3 that there is a deficit of milk in both sub-regions of Asia and that this deficit has grown substantially during the last ten years. The value of total net importation of dairy products to the region as a whole in 1985 was more than US$ 800 million. The net importation of beef was on a much smaller scale.

Table 5.4 shows basic data on the cattle population, milk and beef consumption and trade with individual countries in the region. The big differences between countries in cattle productivity and milk consumption should be noted. Most of the countries are highly deficient in dairy products, with the Philippines, Malaysia and Thailand importing more than 80 percent of dairy products consumed in these countries. Strong dairy development programmes in many countries of the region aim to increase milk production and decrease imports. Importation of Friesian cattle and their semen to establish nucleus herds and for crossbreeding serves this purpose.

5.1.1 China

The total cattle population of China numbered approximately 51 million in 1985. Yellow cattle and their crossbreds predominate, numbering perhaps 50 million. They are used for draught and meat production. The number of dairy cattle is relatively small — 900 thousand in 1985 (Table 5.4). Their average milk production was about 2 700 kg per cow per year which should be considered a good achievement. However, average milk consumption per head of human population per year is one of the lowest in the world (2.6 kg) which explains the large effort China is making toward dairy development. The most popular among dairy or dual purpose breeds are the Black and Whites, developed from imported purebreds or crosses between exotic breeds and native yellows. First importations from the USA, UK, Germany, Canada and Japan took place in about 1870. Since 1945, Holstein-Friesians have been imported from USA and Friesians from the Netherlands since 1950. Small

Table 5.2 **Beef and milk production per caput and cattle productivity in the Far East and Asian CPE countries (1975-85)**

	Production per caput *(kg)* of			Productivity per animal *(kg)*		
	Total meat	Beef & veal	Total milk	Cows milk	Beef & veal	Cows milk
FAR EAST						
1975	3.7	0.8	32.7	13.0	3.6	500
1980	4.2	0.8	35.4	14.1	3.4	519
1985	4.5	0.8	40.3	17.4	4.1	619
ASIAN CPEs						
1975	10.2	0.3	2.9	1.2	4.8	923
1980	14.1	0.3	3.3	1.4	5.2	1 154
1985	19.3	0.4	4.6	2.5	8.6	1 706

Table 5.3 **Net trade in meat and milk in the Far East and Asian CPE countries (1975-85)**

	Total meat		Total milk		Beef & veal	
	('000 MT)	*(million US$)*	*('000 MT)*	*(million US$)*	*('000 MT)*	*(million US$)*
FAR EAST						
1975	-93.0	-74.5	-2 115	-317.3	-30.1	-38.7
1980	-79.1	-89.1	-3 826	-730.4	-8.0	-22.0
1985	-118.6	-104.4	-3 638	-639.0	-5.6	-14.5
ASIAN CPEs						
1975	+ 154.0	+ 183.2	-315	-64.1	-10.3	-3.4
1980	+ 213.1	+ 389.2	-601	-138.0	+ 13.7	+ 10.2
1985	+ 260.3	+ 476.5	-817	-192.6	+ 22.3	-3.7

+ Exports - Imports

Table 5.4 **Cattle production in the countries of Asia and the Pacific in 1985** (from FAO, 1985)

Country	Number of cattle	Beef and veal: No. of animals slaught-ered	Beef and veal: Meat production	Carcass weight	No. of dairy cows	Milk yield	Milk produced available per person	Net imports (milk equivalent)
	('000)	('000)	('000 MT)	(kg/animal)	('000)	(kg/animal)	(kg)	(US$ millions)
Afghanistan	3 750	560	68	121	1 220	500	36.9	—
Bangladesh	36 500	2 300	124	56	4 000	205	8.1	64.1
Burma	9 550	660	79	120	2 290	2 445	15.1	6.1
China	51 286	3 200	280	875	900	2 773	2.4	26.1
Hong Kong	3	187	38	204	1.2	2 362	0.5	15.8
India	182 410	1 115	89	80	28 130	658	24.4	52.3
Indonesia	6 859	863	135	157	196	829	1.0	61.5
Korea DPR	1 100	250	37.5	150	31	2 419	3.7	0.1
Korea Rep.	2 652	750	162	216	196	4 929	23.4	7.0
Malaysia	570	105	12	113	47	680	2.1	94.2
Mongolia	2 374	439	66	150	563	396	116.9	4.8
Nepal	7 050	59	5	85	438	500	13.3	1.5
Pakistan	16 549	1 750	235	134	2 710	886	23.9	34.3
Papua New Guinea	123	14	2	150	1.5	267	—	8.2
Philippines	1 900	541	72	133	11	909	0.2	72.5
Sri Lanka	1 750	200	13	67	430	477	12.7	30.5
Thailand	4 800	762	152	200	13	2 692	0.7	78.3
Vanuatu	100	22	3.5	160	10	204	14.1	0.5

numbers of Friesians were imported in recent years from Australia, Canada, Denmark, Federal Republic of Germany and Japan. According to the newly established Black and White Dairy Cattle Association the yield of 270 000 mature cows was 4 461 kg of milk with 3.3-3.4 percent fat content. Black and White cattle are maintained in the southern, central and eastern part of China.

An FAO mission (Poirier and Shah, 1985) which visited farms related to the fourteen target cities (e.g. Chengdu, Chongqing, Fuzhou, Guangzhou, Guilin, Hefei) found cattle to be Black and White and derived from European or North American Holstein-Friesians. Quality of the animals was very heterogeneous (in terms of size, body capacity, feet, legs and especially udders). In December 1983, 3 818 farms were involved in milk production, with 66 311 animals (female calves, heifers and cows). Since 1981, the number of milking cows has increased at an average rate of 10.3 percent each year with 9 of the target cities having an increase of more than 12 percent each year, 4 having an increase of less than 8 percent, and 1 having an increase close to the average. It is considered that an annual growth of 10 to 15 percent could be sustained. The government is encouraging collective and private farms to establish or increase herds. In the year 2000 they plan to have 10 million Black and White dairy cows.

The use of semen from overseas is limited. All females are artificially inseminated with frozen semen (mainly as pellets). At AI stations, semen is mostly collected in winter and spring in order to increase the quality and rate of fertilization. The management of reproduction in herds is very limited and calving rates show room for improvement. The semen of top-proven bulls which state farmers use is only available from Beijing, Nanjing and Shanghai. Use of top-proven bulls from overseas is considered a key to increasing the present genetic level of the population and to creating young males from the best cows.

Increase in milk output from farms between 1981 and 1984 was, on average, 16 percent per annum. This is likely, over this short time period, to be mainly due to improvement in management, genetic value and feeding conditions since the number of milking cows has only increased by 10 percent. Some data on reproductive performance and milk yield in state and cooperative farms are presented in Table 5.5.

Mastitis is an important problem with an incidence of about 20 percent (this reflects cases with acute mastitis — cows with one to two lost quarters); thus an improved hygiene programme is considered essential (Poirier and Shah, 1985). Feet and leg problems were observed on all farms and are a characteristic of intensive stall-feeding systems. Digestive disorders are very common especially on farms feeding animals with rice straw of low nutritive value and high levels of concentrates fed in a very powdery form. Thus butterfat is low (3.0 to 3.2 percent).

China. This woman tends one of the few pilot Holstein herds in China. The herd, on a People's Commune in northern Hobei, is in the background

China. Feeding cows wich heve been crossbred from the Dutch Friesian

Among calves, pneumonia, arthritis and diarrhoea are common, often as a result of inadequate housing (no bedding straw, dampness and draughts). Other disorders of cows include, principally, retained placenta, metritis and reproduction problems (Poirier and Shah, 1985).

A good example of Black and White cattle production in China has been given by Han and Hong (1981) for Jilin Province. The population of Black and White in Jilin Province, was 13 103 in 1980, mostly in the Changchung area, 17.4 percent being on state farms and 80 percent on collective farms. Average annual milk yield was 3 294 kg (4 675 kg for state farms and 2 989 kg for collective farms) but almost half the lactating cows yielded more than 6 000 kg. Milk fat, protein and lactose percentages were 3.1-3.7, 2.6-3.4 and 4.2-4.6 respectively. Udder index and milk flow rate for 16 cows on a state farm were 41.8 percent (39.3-46.8 percent) and 1.89 kg/min (1.65-2.18 kg/min) respectively. Carcass weight of unfinished cull cows was 325.6 kg, lean meat, bone and fat percentages being 74.6, 23.4 and 2.0 percent respectively. Fertility was not considered satisfactory. On state farms and collective farms respectively, calving interval averaged 425 and 461 days and age at culling 7.7 and 8.5 years.

Table 5.5 **Calving interval, days open and milk yield per cow of Holstein-Friesian cattle in state and collective farms in China** (from Poirier and Shah, 1985)

	Calving Interval *(days)*		Days open		Milk yield per cow in 1984 *(kg)*
	State	**Collective**	**State**	**Collective**	**All Farms**
Hefei	370	390-430	80-100	90-120	4 627
Suzhou	380-420	—	80-120	—	4 032
Wuxi	395	420	110	120	5 277
Hangzhou	390	400	110	120	5 625
Nanchang	380	380	90	90	4 158
Fuzhou	380-420	380-420	80-120	80-120	2 676
Guangzhou	409	540	110	250	3 450
Guilin	380	—	100	—	1 785
Changsha	390	410	100	100	3 676
Chongqing	454	—	168	—	4 169
Chengdu	400-420	—	100-120	—	4 728
Shenyang	380-400	400	80-120	80-120	4 867
Dalian	380-400	380-400	80-115	80-120	5 429
Qingdao	408	402	95	91	4 792

5.1.2 Hong Kong

Cattle breeding in Hong Kong plays a very limited role. There are now about 3 000 cattle with 1 200 dairy cows and average production is close to 2 400 kg per year. This gives only about 0.5 kg of milk per year per person and 99 percent of dairy products is imported (Table 5.4). Despite this fact, the number of cattle in Hong Kong has substantially declined during the last few decades (H.N. Cheng, 1985, personal communication) caused by lack of land.

At present over 90 percent of cattle in Hong Kong are Friesians. The cattle originated from Australia, New Zealand and USA. Seventy-nine percent of all cows are artificially inseminated with semen originating from Canada and USA. A production level of 3 700 kg/head/year was achieved on some farms. The reproductive performance was regarded as fair. However, the heat and humidity of the summer months affect the performance. Disease resistance was also considered to be fair. Subclinical mastitis and footrot are the main problems. Calf survival was 95 percent.

5.1.3 India

In 1985 India had a cattle population of over 180 million, the largest in the world and of which 28 million were dairy cows (Table 5.4). The total milk production in India in 1984-85 was 41.2 million tons. The average availability of cow's milk is 24 kg per person per year but considering that the higher proportion of milk consumption comes from buffaloes, the total availability of milk per person and per year will be much higher. India has 26 breeds of cattle and 7 of buffaloes (Acharya and Bhat, 1984). However, a large proportion of the cattle and buffaloes are nondescript.

The major emphasis in dairy development has been to improve milk production of low producing and nondescript cattle in the areas of large milk consuming markets by introducing temperate dairy genotypes through crossbreeding. These programmes are coupled with the provision of inputs such as compounded concentrate feed, fodder seed, breeding and health advice and milk marketing at remunerative prices.

From 1961 to 1978 India had imported 1 470 bulls and 6 165 females of exotic breeds of which 275 bulls and 1 825 females were of Friesian and Holstein-Friesian strains but most of the others were Jerseys. More recently Friesian has become the most favoured breed among the exotic dairy cattle breeds for crossbreeding. Currently either pregnant heifers or frozen semen of genetically superior bulls are being imported. Twenty three farms of the Holstein-Frieisan breed with a total of 1 270 animals are available for producing Holstein-Friesian bulls which go to the semen freezing network. The majority of these farms have been established under the "Operation Flood" in cooperation with the National Dairy Development Board and with the assistance of the Governments of Australia, Canada, Denmark, New Zealand, Switzerland, UK and USA.

The performance of Holstein-Friesian cattle in India at various institutional farms is given in Table 5.6. It will be seen that the lactation milk yield ranged from 1 909 litres to 3 442 litres, the age of first calving from 29.6 to 35.9 months and calving intervals from 421 to 527 days.

Crossbreeding in cattle involving *Bos taurus* and *Bos indicus* breeds has been in operation both in the organized farms and under field conditions in India since the beginning of the century. A number of reviews on the results of crossbreeding have been published. Most recently Acharya (1986) summarized results of crossbreeding on institutional farms and in villages involving farmers' animals. The crossbreds have been observed invariably to be superior to the indigenous cows and buffaloes in reproduction, milk production and efficiency of milk production as reflected by dry matter intake and cost of milk production. This is also shown by the performance of crosses relative to Indian breeds as summarized in Table 5.7. Further details of crossbreeding results involving a variety of traits were summarized by Katpatal (1982) including a study involving both Friesians and Jerseys, reported by Panda and Sadhu (1983), shown in Table 5.8.

For the organization of genetic improvement programmes, 500 key village blocks and 122 intensive cattle development projects (ICDPs) have been established throughout the country. In addition 45 frozen semen stations have been established allowing a total of 8.38 million inseminations with exotic semen to be made annually. It is intended that the number of inseminations with exotic semen be raised to 12.75 million per annum.

The world's largest cooperative dairy development programme "Operation Flood" will now cover practically all the states and the four union territories. It is intended that the number of farm families covered will rise from 3.48 million to 10 million and the number of primary milk producers' societies from 29 000 to 50 000.

Under farmer management conditions, 305-day lactation milk production averaged 2 286 $\pm$46 kg in Holstein $\times$ local crosses and 1 695 $\pm$35 kg in Jersey $\times$ local crosses in Maharashtra.

In Punjab, the average lactation production was 2 117 kg, 916 kg and 1 416 kg respectively in crossbred cows, local cows and buffaloes in farms predominantly maintaining cows. The average lactation in the Holstein-Friesian halfbreds and 3/4 breds averaged 2 113 $\pm$40 and 2 194 $\pm$44 kg in herds belonging to farmers with land and 1 946 $\pm$56 and 2 001 $\pm$81 kg respectively in the case of herds of landless farmers.

Table 5.6 **Averages standard errors for reproduction and production traits of Holstein-Friesian in India**

Location and origin of cattle	**Age at first calving** *(months)*	**Lactation milk yield** *(kg)*	**Lactation length** *(days)*	**Calving interval** *(months)*	**Reference**
IVRI (Australia and New Zealand)	31.9 $\pm$1.2 (23)	3442 $\pm 244^{1}$ (24)	401 $\pm 18^{1}$ (24)	484 $\pm 22^{1}$ (21)	Bhat *et al.* (1978)
IARI (Australia)	32.5 $\pm$2.8 (15)	3140 $\pm 311^{1}$ (18)	286 $\pm 14^{1}$ (18)	485 $\pm 37^{1}$ (13)	Bhat *et al.* (1978)
H. Pradesh	29.6 $\pm$1.6	3911 $\pm 136^{P}$	347 $\pm 10^{P}$	421 $\pm 15^{P}$	Sadana and Basu (1980)
IVRI (Australian strain)	35.9 $\pm$1.6 (11)	3019 $\pm 117^{1*}$ (11)	—	494 $\pm 27^{1}$ (10)	Kumar and Singh (1979)
IVRI (New Zealand strain)	26.0 $\pm$1.0 (38)	1909 $\pm 117^{1*}$ (37)	—	527 $\pm$21 (30)	Kumar and Singh (1979)
IVRI (Indian born)	35.2 $\pm$1.1 (37)	2296 $\pm 90^{1*}$ (29)	—	514 $\pm$34 (15)	Kumar and Singh (1979)
Hissar (Australian)	31.6 $\pm$0.6 (181)	2889 $\pm 52^{1*}$ (178)	366 $\pm 8^{1}$ (178)	429 $\pm 10^{1}$ (174)	Duc and Taneja (1984 a & b)
Hissar (Australian)	29.9 (119)	2546^{P} (119)	320^{P} (119)	422^{P} (119)	Arora and Sharma (1983)
NDRI Karnal	—	—	—	421 $\pm 15^{P}$ (39)	Sadana and Basu (1983)
Pune	30.1 (304)	—	—	465 (363)	Mangrurkar *et al.* (1985)

Figures within parentheses indicate number of observations
1 = First Lactation * = 305 days or less
P = Pooled over all lactations

Table 5.7 **Reproduction and production characteristics (mean ± s.e.) of Indian cattle and crossbreds** (from NDRI, 1981)

	Age at first calving *(months)*		**Days open**		**Calving interval** *(months)*		**Lactation yield** *(kg)*		**Lactation yield** *(days)*		**Dry period** *(days)*		**Milk yield per day of calving interval** *(kg)*
	mean	**s.e.**	**mean**	**s.e.**	**mean**	**s.e.**	**mean**	**s.e**	**mean**	**s.e.**	**mean**	**s.e.**	
Indian breeds													
Dangi	54	2.0	229	6	17	0.6	615	34	292	11	221	16	1.2
Deogir	47	2.1	181	9	15	0.4	1 423	60	313	8	157	8	3.1
Deoni	53	1.0	163	6	15	0.3	879	23	270	4	172	7	2.0
Gir	47	0.8	160	4	16	0.5	1 403	31	257	4	213	5	2.9
Gaolao	46	0.4	188	4	16	0.3	534	15	295	5	185	7	1.3
Hallikar	45	1.5	298	7	20	0.9	541	6[illegible]	285	10	302	28	0.9
Hariana	59	0.4	166	0	20	0.5	1 136	34	232	4	166	8	1.9
Kangayam	44	0.4	225	6	17	0.3	643	10	212	30	242	6	1.3
Kankrej	47	0.8	212	6	16	0.4	1 850	5[illegible]	351	8	141	9	3.8
Ongole	40	0.4	229	5	17	0.6	613	60	217	8	366	23	1.2
Rathi	40	0.4	256	8	19	0.7	1 931	53	331	4	243	6	3.3
Red Sindhi	42	0.6	146	5	15	0.3	1 605	24	284	2	146	5	3.7
Sahiwal	40	0.2	156	3	15	0.6	1 718	36	283	1	156	3	3.8
Tharparkar	49	0.4	145	8	15	0.8	1 659	53	280	6	145	8	3.7
Nondescript	59	2.5	280	8	19	1.0	534	14	303	6	264	20	0.9
Crossbred cattle (Bos indicus F x Bos taurus M)													
H x F	33	1.1	173	15	15	0.8	3 195	205	340	5	85	15	7.0
H x BS	29	0.9	134	12	14.1	0.6	2 785	163	336	24	115	23	6.6
H x J	33	1.2	134	10	14	0.4	2 868	215	308	4	98	5	7.1
G x J	25	1.8	101	10	13	0.9	2 713	225	324	6	57	6	7.1
G x F	35	0.5	113	11	13	0.3	2 254	97	288	6	105	16	5.7
RS x F	29	0.6	90	10	12	0.4	2 326	94	283	8	84	18	6.3
RS x RD	28	1.2	71	10	12	0.3	2 213	115	267	9	85	17	6.2
RS x J	29	0.9	109	9	14	0.5	1 501	82	305	7	98	12	3.7
R x J	31	1.2	104	11	13	0.5	2 801	96	321	8	60	11	7.4
T x F	34	0.6	119	12	13	0.4	2 600	49	311	17	96	17	6.6
S x F	34	0.8	133	11	14	0.4	2 356	20	294	5	121	12	5.7
S x J	33	0.6	120	15	13	0.6	2 659	29	314	16	92	6	6.7
J x BS	35	5.1	147	77	14	2.9	2 188	607	292	12	141	18	5.0
L x J	31	3.5	218	17	18	0.8	1 151	45	328	4	109	23	3.7

H = Hariana. S = Sahiwal. RS = Red Sindhi, G = Gir. T = Tharparkar. L = Nondescript. R = Rathi. F = Friesian. BS = Brown Swiss. RD = Red Dane and J = Jersey

In Karnataka, crosses of Hallikar, a draught breed, with Holstein-Friesian produced 1 919 kg in comparison with 843 kg in Hallikar.

In Haryana the average lactation milk production was 2 211 kg in Holstein-Friesian crosses compared to 1 033 kg and 674 kg in buffaloes and local cows, respectively.

These figures indicate the superiority of Holstein-Friesian crossbred cows for milk production relative to indigenous cattle and buffaloes under highly varied ecologies and variable management and feeding systems. The potential for milk production in these crossbred cows is much larger than is being exploited under present feeding and management conditions. The Holstein-Friesian crossbreds can yield on an average 4 000 kg per lactation, but present averages are much lower as shown in Tables 5.9 and 5.10. Table 5.9 provides information from a number of farms and shows, in particular, that half-bred Holstein-Friesian x Sahiwal crosses appear to have yielded better in their first lactation than crosses with lower or higher proportions of Holstein-Friesian blood. The same conclusion arises from Table 5.10 based on data from military farms. However, the results of Katpatal (1982) in Table 5.9 for pooled lactations (up to 4) show 3/4-bred cows to advantage. Half-bred crosses were also heavier at various ages than lower or higher grades as seen in Table 5.11.

5.1.4 Indonesia

Indonesia, in proportion to the human population, has only a small number of cattle totalling 6.8 million of which only about 200 000 are dairy cows. Average milk production per cow per year is 840 kg of milk. Milk production per head of human population is very low, about 1 kg per year (Table 5.4).

The dairy cow population is made up almost exclusively of purebred and grade Holstein-Friesian. Breeding of dairy cows is done by AI using imported

Table 5.8 **Milk yield of Holstein-Friesian (HF) and Jersey (J) crossbreds with Hariana and Bengal cattle, India** (from Panda and Sadhu, 1983)

	HF × Hariana	HF × Bengal	J × Hariana	J × Bengal
No. of cows	675	953	624	739
Milk yield (av. 1-6 lact)	2 610	2 183	2 144	1 945

HF vs J: $P \leq 0.01$

Table 5.9 **Least squares means and constants with their standard errors for the fixed effects associated with breed, India (Sahiwal and crossbreds with differing proportions of Holstein-Friesian (HF) blood)** (from Katpatal, 1982)

	(Bhat *et al.*, 1978) from several farms			(Katpatal 1970)
	Age first calving *(days)*	Yield first lactation *(kg)*	First lactation length *(days)*	Yield lactation 1-4 *(kg)*
LS Mean	1 141 ± 6.6	2 006 ± 22.1	287.3 ± 2.3	2 642 ± 18
Sahiwal	67.8 ± 8.4	-394.5 ± 28.5	—	
1/4 HF	7.6 ± 21.4	-159.2 ± 73.96	+1.6 ± 7.6	[1]-265 ± 58
1/2 HF	-93.3 ± 16.2	+187.0 ± 58.7	-5.9 ± 6.0	-239 ± 59
3/4 HF	-21.1 ± 29.9	+90.2 ± 106.0	10.3 ± 10.6	17 ± 44
7/8 HF	10.3 ± 41.3	-36.4 ± 134.5	5.1 ± 13.8	-106 ± 91

[1] 5/16 HF blood

Table 5.10 **Least squares means for first lactation traits in Holstein-zebu crosses at military dairy farms** (from Taneja and Bhat, 1986)

Genetic groups	Age at first calving *(months)*	Lactation yield in 300 days *(kg)*	Lactation length *(days)*	Calving interval *(days)*
3-8 H-5/8 Sh	35 ± 0.4	2 286 ± 68	316 ± 6	424 ± 11
1/2 H-1/2 Sh	33 ± 0.3	2 787 ± 44	326 ± 4	424 ± 7
5/8 H-3/8 Sh	33 ± 0.2	2 654 ± 38	318 ± 4	419 ± 6
3/4 H-1/4 Sh	34 ± 0.3	2 632 ± 45	322 ± 4	439 ± 7
7/8 H-1/8 Sh	34 ± 0.4	2 647 ± 71	388 ± 6	446 ± 11
1/2 H-1/2 Thp	33 ± 0.8	2 760 ± 115	315 ± 11	446 ± 18
1/2 H-1/2 Rs	32 ± 0.6	2 528 ± 94	312 ± 9	392 ± 14
1/2 H-1/2 Gir	32 ± 0.5	2 526 ± 62	313 ± 5	414 ± 11

H, Holstein-Friesian; Sh, Sahiwal; Thp, Tharparkar; Rs, Red Sindhi

Table 5.11 **Least square means of body weight of Sahiwal and Sahiwal-Friesian crossbreds in India at various ages from the data of Taneja (1973)** (Quoted by Katpatal, 1982)

Age	**Body weights**				
(months)	**Sahiwal**	**1/4F**	**1/2F**	**3/4F**	**7/8F**
0	20.4	23.7	24.5	25.7	21.3
3	56.7	62.1	69.6	64.9	21.3
6	107.3	117.6	134.0	120.2	122.3
12	184.1	203.3	220.4	204.9	197.5
24	280.9	297.2	326.8	298.6	292.5

Table 5.12 **Dairy sector development in Indonesia** (from FAO, 1986a)

Year	**Population of dairy cattle** *(1 000 head)*	**Local milk production** *(1 000 tons)*	**Imported milk** *(1 000 tons)*	**Milk consumption** *(1 000 tons)*	**Number of dairy farmers** *(1 000 farms)*
1979	94	72.2	474.2	546.4	52.1
1980	103	78.4	594.3	672.7	55.8
1981	113	85.8	521.1	606.9	59.7
1982	140	116.7	536.0	652.7	64.0
1983	161	142.9	482.4	627.0	66.0
Annual increment	14.5	19.1	1.5	4.2	6.1
1984	169	170.3	540.3	710.6	74.8
1985	208	229.4	517.8	747.2	81.1
1986	254	316.2	468.8	785.0	87.4
1987	293	409.2	418.1	827.3	93.0
1988 (estimate)	329	532.8	336.3	869.1	98.3
Annual increment	18.2	33.1	-11.0	5.2	7.1

and domestic frozen semen from Holstein-Friesian bulls. While semen is provided by the government, the breeding services are either provided by the government or by the dairy cooperatives.

Table 5.12 shows the development of the dairy sector from 1979 with an estimate up to 1988 (FAO, 1986a).

5.1.5 Korea, Republic of

As a result of rapid economic growth in Korea, the rise in income has stimulated a major increase in the consumption of animal protein. The government has given high priority to the development of the Korean livestock industry and dairy cattle have been the most significant element in the livestock sector of the Korean economy. While in 1961 milk production was approximately 1 000 tons, by 1973 over 100 000 tons of milk from domestic production were supplied to 45 milk plants. The dairy cattle population increased from less than 1 000 in 1960 to over 20 000 in 1970 and 20 times that in 1985 (Table 5.13). All the dairy cattle of the country originate from imported stock. The Holstein-Friesian breed accounted for 34 308, Jersey for 573 and Guernsey for 111 of the total number of 36 128 dairy cattle in 1972 (Young Kum Shim, 1982; FAO, 1976a). More details of the Holstein-Friesian population, milk production and importations are shown in Table 5.13. The total number of Holstein-Friesian cattle imported from 1962 to 1985 was 112 680. The principal countries of origin were USA and Canada but animals also came from Australia and New Zealand.

It was reported (FAO, 1977a) that semen production from 12 Holstein-Friesian bulls in Korea in 1976 amounted to 130 000 doses and conception rate of 6 619 Holstein-Friesian cows inseminated in 1975-76 was approximately 73

Table 5.13 **Holstein-Friesian population in the Republic of Korea** (from Park Shin-Ho, 1986, personal communication)

Year	Total number (including replacements)	Total milk produced (received by plants)	Milk production of cows over 2 yr-old	No. of cattle imported
1962	2 406	2 647	2 752	1 776 in 1962-65
1965	6 612	10 685	3 268	
1970	23 624	51 888	3 958	6 841 in 1966-70
1975	85 542	162 926	4 220	15 731 in 1971-75
1980	179 841	457 580	4 532	56 481 in 1976-80
1985	399 000	1 000 000	4 800	31 851 in 1982-85

percent. In 1978, 77 percent of dairy cattle were bred by artificial insemination. The age of first calving in dairy cows was 27-30 months and calving intervals were 409 days for Holstein-Friesian cattle and 443 days for Korean native cows. Calf mortality for Holstein-Friesian was 6.7 percent (Bong Kug Ohh, 1981).

Park Shin-Ho (1986, personal communication) reported that crossbreeding has not been practised in Korea. Nearly all calves born were the results of insemination using frozen semen produced by bulls kept at the AI Centre of the National Livestock Cooperative Federation and the Institute of National Livestock Breeding. However, semen is also imported, at the rate of approximately 20 000 to 30 000 straws per year.

The number of dairy cattle per farm normally is approximately 10, but numbers can range from 5 to 300 head. Virtually all arable land is now devoted to the growing of food grains, resulting in a serious shortage of land for pasture and fodder crops. Most farmers, therefore, feed livestock with by-products of the grain industry and allow animals to graze unimproved hillside or riverbank grass. They purchase commercial feeds which are used heavily due to limited roughage. Treated rice straw is also fed mainly to dry cattle (Park Shin-Ho, 1986, personal communication; Young Kum Shim, 1982). The only major health problem in Korea for dairy cows is mastitis.

5.1.6 Malaysia

Malaysia, with over 15 million people, has only 570 000 cattle with about 50 000 dairy cows which produce on average 680 kg of milk per cow per year (Table 5.4). In spite of large efforts to increase milk production, the country imports about 94 percent of dairy products consumed. A former programme involving purebred Friesians has been discontinued, but the main component of the national dairy breeding programme from the 1960s onward is based on Friesians with F_1 *inter se* mating using selected crossbred sires though other *Bos taurus* breeds were also used (G. Sivasupramaninam, 1985, personal communication). In order to achieve the goals of the Third Malaysia Plan a large-scale importation of crossbred calves was started in 1978 (Sivarajasingam *et al.*, 1982) on a contract basis from Australia and New Zealand. Friesian and Jersey cows were inseminated with Sahiwal semen and the female calves were air-freighted to Malaysia from Australia (6 to 8 months old) and New Zealand (4 to 6 months old). They were kept and reared at the Institute Haiwan, Kluang (COWIconsult, 1980). They were inseminated mostly with imported Friesian semen though some semen from crossbred bulls with 50 percent *Bos taurus* inheritance was also used. The general approach has been in line with accepted ideas and experience for dairying under wet tropical conditions and the introduction of crossbreds with 50 percent *Bos taurus* and 50 percent *Bos indicus* has been successful. Progeny testing schemes and a selection of preferred exotic breeds were recommended — and also particularly from where the bulls or semen of the *Bos indicus* should be taken.

Table 5.14 gives an estimated dairy cattle population in Peninsular Malaysia in 1981. Seventy percent of the total population are Local Indian Dairy (LID) cattle, the crossbreds making up the balance. It is expected that the proportion of Friesian crossbreds will steadily increase in the next few years as most of the Sahiwal × Friesians are heifers (Sivarajasingam *et al.*, 1982).

Table 5.15 provides comparative data on milk yields for cows of various breeds and crosses. The Friesian halfbreds generally produced about 1 600 kg per lactation and the Jersey halfbreds 1 200 kg per lactation (Sivarajasingam *et al.*, 1982). Comparative data for reproductive efficiency in Local Indian Dairy and crossbred cattle in Peninsular Malaysia are given in Table 5.16. As can be seen, the LID and Red Sindhi had late first calving ages (4 years) whereas the Friesian × LID calved about 1.4 years earlier. The number of inseminations per conception, a vital measure of reproductive efficiency, generally seems to be high in Friesian crossbreds relative to LID cows. Certain other characteristics of LID and crossbred cattle in Peninsular Malaysia are given in Table 5.17.

Dairy cattle production in Peninsular Malaysia is still a young industry with a number of gaps to be bridged. The bulk of the animals are reared by smallholders. Under the existing systems, the Local Indian Dairy cattle serve as an excellent genetic resource. With the introduction of *Bos taurus* crossbreds that have higher producing abilities, several problems arise since such animals demand greater care and higher feeding standards. Their response to the hot and humid environment becomes critical since the crosses have a larger heat load to dissipate. The farmers themselves have to be aware of these consequences and adopt appropriate management policies. This awareness has, to date, been slow to come and this, in turn, has resulted in a number of problems such as high mortality, low yields, anoestrous states of

Table 5.14 **Estimated dairy cattle population in Peninsular Malaysia, 1981** (from Sivarajasingam *et al.*, 1982)

Type	**Total**	**Large farms**	**Small holders**	**%**
Local Indian Dairy	72 000	420	71 580	70.0
Sahiwal-Friesian	11 000	3 800	7 200	10.7
Other crosses	19 800	5 800	14 000	19.3

Table 5.15 **Milk yield of local Indian dairy, exotic and crossbred cattle in Peninsular Malaysia** (from Sivarajasingam *et al.*, 1982)

Breed type	**No.**	**Lactation**	**Mean** *(kg)*	**Reference**
LID[1]	594	all (survey)	502	Sivarajasingam *et al.* (1974)
Red Sindhi (RS)	183	five	[2,3] + 835 ± 12	Sivarajasingam (1974)
Friesian (Fr)	209	first	2 171 ± 67	Sivasupramaniam *et al.* (1981)
Friesian (Fr)	119	—	1 342 to 2 06	,,
Grade Friesian	46	all	9.1 daily	Sivarajasingam *et al.* (1974)
50% Friesian x 50% zebu	109	five	[2]1 359 ± 59	Sivarajasingam *et al.* (1975)
50% Friesian x 50% LID	223	first	1 758 ± 72	Sivasupramaniam *et al.* (1981)
75% Friesian x 25% LID	57	first	1 452 ± 103	Sivarajasingam *et al.* (1981)
50% Sahiwal x 50% Friesian	30	first	8.0 daily	Sivarajasingam *et al.* (1982)

[1] LID = Local Indian Dairy

[2] First 120-day milk yield adjusted for age

[3] Includes milk consumed by calf

Table 5.16 **Measures of reproductive efficiency in Local Indian Dairy and crossbred cattle in Peninsular Malaysia** (from Sivarajasingam *et al.*, 1982)

	Breed type	No.	Mean *(days)*	Reference
First calving age	LID[1]	40	1 461 ± 25	Sivarajasingam *et al.* (1974)
	Red Sindhi	53	1 527 ± 30	"
	50% Fr/50% zebu	85	1 005 ± 20	"
	50% Fr/50% LID	58	1 519 ± 61	Sivasupramaniam *et al.* (1981)
	75% Fr/25% LID	58	1 153 ± 43	"
	Friesian	42	906 ± 28	"
Calving interval (days)	LID	227	408 ± 5	Sivarajasingam *et al.* (1974)
	Red Sindhi	156	522 ± 12	"
	50% Fr/50% zebu	74	412 ± 10	"
	LID	113	454 ± 12	Baharin (1977)
	Friesian	88	388 to 530	Sivasupramaniam *et al.* (1981)
Inseminations per conception	Red Sindhi	151.2	2.17 ± 0.14	Sivarajasingam *et al.* (1974)
	50% Friesian/ 50% zebu	132	2.33 ± 0.10	"
	LID	34	2.1	Eusof Jamak *et al.* (1981)
	Grade Friesian	47	2.4	"
	Sahiwal Friesian	74	3.9	"
	Sahiwal Friesian	45	2.7	Mohamas Ngah (1981)

[1] LID = Local Indian Dairy

Table 5.17 **Growth characteristics of Local Indian Dairy and crossbred cattle in Peninsular Malaysia** (from Sivarajasingam *et al.*, 1982)

Trait	Breed type	No.	Mean (kg)	Reference
Birth weight	LID[1] male	16	20.2 ± 0.8	Sivarajasingam (1975)
	LID female	13	19.0 ± 0.8	"
	LID	82	[2]21.8 ± 0.8	Sivarajasingam *et al.* (1982)
	50% Friesian/ 50% zebu	137	26.3 ± 0.3	"
	50% Friesian/ 50% LID	1228	[2]23.2 ± 0.6	"
	50% LID/ 50% Friesian	29	[2]22.4 ± 1.3	"
Yearling weight (kg)	LID (male)	14	162.5 ± 5.5	"
	LID (female)	12	142.8 ± 7.2	"
Average daily gain (kg)				
Birth to 6 months	LID (male)	13	0.32 ± 0.03	Sivarajasingam (1975)
	LID (female)	12	0.28 ± 0.02	"
7 to 12 months	LID (male)	11	0.52 ± 0.04	"
	LID (female)	10	0.38 ± 0.04	"

[1] LID = Local Indian Dairy — [2] Adjusted for sex of calf, dam parity and season within year

reproduction, tick infestation and higher disease incidence. However, as most of the farmers are new to milk production the situation is expected to improve (Sivarajasingam *et al.*, 1982).

Various reports have appeared on the dairy industry in Peninsular Malaysia. These include general data on LID and temperate crossbred animal numbers and productivity (ODA, 1982; COWIconsult, 1980). The performance of Sahiwal × *Bos taurus* crossbreds has been reported on more recently (Kumar and Cheah, 1984; Cheah and Kumar, 1984). By the end of 1984, some 22 713 head had been imported from Australia and New Zealand under the Sahiwal × *Bos taurus* programme referred to earlier. Most of these imported crossbred calves (4 to 6 months old) were raised at holding stations and sold as pregnant heifers to smallholders (Kumar and Cheah, 1984). The *Bos taurus* females, used in the exporting countries, were mainly Friesians but Jerseys and other dairy breeds were also used in the programme (Cheah and Kumar, 1984).

5.1.7 Nepal

This country, with a human population of 16 million, has 7 million cattle. The number of dairy cows is low, below 0.5 million. The average milk yield is 500 kg. With 13 kg of milk available per person per year, Nepal is almost self-sufficient and importation of dairy products is small (Table 5.4). The country is endeavouring to increase milk production.

Nepal. Women farmers carying fodder for feeding of cattle, an example of labour-intensive systems of production

Holstein-Friesian semen (obtained through FAO) has been used for crossing with the local breed, alongside semen of other breeds (Ayrshire, Brown Swiss and Jersey). The Holstein-Friesian crosses were found to perform better than the other crosses (S. Shrest, 1985, personal communication). The Holstein-Friesian crosses on the government farm yielded about 1 500 kg in first lactation rising to 2 200 kg in the third, but falling away rapidly to half that amount by the fifth lactation.

On this farm, the average age of first calving was 764 days and the average calving interval 438 days.

5.1.8 Pakistan

The cattle population of Pakistan is 16.5 million with 2.7 million dairy cows. The average yearly milk yield per cow is close to 900 kg (Table 5.4). Important for Pakistan is the buffalo population of 12.7 million which produces in total 7 million tons of milk in comparison with 2.4 million produced by cattle. Together total milk production amounts to about 100 kg per year per person. Pakistan is taking steps to increase milk production.

Sahiwal cattle and indigenous nondescript cattle were used for crossbreeding and imported Holstein-Friesian frozen semen began to be used in 1973-74 (S.K. Shah, 1985, personal communication). Donated semen was provided under various programmes (Australia — 2 000, China — 20 150, France — 4 000, New Zealand — 37 490, Federal Republic of Germany — 77 500, USA — 4 000 doses). From 1973 to 1984, the total amount of donated semen was 151 140 doses. Insemination was carried out on government livestock farms as well as in rural areas.

In addition, several hundred doses of frozen semen from top proven bulls were purchased from the USA to inseminate elite Sahiwal cows. The half-bred Holstein-Friesian × Sahiwal bulls were used for breeding Holstein-Friesian × Sahiwal cows. Interbreeding using half-bred cattle is part of the policy of the National Livestock Development Committee of Pakistan. This policy specifies that exotic genes in the upgraded cattle population should not exceed 50 percent (S.K. Shah, 1985, personal communication).

The frozen semen from Friesian and Holstein-Friesian bulls used over the 1973-84 period on local nondescript cattle was reported to have given distinctly encouraging results for production, reproductive traits and health. No serious losses were noted for the improved cattle from otherwise common diseases (haemorrhagic septicaemia, rinderpest, blackquarter (clostridium group) or foot-and-mouth disease) in comparison with local cattle.

A first batch of 175 Holstein-Friesian cows was imported during the 1977-78 period and located at the government livestock farm near Quetta, Baluchistan.

Observations made seem to indicate that their performance has been quite satisfactory over the last six to seven years (S.K. Shah, 1985, personal communication). Based on this experience, another batch of 750 pregnant Holstein-Friesian heifers was imported in 1983 (Asian Development Bank project). Some 370 of these pregnant cows were distributed among local farmers for milk production, guidance being provided by the Livestock Department for general management, feeding, health cover etc. Observations indicated that cows and young calves were in good condition at the rural farmer level although farmers indicated low conception rates for cows using AI, and tick-borne red-water fever in young calves. A batch of 100 pregnant Friesian heifers was also imported from the Netherlands in the 1982-83 period to the government-managed farm near Peshawar, North West Frontier Province. The aim was also to produce germ plasm for crossbreeding with local nondescript cattle. Up until now, they have been performing well and maintaining satisfactory general body condition and health (S.K. Shah, 1985, personal communication).

As reported by M.J. Khan (1985, personal communication) in the North West Frontier Province, zones have been broadly classified as tropical plain, irrigated areas and dry, hilly, rainfed areas. The indigenous dairy cattle in these areas have a milk yield of around 900 litres per lactation. For commercial improvement, AI services started in 1978 and by August 1985, 77 200 Friesian crossbred calves had been produced. In the period 1978-80 there were importations of some 2 480 doses of Holstein-Friesian semen from New Zealand and 3 040 doses of semen from Black and White cattle in China. In the initial stages, inseminations were performed on the basis of semen availability. In 1979, a semen production unit was established which allowed a regular insemination programme to be maintained (M.J. Khan, 1985, personal communication).

Under a breeding policy, crossbreeding of nondescript cows will be by use of exotic semen from Friesians in the plain and the irrigated areas and Jersey crossbreeding for hilly and rainfed areas to produce 50:50 native: exotic with selection of F_1 sires to produce subsequent generations.

The surplus exotic bulls are given to progressive farmers for crossbreeding purposes (natural service). The production potentials in the Friesian crossbred progeny have been studied in certain key villages. The analysis of data is said to indicate that these crossbreds reach puberty about one year earlier, and yield twice as much milk as local cows. The male progeny, which are heavier at birth, gave some 40 kg more meat at one and a half years of age than local cattle. The Friesian crosses are popular with farmers (M.J. Khan, 1985, personal communication). However, the weakest part of the programme is the lower fertility in crossbred cows (30 to 40 percent) whilst local indigenous cows show 60 percent. The F_1 females exhibit prolonged oestrus (up to three days) and the bulls produce low quality semen.

In 1985, 140 pregnant Friesian heifers were imported from Australia by a private breeder some 20 km from Lahore (Punjab Province) and 50 Friesian (as well as 50 Jersey) cows were imported from Denmark by another private breeder some 280 km west of Lahore (S.K. Shah, 1985, personal communication). The Punjab Province has a hot and humid climate with summer temperatures frequently in the 40°C to 46°C range. The North West Frontier Province, where Friesian herds are also sited, has a similar sub-tropical climate with high temperatures around 44°C and relative humidities ranging from 20 to 100 percent. The winters are cold but temperatures rarely drop below freezing point. Rainfall mostly occurs during the summer months (50 to 60 percent) and ranges from 750 mm to 1 300 mm. There is abundant green roughage for the animals and supplementary concentrates are provided. Health control measures are considered adequate (S.K. Shah, 1985, personal communication). The Province of Baluchistan, where the largest Holstein-Friesian herds are to be found, experiences very cold winters (with minimum temperatures ranging from −2.5°C to −12.5°C) and appreciably hot summers (with maximum temperatures of 41°C to 44°C). Rainfall is scanty and erratic and usually occurs in the winter (71 mm to 250 mm per annum). The terrain is mainly mountainous and has little natural edible vegetation for large ruminants. Exotic animals are thus fed indoors on fodder and forage (lucerne) cultivated in orchards (green or as hay) and given conventional concentrates. Animals are bred using AI with semen from bulls bred on the farm.

A number of reports provide information from experimental stations and some results on body weights of Friesian-Sahiwal crosses are shown in Table 5.18 for a herd at the Livestock Production Research Institute (LPRI) at Bahadurnagar and Table 5.19 for a Friesian crossbreeding trial with nondescript cattle under conventional rural management in the Sheikhupura district. The cattle at the Institute became heavier particularly at the later ages than those under rural management, but variability of weight (as shown by the coefficient of variation) was also lower in the institute herd.

Crossbred cattle in the LPRI herd (50 and 75 percent Friesian, Jersey or Swedish Red and White with Sahiwal) all previously vaccinated against foot-and-mouth disease, were examined during outbreaks of the disease in 1976 and 1977. All the crosses showed greater losses in body weight and a bigger reduction in milk yield than did the local Sahiwal (Kazimi and Shah, 1981).

The reproductive and productive performance of Holstein-Friesian crosses with the Sahiwal at LPRI is shown in Table 5.20 for three proportions of Holstein-Friesian blood. The comparison has to be treated with some caution since the half-breds are averaged over 5 lactations and the other grades over only 3 or 4. However, since milk yield appeared to increase up to the 3rd lactation and declined thereafter (Haider *et al.*, 1984) it seems appropriate to deduce that the performance of the

F_1 generation was better in most respects than that of lower or higher grades. A similar conclusion may be drawn from Table 5.21 which provides information on crosses of Holstein-Friesian with nondescript cows (first lactation only).

The yields shown were considered quite satisfactory and are seen to be, if anything, better in the rural extension area but lactation lengths were also long. Calving intervals were clearly longer than desired at both locations. A

Table 5.18 **Body weights and age at first calving of Friesian × Sahiwal F_1 crossbreds at LPRI, Bahadurnagar, Pakistan** (from Zaman *et al.*, 1983)

	Females				Males			
Liveweight *(kg)*	**No.**	**Mean**	**s.e.**	**COV**[1] *(%)*	**No.**	**Mean**	**s.e.**	**COV** *(%)*
Birth	144	25	0.2	0.96	141	24	0.3	1.05
3 months	120	65	0.9	1.45	12	76	3.1	4.12
6 months	110	111	1.9	1.69	12	124	3.3	2.63
9 months	109	162	2.8	1.74	12	205	5.0	2.45
12 months	109	213	3.5	1.62	12	266	8.1	3.03
15 months	109	266	2.7	1.02	12	307	8.4	2.75
18 months	109	310	4.2	1.35				
24 months	102	372	3.0	0.81				
at first heat	100	255	2.7					
at first calving	100	370	2.9					
Age at first heat	100	437	5.0					
Age at first calving	100	770	6.3					

[1]COV = Coefficient of variation

Table 5.19 **Body weight, age and weight at maturity and first calving in Friesian × nondescript crossbreds under rural management at Sheikhupura district, Pakistan** (from Zaman *et al.*, 1983)

(Age in days and weight in kg)

Liveweight at	**No.**	**Mean**	**s.e.**	**COV**[1] *(%)*
3 months	41	69	2.2	3.19
6 months	50	108	3.9	3.60
9 months	52	138	5.4	3.92
12 months	40	170	7.3	4.27
15 months	55	222	7.2	3.22
18 months	56	255	7.1	2.79
24 months	59	304	8.0	2.64
at first calving	52	340	7.1	2.10
at first heat	53	215	6.6	3.08
Age at first heat	53	457	16	3.56
Age at first calving	52	888	21	2.42

[1]COV = Coefficient of variation

report of work in the same area of the country initiated in 1974 on progeny testing with imported frozen Holstein-Friesian semen has been published by Shah and Chaudhry (1979).

Studies on the adaptability of small groups of crossbred cows (5 each of Friesian × Sahiwal, Jersey × Sahiwal, and pure Sahiwal) were carried out at LPRI in the summer of 1979 (Shah *et al.*, 1980) and showed that in spite of the heat stress experienced, the F × S crosses in particular showed a highly significant increase in milk yield relative to Sahiwal. Use of 75 percent crosses the following year again showed the crosses to give more milk than Sahiwal in spite of severe heat stress (Shah *et al.*, 1980).

Shah *et al.* (1980) also found that crossbred bullocks (both 50 percent and 75 percent Friesian) in the summers of 1978 and 1980 showed severe heat stress but still performed satisfactorily as draught animals before the plough.

5.1.9 Philippines

This country has only 1.9 million cattle but 2.9 million buffaloes. While cows produce about 10 000 tons of milk per year, 18 000 tons come from buffaloes. In total this gives about 0.5 kg of milk produced per person and year. As a result the country has to import 98 percent of dairy products for its population of 49 million people. Efforts toward establishing its own milk production are being undertaken.

Table 5.20 **Productive and reproductive performance (weighted averages of several lactations) of Holstein-Friesian × Sahiwal crossbreds at three levels of HF blood at LPRI, Bahadurnagar, Pakistan** (from Zaman *et al.*, 1983; Haider *et al.*, 1984)

	Proportion of Holstein-Friesian		
	25%	50%	75%
	mean ± s.e.	mean ± s.e.	mean ± s.e.
No. of lactations	3	4	5
Age at first heat (days)	569 ± 5.6 (32)	474 ± 12.0 (49)	437 ± 5.0 (108)
Weight at first heat (kg)	231 ± 14.4 (32)	230 ± 4.0 (49)	255 ± 2.7 (108)
Age at first calving (days)	1 093 ± 29.3 (29)	904 ± 23.0 (43)	770 ± 6.3 (100)
Weight at first calving (kg)	359 ± 9.4 (29)	362 ± 9.0 (42)	370 ± 2.9 (100)
Total milk yield (kg)	1 418 ± 89.0 (46)	2 181 ± 87.0 (98)	2 816 ± 48.0 (325)
305 days' milk yield (kg)	1 407 ± 89.0 (46)	2 134 ± 83.4 (98)	2 587 ± 43.0 (325)
Lactation length (days)	216 ± 10.8 (46)	280 ± 5.2 (98)	324 ± 5.0 (325)
Dry period (days)	220 ± 18.4 (30)	199 ± 16.4 (72)	151 ± 6.2 (299)
Calving interval (days)	429 ± 17.1 (30)	479 ± 18.2 (71)	479 ± 7.8 (300)
Service period (days)	150 ± 17.25 (29)	197 ± 9.0 (68)	196 ± 7.9 (298)
Gestation period (days)	284 ± 1.36 (57)	282 ± 0.84 (108)	282 ± 0.52 (373)
Services per conception	2.70 ± 0.30 (56)	2.45 ± 0.34 (109)	2.78 ± 0.08 (383)
Fat content (%)[1]	4.3 ± 0.10 (21)	4.1 ± 0.10 (27)	4.23 ± 0.04 (100)
Solids not fat content (%)[1]	7.98 ± 0.10 (21)	7.8 ± 0.10 (27)	8.03 ± 0.05 (100)

Number of observations is given in parentheses

[1]1st lactation only

Table 5.21 **Productive and reproductive performance (1st lactation only) of crosses of Holstein-Friesian × nondescript local cattle, at three levels of HF blood in the rural extension area of Sheikhupura District, Pakistan** (from Zaman *et al.*, 1983; Haider *et al.*, 1984)

	Proportion of Holstein-Friesian		
	25% mean ± s.e.	50% mean ± s.e.	75% mean ± s.e.
No of lactations	1	1	1
Age at first heat (days)	717 ± 40.6 (22)	670 ± 50.0 (29)	456 ± 16.3 (53)
Weight at first heat (kg)	—	—	215 ± 6.6 (53)
Age at first calving (days)	1 090 ± 55.3 (19)	995 ± 61.0 (25)	888 ± 21.5 (52)
Weight at first calving (kg)	—	—	440 ± 7.7 (52)
Milk production per lactation (kg) (estimated)	1 491 ± 184 (16)	2 458 ± 154 (22)	[1]2 695 ± 122 (64)
Lactation length (days)	364 ± 35 (11)	374 ± 25 (12)	400 ± 16.2 (58)
Dry period (days)	149 ± 25 (10)	134 ± 45 (9)	134 ± 15.1 (54)
Calving interval (days)	558 ± 39 (13)	499 ± 42 (9)	550 ± 268 (54)
Service period (days)	243 ± 47 (10)	222 ± 39 (11)	268 ± 268 (54)
Gestation period (days)	280 ± 1.53 (19)	281 ± 0.90 (25)	282 ± 0.90 (54)
Services per conception	2.52 ± 0.47 (19)	2.44 ± 0.35 (25)	2.27 ± 0.18 (56)

Number of observations is given in parentheses
[1]305-day yield

A total of 3 475 Friesians were imported to the Philppines from 1945 to 1985. Countries of origin were the USA (493), New Zealand (2 189), Australia (633), India (106) and Japan (54). From 1972 to 1985, 49 993 doses of semen were imported with countries of origin being the USA (25 793 doses), Netherlands (10 000 doses). New Zealand (10 000 doses) and Canada (1 200) (F.C. Bandong 1986, personal communication). Castillo *et al.* (1979) reported that 114 Friesian cows, imported in 1977, were 18-24 months of age and 3-6 months in-calf. After a quarantine period the cows were distributed to farmers. The milk yield of these Friesian cows was 2 785 kg in the first lactation (in 334 days). In the second lactation the milk yield of 53 cows was 2 781 kg in a lactation length of 255 days. Some farmers allowed the calf to stay with the dam especially if the calf was female. Hence, the milk recording started at an average of about two weeks after calving.

Lopez *et al.* (1985) reported on a cooperative project between the provincial government of Laguna and the University of the Philippines at Los Banõs (UPLB)-Laguna dairy demonstration farm. The aim was to establish a commercial dairy farm in Laguna. The dairy farm became operational in 1978 when 12 Friesian cows from Magnolia farm were transferred and recorded for milk production and reproductive performance. There were also a number of crossbreds (Friesian × Red Sindhi) transferred from the Dairy Training and Research Institute (DTRI). The average milk production from 1978 to 1982 was 6.3 litres/cow/day and 6.9 litres/cow/day in 1983-1984. Holstein-Friesian × Red Sindhi crosses yielded an average 1 963 litres milk in 265 days and some Australian purebred Friesians 1 815 litres in 300 days. Other pure (Magnolia) Holstein-Friesians have still to complete lactations.

Roughage feeding comprised mostly fresh grass *ad libitum,* supplemented with sugarcane tops and rice straw (sprinkled with molasses). Silage was given at the rate of 3-5 kg/day for each lactating cow during dry periods. Salt and mineral supplement were also provided. Concentrates are provided at the rate of 1 kg per 2 kg of milk.

In this study the authors stated that the calving interval of crossbreds (HF × RS) was shorter than purebreds. The (Magnolia) Holstein-Friesian encountered reproductive problems.

Ocampo *et al.* (1979) reported that there was an attempt under the DTRI scheme to introduce dairy-type animals in the "backyard" sector. Six pregnant Friesian grade "culls" were sold to the farmers at meat price. They were managed under the typical backyard system, maximum forage, minimum of concentrate (1 kg/day) irrespective of production level. The only major deviation was the additional attention and twice-a-day milking. The results in this study were better than in DTRI. The milk production/lactation was 1 622 litres compared to 1 360 litres at DTRI.

Philippines. Imported Dutch Friesian cattle being fed in confinement

Philippines. Friesian herd in a coconut plantation, grazing on a cover crop of indigenous legumes

A report by Devkota (1982) on Holstein-Friesian cattle in semi-commercial conditions at Batangas showed that mean 305-day milk production for the first to fourth lactations was 2 953 kg, the age at first calving was 920 days and the calving intervals 426 days. Considering all farms together productive and reproductive performances were considered satisfactory, though on average, all farms were losing financially.

The adaptability and breeding efficiency of imported American Brahman (AB) and Holstein-Friesian (HF) cattle under pasture conditions at the Central Mindanao University (CMU), Musuan, Bukidnon was studied by Casinabe and Boloron (1981). They showed that heat tolerance coefficients and rectal temperatures of these two breeds did not differ significantly. Significant differences in favour of AB when compared to HF were found in the number of cows dying and the number of calves produced and calf mortality.

The performance of Philippine-born Holstein-Friesian and Jersey cattle of a dairy herd (Baguio Dairy Farm) reported by Hermosura and Mordeno (1982) showed average daily milk production and average lactation periods to be Holstein-Friesian: 8.1 kg and 262 days; Jersey: 6.5 kg and 242 days. Percent conception rates were — Holstein-Friesian, 40.6; Jersey, 64.0; average calving intervals (years) — Holstein-Friesian, 1.7; and Jersey, 1.1; calving percentages — Holstein-Friesian, 69.4; Jersey, 92, and the heat tolerance coefficients — Holstein-Friesian, 92.8 and Jersey, 92.4.

5.1.10 Sri Lanka

This country with 16 million people has 1.75 million cattle and 0.95 million buffaloes. Cattle produce about 200 000 tons of milk annually and buffaloes 64 000 tons. The number of dairy cows in Sri Lanka is only 430 000 with an average yield of 477 kg per annum. Almost half of the total dairy products consumed in this country are imported and steps are being taken in Sri Lanka to improve milk production (Table 5.4).

The local Sinhala (Lanka) which constituted about 95 percent of the national herd was a poor producer of milk giving about 440 kg of milk in a 189-day lactation under good management (Jalathge, 1969). Jalathge also reported that in up-country zones, upgrading the local Sinhala to the Friesian or Ayrshire might with advantage have continued almost indefinitely, but that in low and mid-country zones, the evidence available seemed to indicate that half-bred and in some cases 3/4-bred animals were the best producers but in no instance was the 3/4-bred animal superior to the half-bred.

Results with Friesian-Sinhala, Jersey-Sinhala and Sinhala cattle are given in Table 5.22 (Buvanendran and Mahadevan, 1975). Progeny from Friesian crossbreds

Table 5.22 **Production characteristics of Sinhala zebu and crossbreds with Friesian or Jersey at Karagoda-Uyangoda in Sri Lanka** (from Buvanendran and Mahadevan, 1975)*

Breed	1956-66			1968-73		
	Milk yield *(kg)*	Lactation length *(days)*	Calving interval *(days)*	Lactation yield *(kg)*	Lactation length *(days)*	Calving interval *(days)*
Friesian × Sinhala:						
F_1		1 573	327	1 482	387	442
F_2		987(34)	302(34)	981	333	421
F_3		—	—	957(30)	332(30)	438(30)
Jersey × Sinhala:						
F_1		1 215	313	1 076	377	453
F_2		809(34)	273(34)	—	—	
B_1		—	—	948	344	407
Sinhala		570	224	234(24)	186(24)	467(24)

Sources: 1956-66 data, Wijeratne (1970); 1968-73 data, Buvanendran (1975, unpublished)
Note: Means based on more than 50 observations except where shown in parentheses

produced more milk than those of the corresponding Jersey crossbreds. (The choice can be governed by meat price and ecological region.) Interbreeding of F_1 crosses to produce F_2 led to a marked decline in milk production but there was no further decline of any importance to F_3 (Buvanendran and Mahadevan, 1975).

Tilakaratne (1980) reported that the average milk yield of Friesians was 3 078 kg in 303 days. Whilst F_1 crosses of Friesian and Lanka produced 1 573 kg in 327 days with a calving interval of 373 days, corresponding results from F_2 crosses were less good with 987 kg milk in 302 days and a longer calving interval of 448 days but both crossbred types yielded more than the 570 kg of Lanka cows (in 224 days), with a calving interval of 391 days.

Some information has been provided based on research in Friesian reproduction which was carried out in Sri Lanka (R.A. Ranasinghe, 1986, personal communication): age at first calving (months): Friesian — 32 (A.P. & H. Bulletin), Friesian — 40 (Jalathge, VRI report), and F_1 Friesian × Lanka — 39 (Mahadevan, 1970).

The incidence of perinatal mortality from data collected from various sources was 1.12 percent for Friesian × Lanka F_1, 1.74 percent for F_2 and 2.44 percent for F_3.

The number of services per conception was 1.46 (R.A. Ranasinghe, 1986, personal communication).

5.1.11 Thailand

With a human population of 51 million there are only 4.8 million cattle but 6.1 million buffaloes in Thailand. The total production of cows' milk is 38 000 tons plus 7 000 tons of buffalo milk which gives nearly 1 kg of milk per annum per person. As a result 90 percent of the dairy products consumed in the country are imported (Table 5.4).

In spite of the lack of tradition in milk consumption the demand is growing rapidly. There are already 13 000 dairy cows in Thailand with an average milk production of 2 692 kg per annum.

Sri Lanka. Friesians heifers grazing on clover-grass pasture

Artificial insemination of dairy cattle in Thailand started in 1956. Native cows were inseminated with frozen Brown Swiss semen from the USA in a programme to upgrade stock (T. Bhannasire, 1985, personal communication).

In 1961 a scheme for the importation of Friesian bulls was initiated. Twelve bulls were imported from Hong Kong (between 1961 and 1970) and 2 each from Italy (1962) and Australia (1963). Fresh semen was used from 1956 and later frozen semen was introduced in 1979. A national AI centre was established in Pathumthani Province over the period 1975-79 under a Thai-Netherlands AI project. This centre produces deep frozen semen of cattle and buffalo which has been used since 1979 by the Department of Livestock Development (DLD).

The breeding programme has involved DLD responsibility for the use of Holstein-Friesian (HF) bulls. The limit for exotic inheritance was 75 percent. It is planned to stabilize exotic inheritance between 62.5 and 75 percent.

Progeny testing of 62.5 percent and 75 percent upgraded bulls, based on milk yield, has been carried out since 1977 (T. Bhannasire, 1985, personal communication). These crossbred bulls were used to provide semen for deep freezing and then for subsequent insemination to produce progeny from upgraded cows (T. Bhannasire, 1985, personal communication). The pure Holstein-Friesian line was derived from imported frozen semen from proven sires from various countries.

The first importations (1976-79) comprised 23 000 doses from the Netherlands, followed by more than 32 000 doses from the USA, 13 000 from Canada and a further 8 000 doses from several sources including Israel, New Zealand, Switzerland and FAO donations.

Under Thailand conditions, crosses with Holstein-Friesian had higher temperatures which caused significant increases in respiration rates and a fall in blood haemoglobin. Lower environmental temperatures tended to increase red cell numbers, body weight, milk yield and voluntary fodder intake, but a decreased water rate (Prajaksakul, 1981).

Experiments were carried out with crossbreds of different types including 266 HF crossbred cows over the period 1960 to 1985. Heat detection for insemination was made by using a vasectomized bull. Upgraded cows with 75 percent HF level or more showed problems of conception. Calves with more than 75 percent HF blood were weak, difficult to handle, had thick hair and exhibited fast panting, all leading to poorer survival. The problem was usually solved by leaving calves with the cows at pasture. Tick-borne disease appeared between one and six months of age (T. Bhannasire, 1985, personal communication). Milk production results from these various crosses are shown in Table 5.23.

Some 190 milk records of some 181 randomized Black and White crossbreds in the Ratchaburi Province were studied between 1979 and 1980 (T. Bhannasire,

1985, personal communication). In some lactations (305 day) milk yields of between 3 000 to 3 500 kg were achieved (Sekasiddhi *et al.*, 1981). Earlier reports from the same province on 62 crossbred dairy cows (including HF) for the period 1975-77 showed milk yields some 1 000 kg lower (Seksit *et al.*, 1977).

Records of 970 crossbred dairy cows from 160 private herds in the Chiang Mai Province for the period 1973 to 1979 were analysed by Seksit et al. (1980). Conception rate was found to be highest during the months of August, September and October with the highest in October (44.3 percent ±12.7 percent) and lowest in May and November (28.8 percent ±13.8 percent and 27.7 percent ±5.3 percent respectively). This covered Holstein-Friesian as well as Brown Swiss and Red Sindhi crossbred cattle. Seksit *et al.* (1980) also studied 247 crossbred and 27 native

Table 5.23 **Milk production of crossbred cattle at Tabkwang Livestock Breeding Station, Thailand** (T. Bhannasire, 1985, personal communication)

Crossbred	305 day milk yield *(kg)*		100 day milk yield *(kg)*		Lactation fat yield *(kg)*		Fat content *(%)*	Lactation length *(days)*	
	mean	s.e	mean	s.e.	mean	s.e.		mean	s.e.
HRIN	2 246	367	825	29	60	6	3.89	221	18
H (x)	1 943	276	718	122	53	13	3.72	221	15
HIN	1 882	297	706	106	52	8	3.67	218	24
RHN	1 791	139	624	31	47	3	3.62	199	18
R	1 459	165	543	64	39	7	3.56	198	6
IN	1 048	57	450	40	27	0.5	3.54	196	6

R = Pure Red Sindhi
IN = 1/2 Australian Illawara Shorthorn (AIS) X 1/2 Native
RHN = 1/2 Red Sindhi X 1/4 Holstein-Friesian X 1/4 Native
HIN = 1/2 Holstein-Friesian X 1/4 Australian Illawara Shorthorn X 1/4 Native
H (X) = 1/2 Holstein-Friesian X 1/2 (Brown Swiss, AIS, zebu, Native)
HRIN = 1/2 Holstein Friesian X 1/4 Red Sindhi X 1/8 AIS X 1/8 Native

cows for age at first calving and calving interval at Chiang Mai Artificial Insemination Station, Chiang Mai Province. The calving interval of Friesian crosses was shorter than that of Brown Swiss crosses or native cows, and the Friesian crosses were also more fertile.

Another report (quoted by T. Bhannasire, 1985, personal communication) on 18 male and 20 female Holstein-Friesian × Thai local cattle from four sites (Lopburi, Saraburi, Pitsanulok and Ayudhaya) showed an average daily weight gain of just under 490 g/day.

Retrospective studies by Sekasiddhi *et al.* (1981) on 206 crossbred dairy cows in Ratchaburi Province (with the AI centre first established there in 1960) for the period 1978 to 1980 indicated that the 50 percent crossbred cow significantly outperformed other groups in terms of calving rates — Holstein-Friesian crossbred bulls having recently been used in addition to purebred bulls. The number of inseminations per calving was 2.9. The percentage of calving from 1 to 3 inseminations was around 69 to 70 percent but declined to 22-25 percent when more than 3 inseminations were required.

A dairy development programme has been started by the government with a budget of 80 million Baht and 3 130 crossbred dairy cows were imported from New Zealand in 1981 (FAO, 1986a).

5.1.12 Vanuatu

This country, with 142 000 people and 100 000 cattle has 10 000 dairy cows with an average production of 204 kg per annum per cow (Table 5.4).

Small numbers of Black and White cattle were introduced in Vanuatu over many years. They were used for milking cows in missions and expatriate communities but are no longer kept for this purpose and have been absorbed into the national beef herd. There they have been crossed with Charolais, Illawara, Limousin and Hereford among others (K. Bazeley, 1985, personal communication). More recently Friesian cattle have been imported for both the commercial dairy herds situated near the capital, Vila. Heifers and a bull were imported from New Zealand. The first herd in a well-run dairy unit consisted of 50 Illawara and 25 Friesians. These are run as two groups one with a bull and the other served by AI. Though production levels have not been recorded a good Friesian yields 20 litres while the local Illawara produces up to 15 litres daily. Friesian cow lactations often extend to only 4 to 5 months with many cows drying themselves off in the hot months of January and February. Reproductive performance has not equalled the local Illawara herd which is exceptionally fertile (K. Bazeley, 1985, personal communication). Friesian semen for AI has been imported from Noumea, New Caledonia. Conception rates to AI have been around 40 percent which is a local average (though little

AI is used in Vanuatu). The use of Illawara semen instead of Friesian is planned due to the perceived lower heat tolerance and fertility of the pure Friesians. The other herd of 50 pure Friesians has similar problems. The cattle often shelter all day in the shade so that dry matter intake is likely to be reduced at this time.

5.1.13 Other Pacific Islands: Fiji, Western Samoa and Tonga

It is reported that there are some 62 000 dairy cattle of various breeds and ages — of these about 15 400 are Black and White cattle (predominantly Friesians). The tabulation below provides a breakdown showing that breeding cows represent more than 60 percent of the total herd.

Country	Black and White cattle nos.	
	Cows	*Total*
Fiji	8 600	14 000
Western Samoa	200	400
Tonga	450	1 000
Total	9 250	15 400

Other islands which are reported to have a small number of Black and White cattle are Papua New Guinea, Solomon Islands, and New Caledonia (M.Z. Hussein, 1986, personal communication).

5.2 NEAR EAST AND NORTH AFRICA

For ecological reasons, Near East countries will be grouped with North Africa.

Arid and semi-arid areas dominate this region (Figure 5.2). The better areas for agriculture and intensive animal production are along the Mediterranean coast and along the river valleys. This is reflected in the distribution of cattle which was shown in Figure 1.2.

In the Near East and North Africa a substantial increase in cow numbers and total beef and milk production has occurred during the last ten years (Table 5.24). However, due to the increase of the human population, the production of beef and milk per person has increased very little (Table 5.25). As a result, importation of beef and dairy products increased from 1975 to 1985 four and six-fold respectively. Milk production in the region has shown considerable progress during the last two decades. Large numbers of Friesian heifers, mainly from Europe, have been imported. For example, in Saudi Arabia up to 1985, 73 percent of imported cows

Fig. 5.2 Near East and North Africa — main agro–vegetation zones

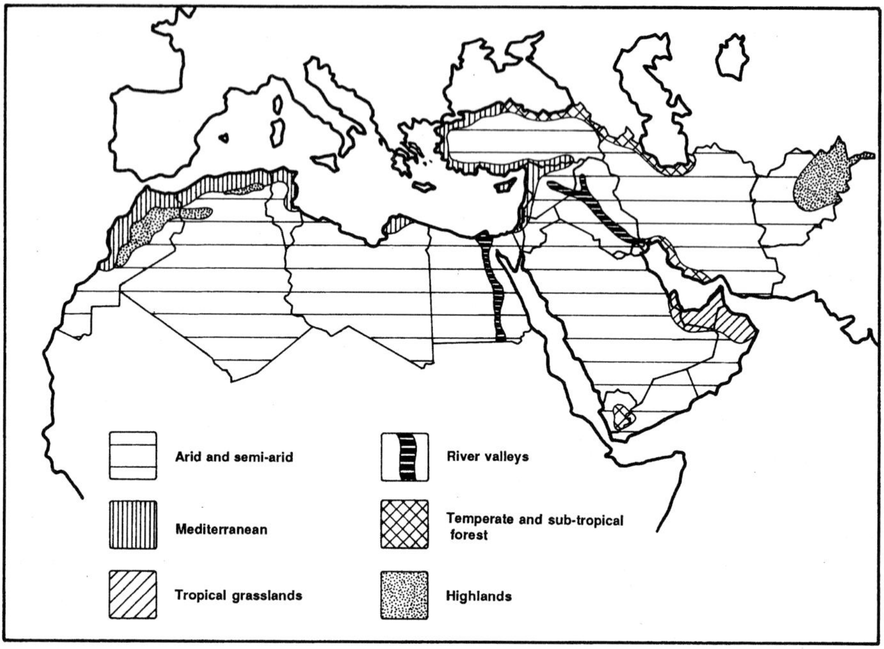

were Holstein-Friesians and 25 percent were Friesians. Black and White cows are considered to be adaptable to the hot and arid conditions in the area (M.Y. Yamoor, 1985, personal communication). AI is used extensively, particularly with Friesian semen for upgrading local cattle.

Milk production is based on intensive fodder production often with heavy investments in buildings and equipment. Future dairy development in the region will depend on finding the right policy for encouraging production, including price incentives and protection against uncontrolled milk powder imports (FAO, 1986b). The Near East and North Africa presently import about 40 percent of the total milk and dairy products consumed resulting in a large net trade deficit in these products (Table 5.26). In Table 5.27, the basic data for cattle production in the region are shown. The largest cattle population was in Iran followed by Morocco. The highest average milk yield per cow was achieved in Cyprus.

Several countries in the region have relatively large numbers of Friesians and their crosses resulting from numerous imports dating from 1942 and in the case of Iraq an artificial insemination programme for upgrading cattle, dating from 1968.

Table 5.24 **Total number of cattle and milk and beef production in the Near East and North Africa (1975-85)**

	No. of cattle	No. of cows	Total meat	Beef & veal	Total milk	Cows milk
	(million)			*(million metric tons)*		
1975	46.8	11.6	2.9	0.8	13.0	7.2
1980	52.0	14.1	3.4	0.9	15.8	8.9
1985	56.7	15.2	4.5	1.2	17.3	10.0
Index	+2.4	+3.1	+4.7	+3.9	+2.9	+3.4

Table 5.25 **Beef and milk production per caput and cattle productivity in the Near East and North Africa (1975-85)**

		Production per caput *(kg)*			Productivity per animal *(kg)*	
	Total meat	Beef & veal	Total milk	Cows milk	Beef & veal	Cows milk
1975	15.5	4.4	70.3	39.0	17.1	612
1980	16.3	4.5	74.5	41.9	17.3	631
1985	18.5	4.8	71.8	41.6	21.2	658

Table 5.26 **Net trade in meat and milk in the Near East and North Africa (1975-85)**

	Total meat		Total milk		Beef & Veal	
	('000 MT)	*(million US$)*	*('000 MT)*	*(million US$)*	*('000 MT)*	*(million US$)*
1975	-236.4	-318.4	-1 423.9	-383.7	-50.9	-73.8
1980	-930.9	-1 572.1	-4 324.3	-1 482.3	-217.4	-414.3
1985	-1 045.1	-1 478.1	-5 614.9	-1 580.5	-384.0	-610.5

+ Exports -Imports

Table 5.27 **Cattle and cattle production in 1985 in countries of North Africa and the Near East** (from FAO, 1985)

Country	**No. of cattle**	**Beef and Veal**		**Carcass weight per animal**	**No. of dairy cows**	**Milk yield per animal**	**Milk produced available per person**	**Net imports (milk equivalent)**
		No. of animals slaughtered	**Meat production**					
	('000)	*('000)*	*('000 MT)*	*(kg)*	*('000)*	*(kg)*	*(kg)*	*(US$ million)*
Algeria	1 450	390	45	115	560	982	28.1	381.0
Cyprus	43	18	4	228	17	3 118	82.2	3.9
Egypt	1 825	960	142	136	955	686	20.5	250.6
Iran	8 200	1 833	165	90	2 300	717	38.1	274.0
Iraq	1 450	310	34	110	380	750	18.9	161.2
Jordan	35	18	2	90	20	1 000	5.7	59.9
Kuwait	90	55	11	200	13	2 500	22.1	60.7
Lebanon	45	105	14	135	37	2 297	33.7	39.1
Libya	200	220	45	205	33	2 030	18.6	104.8
Morocco	3 300	850	120	141	1 300	638	38.3	34.6
Qatar	5	4	—	120	4	1 535	21.6	18.0
Saudi Arabia	550	220	24	110	270	1 259	32.1	287.1
Syria	770	275	30	110	330	1 600	57.1	78.4
Tunisia	600	200	29	147	202	1 317	38.1	40.4
Turkey	17 300	4 200	230	55	5 100	667	69.0	20.0
U.A. Emirates	30	11	3	250	10	495	3.6	67.0

Generally speaking, Friesian herds are maintained successfully in the Near East and North Africa through improvements in local systems of management. These improvements include maintenance of high levels of nutrition and strict control of tick-borne diseases and the use of vaccines to prevent diseases of major importance (e.g. foot-and-mouth and rinderpest) as well as the use of artificial insemination and the use of semen from progeny tested sires.

It has been shown in the region that provided high standards of management are applied, it is possible to maintain exotic breeds of cattle, such as Friesians, at a high level of productivity and fertility even under the hottest and driest conditions (F.I. El-Dessouky, 1985, personal communication).

The region is, on the whole, characterized by the low proportion of the total area of each country classed as arable soil (frequently around 6 percent due to a preponderance of mountain ranges or deserts or both as against over 30 percent of various West European countries). The region has around 15 million cattle and buffaloes and nearly 90 million head of sheep and goats, with over 5 million of these cattle and 30 million of the sheep in three North African countries, Algeria, Tunisia and Morocco (Doumandji, 1982).

As pointed out by Doumandji (1982) cattle breeding is a relatively recent and limited practice, though not marginal, compared with the thousands of years during which the smaller farm animals have been kept. In North Africa the most successful cattle breeds are Black and White, Red and White and notably Montbéliarde, though others are considered to have potential. With a premium on land and the need to irrigate for high green fodder production in most areas outside a narrow coastal belt, dual-purpose cattle are favoured since, in addition to the milk yield, meat prices make the slaughter of young bulls, and culled cows an attractive proposition.

Veterinary protective measures are considered most important in the region when establishing new production units in view of the endemic nature of various serious infectious diseases such as tuberculosis, foot-and-mouth disease, brucellosis, anthrax (North Africa) and rinderpest (Middle East).

Table 5.27 provides an overview of cattle numbers and production in the region.

5.2.1 Algeria

This country has a shortage of arable land. Dual-purpose animals are considered the most appropriate type (Doumandji, 1982). There are nearly 1.5 million cattle in Algeria of which one third are dairy cows with an average production of nearly 1 000 kg of milk. Algeria is a large importer of dairy products (Table 5.27).

In the modernized sector of cattle production, cows can yield about 3 000 kg but this modern sector accounts for only 20 percent of the total cow population (Doumandji, 1982). Black and White cattle play an important part in this modern sector. Crossing of European breeds with local cattle is considered useful to provide adaptability to and protection from extremely hot, long summers and the frequent sirocco winds (dry), sometimes loaded with sand.

The results on the performance of Holstein-Friesian cattle of Canadian origin over a four-year period (1977 to 1980) at the IDEB Livestock Experimental Station, Baba-Ali are shown in Table 5.28. The purpose of the project was to study the introduction of this high yielding type of cattle and their adaptation to the new environment, using a herd of 45 cows. The results obtained were considered encouraging (M.T. Benyoucef, 1986, personal communication).

5.2.2 Cyprus

In the small cattle population of this country (less than 50 000 head) Black and White Friesians predominate. In 1984 the Friesian population numbered 34 641 cows (C.S. Serghiou, 1985, personal communication). No distinction is made between Holstein and the UK or Israeli Friesian. The Friesians are used as a dual-purpose breed for the production of both milk and meat. In 1984, 7 546 AI first services were performed in a population of 15 450 females (mostly Friesian). Semen continues to be imported each year its origin varying from year to year. Since 1978 nearly 90 000 doses have been imported (more than half through the FAO semen donation scheme), Canada being the largest source, followed by Finland, New Zealand (in 1985), the USA, and smaller quantities from Israel and Greece (C.S. Serghiou, 1985, personal communication).

Management of the cattle is based on minimum quantities of roughage and large quantities of concentrates. The climate is typically Mediterranean. The level of productivity is satisfactory although the high use of concentrates causes low fat content. The trend in milk production and fat percentage over the years is shown in Table 5.29.

Reproductive performance is considered satisfactory and calf mortality is less than 5 percent, but Cyprus is not regarded as a high disease risk area and is free of most epidemic and contagious diseases (C.S. Serghiou, 1985, personal communication).

The Friesian cow is very popular with farmers and it is in fact the only exotic breed that has been established in Cyprus. Because of this popularity it is gradually replacing the local cattle population. Simmental, Jersey and Aberdeen Angus breeds have been imported at various times but have not established themselves.

Table 5.28 **Performance of Holstein-Friesian cows of Canadian origin in Algeria (1977-80)** (from M.T. Benyoucef, 1986, personal communication)

	1st lactation[1] n = 14	2nd lactation[1] n = 10	3rd lactation[2] n = 21	4th lactation[2] n = 23
Age at calving (days)	944	1 364	1 796	2 187
Lactation length (days)	318 ± 17	326 ± 24	308 ± 11	310 ± 10
Dry period (days)	—	102	106	8
Calving interval (days)	—	420	432	39
Days open	—	138	150	109
Total milk production (kg)	4 834 ± 552	5 477 ± 841	5 980 ± 1 078	6 420 ± 798
FCM (kg)	4 484	5 222	5 602	6 002
Fat yield (kg)	170	202	214	229
Fat content (%)	3.52	3.69	3.58	3.57

[1] Females born at CNRZ (Baba-Ali)

[2] Females imported from Canada

With the first Friesian cattle imports (mostly from Israel and UK) attempts were made to upgrade local cattle. This procedure was, however, abandoned at an early stage when it was found that the pure Friesian proved to have no problems in its adaptation to the local environment, provided management and feeding were improved.

It is considered that the Friesian breed regardless of strain will now stay in Cyprus and is likely to replace the local cattle, which are used mainly as draught animals. Future plans involve the establishment and operation of local semen production facilities using young bulls. This will most probably involve the selection of dams of young sires which would require performance testing for growth and growth efficiency locally followed by the insemination of these dams with high quality imported semen from progeny tested bulls (C.S. Serghiou, 1985, personal communication).

A number of studies, necessary as a result of the growing Friesian population, have been carried out in Cyprus on the replacement of imported feed, particularly barley (Economides *et al.*, 1982). A series of trials on substitutes included the effect of dried citrus pulp and grape marc on milk yield and composition (Economides, 1974), soybean meal and urea in diets of growing heifers (Georgiades *et al.*, 1979), and sorghum, wheat and maize as a barley substitute in ruminant diets (Economides *et al.*, 1982).

5.2.3 Egypt

During the last 10 years the cattle population in Egypt has fallen from 2.1 to 1.8 million but numbers of buffaloes increased from 2.2 to 2.4 million. Average milk yield per cow is below 700 kg per year. As a result, production of milk per head of the human population is low and imports of dairy products high (Table 5.27).

Importation of Friesian cattle by Egypt started in limited numbers in the 1930s in an attempt to improve milk production (Ragab *et al.*, 1973). Following this,

Table 5.29 **Milk production and fat percentage of Cypriot Friesian cattle** (from C.S. Serghiou, 1985, personal communication)

	1965	1970	1973	1975	1980
305-day milk (kg)	4 075	4 577	5 097	5 033	5 476
Fat (%)	3.53	3.72	2.97	3.22	3.37

the Tahreer province imported substantial numbers of Dutch and Danish Friesian heifers and cows — 1 442 head — from 1954 to 1964. A total of 948 Friesian in-calf heifers and 19 bulls were imported in the period 1959 to 1961 from the Netherlands alone. The present number of Black and White Holstein-Friesians in the country is thought to be in tens of thousands (E.S.E. Galal, 1985, personal communication).

Artificial insemination for state farms is primarily based on home-produced frozen semen from unproven bulls. However, some farms are using imported semen (from Netherlands, Denmark, UK, USA, Canada and Israel). Management of cattle is not always adequate, nor the environment satisfactory for optimum performance.

In the last ten years some relatively large specialized dairy farms have been established which utilize intensive systems. Smaller herds of 15 to 40 cows are also organized near urban centres. Machine milking is replacing hand milking. The production levels are considered relatively low with not very accurate estimates available from commercial Friesian herds. Estimates for state herds are, for a 305-day milk production, 2 500 to 4 000 kg; age at first calving, 27 to 32 months; calving interval, 13 to 16 months; calf mortality, up to 10 percent and adult mortality, 3 percent. Disease resistance is distinctly lower than for local cattle or crossbreds. In some of the large specialized farms, milk production is 5 000 to 5 500 kg, age at first calving 28 months, calving interval 13 to 14 months, calf mortality 2 to 8 percent and adult mortality 1 to 2 percent (E.S.E. Galal, 1985, personal communication).

The annual average milk production of a local cow (305 days) is 700 kg at 4 percent butterfat (Wasfy, 1982). Studies carried out on 1 126 Friesian cows in the Tahreer Province (reclaimed Nile Delta land) over 4 lactations (a total of 2 631 lactation records) shows a yield of 2 804 kg (305 days) with milk yield increasing from the first to the fourth lactation (Ragab *et al.*, 1973). Records of 178 Friesian and 207 crossbred cows with 1 403 lactation records (up to 10 successive lactations) indicated that milk yield increased up to the fifth lactation then declined, whilst the calving interval and service period tended to decrease with age (Abdel-Ghani and Fahmy, 1966).

Some aspects of the reproductive performance of Friesian cattle under semi-arid conditions showed age at first calving: 33.5 months; number of services per conception: 1.9 for heifers and 2.6 for cows; calving interval: 357 days for regular calvers and 524 days for irregular calvers whilst gestation length was 276.3 days for heifers and 275.8 days for cows (Mohamed *et al.*, 1985).

Friesians are becoming very popular with farmers for both milk and beef production and are usually used for grading-up of the local cattle (E.S.E. Galal, 1985, personal communication). However, the buffalo is a strong competitor to Friesians in Egypt, with a butterfat content in its milk in excess of 7 percent and a 1 700 kg milk yield per lactation in addition to its higher tolerance of disease and harsh environment (E.S.E. Galal, 1985, personal communication). Friesian and buffalo herds were compared over a period of ten years (Fahmy *et al.*, 1975). The herds had varying numbers over the period (Friesian from 498 to 623; buffaloes from 775 to 929). The Friesian produced 2 718 kg per year per milking cow and 2 065 kg on a total herd basis whilst the figures for buffaloes were 1 745 kg and 881 kg.

Heat tolerance of Friesians and buffaloes has been studied by Abdel Aziz *et al.* (1979). Kotby *et al.* (1978) found the degree of response to varying environmental temperature to vary with origin of the animals. Friesians imported from New Zealand were found to be the least heat sensitive whilst those from UK were the most heat sensitive.

A study on the life span of nearly 1 000 imported Dutch Friesians in one herd was reported by Fahmy *et al.* (1978). These animals were imported during 1959-61 and the study extended to 1972 by which time the animals were no longer in the herd. Disposal age varied from 3 to 14 years of age. Nearly 80 percent had been eliminated by the age of 6. Half the imported animals were slaughtered because

Table 5.30 **Estimates of average lactation yields of Black and White cows of different origins in Iran** (from Mason, 1985)

Breed type	**Lactation length** *(days)*	**No. of cows**	**Milk yield** *(kg)*	**Fat** *(%)*	**Fat yield** *(kg)*
Iranian Friesian	312	240	4 456	3.8	169.3
Bulgarian Friesian	310	85	4 125	3.97	163.7
German Friesian	319	165	4 432	3.85	170.6
Red Danish	315	180	4 867	4.16	202.4
Danish Friesian	316	383	5 214	3.8	198.1
British Friesian	312	579	5 581	3.61	201.5
Dutch Friesian	312	856	5 805	3.56	206.6
Israeli Friesian	310	1 520	6 515	3.4	221.5
Holstein-Friesian (USA and Canada)	312	208	6 740	3.43	231.2

of brucellosis, 13 percent as a result of tuberculosis and 13 percent due to calving troubles and sterility. The average productive life span ranged from less than one year to as much as 11 years with an average of 2.25 years, and an average lifetime milk production over that period of 4 670 kg (Fahmy *et al.*, 1978).

5.2.4 Iran

Of a total cattle population of 8.2 million the number of cows is 2.3 million. The estimated number of crossbreds is 250 000 and of exotic breeds 165 000. The average milk yield per cow is slightly over 700 kg per year. Importation of dairy products is high (Table 5.27). Importation of Friesian cattle has been going on since the 1950s with large-scale importations starting in the early 1970s, the most recent of which was of 16 000 registered Holsteins from Canada. Smaller numbers and semen are still being imported.

There are various estimates of the present numbers and yields in departmental reports. Most large-scale dairy farms (private as well as those belonging to various institutions) keep herd records. Government breeding farms keep individual records. The official milk recording and bull selection scheme covers only the 16 000 Canadian Holsteins which are mostly in private herds.

The estimated numbers of cattle of different breeds and crossbreds are:

Breed	No.
Friesian	125 000
Brown Swiss	25 000
Red Danish	7 500
Jersey	7 500
Friesian crossbreds	120 000
Brown Swiss crossbreds	100 000
Other crossbreds	30 000

Estimates of lactation milk yields and fat percentage for Black and White cattle of different origins are shown in Table 5.30.

5.2.5 Iraq

There are nearly 1.5 million cattle in Iraq with 380 000 dairy cows. Milk yield per cow is 750 kg per year, and total production accounted for only about one third of the milk and milk products available in 1984 in the country (Table 5.27).

The climatic conditions in Iraq are characterized by a long dry and hot summer and cooler winter. The solar radiation level is very high especially at midday.

There are now approximately 11 000 pure Friesians and 300 000 crosses in Iraq as a result of imports dating from 1942 and an artificial insemination programme and improvement policy since 1968. Attempts have been made to ensure high standards of feeding, management, disease prevention and control, forage production and conservation, breeding and reproduction, milk handling, recording and processing as well as marketing.

There are seven dairy cattle farms with 800 adult cows, each maintained by the State Organization for Animal Production and a further eight farms belonging to the State with a total of 5 600 Friesian cows. The pure Friesians were imported from UK, Netherlands, Federal Republic of Germany, France, Denmark, Canada, USA and Australia. Frozen semen has been used since 1974 and since 1975 semen has been stored in straws. The highest number of inseminations was in 1977 with 151 713, but by 1984, the number of AI services had declined to 57 336 (F.I. El-Dessouky, 1985, personal communication).

Cattle are generally maintained in semi-open sheds with yards whilst in the larger state farms they are kept in closed or semi-closed systems. The feed includes concentrates, wheat straw and a substantial amount of green lucerne or hay. Heifers are mated at 18 months of age (300 to 350 kg). All animals are kept under veterinary supervision, vaccinated against common infectious diseases and tested for tuberculosis and brucellosis. Calves are weaned at 3.5 to 4 months of age (100 to 120 kg). The cow life span in Iraq is four lactations (about seven years), the annual culling rate is 20 percent (mainly due to infertility problems), the mortality rate in adults is considered low (2 to 5 percent) but that of calves is high (nearly 20 percent). The average 305-day milk yield varies with farms from 2 500 kg to 4 000 kg for purebreds and about 2 500 kg for crossbreds. (F.I. El-Dessouky, 1985, personal communication).

Iraq is adopting a grading-up system involving Friesians to improve its local cattle. Such crosses are very popular with breeders and farmers. Although there is an efficient AI network, some breeders prefer to have their cows serviced by their own purebred or crossbred bull. An extensive recording system (dairy herd improvement service) was initiated with FAO assistance in 1966 and discontinued in 1971 because of operational difficulties (F.I. El-Dessouky, 1985, personal communication).

A commendable number of scientific papers have been published on many aspects of the performance of Friesians and their crosses in Iraq, and, to a degree, these results can be taken as representative of the region. Among these studies, Asker *et al.* (1965a) analysed 440 lactation records of pure Holstein-Friesians or Friesian crosses (Holstein-Friesian bulls mated to native cows) with 50, 75, and 87.5 percent Friesian blood (86, 60 and 10 animals respectively). The main factors affecting milk production were found to be age, month of calving and lactation period. Results showed that native cattle were inferior to other groups in most of the economic dairy traits and that the Ayrshire and its crosses, which were also studied, were inferior to the Friesian and its crosses. For the Friesian crosses increasing Friesian blood level decreased yield (Asker *et al.*, 1965b).

Some of the genetic and non-genetic factors associated with milk yield in Friesian and native (Jenubi) cows and their crosses at the Abu-Ghraib farm were studied for the 1969 to 1976 period (Al-Rawi and Said, 1980a), and statistical analysis of 704 lactation records were used to cover seven breeding groups (purebred, 1/2, 3/4, 7/8, 15/16, 31/32 Friesian and indigenous cattle). A subset of 480 lactation records of only purebred Friesians was analysed separately to investigate the effect of country of origin on milk production. Several non-genetic factors influenced milk yield. Month and year of calving effects were thought to be due to fluctuations in temperature, nutrition, disease and management. Breeding groups differed significantly in milk yield. The 3/4 Friesian blood level was superior in milk yield at all stages of lactation. The total average milk yield (LSM) was 3 566 ±106 kg. With an increase in Friesian blood levels beyond 3/4 there was a decrease in milk yield (Al-Rawi and Said, 1980a). In these studies, in contrast to others, no significant difference was found in milk production in Friesians of different origin (UK, Denmark, Netherlands, France, Federal Republic of Germany or native-born) with an overall milk yield (LSM) of 3 211 ±69 kg. Further studies of 217 purebred Friesian cows at the same experimental farm for the period 1969 to 1975 showed differences between strains in age at first calving and lactation length (Table 5.31).

Table 5.31 **Least squares means (± s.e.) for Friesians of different origin for age at first calving, lactation length and first lactation yield in Iraq** (from Al Rawi and Said, 1980b)

Origin	No.	Age at first calving *(month)*	Lactation length *(days)*	Lactation yield *(kg)*
UK	24	29 ± 0.9	312 ± 24	2 816 ± 196
Denmark	19	28 ± 1.1	302 ± 27	2 421 ± 220
Netherlands	4	25 ± 2.3	380 ± 59	3 230 ± 480
France	107	31 ± 0.4	362 ± 11	3 178 ± 93
F.R. Germany	19	28 ± 1.1	372 ± 27	3 073 ± 220
Native born	44	35 ± 0.7	339 ± 18	2 661 ± 145

Talo (1978) examined first lactation records at Abu-Ghraib of Friesian heifers imported from France (176), Denmark (122) and United Kingdom (41) as well as locally raised purebreds (59) and 1/2 to 63/64 grade Friesians (190) for reproductive traits and milk yield. Both Friesians of UK and French origin did better than grades and, in contrast to earlier findings (Ali, 1977 and Eltawil *et al.*, 1977), 15/16 and higher grades were superior to lower grades in total and 305-day milk yields, length of dry period and services per conception.

Al-Ani (1980) estimated heritability and phenotypic and genetic correlations from 1 352 records of Friesians and their crosses for the period 1948 to 1978 from three farms (Abu-Ghraib and Musaiyib farms of the Ministry of Agriculture and the College of Agriculture farm, Baghdad) and showed highly significant ($P < 0.01$) effects of sire and grade of Friesians on milk production. Babona (1981) analysed 1 102 first and second lactation records produced by 773 purebred Friesian cows at three state farms of the Ministry of Agriculture and Agrarian Reform imported as pregnant heifers from the Netherlands (April 7th farm) and Denmark (Alkhaliss and Alwahda farms). He reported that sires had a significant effect ($P < 0.01$) on all traits studied. Herds differed in total milk yield ($P < 0.01$) and lactation length ($P < 0.05$) whilst calving season significantly affected ($P < 0.01$) total milk yield, lactation length, days open and calving interval. Milk yield varied with season of calving being best for autumn (4 353 kg) followed by summer (4 122 kg), spring (4 104 kg) and winter (3 867 kg). Al-Rawi and Said (1980c) analysed sources of variation in the performance of 322 purebred and crossbred Friesian calves born at Abu-Ghraib farm between 1969 and 1976. Prime effects noted were that high-grade (above 15/16) Friesians showed good growth rates although birth weights of calves born in Iraq were, on average, lower (6 to 7 kg) than their contemporaries in the country of origin.

Lactation records (674) obtained between 1942 and 1978 for progeny of imported and native born Friesian bulls (41) at the Abu-Ghraib and Musaiyib farms indicated that the origin of the bulls did not affect the traits of economic importance (Al-Rawi and Al-Ani, 1981).

A series of reports has also appeared dealing with various factors affecting semen characteristics of imported purebred and locally born purebred and crossbred Friesian bulls. Seasonal variation in ejaculate characteristics of Friesian bulls was not found to be significant at Abu-Ghraib AI Centre (established in 1965) although deterioration in semen characteristics was noted in Iraqi born bulls (Dessouky and Juma, 1968) and later confirmed (Juma and Dessouky, 1969) though El-Dessouky *et al.* (1977a) drew attention to adverse environmental effects on reproductive capacity and semen characteristics of imported Friesian bulls at Abu-Ghraib.

In a comparison of three climatic regions in Iraq (southern, central and northern) the cooler northern region is said to favour Friesian semen quality and make that area suitable as a site for semen production (Al-Chalabi *et al.*, 1978).

Crossbred Friesians at Abu-Ghraib, near Baghdad, were found to gain weight more efficiently for beef production than two pure native cattle breeds (El-Dessouky *et al.*, 1973). In feeding experiments at Mosul in northern Iraq with 58 yearling bull calves, 8 Friesians were compared with local breeds (Restaki, Janoubi, Karadi, Sharabi) and water buffalo, and the Friesians and water buffalo were found to gain at a significantly higher rate ($P<0.01$) than the other breed groups (Dahal, 1976).

Dabdoub (1975) reported on the seasonal effect of environment (temperature and humidity) on rectal temperature, respiratory rate and various haematological parameters in 7 purebred and 7 crossbred Friesians and concluded that the crossbred animals were more tolerant. A larger study in central and northern Iraq from 1978 of 384 imported and local pregnant Friesian heifers from various sources (USA, Canada, Denmark and local), distributed over various government farms, showed that animals from the different sources differed in body temperature and respiration rate as indicators of heat stress (Ablahad, 1980). Another study showed that the indigenous and crossbred animals were less affected by heat stress than purebred and high-grade Friesians (Baghdasar, 1980).

In an investigation on 81 Holstein-Friesians imported from Canada and USA, coat clipping and water sprinkling were shown to have a major beneficial effect on daily milk yield (Al-Tassim, 1982). It was also found experimentally (54 heifers) that the use of air coolers and fans with slight feeding adjustment resulted in a marked increase in milk yield and body weight (El-Dessouky *et al.*, 1977b).

Jajo (1984) found, as would be expected, that sweating rates were highest at high temperatures ($P<0.01$) but were not related to overall 305-day milk yields. However, during a 15-day trial with Friesian cows those with the higher milk yields also had the higher sweating rates ($P<0.05$).

In an earlier study at Abu-Ghraib, Kassir and Juma (1968) using artificial insemination records on 108 purebred and crossbred Friesians (January 1962 to September 1967) and 443 birth records (1961 to 1965) found that calf losses in purebreds were twice those in crossbreds. Over 80 percent of the calf losses occurred prior to weaning. Digestive troubles and tick fever were the major causes of calf mortality and twinning and abortion occurred only among the Friesians. The crossbreds showed a lower incidence of still-births.

The occurrence of tick fever (*Theileria annulata*) in purebred Friesians and crossbreds over the period 1961 to 1965 at Abu-Ghraib was reported by Yousif (1969). The cases occurred between March and October, the known tick activity season, and infection was detected in 44 percent of purebred adult Friesians while only 2.8 percent of the crossbred adults were found to be infected. Mortality from theileriasis was 29 percent in purebred and only 0.3 percent in crossbred adults (Yousif, 1969).

Infertility problems in 1 155 native and 372 crossbred cows from data collected weekly from a slaughterhouse (Rasafa) over a year seemed to indicate that percentages of retained corpora lutea and cystic ovaries were lower in native cows (4.13 and 2.80 percent respectively) than in the crosses (7.27 and 5.09 percent) (El-Dessouky *et al.*, 1969).

The incidence of mastitis in purebred Friesian (Danish, French, Israeli and West German origin) and crossbred cattle (× native: 3/4, 7/8, 15/16, 31/32, 63/64 Friesian) was analysed using records for cattle at Abu-Ghraib (Ali, 1977) and the effect of grade was found to be significant.

5.2.6 Jordan

There are only 35 000 cattle in Jordan with 20 000 dairy cows of which average yearly milk yield is 1 000 kg. Importation of milk and milk products account for most of what was available for consumption in 1984 (Table 5.27).

Friesian cattle were introduced in small numbers into Jordan at the beginning of the 1950s. According to Lubbadeh (1985) (as personally communicated by M. Duwayri, 1985) most exotic Friesian cattle in Jordan are of Dutch origin. There were large fluctuations in numbers resulting from importations of heifer calves from Europe and their re-export to neighbouring Arab countries. The number of Friesian cattle increased from 6 450 in 1975 to 10 920 in 1982.

The purebred Friesian population is maintained by these imports or by mating females to Friesian bulls or by artificial insemination using imported frozen Friesian semen mainly from Germany (Lubbadeh, 1985). Cattle breeding is concentrated in three districts: Amman (4 percent), Irbid (31 percent) and the Jordan Valley (17 percent). The cattle are raised mainly in specialized dairy farms primarily in fertile, cold districts (e.g. Irbid) or in dry desert districts (e.g. Dhalil) where they have acclimatized relatively easily although they require special care in view of their susceptibility to local diseases. The cows have reduced production as is usual with temperate breeds in hot, low desert conditions.

The sex distribution in the Friesian population is not optimal. Although the number of bulls has fallen in recent years with a markedly increased use of artificial insemination, the number of bulls still remains high due to small farms retaining one bull for every 5 to 10 cows. The proportion of cows in milk at 67 percent is also lower than the herd optimum of 75 percent (Lubbadeh, 1985).

Table 5.32 provides average weight gains of Friesian heifers from birth to one year of age and slaughter weights and ages. The age at sexual maturity was around 15 to 18 months (250 to 300 kg body weight). The table also shows that calf mortality at Ad-Dhalil station was around 5 percent. Commercial farms tend to have very variable calf losses from none in some years to 30 percent in others; a national average is regarded as 10 to 15 percent for suckling calves (Lubbadeh, 1985).

Table 5.33 provides data on calving intervals and age at calving for Dutch Friesians for the Deir'Ala station (1965 to 1974) and for American Holstein-Friesians at the same station at a later period (1975 to 1982). Although age at first calving was lower for the females of American origin they had longer calving intervals.

Estimated milk production for commercial farms per cow is around 4 000 kg on large farms and 5 000 kg on smaller farms where cows are likely to have more individual care (lactation length 250 to 450 days). The average 305-day milk production for Friesians of Dutch origin was 3 733 kg (Al-Kilani, 1981, quoted by Lubbadeh, 1985). A study by the Directorate of Animal Production and Health of

Table 5.32 **Productive characteristics of Friesian calves in Jordan**[1] (from Lubbadeh, 1985)

	Ad-Dhalil station	**Deir'Ala station**	**Other Breeding Farms**
Weight at birth (kg)	36.1	40	35
Weight at weaning (kg)	94	110	100
Weight at one year of age (kg)	250	300	250
Mortality (%)	5.2		
Daily gain to one year (g)	586	712	589
Slaughter age (months)	6-12	6-12	3-9
Weight at slaughter (kg)	250	300	150-200

[1] Data obtained, in part, from reports and study made by the Directorate of Production and Animal Health, and mostly from personal interviews.

110 farms with 1 837 Friesian cows indicated an average milk production of 4 000 to 4 500 kg (280 to 400-day lactation) for these farms (Lubbadeh, 1985). Records for 20 Friesian heifers at the Ad-Dhalil farm (University of Jordan) indicated an average of 4 220 kg (294 to 373-day lactation) and a 305-day milk yield of 4 167 kg. Their dams at the same station gave 3 000 kg average-indicating higher feed and care levels for the young cows in milk. Higher feed and care levels also provided higher yields in the Jordan Valley (hot region) at the Deir'Ala station for both Dutch Friesians and American Holstein-Friesians as shown in Table 5.34. Although, as might be expected, the milk production from the cows of American origin was higher, it should be noted that the two groups are not contemporary and that changes in feeding and management might contribute to the difference.

Butterfat content in Friesian cows in Jordan is usually in the range of 3.5 to 4 percent depending on the farm and feed used (Lubbadeh, 1985). For the Deir'Ala station (1972) it ranged from 3.2 percent (morning) to 4.66 percent (evening).

In the Friesian herd on the Deir'Ala station an average of 3.4 calvings per cow was obtained. The number of inseminations per conception is reported nationally to range from 1 to 6 and a survey team visiting cooperative farms found an average of 3, whilst a 1975 report for the Deir'Ala station stated that the average for Friesian cows of American origin was 3.2 and that for those of European origin, 1.47. A 1982 report of the Directorate of Production and Animal Health gave national

Table 5.33 **Calving interval and age of calving of Friesians of different origin at Deir 'Ala Station, Jordan** (from Lubbadeh, 1985)

	Dutch origin	American origin	
	Imported	Imported	Born at Deir'Ala
Calving interval (days)			
first to second calving	262 (22)	470 (40)	386 (33)
second to third calving	400 (13)	430 (36)	377 (20)
third to fourth calving	380 (11)	418 (28)	428 (15)
Age at first calving (mths)	36 (30)	27 (45)	26 (43)

Source: Records of Deir'Ala station for the period 1965-74
No. of cows given in parentheses

Table 5.34 **Milk yield of Friesian cows of Dutch and American origin at Deir 'Ala Station, Jordan** (from Lubbadeh, 1985)

Lactation	Number of cows	Lactation length *(days)*	Total milk production *(kg)*	Actual or estimated milk production for 305 days *(kg)*
Dutch Friesians (1965-74)				
First	30	254	2 212	2 429
Second	22	297	3 256	3 326
Third	13	299	3 458	3 660
Fourth	11	280	3 560	3 724
Fifth	6	326	4 553	4 240
Average first to fifth lactation	—	291	3 408	3 476
American origin (1975-82)				
First	86	344	4 170	3 745
Second	71	291	4 059	4 328
Third	55	318	5 254	5 105
Fourth	53	326	5 005	4 938
Fifth	28	321	5 326	5 157
Average first to sixth lactation	—	317	4 948	4 878

Source: Records of Deir' Ala station

rates of 1.6 average (1 to 3 range). It is estimated for about 7 500 adult cows or farms that approximately 31 percent are artificially inseminated and 69 percent served naturally (Lubbadeh, 1985).

Feeding patterns have been considered in some detail by Lubbadeh (1985). Feed is said to represent 70 percent of milk production costs in Jordan due to the high prices of feed (Al-Kilani, 1981, quoted by Lubbadeh, 1985) with many of the feed components now being imported.

5.2.7 Kuwait

There are 90 000 cattle in Kuwait with 13 000 cows yielding an average of 2 500 kg milk per year, but imports of milk and milk products are very substantial (Table 5.27).

A dairy project was launched in 1960 with the importation of Friesians from UK and New Zealand, locally born Friesians also being present in Kuwait (Younis *et al.*, 1976). The latest information according to Bahman (c. 1983) shows the total number of cows in Kuwait to have virtually doubled from 8 435 in 1975 to 13 956 in 1979. The most numerous breed is the Friesian, some 90 percent (12 000) of the total number of cows. Milk cows are mostly located in the Sulaibiyah area 25 km from the city of Kuwait primarily for fresh milk production which is processed by a nearby dairy plant (Bahman, c. 1983). Friesians were imported over an extended period (1960, Dutch; 1962, Australian; 1972, UK; 1978, German). Other breeds were imported in the early 1960s (Jersey, 1960; Red Danish, 1960; Red Sindhi, 1961). Semen for artificial insemination was also imported over an extended period (UK, 1965; USA, 1973; Canadian, 1973; Iraqi, 1980).

The productivity of cows under experimental conditions in Kuwait is shown in Table 5.35. Under similar conditions, there was lower production for other breeds [Jersey: 2 900 kg; Red Danish: 3 230 kg (2 lactations); domestic Red Sindhi: 1 173 kg (3 lactations)]. In another trial, the effects of climatic conditions on small numbers of Friesian cattle of Dutch and UK origin were investigated and milk yields of 4 090 kg (3.8 percent fat) and 3 890 kg (3.4 percent fat) respectively recorded over 4 lactations. In yet a further trial, imported Friesian cattle of Dutch, German and Australian origin were compared. In general average yields of around 4 000 kg were achieved in first lactation with some indication that country of origin of the cattle played a part in determining yield (Bahman, c. 1983).

A report on the Friesian herd of the Kuwait Dairy Company farm with

Table 5.35 **Production of Friesians in Kuwait** (from Bahman, c. 1983)

	Milk yield *(kg)*	**Fat** *(%)*
1st lactation	3 700	3.9
2nd lactation	3 820	4.0
3rd lactation	4 300	3.5
4th lactation	3 780	3.6
Average	3 900	3.8

403 records collected between 1962 and 1966 for both imported cattle (UK: 227; New Zealand: 89) and others born locally (87) showed that the number of services per conception was 2.23 and the source of the animals significantly affected this ($P < 05$) — those from New Zealand required significantly fewer services and it was argued that this was due to those from New Zealand being more adaptable to climatic conditions (Younis *et al.*, 1976). With an average gestation length of 272 days, the month of calving significantly ($P < 0.05$) affected gestation length. The shortest lengths were from conceptions in April, May and June, the hottest months, when the cows were heat stressed (Kotby *et al.*, 1975, quoted by Younis *et al.*, 1976). The average birth weight was 32.5 kg with the lightest calves delivered by New Zealand Friesians and the heaviest by British Friesians, the difference being highly significant ($P < 0.01$) (Younis *et al.*, 1976).

The data obtained by Bahman (c. 1983) through a field survey of farms in the Sulaibiyah area (8 100 cattle with an average of 450 per farmer) showed the average weight at birth to be 25 kg, average weight at one year 200 to 250 kg, average weight at time of sexual (breeding) maturity 300 or 320 kg, age at first calving 27 to 30 months, lactation length 280 to 350 days, average milk yield 3 000 kg, gestation length (9 months ±10 days), percentage of milk fat (3.2 to 3.8 percent), average number of lactations 3 and services per conception 2.5 to 3, with insemination either by natural service or artificial insemination. Annual replacement of animals in the herds ranged from 25 to 30 percent. Mortality rates amounted to 12 to 15 percent of the newly born calves. Bahman (c. 1983) provided data on males showing birth weight to be 30 kg, weaning weight 117 kg, weight at one year 250 kg, daily gain 0.72 kg, age at slaughter 9 to 12 months, and slaughter weight 150 to 250 kg. The mortality rate at birth for males was given as 15 percent on good farms and 70 to 80 percent on inferior ones.

5.2.8 Lebanon

There are 45 000 cattle in Lebanon including 37 000 cows which yielded an average of 2 300 kg of milk annually. Production is supplemented by large amounts of imports (Table 5.27).

The Friesian breed was introduced into the Lebanon some thirty years ago for crossbreeding with local Baladi cattle using artificial insemination (R. Edriss, 1985, personal communication). The Baladi breed is a small animal giving less than 1 500 kg milk per year with a fat content of 5-6 percent.

During the first ten years of breeding (1950 to 1960) many problems occurred with the F_1 (Baladi × Friesian) due to calving problems. However, grading-up later removed these problems and farmers were satisfied with the results. Currently farmers depend solely on imported Friesians (R. Edriss, 1985, personal communication).

In a report from the Institut de Recherches Agronomiques Liban (Choueiri *et al.*, 1966) data available for a Friesian herd derived from cows from the Netherlands over the period 1951 to 1966 were used. The average number of services per conception (622 cows) ranged from 1 to 9 with an average of 1.53. Calving interval (267 cows) ranged from 294 to 390 days. Age at first calving ranged from a minimum of 586 to a maximum of 1 529 days with an average of 848 days. The interval between calving and first post-partum service (267 cows) ranged from 40 to 305 days with an average of 102 days. Monthly distribution of calving (405 normal calvings) indicated that the highest rates were between November and June, with a December peak, whilst the rate between August and October was lowest.

Milk yield for 314 lactations is quoted in Table 5.36 (Choueiri *et al.*, 1966). The high milk production reflects good feeding and management conditions.

5.2.9 Libya

This country has 200 000 cattle with only 33 000 dairy cows. Average annual milk production per cow is slightly over 2 000 kg. Domestic production, however, accounted for only about a fifth of production plus imports of milk and dairy products combined in 1984 (Table 5.27).

Over the period 1971 to 1983 more than 12 000 purebred Friesian cattle were imported by the public sector and more than 21 000 by the private sector

Table 5.36 **Milk yields and lactation lengths of purebred, imported Friesians in the Lebanon** (from Choueiri *et al.*, 1966)

Lactation	Number of lactation records	Milk yield *(kg)*	Lactation length *(days)*
First	99	5 337	305
Second	72	5 210	287
Third	51	6 105	291
Fourth	55	6 515	298

in farming. By 1985 the total had risen to around 50 000. The countries of origin were UK, Canada, Denmark, Netherlands and Federal Republic of Germany. In addition, between 1979 and 1984, some 44 000 doses of semen were imported from the Netherlands, UK and Denmark in 1979 and the Federal Republic of Germany (the major supplier) and New Zealand thereafter (Al Hady Al-Zweik, 1985, personal communication). The cattle are bred in a moderately warm climate and there is a small difference between the north and south in the country as regards temperature and humidity.

Average milk yield for a public sector project involving 20 stations was 3 800 kg in 1975 and 5 700 kg in 1982 in 300 days' lactation. Mortality rates for suckling calves were 12 to 14 percent; for growing calves, 4 to 8 percent; and for adult cattle, 1 to 3 percent.

Although Friesian cattle were found to have inadequate resistance to bovine pleuropneumonia in the region they had proved to be successful in farmers' dairying programmes. No attempts at crossbreeding Friesians with domestic cattle are reported.

5.2.10 Morocco

With 3.3 million cattle including 1.3 million dairy cows, Morocco is the largest beef and milk producing country in North Africa, and comes closer to providing from its own production most of the milk and dairy products consumed than any other country in this region (Table 5.27).

Importations of purebred cattle for livestock improvement consisted of Black and White Friesians (90 percent), Holstein (5 percent), Red and White (3 percent) and Tarentais (2 percent) and numbered 160 000 of which 110 000 were adult cows. By 1984 imported animals constituted 7 percent of the total cattle population.

An official policy to improve milk production was initiated in 1970, and in 1973, formal registration of purebreds was started by the Ministry of Agriculture. Crossing of imported breeds with local ones, e.g. the predominant Brown Atlas and the Blond Oulmès, was used. The Brown Atlas was primarily used for crossing with exotic breeds. Crossbred cattle constituted 15 percent of the total cattle population in 1984 but the majority (78 percent) were local cattle (Benlekhal, 1985). The total cattle population has been declining since 1971.

Crossbreeding was facilitated by the introduction of artificial insemination in 1950. In 1973 frozen semen was introduced. The AI centres produce 500 000 frozen semen doses per year. Embryo transfer was started in 1981 on a small scale.

Table 5.37 **305-day milk yield (mean and s.d.) of different cattle breeds in Morocco official records** (from Benlekhal, 1985)

		Lactation				
Breeds		1	2	3	4	**Total**
Black & White	nos.	1 241	671	497	484	2 893
Friesians	(kg)	3 204 ± 320	3 430 ± 742	3 703 ± 831	4 139 ± 884	3 499 (3.80%)[1]
Holstein	nos.	516	328	131	94	1 069
	(kg)	4 277 ± 727	4 312 ± 859	4 827 ± 859	5 286 ± 922	4 444 (3.40%)
Red & White	nos.	109	140	581	51	358
	(kg)	3 076 ± 511	3 396 ± 403	3 665 ± 645	4 061 ±628	3 437 (3.95%)
Tarentaise	nos.	71	41	22	11	145
	(kg)	2 484 ± 320	2 717 ± 381	3 250 ± 391	3 654 ± 994	2 756 (3.82%)

[1] Average fat percentage

The performance of Moroccan Holstein-Friesian cows in relation to breeding and milk production has been reported (Boujenane and Maty Ba, 1986). In this study 1 827 records obtained from 807 Holstein-Friesian cows raised in 24 herds and located in four agricultural areas were analysed. The following averages were reported on first calving (29.5 months), service period (139.4 days), dry period (76.2 days), 305-day milk yield (3 345 kg), fat (3.71 percent) and fat yield (123.6 kg). The average calving interval was reported as 411.2 days and the average length of lactation was given as 338 days. The author concludes that the crossing of local breeds with Friesians would be beneficial.

Milk yield of cattle of different origin is shown in Table 5.37. Holsteins had the highest yields but not significantly higher than the yields of Black and White Friesians and Red and White European cattle. Crossbreeding of the local cattle with Friesians produced an increase of yield by about 1 000 kg of milk as shown in Table 5.38 on both livestock stations and on ordinary farms.

There is need for protection against brucellosis, tuberculosis and foot-and-mouth disease (Benlekhal, 1985).

Table 5.38 **Milk production (mean and s.d.) of local cattle and crossbreds (local × Black and White Friesians) in Morocco** (from Benlekhal, 1985)

	Lactation length *(days)*	Milk yield *(kg)*	Source
1. Government stations			
Local Brown Atlas	180 ±82	829 ±263	1980 — Moghrane Farm
Local Blond Oulmès	120	829 ±479	1980 — Moghrane Farm
Crosses	180	1800	El Jadida
2. *Smallholder farms*			
Local cattle	—	600-700	Tadla Irrigated Plain — 1975 survey
Local cattle	180-210	620	Tessaout Plain 1976 survey
Crosses	180-240	1 520	Tadla (Khiar) 1976 survey

5.2.11 Qatar

There is one private farm which raises Black and White cattle (M. Motassem, 1985, personal communication). It was established in 1973 through Australian imports and holds 1 800 stock (1 000 milking cows). Artificial insemination with semen from Kuwait was tried in the first year. However, as a result of numerous difficulties, natural mating was used subsequently. Three hundred pregnant heifers were recently imported from the Netherlands. A loose housing system is employed day and night for all stock other than dairy cows which are housed in closed barns fitted with air conditioners used during the hottest period of the day (09.00 to 16.00 hours from May to September) when the temperature reaches 45°C (95 percent relative humidity) during the summer. Calvings take place in the autumn and winter months. Mechanical bucket milking is used. The lack of roughages results in a heavy reliance on concentrate feeds. All male animals are sold for slaughter at 3 to 5 months of age. The average milk yield is 3 200 kg over a 290-day period. Fat content is low at 2.9 percent. Calf weights at birth are 25 to 35 kg.

5.2.12 Saudi Arabia

This country with a vast territory has only just over 0.5 million cattle with 270 000 dairy cows. The average annual milk production per cow is 1 260 kg and importation of milk and milk products is substantial (Table 5.27).

The biggest Black and White cattle dairy farm had 14 500 animals on site in 1984 (Zeinstra, 1984). Construction was started in 1979 and completed in 1982. It is planned to increase the stocking rate to 11 500 milking cows and the appropriate number of young stock. By 1983 some 7 500 American Holstein-Friesian and Dutch Friesian animals had been imported. However, the plan now is that the farm should reach full capacity from its own stock by 1988 following a reorganization of plans (Zeinstra, 1984).

Semen has been imported from USA top selected sires and used at random in the herd. Natural service is used for difficult breeders only. Under a modified plan 2 500 heifers born in the project were due to calve in the first months of 1985. The minimum age and weight of heifers at the first insemination were 15 months and 340 kg respectively (J.S. Zeinstra, 1984, personal communication). Efforts are made towards concentration of calvings in the first five months of the year.

The farm produced in 1983 nearly 38 million litres of milk from 6 600 adult animals with the average yield being 5 900 kg. The variation among cows is high, from 2 000 to 3 000 kg of milk. Some feeding problems have still to be solved as well as the problems of overcoming heat stress.

Calves are born in the dairy production units (holding 2 500 cows, instead of the preferred 1 000, with 120 to 160 cows in a production group) and then transported to the central rearing section on the second day. Mixed colostrum is fed for three days, followed by milk replacers (40 kg per head), concentrates and good lucerne hay from the second week of life. Mortality during the first twelve weeks of age was low at 3.6 percent. It is considered that open hutches for calves provided with a sunshade are superior to a closed building (J.S. Zeinstra, 1984, personal communication).

Major health problems are foot-and-mouth disease, metritis and nephritis. The former (FMD) occurs despite vaccination against four types of the virus. This infection is considered to be due to a lack of protection by vaccination up until now against all the types and subtypes present in the country. American Holstein-Friesians are reported to be more susceptible to the disease than Dutch Friesians. Metritis is found in cows which calve in the hot period and they respond only slowly to treatment. It is suspected that many of the nephritis infections are derived from the uterus due to low production of urine in hot weather which reduces the flushing effect. Kidney infections were observed in 30 percent of cows during the post-partum period in 1983.

Tunisia. Friesian cattle fed with local roughages

The negative energy balance during the first 100 days of lactation commonly found under temperate conditions, with the body reserves making up the balance of requirement to support production, is complicated under heat stress when dry matter intake drops by up to 25 percent. The resulting additional energy deficit is considered to lead to metabolic disorders such as ketosis, acetonaemia and fatty liver syndrome since in 40 percent of all post-mortem examinations in 1983 these diseases were diagnosed. The basic ration is considered insufficient to maintain the animal so the quality of the concentrates is the key factor in production and even survival. The best quality roughage is stored to meet the most severe stress from May to July and an exceptionally energy-rich concentrate is fed to bridge the drop in intake during this period. The standard feeding ration consists of greenchop, hay and concentrate (65 to 70 percent of diet). The animals are fed in groups eight times a day (J.S. Zeinstra, 1984, personal communication).

5.2.13 Syria

Cattle (770 000) and cattle production play a secondary role in Syria, after sheep and goats. According to the Syrian Statistical Abstract (1984), the agricultural census in 1981, showed the total number of cattle to be 767 000 (including 512 000 dairy cows), 13.4 million sheep and 1.2 million goats. The local Shami (Damascus) cattle is of zebu type. It is characterized by good adaptation and production.

Up to 1979, 95 000 purebred, exotic European cattle (Friesian) were available in Syria and 6 000 newly imported purebred animals were distributed in 1978 to the peasants in the most favourable climatic regions. Dairying in Syria is undertaken in both the public and private sectors. Cattle produced 562 000 tons of milk in 1981. Milk production continued to increase by 7 percent each year over the period 1967-1977 (FAO, 1980a).

Khiara state farm, where 200 imported pregnant Friesian heifers were received in 1974, reported losses of 0.5 percent among the imported heifers, 8 percent calf mortality and an 85 percent calving rate. At Ek-Rakka state farm the losses were nearly the same but on other, less well-managed state farms, losses of young calves born from imported pregnant heifers reached a level of 20-25 percent (FAO, 1976b).

De Haen and Langholz (1981) have reported that in smallholder Friesian cattle units, average milk yields were 3 200, 3 370 and 3 440 kg per year under the feeding regimes based on maize/vegetable wastes, grain and legumes, and lucerne or lucerne/green maize. It was concluded that at current prices intensive dairying with Friesian cattle by smallholders was fairly successful and profitable. The authors also quote the following average production levels:

	Nos.	Lactation (days)	Milk yield (kg)	Calving interval (days)
Smallholder Friesian	322	293	3 151 ± 847	380
Smallholder Shami	1 010	284	2 995 ± 975	383
State farm Friesian	390 (per farm)		2 860 (2 080-3803)	463

FAO (1982a) reported 3 037 Shami cows and also more than 3 000 Friesian and crossbred cows in four Farmer Breeders' Associations in Syria. The average milk yield per cow at these Associations ranged from 2 700 to 3 300 kg in 305 days. Calving intervals were on average longer than 12 months and the first calving normally took place at 27-30 months of age. Natural service was generally used for the first time, with AI on subsequent heat. The National Breeding Programme for cattle is based on crossing Shami cattle with Friesian cattle until selected AI bulls of the Shami breed are available that can compete with the Friesian.

Approximately 6 000-7 000 inseminations were carried out per year throughout the country. It was shown, in 1980 data, that 6 308 cows were inseminated and 4 144 (66 percent) cows were pregnant. Notwithstanding, there is still a need to maintain the breeding population of the imported Friesian, hence 200 000 doses of Friesian semen were imported (FAO, 1982a).

Infectious diseases affecting milk production in Syria, reported by FAO (1976b), were brucellosis, mastitis and foot-and-mouth disease. Brucellosis must be regarded as the most important disease confronting the further development of milk production. In some areas, blood tests carried out in the laboratory showed that 30 percent or more of the cows in the herd were infected with brucellosis.

Twenty-five cooperatives, primarily for dairying, owned 2 100 adult cows in 1971 and four more cooperatives were established by distribution of 200 Friesian heifers in calf. Modern milking equipment and artificial insemination are being introduced. Average production of native cattle, apart from the Shami, is low, at around 500 kg/animal/year (Motassem, 1972). These cattle breeding cooperatives are still operating and expanding their membership.

5.2.14 Tunisia

There are 600 000 cattle in Tunisia, including 200 000 dairy cows of which the average annual milk yield is 1 300 kg. About the same quantity of milk and milk products of cattle was imported in 1984 as produced (Table 5.27).

The local breed in Tunisia is of the Iberian type. The milk yield of local cattle at Beja Station of the Institut National de la Recherche Agronomique was 500 kg in first lactation and 840 kg in the second. Crossing with Tarentais doubled the milk yield and crossing with Brown Swiss increased yield 2.6-fold.

Friesian cows numbering 23 000 were either originally imported from the Netherlands or born locally from imported animals mainly on state farms (Hall, 1977). In 1984 the total number of dairy cows was 330 000 of which 50 000 were purebred Friesian or Holstein-Friesians, the rest being of local breeds or crossbreds.

Brucellosis, foot-and-mouth disease, tuberculosis and blackquarter are problems which require preventive health measures (Romdhane, 1982).

The milk yield of Black and White cows in different production sectors is shown in Table 5.39 for the period 1977 to 1982 and that for Holstein-Friesians for the period 1978 to 1982 in Table 5.40 (Brahmia and Ben Mehdi, 1982). The average annual increase in milk production of registered Black and White cows between 1977 and 1983 was 128 kg and for cows which exceeded 305 days of lactation the increase for 305-day lactations was 92.8 (Tunisia, 1983).

Morocco. Black and White cattle drinking water from a pond in the Guelmin region

Table 5.39 **Milk performance for Black & White cows in different sectors in Tunisia in 1983**

		OTD[1]	OEP[2]	CFPA[3]	UCP[4]	State	Private	Total & Average
Milk yield	no.	2 853	161	334	1 425	4 773	950	5 723
per cow (kg)	Av.	4 578	3 936	3 431	3 175	4 057	3 858	4 024
Days in milk	Av.	305	311	323	302	306	300	305
Milk yield (305 days)(kg)	Av.	4 673	4 058	3 596	3 299	4 160	4 074	4 146
Age at 1st	no.	909	22	93	363	1 387	137	1 524
calving (mths)	Av.	28	31	30	31	29	30	29
1st calving over 30 mths	(%)	12	32	39	43	22	31	23
Duration of	no.	1 748	135	194	907	2 984	534	3 518
dry period (days)	Av.	74	74	71	81	76	79	77
Cows dry over 80 days	(%)	29	27	22	36	31	37	32
Cows dry up to 50 days	(%)	21	34	29	27	23	23	23
Calving interval more than 14 mths	(%)	16	22	24	26	20	20	20

[1] OTD — State Lands Office
[2] OEP — Livestock and Pasture Department
[3] CEPA — Agric. Schools
[4] UCP — Agric. Cooperatives

5.2.15 Turkey

Approximately 65 percent of Turkey's milk production comes from cows and buffaloes. The estimated average milk yield is low, about 670 kg per lactation for cows. The total cattle population numbered around 17 million of which 5 million were dairy cows (Table 5.27).

The policy for cattle genetic improvement in Turkey is to establish and upgrade Friesian and Jersey cattle and dual-purpose cattle (for milk and beef) based on the Brown Swiss (FAO, 1976c).

Table 5.40 **Milk production of Holstein-Friesian cows in Tunisia** (from Brahmia and Ben Mehdi, 1982)

	Full lactation			
Year	No. of cows	Days in milk	Milk yield (kg)	Milk fat (%)
1978[1]	122	308	3 380	3.45
1979[2]	126	348	4 623	3.36
1981[3]	210	318	5 101	3.33
1982	241	319	4 988	3.60

[1] The results are solely for animals in the first lactation.
[2] Animals having reached the second lactation.
[3] Herd having reached adult stage (fourth lactation)

The native breeds in Turkey are Native Black, Grey Steppe and South Anatolian Red. It was estimated that in 1975 these breeds accounted for 37 percent, 7 percent and 4 percent respectively of the cattle in Turkey. The milk yield of the Native Black in villages is 500-800 kg/lactation. Approximately 8 percent of this breed, numbering over 5 million (1.8 million milking cows), was mated to Jersey, Brown Swiss, or Friesian bulls. Some 930 000 Brown Swiss crosses and 250 000 Friesian crosses in the country came from the Native Black cows.

Grey Steppe is the biggest of the native breeds. The milk yield of this breed is estimated to be between 1 000 and 1 600 kg/lactation. There were 963 000 of this breed (345 000 milking cows) in 1975. Since 1930 grading of the Brown Swiss has produced the Turkish Brown of which there were (including Brown Swiss) approximately 930 000 in 1975. Government policy is to continue this grading. The South Anatolian Red breed numbering about 580 000 (210 000 milking cows) in 1975, has four varieties: Halep, Kilis, Cukurova and Dörtyol. Halep and Kilis are the more productive, averaging 3 000 and 2 700 kg milk respectively at two state farms.

The South Anatolian breed is said to be well adapted to the hot climate and poor nutritional conditions, and is resistant to local diseases. Experimental crossing of this breed with the Friesian increased milk yield and liveweight in the F_1 but backcrossing to the Friesian produced no further improvement.

Table 5.41 **Numbers and performance of various types of cattle in Turkey under farm conditions in 1975** (from Yalçin, 1977)

	Number of records	Milk yield (kg)	Lactation length (days)	Fat (%)
Native Black (NB)	52	1 457	301	4.6
F_1 (NB × Jersey)	45	2 320	317	4.5
R_1 (NB Backcross to Jersey)	25	2 745	312	4.6
Grey Steppe (GS)	13	1 368	250	—
F_1 (GS × Brown Swiss)	13	1 849	259	—
South Anatolian Red (SAR)	54	1 881	220	3.4
F_1 (SAR × Friesian)	50	2 955	279	3.5
R_1 (F_1 × Friesian)	6	2 956	292	3.6
Friesian Purebred	73	3 376	292	3.6

Yalçin (1977) has extracted information from specific studies on the performance of crossbreds as shown in Table 5.41.

Overall mortality of animals in the national herd was estimated to be 8 percent and calf mortality was close to 10 percent. Veterinary services were on the whole insufficient. Important diseases were foot-and-mouth disease, rinderpest, tick-borne diseases and brucellosis (FAO, 1976c).

FAO (1976c) reported that artificial insemination was free of charge to farmers and that fresh semen was used. When natural service was employed, the calving percentages of indigenous and upgraded cattle were said to be within the ranges of 55-65 and 65-75 percent, respectively.

Forde (1984) reported that the number of inseminations in 1984 was reduced to half the 1983 level and that semen production was approximately 500 000 doses, sufficient to supply only western Turkey. Forde (1984) also reported that estimated pregnancy rate, based on 272 inseminations, was 64 percent in the autumn of 1984.

5.2.16 United Arab Emirates

This country has only 30 000 cattle with 10 000 dairy cows. Nearly all the milk and dairy products available for consumption are imported (Table 5.27).

The basic study on Black and White cattle was that on British Friesians by Ansell (1974). The average yield of 19 UK-born Friesian heifers in a four-year study was 3 252 kg for the 305-day first lactation (Ansell, 1976) whilst the second lactation achieved yields of 4 569 kg (Ansell, 1985).

Suppression of spermatogenesis and early foetal reabsorption were noted during the first year of the study in that no successful conceptions took place between the months of June and October although libido and manifestations of heat were not seriously affected. It was also found that calving in the hot season subjected animals to unacceptable stress conditions. Gestation periods were shorter by approximately ten to fourteen days for summer calvings on average, birth weights of calves were significantly reduced and calves born during the height of the summer were found to be as much as 56 days premature in some cases. The shortened gestation periods adversely affected the dry periods prior to parturition. Thus seasonal calving and close supervision of parturition are considered essential under climatic conditions as extreme as those encountered in the United Arab Emirates with matings limited to the months of February, March and April to avoid the problems of the summer months (Ansell, 1976).

With care in the first ten days of life (or longer for premature calves) including artificial cooling to avoid extreme hyperthermia, heat stress was well tolerated, posed few problems and resulted in no calf losses or calf scours in the 94 born during the study (Ansell, 1976). Various stress symptoms were encountered at the upper limit of heat tolerance such as open mouth breathing, high respiratory rates, second phase breathing, body splashing, refusal to lie down, huddling and high rectal temperatures. These were avoided if relatively low-cost specific steps were taken to mitigate climatic stress including showering, provision of shelter, appropriate feeding schedules and adequate water supply.

The principal conclusion of these studies, although based on small numbers of animals, was that with high management levels successful dairy development was possible in these climates (high ambient temperatures and humidities) with European breeds (Ansell, 1976).

5.3 AFRICA SOUTH OF THE SAHARA

In this chapter we deal with African countries, except North Africa which in FAO's regional distribution belongs to the Near East and North Africa region. There are about 140 million cattle at present in Africa south of the Sahara, of which 18 million are dairy cows. The distribution of cattle in Africa (see Figure 1.2) depends mainly on agroecological zones which are shown in Figure 5.3.

Fig. 5.3 Africa south of the Sahara — main agro–vegetation zones

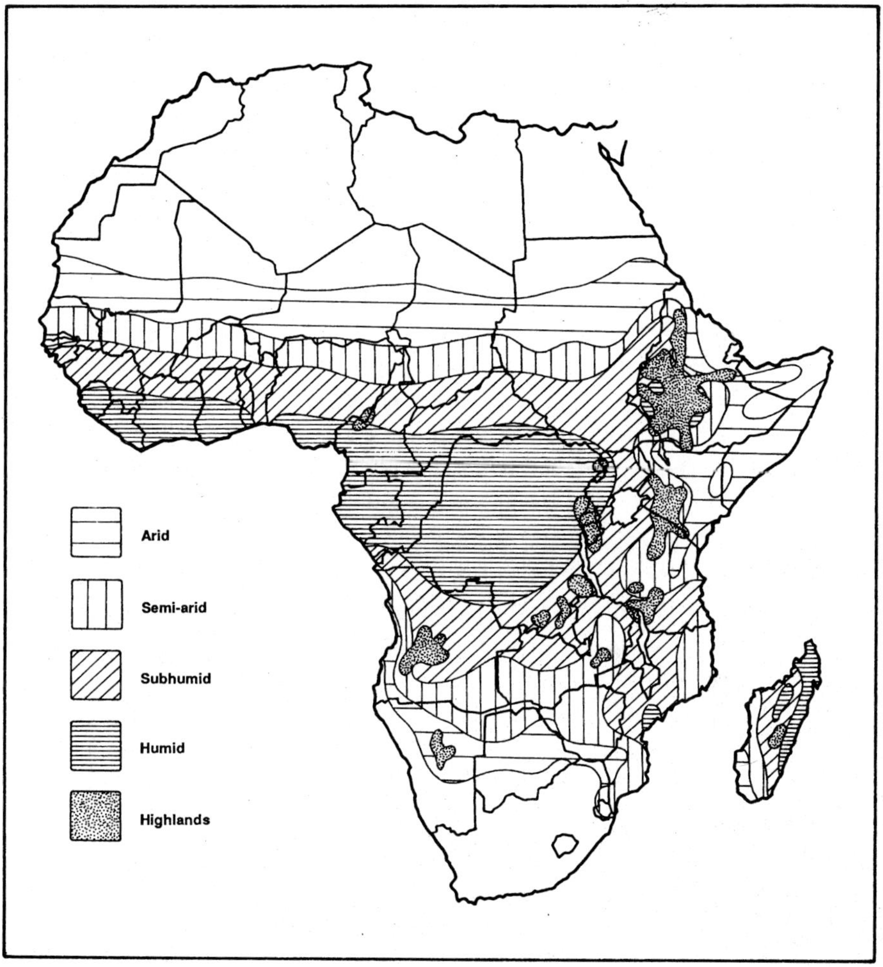

Six main ecological zones can be distinguished in Africa:

1. *Arid zone* with rainfall up to 400 mm per annum. The zone is made up entirely of rangeland. Nomadic and transhumant systems dominate here with extensive grazing.

2. *Semi-arid zone* with rainfall of 400-600 mm per annum. Two types of livestock system exist in this zone, namely, pastoralism and agropastoralism.

The former is characterized by the transhumance and semi-nomadism while in the latter, there is much dependence on crop residues for livestock feeding. Millet and sorghum are the chief crops cultivated.
3. *Dry sub-humid zone* with rainfall of 600-800 mm per annum. Farming by smallholders dominates with millet, sorghum, maize and groundnuts being cultivated. Range grazing is based on wooded grassland (savannah) or woodland.
4. *Moist sub-humid zone* has a rainfall of 800-1 200 mm per annum. The zone is widespread and diverse with subsistence smallholder farming based on a rotational use of fallow land. Tree crops including tea and coffee are important and maize and sorghum are the dominant cereals. Range grazing depends on tree and bush clearing. Poor dry season nutrition is the major limiting factor to cattle productivity. The presence of tsetse fly is also an important constraint on the keeping of cattle.
5. *Humid zone* has a rainfall of about 1 200 mm per annum with a mean temperature of 27-32°C and high relative humidity (80-90 percent). This zone has dense forests in the coastal belt with derived savannah in the low humid areas of the north. The entire zone is infested with tsetse fly and therefore livestock production is limited to trypanotolerant breeds. Cattle are of minor importance in this area, trypanotolerant sheep and goats being the dominant ruminants.
6. *Highlands* which include the land area above 150 mm cover only about 4 percent of the continent but are important for livestock production. They support about 20 percent of cattle and sheep and 10 percent of goats. Tropical and temperate pasture and forage crops coexist and exotic breeds of cattle easily adapt to this area.

The majority of cattle in Africa are distributed in the arid and semiarid Sudano-Sahelian belt in sub-humid and mountainous East Africa and in sub-humid and semi-arid southern Africa.

Twelve million square kilometres of Africa are desert and three million dense rain forest — both quite unsuitable for rearing cattle. The rest, about 15 million square kilometres, are wooded steppe and savannah, and are theoretically suitable for grazing cattle; unfortunately, due to trypanosomiasis (transmitted by the tsetse fly), about 7 million square kilometres of this area are practically unused by cattle, except by a small number of trypanotolerant animals (Figure 5.4). It has been estimated that when cleared of tsetse fly, this area would provide suitable grazing for an additional 120 million cattle, which would allow almost a doubling of the number of cattle in Africa.

The indigenous cattle of Africa make up almost the entire population of the region. The main groups are East African Zebu, West African Shorthorn, Sanga (crosses between *B. taurus* and *B. indicus*), N'dama and Africander.

Fig. 5.4 Tsetse–infested areas and cattle distribution

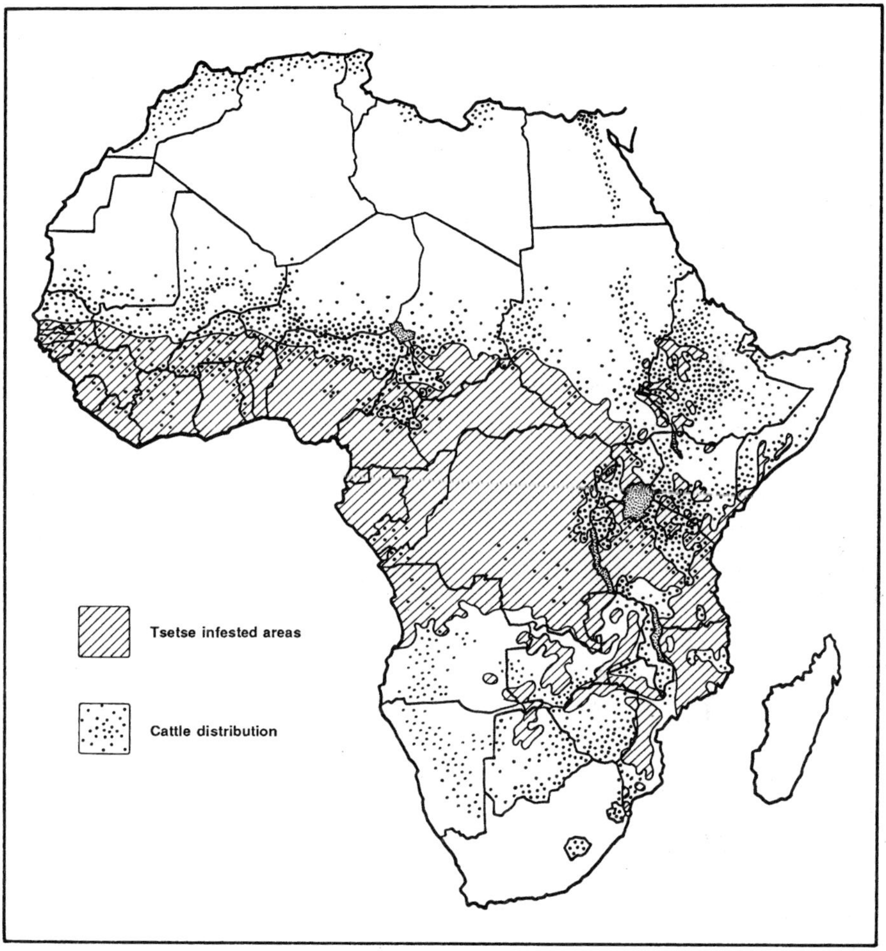

Exotic breeds have been introduced, particularly into the drier tropics or the cooler highlands. However, the proportion of pure and high-grade exotics is unlikely at present to exceed one percent of the cattle population, except in Zimbabwe and Kenya. The main exotic breeds introduced are the Friesian and Jersey.

Table 5.42 shows the number of cattle and their production in African countries. It is evident that the proportion of milking cows is relatively low and, in general, productivity is also low.

Table 5.42 **Cattle production in Africa south of the Sahara, the production in relation to the size of human population and the net trade in milk and milk products for 1985** (FAO, 1985)

Country	No. of cattle	Beef and Veal		Carcass wt. per animal	No. of dairy cows	Milk yield	Milk available per person	Net imports (milk equivalent)
		No. slaughtered	Meat production					
	('000)	*('000)*	*('000 MT)*	*(kg)*	*('000)*	*(kg/animal)*	*(kg)*	*(US$ millions)*
Angola	3 360	360	54	150	295	502	16.9	44.5
Botswana	2 700	225	42	185	275	350	86.7	7.8
Burkina Faso	2 800	255	24	95	560	200	16.1	12.2
Cameroon	3 642	330	50	150	93	500	4.8	10.6
Cent. Afr. Rep.	1 800	180	25	140	42	110	1.9	1.7
Chad	3 400	210	25	120	340	270	18.3	8.2
Côte d'Ivoire	843	310	42	137	135	121	1.6	49.5
Ethiopia	26 000	1 950	213	110	2 723	219	13.7	53.0
Ghana	820	100	11.5	115	123	55	0.5	4.2
Guinea	1 800	180	18	100	225	185	6.9	3.1
Kenya	12 000	1 560	187	120	2 300	478	53.4	3.2
Liberia	42	31	4	125	6	100	0.5	4.6
Madagascar	10 400	1 071	137	128	58	700	4.1	3.2
Malawi	920	92	19	202	86	460	5.8	3.2
Mali	5 800	296	36	122	580	200	14.4	5.1
Mauritius	60	5	1	206	10	2 500	23.8	15.4
Mozambique	1 330	245	37	150	380	170	4.7	13.7
Namibia	2 000	202	37	183	165	684	43.9	—
Niger	3 530	340	39	115	540	200	17.7	5.6
Nigeria	12 000	1 750	221	126	1 200	295	3.7	80.7
Rwanda	660	140	14	104	136	529	11.9	6.7
Senegal	2 200	300	41	138	220	360	12.3	18.0
Somalia	3 600	380	42	110	400	350	30.1	7.3
South Africa	12 733	2 900	593	205	920	2 826	80.3	0.4
Sudan	20 000	2 000	330	165	3 500	486	78.9	26.6
Tanzania	14 000	1 500	153	102	2 650	160	18.8	11.0
Uganda	5 200	624	94	150	1 040	350	23.5	4.7
Zaire	1 350	160	24	150	8	855	0.2	23.1
Zambia	2 600	216	35	160	229	300	10.4	3.0
Zimbabwe	5 800	530	80	150	115	1 696	22.2	0.6

Table 5.43 **Cattle numbers and beef and milk production in Africa south of the Sahara** (from FAO Agrostat)

	No. of cattle	No. of cows	Total meat	Beef and veal	Total milk	Cows' milk
	(million head)			*(million metric tons)*		
1975	126.6	15.3	3.8	1.6	6.6	4.9
1980	137.4	17.0	4.6	1.9	7.6	5.8
1985	140.3	18.0	5.1	2.0	8.3	6.4

Table 5.44 **Beef and milk production in relation to the size of the human population (per caput) and productivity of cattle in Africa south of the Sahara** (from FAO Agrostat)

	Production per caput (kg) of				Productivity per animal	
	Total meat	Beef & Veal	Total milk	Cows' milk	Beef & Veal	Cows' milk
1975	11.5	5.0	19.7	14.8	13.4	320
1980	11.9	5.0	19.7	15.0	13.8	341
1985	11.3	4.4	18.5	14.1	14.3	356

It can be seen from Table 5.43 that during the last ten years, beef and milk production have increased mainly due to an increase in the cattle population. The production of beef and milk achieved for each person of the human population and the productivity of cattle have remained practically unchanged during the last ten years (Table 5.44). As a result the continent has become more and more dependent on imported dairy products and beef (Table 5.45).

Africa imports at present a third of the milk and dairy products consumed. The figure for West Africa is as high as 81 percent (FAO, 1986b). This highlights the need to develop dairying in Africa.

Imports of Friesian cattle to Africa were made mainly during this century to increase milk production. Small numbers were imported which were bred pure and also used for crossbreeding mainly in East Africa where the mild climate, especially at the higher altitudes, is suitable for milk production. The dairy industry in West Africa is just beginning to develop. It is generally believed that Friesian cattle, when judiciously used for upgrading of the local cattle, can play a major role in the development of milk production in Africa. This section gives results of adaptability and productivity of Friesian cattle and their crosses in some countries of Africa.

5.3.1 Botswana

The milk output of the dairy cows in Botswana, which represent only about ten percent of the total cattle population in the country, nonetheless amounted to the largest quantity (87 kg) per person of the human population in 1985.

Table 5.45 **International net trade in meat and milk (in milk equivalents) in Africa south of the Sahara** (from FAO Agrostat)

	Total meat		**Total milk equivalent**		**Beef & Veal**	
	('000 MT)	*(million US$)*	*('000 MT)*	*(million US$)*	*('000 MT)*	*(million US$)*
1975	+ 43.1	+ 47.5	-1.717	-432.2	+ 48.6	+ 59.6
1980	-75.0	-153.3	-3.419	-1 003.9	-33.2	-72.3
1985	-136.9	-200.9	-4.610	-979.4	-51.3	-80.3

+ Exports — Imports

At present levels of availability, therefore, the milk produced in Botswana provided in 1985 more than 80 percent of the total milk and dairy products (including the imports) (Table 5.42).

Dairying started originally as a by-product to cattle ranching. Milk surplus to the needs of calf rearing was produced seasonally from low-yielding Tswana cows or their crosses (FAO, 1982b).

The Botswana Ministry of Agriculture (1980) reported that Botswana imported 30 Friesian heifers from South Africa in 1978. The dairy cows were zero-grazed and given sorghum, silage and brewer's grains. Concentrates were given at the rate of 0.4 kg/kg milk produced. Yields under such an intensive system of management are shown in Table 5.46. Calves were bucket-fed after a 48-hour suckling period and weaned at 12 weeks old, after which they were fed silage-hay and brewer's grains. Breeding age was 15-18 months. Vaccinations were given for blackquarter, anthrax, botulism and brucellosis (heifer calves at weaning). The herd was sprayed with acaricide weekly during summer and every fortnight during winter to control tick-borne diseases.

Purebred Friesian cows had lowered appetites during the hottest months of the year. This can be overcome by adopting intensive management systems through proper feeding practices and the provision of shelters (FAO, 1982b).

The limited number of observations available in the country suggests that crossbred Friesian-indigenous animals can be kept if feed and water supplies, as well as health control, are at a satisfactory level.

5.3.2 Cameroon

Domestic production of milk from cattle accounted for less than half the share of the total milk consumed amounting to 8 kg per head of the human population in 1985, the balance being met by imports of milk and dairy products. As also shown in Table 5.42, the proportion of the total cattle population used for dairying is quite small (2.5 percent).

The policy of the Government is to stimulate local dairy production in all parts of the country in order to increase production of animal protein and also save foreign exchange on the importation of dairy products (Ndumbe, 1981).

More than 99 percent of the national herd is made up of zebu cattle which seem to be well adapted to the environment and resistant to most diseases. The only treatment that they receive is vaccination against rinderpest and contagious bovine pleuropneumonia (FAO, 1980b).

Since 1974, the Institute of Animal Research (IRZ) has intensified research in dairy production and development with pure exotic dairy breeds (Holstein,

Table 5.46 **Milk production of 30 imported Friesian cows in Botswana**

	No.	Mean
Lactation yield average (kg)	21	4 590
Lactation length (weeks)	21	45.6
Lactation yield to 44 wks (kg)	21	4 285
Calving intervals (days)	13	362

Table 5.47 **Percentage distribution of breeds in the northern region[1] of Côte d'Ivoire in the years 1979-85**

	Baoulé	N'dama	Zebu crossbreds[2]	Zebu
1979	73	23	3	1
1982	67	27	4.5	1.5
1985	56	25	16	3

[1] The cattle population in the northern region was 20 animals/km² and in the southern region 0.1 animal/km² (Government of Côte d'Ivoire, 1986)

[2] Zebu crossbred mean Baoulé × zebu and N'dama × zebu

Jersey and Montbéliard). The studies showed that Holsteins produced an average of 12.5 kg of milk per day (Ndumbe, 1981).

5.3.3 Côte d'Ivoire

In 1985, as can be deduced from Table 5.42, more than 90 percent of cow's milk and dairy products available in Côte d'Ivoire were imported with a substantial import bill equivalent to nearly US$50 million. Diseases including those transmitted by tsetse fly pose a serious problem to development of the dairy industry.

Two types of cattle production systems exist in Côte d'Ivoire — the sedentary and transhumant systems. The first covers about 66 percent of the total cattle population. Breed distribution is 47 percent zebu × *Bos taurus* crossbreds, 38 percent Baoulé and 15 percent N'dama. Due to the presence of tsetse fly in the south of the country, cattle are mainly located in the north. The breed distribution in the northern region of Côte d'Ivoire between the years 1979-85 (Table 5.47) shows a decrease of the Baoulé population and an increase in zebu crossbreds (Hoste, 1986).

Serres *et al.*, cited by Mongodin (1985), reported that Australian Friesian-Sahiwal cows (offspring of male zebu-Sahiwal × female Friesian) could produce on average 3 000-4 000 litres of milk per cow per year in a low altitude humid tropical zone at Kairi station even in the third or fourth generation after importation.

Hoste (1986) advocates caution in the use of European breeds (including Friesian) for improving the national herd in Côte d'Ivoire because of some serious health problems, the main one (Touré, 1986), being trypanosomiasis. Other important diseases in Côte d'Ivoire include contagious bovine pleuropneumonia, anthrax, brucellosis, pasteurcllosis and foot-and-mouth disease. These diseases, together with the climatic conditions, pose serious limitations on the possibilities for imported exotic breeds and their crosses.

5.3.4 Ethiopia

Ethiopia has the largest cattle population in the African continent with 26 million animals. Approximately 10 percent are milked, and provide the human population with an average of about 14 kg milk per person annually. As seen from Table 5.42, the average yield per cow was only 220 kg per year and imports of dairy products were substantial. In 1985, domestic production accounted for about two thirds of the total milk and milk products consumed.

Dairying is an integral part of traditional cattle production in Ethiopia. In the highlands, where about 70 percent of the human and livestock population live, cattle of various local breeds are owned by most peasant farmers mainly for draught purposes but are also milked to provide sustenance for the families and cash income mainly from the sale of butter. In the lowlands some nomadic pastoralists subsist almost entirely on milk and livestock trading.

Improvement schemes for milk production started in 1967 with the setting up of the Chilalo Agricultural Development Unit (CADU) to promote small-scale dairying. This Ethio-Swedish integrated project was located in the Arussi province (known since 1975 as the Arussi Rural Development Unit — ARDU) situated in the highlands. Indigenous females comprising chiefly the local Arussi (East African

Shorthorned Zebu) and some Boran, Fogara and Barca were initially crossed to the Friesian and Jersey and F_1 crossbred females sold to participating farmers (Brännäng *et al.*, 1980). However, the use of Jersey was discontinued early in the programme, as the farmers showed preference for the Friesian crosses and disliked the Jersey crosses due to their small teats which were inconvenient for milking. The Friesian crosses were also liked because of their large size. Thus, the programme continued with Jersey and Friesian half-breds being graded up to the Friesian only. The different breed combinations were tested at Asella and data on their performance were also obtained from farmers.

The results from Asella station are presented in Table 5.48 and those from 232 crossbred animals sold to small farmers are shown in Table 5.49. At Asella, although there were differences between breed groups for individual traits, in terms of overall productivity, both Jersey and Friesian half-breds were similar, as were all grades of Friesian. The production level on the smallholdings was similar to that at Asella station and the genotypes also ranked in the same order as those at Asella (Kiwuwa *et al.*, 1983).

The Arussi experiment and other observations led to the implementation of further dairy development plans for Ethiopia. Today there is in Ethiopia a growing, modern milk production sector, although this is restricted to the highlands, where the climate is favourable for dairy development and to around the big cities where a good market exists for dairy products. This dairy sector is based mainly on Friesian crossbreds in 98 cooperative dairy farms, 18 state dairy farms and many privately-owned farms, mainly near Addis Ababa. In addition, large numbers of crossbred cows are held by urban dwellers.

The cooperative farms own usually 10 to 20 dairy cows, mainly crossbred Friesians with 50 to 75 percent Friesian blood. The average milk production on cooperative farms is 1 121 kg per cow per year with average yields of over 2 000 kg achieved on some farms. The best production for Friesian crosses was reported by the Welayita Agriculture Development Unit (Southern Sheva province) where average milk yield per cow in 1980-81 was 2 889 kg. Yields of this order, and even higher for individual cows, were also achieved in a largely purebred herd of Friesians (mainly of Kenyan origin) maintained in the early 1970s at the Alemaya College of Agriculture under an intensive system of feeding and management which was unlikely to have been economic (Wiener, 1972).

The average calving interval for Friesian crossbreds on cooperative farms was 430 days and mortality 12.5 percent.

The state dairy farms which before nationalization in 1974-75 operated as private enterprises, owned, in 1983, 4 700 cattle (Friesian crossbreds) of which 1 800 were milking cows. The average milk production was 9.3 kg of milk per cow per day which amounted to an annual production of about 3 000 kg. However, the calving

Table 5.48 **Least squares means for different traits in the crossbreeding experiment at Asella Farm, Ethiopia** (from Kiwuwa *et al.*, 1983)

Genotype	Age at first calving*	Calving interval**	Breeding efficiency**	Post partum weight**	Lact. milk yield**	Total FCM** (4%)	Fat content**	Annual FCM yeld per unit metabolic weight of cow**
	(mo)	*(days)*	*(%)*	*(kg)*	*(kg)*	*(kg)*	*(%)*	*(kg)*
Arussi	34	439	68	236	809	805	5.5	12
Zebu		459	64	309	927	922	5.3	11
1/2 F 1/2 A	34	427	95	307	1 977	1 942	4.5	23
1/2 F 1/2 Z	35	458	97	328	2 352	2 367	4.4	24
3/4 F 1/4 A	34	464	101	324	2 374	1 958	4.3	23
3/4 F 1/4 Z	34	475	98	325	2 356	2 246	4.2	23
7/8 F 1/8 L	36	525	98	336	2 318	1 718	4.1	(17)
1/2 J 1/2 A	34	403	—	269	1 741	1 942	5.1	26

A — Arussi, F — Friesian, Z — Zebu, L — Local, J — Jersey

*Significant influence of the genotype ($P < 0.05$)

**Highly significant influence of the genotype ($P < 0.01$)

Table 5.49 **Least squares means for total lactation yield and lactation length for crossbred cows in smallholdings — Arussi Province, Ethiopia**

Genotype	Total lactation milk yield** (kg)	Lactation length** (days)	Calving intervals (days)	Annual milk yield (kg)
1/2 J 1/2 A	1 353	301	411	1 265
1/2 F 1/2 A	1 613	353	546	1 398
1/2 F 1/2 Z	1 985	313	422	1 800
3/4 F 1/4 A	2 317	391	428	2 033

** Highly significant influence of the genotype ($P < 0.01$)

Table 5.50 **Preliminary results of the performance of Friesian and other crosses in experimental trials at the Institute of Agricultural Research, Ethiopia**

	Liveweight at one year[1] (kg)		Services per conception[2]		Calving interval[3] (days)		1st lactation milk yield[4] (kg)	
Sire Breed	*No.*	*mean*	*No.*	*mean*	*No.*	*mean*	*No.*	*mean*
Friesian	135	220.6	38	1.58	9	392	18	2 140
Jersey	145	193.3	58	1.54	21	421	30	1 691
Simmental	129	224.8	48	1.66	12	453	22	1 942
Zebu[5]	—	—	491	1.74	155	507	—	—

[1] All 4 stations (from Kebede and Galal, 1982) [2] At Holetta and Bako only (the wetter stations) (from Tegegn *et al.*, 1981)
[3] From Holetta and Bako only (the wetter stations) (from Galal *et al.*, 1981)
[5] All 4 stations F_1 crosses only (from IAR, 1980)
[6] Average of 3 zebu breeds

interval was high at 442 days and calf mortality was 22 percent.

The breeding policy of the Ethiopian Government is to upgrade dairy cattle to Friesian. For this purpose two AI centres exist, one in the Arussi Rural Development Unit (3 800 inseminations per year) and the other in Addis Ababa (1 400 inseminations per year). The Government supports the distribution of halfbred Friesian heifers and of purebred or nearly purebred Friesian bulls to cooperatives and private farms. This leads to the rapid increase of Friesian blood in dairy herds. At present there is no clear programme to stabilize the percentage of Friesian blood at a level suitable to existing conditions.

A large project initiated by the Institute of Agricultural Research in 1974 with FAO participation should however provide useful information on this point since it involves the production of first crosses, backcrosses and *inter se* mating among the crosses (F_2). This will show whether additional Friesian blood (grading-up) or a stabilized proportion is more advantageous (Wiener, 1972). Initial results from still relatively small numbers from that experiment, which involved three exotic and four zebu breeds at four locations, differing in climate and feeding practice, are shown in Table 5.50. Friesian crosses yielded more milk and had a shorter calving interval than the others but early results from the publications quoted suggest that the relative merit of the crosses may differ between the four locations. McKonnen (1983) also provided results of comparisons of purebred Holstein (of Cuban origin) and Barca and Boran crosses with the (Cuban) Holstein bulls at Debre Zeit Junior College of Agriculture in the highlands south of Addis Ababa. Total first lactation yields for these three groups were 3 768 kg (31), 2 468 kg (12) and 1 951 kg (33) respectively (number of heifers in parentheses). Calving intervals were not significantly different and age at first calving was slightly lower (though very variable) for the Holsteins (27 months) than for the crosses (28 and 30 months).

5.3.5 Guinea

The dairy cattle population of Guinea, numbering approximately 225 000, produces about 7 kg of milk per head of the human population. This accounts for about three quarters of cows' milk and dairy products available in 1985. Imports made up the rest (Table 5.42).

In Guinea with the exception of about 5 percent of cattle that are zebu crosses, all other cattle are of the N'dama breed. A number of imported purebred cattle are kept on government breeding stations. The most important exotic breed is the Red Steppe, imported from USSR. Some Jersey, Ayrshire and Friesian cattle imported from different countries are also kept at the Institut de Recherches Agronomiques de Foulaya (FAO, 1980b).

Crossbreeding to increase milk production is not developed. Contagious bovine pleuropneumonia is still a major problem in Guinea. Anthrax, blackquarter and pasteurellosis are also major diseases. Trypanosomiasis is generally not a serious problem among cattle due to the trypanotolerance of the N'dama (FAO, 1980b).

5.3.6 Kenya

The dairy industry in Kenya is probably one of the most advanced among the African countries. This was achieved by crossbreeding with European breeds and by favourable climatic conditions and good management. The production of cows' milk, as shown in Table 5.42, was sufficient to provide 53 kg per head of the human population in 1985 and net imports of dairy products were relatively small in both absolute and proportionate terms.

The history of European cattle breeds in Kenya goes back to the beginning of the twentieth century. The first Friesians were imported from South Africa to the Government Stock Farm at Naivasha in 1908 and the next importation took place in 1920. In 1937, some bulls and cows were imported from the USA and the Netherlands and artificial insemination was started in Naivasha in the late 1930s. In 1946 the Friesian Cattle Breeders' Society was founded. Importation of semen started in 1962 (M. Nevill, 1986, personal communication). Up to 1975, the numbers of imported cattle amounted to 1 865 males and 4 244 females. Of the imported females 73 percent were Friesian (largely because of a big importation from the Netherlands in 1971-75), 9 percent Ayrshire and the remainder Jersey (6 percent), Guernsey (6 percent), Red Poll (3 percent) and Shorthorn (3 percent). Among the imported males, Friesians accounted for only 22 percent and the other breeds 23, 14, 15, 12 and 6 percent respectively. The introduction of acaricides in the 1940s allowed the control of tick-borne diseases to be effected so imported animals could survive better (Stotz, 1979).

According to FAO statistics, the estimated number of cattle in Kenya in 1984 was 12 million including 2.8 million dairy cows which produced 1.3 million tons of milk. The average production per cow per year was 482 kg milk. The 1983 Annual Report of the Livestock Development Division, MOLD, gives the following breed composition — 77 percent local zebu (mainly SEAZ Small East African Zebu), 3 percent grade beef and 20 percent grade dairy cattle. This last group comprises grades with a high level of blood of cattle of European origin. Among grade cattle, 27 percent were identified by type appraisals as Ayrshires, 25 percent as Friesians, 16 percent as Guernseys and 8 percent as Jerseys (Kenya, 1980a). Smallholdings (less than 20 ha of farm land) own 80 percent of dairy cattle. In 1980, there were 1 800 large-scale dairy herds with an average size of 154 cows.

Milk yield of grade cows on small farms varied from 1 747 to 2 466 kg per cow per year while zebu cows yielded some 300 kg of milk (FAO, 1981).

The total Friesian population in the 1980s was put at 530 000, or 26 percent of the total dairy cattle, with 240 000 milking cows and 10 000 breeding bulls. Over the years some 108 Friesian or Holstein-Friesian breeding bulls were imported from the following countries: South Africa (in 1908 and 1920), USA (in 1973), and the Netherlands (in 1937). Later importations mainly of heifers were from Germany, the Netherlands, Sweden, United Kingdom, Canada and Denmark though some bulls were also imported. The strategy was for large-scale farms to raise purebred cattle. The smallholders upgraded local breeds to Friesian as grade animals were found to be more adaptable to the climate and resistant to disease whilst purebreds demanded good management (L. Enemark, 1986, personal communication). The semen source for smallholders is mainly local whilst large-scale farms in a dairy development programme with 1 875 cows showed an average yield of 3 117 kg milk with a butterfat level of 3.78 percent. The average calving interval was 430 days for these herds (L. Enemark, 1986, personal communication).

The cattle breeding policy for the different ecological zones in Kenya is illustrated in Table 5.51 (Meyn and Wilkins, 1974). Dairying is concentrated in the Afro-Alpine equatorial and humid to semi-arid regions (ecological zones I-III) with approximately 600-900 mm annual rainfall and suitable for forage production.

Cattle production systems in smallholdings in Kenya are described in Table 5.52 and performance characteristics in Table 5.53.

The Livestock Recording Centre situated in the National Animal Husbandry Research Station at Naivasha provides services for the National Breeding Programme. Milk recording, which started in 1949, comprised, in 1980, 13 715 cows in 130 herds (Kenya, 1980b). In milk recorded populations, 41 percent were Friesian grades. The number of progeny tested bulls has varied from 18 to 30, Friesians numbering 5 to 6. In 1980, 1 528 head of cattle were registered with the Kenya Stud Book (nearly 50 percent Friesians).

Artificial insemination plays an important role in the upgrading process. In 1979, the total number of AI services was 548 741 (91 percent with semen from crossbred bulls and 9 percent from zebu) (Stotz, 1979).

The major diseases in dairy cows were tick-borne diseases (East Coast fever). The latter was responsible for half of the total calf mortality in some areas. The other important diseases were foot-and-mouth disease and trypanosmomiasis. Brucellosis was also hazardous to human health because a high proportion of the milk consumed was unpasteurized. Mastitis was reported to be up to 50 percent in dairy cows in some areas (FAO, 1981).

Table 5.51 **Ecology and cattle breeding policy for commercial and subsistence farmers in Kenya** (from Meyn and Wilkins, 1974)

Ecological zone[1]	Climate	Approximate annual rainfall *mm*	Percentage of surface area	Stocking rate[2] *(Stock units/ha)*	Breeding policy Subsistence smallholders	Breeding policy Commercial large-scale
I to III	Afro-alpine equatorial, humid to semi-arid	900	21	2	Dairying[3] European breeds	(i) Dairying[3] (ii) Dual-purpose European breeds[4]
IV	Semi-arid	600-900	9	3-4	Subsistence beef[5]	(i) Dairying[5] Ranching[6]
					European x zebu crosses	(ii) Beef
V	Arid	400-600	50	4	Subsistence: beef-Sahiwal	Beef-Boran
VI	Very arid	under 400	20	15	Subsistence: beef-indigenous zebu	

[1] From Pratt *et al.*, 1966

[2] Stock units, equivalent to 450 kg liveweight

[3] Ayrshire, Friesian, Guernsey, Jersey

[4] Simmental, Brown Swiss, Red Poll

[5] Crisscrossing: Friesian, Simmental, Brown Swiss, Red Poll with Sahiwal and Boran

[6] Commercial crosses: Charolais, Simmental, South Devon, Sussex, Hereford, Friesian, with Boran and Sahiwal

Table 5.52 **Cattle production systems in smallholdings in Kenya** (from Stotz, 1983)

	Production systems			
	A	**B**	**C**	**D**
Cattle breed	Zebu cattle	Cross cattle	Grade cattle	Grade cattle
Calf rearing	suckling of male + female calves for 5-7 months	suckling of male + female calves for 5-8 months	bucket feeding 16 weeks whole milk	bucket feeding 16 weeks whole milk
Main forage source	Kikuyu Star grass	Kikuyu Star grass	Kikuyu Napier grass	Napier grass
Feeding system	permanent grazing	rotational grazing	rotational grazing + stallfeeding	stallfeeding
Disease control measures	none	dipping	dipping + drenching	dipping + drenching
Concentrate and mineral feeding	none	minerals only	minerals + concentrates 450 kg/cow/year	minerals + concentrates 1 000 kg/cow/year

Table 5.53 **Performance of cattle in different production systems in Kenya** (from Sloane, 1985)

	Production systems[1]			
	A	B	C	D
Lactation milk yield (kg)	600	1 400	2 200	2 800
Calving interval (months)	18	17	16	14
Cow mortality (%)			6	
Calf mortality(%)	25	20	15	
Young stock mortality (%)		7	8	

[1] See Table 5.52

Mutiga *et al.* (1981) reported 27 percent of bovine dystocia in pure Friesians, compared to 18 percent in Guernsey, 14 percent in Ayrshire, 13 percent in Jersey and 11 percent in zebu.

In 1962, the National Sahiwal Stud at Naivasha was formed, based on imported animals from India and Pakistan. The herd had 1 300 Sahiwals including 500 cows. The average lactation milk yield was 1 630 kg in 1972 (Meyn and Wilkins, 1974). Introduction of Sahiwals was aimed at crossing with European grade cattle in the

Saudi Arabia. Friesian calves in individual pens at the Al Safi farm

Table 5.54 **Milk yield (kg) (mean and s.d.) of Friesians (F), Sahiwal (S) and their crossbreds (F × S) in extensive grazing conditions at Naivasha, Kenya** (from Mwandotto, 1980)

	Breed type		
	F	S	F × S
First lactation	2 221 ± 361	1 700 ± 385	2 590 ± 624
Second lactation	2 597 ± 724	1 745 ± 480	2 738 ± 651
Third lactation	2 444 ± 518	1 669 ± 477	2 163 ± 856

semi-arid and coastal areas and with indigenous zebu in very arid areas. Table 5.54 shows the crossbreeding results of Friesians with Sahiwals. These results from a well-managed farm show that crossbreeding improves the milk yield of the Sahiwal zebu. A heterosis effect, especially in the first lactation, seems to be evident. Crossbreds exceeded pure Friesians in the first two lactations and the average of both parental breeds by 32 percent in the first lactation, by 26 percent in the second, but by only 5 percent in the third.

Beef production in Kenya is based on local zebu cattle. There are 800 large farms/ranches with the number of beef animals averaging 500 head. In smallholdings, beef production is a by-product of dairying (Kenya, 1980a).

Crossbreeding the zebu with Friesians also proved beneficial for beef production. Tonn (1974) presented data showing that F_1 Friesian x Boran steers were 3 months younger at slaughter and weighed 8-15 percent more than Boran steers. Crossbred Simmental × (Friesian × Sahiwal) were higher in birthweight by 39 percent and in weaning weight by 22.6 percent than Sahiwals. The same was true of Simmental × (Friesian × Boran) crossbreds with birthweight and weaning weight higher than for Boran by 19 and 24 percent respectively (Kamau, 1975).

Mwandotto (1980 and 1981) found only a 1 percent difference in birth weight between Friesian × Sahiwal or Friesian × Boran calves and the purebred zebu. In daily gains, the Friesian crossbreds were superior to Sahiwal by 20 percent and to Boran by 12 percent.

Kiwuwa (1974) studied production characteristics of purebred Friesian and Jersey cattle and their grades (with more than 75 percent *Bos taurus* blood) using data from the Central Registry Office of the East African milk records of Nakuru for the period 1957-66 (a total of 3 897 records from 86 farms). He concluded that milk yield was higher in Friesians than in Jerseys but calving intervals were shorter in the latter. Milk yields per day of calving interval were quoted as follows: Friesian 7.0, Friesian grades 7.5, Jersey 6.2, Jersey grades 5.6; when expressed in terms of fat-corrected milk the corresponding figures were 6.6, 7.2, 7.3 and 6.4 respectively.

5.3.7 Liberia

The animal population of Liberia is small mainly due to the presence of the tsetse fly. Only 4 070 out of a total of about 170 000 agricultural holdings keep cattle. The average number of cows per holding is 3.5 (Provost, 1977) and the total number of cattle in the country is 42 000. Production of milk was sufficient to supply only 0.5 l for each person in the country (Table 5.42) and imports of dairy products play a significant role and accounted for 90 percent of the total milk and dairy products in 1985.

There are two main types of cattle in Liberia: N'dama and West African Dwarf Shorthorn called Muturu. The milk yield of both breeds is low and they produce about 450 litres of milk per lactation (FAO, 1979a).

In the late 1950s, first efforts to introduce exotic breeds to improve cattle production were made. Artificial insemination was used to cross N'dama with Brown Swiss, Jersey and Santa Gertrudis. Later at Suakoko, N'dama cattle were crossed with six breeds: Holstein-Friesian, Jersey, Aberdeen Angus, Brahman, Santa Gertrudis and Brown Swiss brought from USA in 1971. The imported breeds and their crosses have not survived because of trypanosomiasis and streptothricosis.

The second attempt was made in 1973 when 20 in-calf Holstein-Friesian heifers were imported from the USA and studied for their adaptability to the Liberian climate, heat tolerance and the effect of tropical diseases on their performance. The average birth weight of the Holstein calves was 34 kg. The heifers matured at about 17-18 months of age with a liveweight of 320-340 kg. The age of first calving was about 26-28 months and milk yield was approximately 2 700 kg per lactation in 1977-78 (12 cows) (EEC, 1986).

Provost (1977) reported that in 1974, 104 cows, 4 bulls, 39 yearlings and calves of the Holstein-Friesian breed were imported from USA.

The disease situation in Liberia is a critical element in cattle production. The imported cattle have to be given preventive treatment against trypanosomiasis and immunized against rinderpest, brucellosis and foot-and-mouth disease, if they are to survive and produce.

5.3.8 Madagascar

Although the dairy cows represent less than one percent of the total cattle their average yield at 700 kg per cow per year (Table 5.42) is among the highest in the region. In consequence, Madagascar produced about two thirds of the total milk and dairy products consumed in the country in 1985. However, this has to be seen in the context of a low level of availability (approx. 6 kg milk equivalent per person in 1985).

The most important indigenous breed in Madagascar is the Malagasy zebu, other breeds of some importance being the Brahman and the Rana. The Rana is a cross between the local zebu and imported dairy breeds (mainly Friesian and Normandy, plus some Sahiwal and Red and White Norwegian). The milk yield of the local zebu cow is about 300 kg per year. Crossbreds (mainly zebu × Friesian and zebu × Normandy), if adequately fed, produce 1 200-1 500 kg of milk per year (5-6 kg daily over a 7-month period). In a number of modern holdings, with more systematic breeding and adequate feeding, yields between 2 500 and 3 000 kg per year have been obtained from crossbreds (Krostitz, 1974).

Robert (1979) reported that Friesian cows produced an average of 5 122 kg milk in a 305-day lactation. A crossbred of indigenous and Friesian produced 2 200 litres of milk per lactation, twice that of indigenous cows which produced only 900 litres of milk per lactation.

The Government has been engaged in the production of zebu × Friesian crossbreds in the Sakay area. The Bureau Central Laitier (BUCL) provides free AI service to dairy farmers in the Tananarive area. Approximately 5 000 dairy cows were inseminated with semen from Friesian bulls annually in 1976 (FAO, 1976d).

The results of crossbreeding of Madagascar zebu with French Friesians are presented in Tables 5.55 and 5.56. Friesians had problems with adaptation as shown by high mortality and delayed first calving (France, 1978).

Friesians were also used for commercial crossbreeding with local zebu for meat production. Experimental data show an improvement in daily gain and carcass quality of crossbreds compared to purebred Malagasy zebus (Serres, 1974).

5.3.9 Malawi

The total cattle population is below one million, less than a tenth of which are dairy cows. The net imports, equivalent to US$3.2 million in 1985, accounted for about a quarter of the total milk and dairy products in the country, but availability for the human population of milk produced in the country was less than 6 kg milk per person and around 8 kg in 1985 if the imports were included (Table 5.42).

Table 5.55 **Body weights (kg) at various ages of males and females of French Friesians (FF) and their crosses with local zebu — Kranjasoa Centre, Madagascar**

	Females			Males			
Age *(months)*	Pure-bred FF	3/4 FF	1/2 FF	Pure-bred FF	3/4 FF	1/2 FF	Pure zebu
6	130	130	100	150	141	118	103
12	188	190	184	215	222	187	134
24	282	313	325	394	390	354	182
36	416	418	388	579	531	461	270

Table 5.56 **Breeding performance and milk yield of French Friesians (FF) and their crosses with local zebu — Kranjasoa Centre, Madagascar**

	Pure French Friesian	3/4 FF	1/2 FF
Age at first calving (months)	38	35	32
Calving interval (days)	381	—	434
No. of services/ calving	1.15	2.39	1.52
1st lactation			
Milk yield (kg)	2 877	1 746	1 557
Length (days)	338	266	291
3rd lactation			
Milk yield (kg)	3 615	—	2 006
Length (days)	381	—	314

Malawi government policy is to use the zebu-Friesian crossbred in dairy development schemes. The use of Friesian semen is by far the most popular. There were two semen importations from USA in the years prior to 1977. The effectiveness of AI in Malawi is shown in Table 5.57.

Production by different breeds and crosses in Malawi milk recording schemes in the southern region (1976-77) is shown in Table 5.58 as given by Allen and Kristensen (1977).

Most of the innovations in dairy herd farms in Malawi were tested at Chitedze Research Station which had close links with smallholder dairy operations. Bwemba Livestock Centre served as the main station from which crossbred heifers were issued to smallholder dairy farmers in the Central and Northern Regions. Local cows (believed to be crossbreds of Sussex, Brahman and Africander-cross cattle) were mated to Friesian bulls to produce 1/2 Friesians. The female yearlings were inseminated with imported Friesian semen and were sold to the smallholders 3-4

Table 5.57 **Results of the use of AI in Malawi [from Department of Animal Health and Industry, Malawi** (FAO, 1977b)]

	Calving percentage of cows inseminated	Services per calf born
Rural area	57	2.3
Government farms	63.5	2.2
Private estates	64.4	1.9

Table 5.58 **Milk production and lactation length in milk recording schemes in the southern region of Malawi** (from Allen and Kristensen, 1977)

Type	No. of cows	Lactation yield *(kg)*	Lactation length *(days)*
High grade Friesian (commercial farm)	25	1 726	271
High grade Friesian (small farm)	9	1 822	364
3/4 Friesian	7	2 238	357
1/2 Friesian	94	1 728	354
Grade Friesian	4	1 526	339
Malawi zebu	5	541	237

weeks after calving. In this way, the smallholders obtained 3/4 to 7/8 Friesian cows. The 15/16, 31/32 and higher grade Friesians were retained for milking or sold to commercial dairy farmers. Agyemang and Nkhonjera (1986) analysed the results from these operations and they are shown in Table 5.59. It appears that for milk yield 3/4 Friesian produced more than 1/2 Friesian, but higher grades brought no further benefit. In terms of age at first calving and other aspects of reproduction, half-breds were either better or similar to higher grades.

5.3.10 Mauritius

In 1983, a livestock census showed that there were 1 752 Friesian cattle in the country out of a total cattle population of 58 000 of which about 10 000 were milking cows. In relation to the size of its population Mauritius imports a substantial amount of milk and dairy products (three quarters of the total in the country in 1985) although the average yield per cow from its small dairy population is very high and provides 25 kg milk per person (Table 5.42).

Black and White Friesian cattle were imported to up-grade local Creole cattle which produce about 500 kg milk per lactation (K. Lutchmeenaraidoo, 1985, personal communication). Artificial insemination has been carried out, the semen being imported from Kenya and Australia.

In 1974 and 1976, about 1 400 live cattle, mostly heifers of Friesian, and Hereford × Friesian breeds were imported from New Zealand in a programme to enhance milk production in rural areas. In 1983 and 1984, the importation of pure Friesian heifers from Australia numbered 165 and 190 respectively. These were for the government livestock breeding centres to increase production of breeding stock for sale to farmers.

The milk production level of Friesian cattle and their crosses on satisfactorily-managed government farms has ranged from 2 000 to 3 000 litres (300-day lactation). The AI conception rate was around 60 percent and the age at first calving varied between 18 and 24 months.

5.3.11 Mozambique

Mozambique has 1.33 million cattle including 380 000 dairy cows with an average milk production per year of 170 kg. In 1985, Mozambique produced from its own dairy cow population about half the milk in the country and imported the other half. Milk produced amounted to about 5 kg per head of the human population (Table 5.42). Indigenous cattle are the East African Shorthorned Zebu with low milk production. Some fairly large dairy farms existed before Independence (1975) around the cities of Maputo and Beira.

In the region of Umbeluzi in the tropical zone (where the AI Centre is located) Holstein-Friesians are found. They are inseminated with frozen semen imported from Canada. In this region, Dutch Friesians based on imported stock from South Africa and Zimbabwe are also kept. Alberro (1980 and 1981) described the dairy herd at the Livestock Station of the University of Marcaneta (tropical coastal belt of Mozambique). In this herd mainly three types of dairy cattle were kept for milk production: Holstein-Friesian, Dutch Friesian and Friesian × Africander. Milk yield in the first lactation was highest in Holstein-Friesians, which exceeded the herd mean of 2 869 kg by 380 kg (Table 5.60). They were followed by Dutch Friesians. The lowest yield was observed in the Friesian × Africander group which also showed an older age at first calving, poorer milk let-down, shorter lactations and more udder wear. No significant differences were found in calving difficulties, disease and mastitis incidence, parasitism and mortality. The author concluded that acclimatized exotic breeds can produce milk efficiently in the not-too-harsh areas of the tropics and sub-tropics if they are properly managed and fed.

5.3.12 Nigeria

Ten percent of the 12 million cattle in Nigeria were classified as dairy cows in 1985. Their milk production was equivalent to 3.5 kg milk per head of the human population approaching 100 million. Although the import bill for milk and dairy products in 1985 was not high among African countries in relation to the size of the human population in absolute terms it was the highest in 1985 (US$ 80 million) (Table 5.42). This in itself highlights the desirability of dairy development, let alone the prospect of increasing the amount available per person.

In Nigeria, 50 percent of the cattle are transhumant. There are four types of indigenous zebu: White Fulani (51 percent), Red Bororo (14 percent), Adamawa Gudali (11.5 percent) and Sokoto Gudali (11.5 percent). The most important factor determining the presence of cattle is the incidence of trypanosomiasis transmitted by the tsetse fly. African *Bos taurus* (trypanotolerant) can be found in small sedentary herds in the south of the country.

The number of purebred Black and White cattle in Nigeria is considered to be small and not likely to exceed more than a few hundred (V. Buvanendran, 1985, personal communication). Almost all are reported to be held in institutional herds (state farms, research institutes and universities).

Table 5.59 **Least squares means for performance records on smallholdings and two stations in Malawi of various grades of Friesian cattle** (from Agyemang and Nkhonjera, 1986)

	Smallholders		Chitedze		Bwemba	
Age at first calving (months)						
1/2 Friesian (F)	36.7[a]	(110)	33.1	(7)	—	
3/4 F	40.1[b]	(55)	42.7	(39)	—	
7/8 F			45.8	(24)	37.0	(39)
15/16 F					37.2	(47)
31/32 F					39.5	(29)
Pure Friesian			45.5	(10)	36.9	(30)
Calving interval (days)						
1/2 F	488	(432)	407	(49)	—	
3/4 F	482	(145)	484	(116)	—	
7/8 F			513	(17)	414[a]	(45)
15/16 F					392[a]	(200)
31/32 F					502[b]	(30)
Pure Friesian			498	(26)	419[a]	(63)
Days open						
1/2 F	216	(441)	130	(49)		
3/4 F	211	(150)	205	(116)		
7/8 F			235	(17)	124[a]	(45)
15/16 F	—		—		116[a]	(200)
31/32 F	—		—		203[b]	(30)
Pure Friesian	—		219	(26)	140[a]	(64)

Table 5.59 **(continued)**

Lactation yield (kg)						
1/2 F	1 953^{a}	(554)	1 677^{a}	(52)	—	
3/4 F	2 424^{b}	(227)	2 345^{b}	(151)	—	
7/8 F			2 495^{b}	(42)	2 641	(44)
15/16 F	—		—		2 484	(187)
31/32 F	—		—		2 720	(33)
Pure Friesian	—		2 366^{b}	(35)	2 790	(62)
Lactation length (days)						
1/2 F	382	(554)	307	(52)	—	
3/4 F	401	(227)	331	(151)	—	
7/8 F	—		365	(42)	280	(44)
15/16 F	—		—		283	(187)
31/32	—		—		320	(33)
Pure Friesian	—		321	(35)	290	(62)
Dry period (days)						
1/2 F	116	(403)	116	(47)	—	
3/4 F	98	(141)	145	(115)	—	
7/8 F	—		157	(17)	79	(35)
15/16 F	—				105	(151)
31/32	—		—		132	(27)
Pure Friesian	—		179	(25)	133	(50)

Notes: For each characteristic, means, within columns, followed by different superscripts differ significantly (P (((((0.05)

Numbers in brackets are numbers of cows recorded for data analysis.

The years recorded for each data set are not identical but are from 1970-1983.

Table 5.60 **Performance characteristics in the first lactation of Holstein-Friesian, Dutch-Friesian and Friesian × Africander heifers in Mozambique** (from Alberro, 1980)

	Holstein-Friesian (30)[1]		Dutch-Friesian (25)		Friesian x Africander (20)	
	Mean	**s.d.**	**Mean**	**s.d.**	**Mean**	**s.d.**
Age at calving (months)	30	2	31	2	36	2
Milk yield — 305 days (kg)	3 249	788	2 690	615	2 548	555
Lactation length (days)	330	43	320	40	296	30
Milk yield per day (kg)	10.6	2.6	8.8	2.0	8.3	1.8

[1] Number of cattle in parentheses

Kenya. Friesian cattle on rape-mustard seed pasture at Egerton Agricultural College. Photo Kenya Information Services

Table 5.61 **Lactation records of F_1 Friesian × White Fulani zebu females, and contemporary pure White Fulani zebu females under identical management at Vom, Nigeria** (from Armour *et al.*, 1961)

Breeding	**No. of animals**	**Age at first calving** *(months)*	**Calving interval 1st to 2nd lact.** *(days)*	**Yield 1st lactation** *(kg)*	**Yield 2nd lactation** *(kg)*	**Length 1st lactation** *(days)*	**Butterfat** *(%)*
Cross Friesian-zebu	11	32	373	1 688	1 679	300	5.3
Pure zebu	11	47	415	855	[2]867	258	6.0

[1] From White Fulani dams mated to Friesian bulls
[2] Only 6 second lactations

Table 5.62 **Crossbreeding results of Friesian × White Fulani at Vom, Nigeria** (from Knudsen and Sohael, 1970)

Genotype	**Age at first calving** *(months)*	**Milk[1] yield** *(kg)*	**Lactation length** *(days)*	**Calving interval** *(days)*	**Liveweight at 24 months of age** *(kg)*
White Fulani	45	837	246	367	241
1/2 Friesian	31	1 690	271	358	314
3/4 Friesian	30	1 898	262	348	456
Friesian	29	2 434	281	368	473

[1] All lactations

Importations have been chiefly from Europe (UK, Netherlands and Germany) whilst semen has been imported from the USA as well as Europe and a batch was provided in 1979 under the FAO Donation Scheme. The AI scheme in Nigeria is in its infancy and, at present, the number of animals inseminated annually is reported not to exceed one thousand. The semen used is either pure Black and White or crossbred Black and White × White Fulani. The former is used in institutional herds whilst the crossbred semen is used in settled herds kept semi-intensively. The performance of purebred Black and White cattle in institutional herds is variable. In general, reproductive rate, livability and milk production are better in herds kept on plateau where temperatures are milder (V. Buvanendran, 1985, personal communication). Results from earlier work on the Vom farm (plateau 1 280 m) indicated that zebu x Friesian F_1 crossbreds yielded 138 percent (first lactation) and 168 percent (second lactation) more than the pure White Fulani zebu (Armour *et al.*, 1961). The results are summarized in Table 5.61.

Another study was made at Vom with the use of imported British Friesian bulls and heifers which allowed purebreds to be supplied to private and government farms in Nigeria (Knudsen and Sohael, 1970). A study covering 210 animals of various grades of Friesian × White Fulani was carried out. In general it was found that halfbred Friesian cows had an average milk yield of 1 600 kg per lactation — over twice that of zebu. The superiority of 3/4 Friesians over zebu amounted to 127 percent. Crossbreds calved 15 months earlier than zebu and had longer lactation by 20 days. The liveweight at 24 months of age of F_1 heifers was higher by 30 percent and of backcrosses by 90 percent over the White Fulani females (Table 5.62).

In the northern Guinea savannah where temperatures are higher, animals which are stall-fed and receive high levels of concentrate perform better than those out on pasture during the day. The performance of crossbreds has been studied to a greater extent than purebreds. In institutional herds, the crosses appear to be well adapted judging by reproductive performance and livability as well as lactation performance (V. Buvanendran, 1985, personal communication; Buvanendran *et al.*, 1981; Armour *et al.*, 1961; Knudsen and Sohael, 1970). Using the herd at Shika (640 m), a study was made of different grades from (1/2 to 7/8) of Friesian × White Fulani based on 12 years' milk records (1967 to 1978). Under the relatively good conditions of feeding and management in this herd, milk yield was found to increase progressively with increases in the proportion of Friesian genes. Lactation length also showed a similar trend, with a difference between 1/2 and 7/8 grades of 41 days. Calving intervals averaged approximately 390 days and did not differ among grades. Calf mortality was high, varying from 16 to 29 percent. The optimum genetic contribution from the Friesian for this environment was considered to be higher than 50 percent (Buvanendran *et al.*, 1981). Some data on the relative mortality of crosses and indigenous breeds given in Table 5.63 show the crosses to be similar to the indigenous zebu.

Crossbreds, however, seem to be more susceptible to diseases. Of particular importance are anaplasmosis, heartwater, streptothricosis and infectious kerato-conjunctivitis. Ticks are responsible for the first three diseases and as crossbreds are more prone to tick infestations, ticks may partly explain the greater incidence of tick-borne diseases in crossbreds. Some results are shown in Table 5.64.

Although statistics are not available, Friesian crosses are reported to be much more susceptible to the skin condition, streptothricosis (Kirchi), than indigenous breeds. In fact, in an ILCA study where Friesian crossbreds were kept alongside indigenous cattle in Fulani herds, streptothricosis was identified as the major disease problem. In order to minimize tick burden, the Fulani farmer de-ticks his animals manually. While this practice was effective in indigenous breeds, it did not control ticks adequately in Friesian crosses in the above study, with the result that many succumbed to Kirchi and anaplasmosis (V. Buvanendran, 1985, personal communication). In another study (Sekoni, 1983), it was reported that a Friesian bull affected with Kirchi had severe scrotal lesions which induced failure of spermatogenesis and caused azoospermia. However, indigenous animals with scrotal Kirchi did not show such adverse effects on semen quality. Friesian cattle imported from UK in 1980 were infected by bovine anaplasmosis and babesiosis (Ajayi *et al.*, 1982).

Nutrition is an important factor influencing performance of the crossbred as is made clear from a genotype and nutrition interaction study on growth rate (Zemmelink *et al.*, 1973). This study showed that on range feeding, White Fulani gained faster than the crossbred. However, in confined management, crossbreds grew faster but only in instances where the quality of the ration supported a daily gain of more than 0.45 kg in the White Fulani (Habers *et al.*, 1972).

Breinholt *et al.* (1981) reported observations on 14 Holstein-Friesian cows under lowland wet tropical conditions at the University of Nigeria farm at Nsukka (430 m). Results indicated that grazing time was significantly lower during the dry season than during the wet season. Grazing time, ambient temperature, hours of sunshine, milk yield and level of concentrates fed were all interrelated.

Experience in Nigeria with Black and White cattle as purebreds or crossbreds is considered to be rather limited. However, there is a growing interest in dairying and, with the increasing number of nomadic farmers settling down and adopting a system of integrated crop and livestock production, crossbreeding for dairying is considered likely to assume greater importance in the future (V. Buvanendran, 1985, personal communication).

Table 5.63 **Calf mortality in different breeds in Nigeria** (from Jagun, 1982; quoted by V. Buvanendran, 1985, personal communication)

Breed	No. of calves born	Percent mortality
Friesian	16	26.7
Crossbreds	551	11.1
White Fulani	316	12.5

Table 5.64 **Relationship of breed to clinical bovine infectious kerato-conjunctivitis in Nigeria** (Umoh and Adewuyi, 1984)

Breed	Proportion affected *(%)*
Friesian cross	15.0
Charolais cross	32.9
Sokoto Gudali	4.9
White Fulani	9.3

5.3.13 Somalia

Somalia is similar to many countries in this region of Africa with about ten percent of its cattle population classed as milking cows and an average milk production of 350 kg per cow per year. This provides, as shown in Table 5.42, about 30 kg per person of the human population, and imports of milk and dairy products added perhaps half as much again in 1985.

Native cattle are the East African Shorthorned Zebu (Jiddu, Boran, Gherra and Abgal). Three quarters of all the cattle are kept in the nomadic or semi-nomadic pastoral system in the parts of the Juba and river valleys which are not infected by tsetse fly. The rest of the cattle are owned by farmers and urban dwellers. Calf mortality is high (25 percent) even in good years.

Friesian and Guernsey cattle were introduced through an AI programme around 1965. Some Friesian bulls and cows were imported from Kenya and Kuwait. Since 1976, the AI Centre in Mogadishu has been in operation. In 1978, some 3 000 AI services were carried out with semen from Friesian bulls (Halpin, 1978).

Aspects of reproduction of zebu cattle and crossbreds with European cattle were studied by Aria and Cristofori (1980). Friesians × Borans and Friesians × Gherra showed a lower age at puberty (by 7 months) and longer oestrous cycle than the native zebus. Giorgetti *et al.* (1983) found that Friesian × Boran (or Gherra) crossbreds had higher milk yield but lower fat content than zebu cows.

5.3.14 Tanzania

Tanzania has an estimated cattle population of 14 million of which 19 percent are milking cows. Milk production per cow is very low — 160 kg/year (see Table 5.42) — but amounts to more (at 19 kg) per head of the human population than in many of the other countries of the region. Thus US$ 11 million spent on imports in 1985 (Table 5.42) added little more than a further ten percent to the available supply of milk and dairy products in that year. Although all statistics have to be treated with caution, data show that in the last 20 years the number of cattle has increased by 60 percent (FAO, 1967). The national herd is dominated by Tanzanian Shorthorn Zebu. Dairy breeds used for upgrading are Friesians, Ayrshires, Jerseys and Sahiwals. In 1978, the estimated number of grades was 29 000 females, 50 percent of them owned by smallholders. Because of inadequate environment, the yield of grades is low (1 500 kg/lactation). Artificial insemination, introduced in 1964, is still in the early stages of development (FAO, 1979b). According to the official livestock policy (Ministry of Livestock Development, June 1983) the major thrust for increased milk production will depend on the more widespread use of dairy cattle with higher milk production potential and the introduction of management systems to exploit that potential. Some further importations of purebred exotic cattle will be undertaken, particularly to stock large farms. The main task is to upgrade indigenous cattle through an extensive crossbreeding programme, including the use of heifer breeding units, artificial insemination, bull centres, breeding stations (natural mating) and distribution of graded bulls and heifers to smallholders and villages.

5.11 Tanzania. Young cows, including Friesian and Friesian-zebu crosses

One of the state farms, Kizimbani, has a grade Friesian and a grade Jersey herd, based on cattle imported in the late 1960s from Kenya. AI with Friesian semen from Kenya and recently from Switzerland is used. Introduction to small farmers of grade Friesians and Jerseys was unsuccessful because of their susceptibility to heat stress, feed scarcity and diseases (East Coast fever and lumpy skin disease), although Jerseys showed more resistance. Milk production of cows with two or more lactations in 1982-83 was 1 870 kg for Friesians and 2 225 kg for Jerseys (adjusted figures for days in milk, calving season, days dry and days open). Calving intervals were 495 days in Friesians and 449 days in Jerseys (C.H. Pedersen, 1985, personal communication).

5.3.15 Uganda

The cattle population in Uganda in 1985 numbered 5.2 million head, of which 1.04 million were milking cows and this provided for nearly all the milk available for consumption and processing in 1985. The imports, equivalent to US$4.7 million in that year (Table 5.42), provided only about a further six percent. Indigenous breeds are East African Zebu, Ankole Long Horn and Bukedi and a variety of crossbreds. Over 90 percent of cattle are owned by smallholders with an average of 2.3 hectares of land. The average milk yield of indigenous cattle is 1.4 kg daily (FAO, 1975).

Grasslands in Uganda are very suitable for livestock in areas freed from the tsetse fly. An early study covering the period 1962-66, although based on relatively small numbers, showed that very acceptable levels of milk production were attained by a commercial herd in the vicinity of Entebbe with annual yields of 2 917 kg (9 Friesians), 2 271 kg (50 Guernseys) and 2 050 kg (15 Jerseys) (Marples and Trail, 1967). The national dairy herd is concentrated on dairy stock farms and experimental stations. They are mainly operated by the Government and some are privately owned. These farms have exotic dairy breeds and crossbreds with zebu and make up about 1 percent of the total cattle population. The exotic breeds are mainly Friesians but also include some Dutch Red and Whites, and Ayrshires. Dairy stock farms are aimed at multiplication, grade animals being sold to smallholders.

This programme was in operation in the early 1970s, at which time commercial milk production was growing. Since 1960, some 10 000 exotic dairy cattle had been imported (FAO, 1975) from Kenya, UK, the Netherlands, France, Canada and USA. The Friesians and their grades at present amount to 80 000. Artificial insemination is used with semen from Europe (A.P. Nakate-Lusejera, 1986, personal communication).

The chosen method for improvement of dairy cattle in Uganda is by upgrading, using exotic sires of suitable breeds. Performance of Friesians has been tested with good results. F_1 animals produced four times more milk than the indigenous zebu cows. To introduce crossbreeding countrywide, 8 000 bulls would be needed and government farms should concentrate on production of exotic bulls (FAO, 1975).

5.3.16 Zambia

Not atypically, close to 10 percent of the cattle population of the country is classed as dairy cows (Table 5.42), the rest belonging to a traditional, relatively poorly-managed sector of smallholdings. The milk produced amounts to only 10 kg per head of the human population and this represented 85 percent of the total consumption in 1985.

Native cattle breeds are the Angoni (Shorthorn Zebu), indigenous to the eastern part of the country, Barotse in the western part and Tonga in the southern part. The latter two breeds represent the Sanga type. These breeds have an average lactation length of 4 months and a daily milk yield of 2 litres (Zambia, 1977).

The commercial sector (15 percent of the cattle population) comprises fenced ranches for beef production and dairy farms. The dairy industry is based on Friesians imported as live cattle or crossbreds obtained by AI with semen mainly from Europe, Kenya and South Africa. Crossbreds from Friesian × indigenous females are distributed among small farms to encourage milk production in rural areas (Daplyn *et al.*, 1975).

There was a severe reduction in the number of dairy herds, from 19 871 in 1964 to 12 411 in 1975. This was because of the decreased profitability of milk production in relation to competing enterprises (Abdalla, 1978).

Hultnäs (1973) reported that Zambian Cattle Development Ltd established four dairy production units with 400 Friesian cows per unit. In 1974, there were some 120 large-scale, expatriate-owned dairy farms which produced about 1.6 million litres of milk per year, the basic stock of these commercial herds being Friesian, but with a number of crossbred herds (Teferra, 1983).

F.M. Kaluba (1986, personal communication) reported that the total number of Friesian cows in Zambia in 1984 was 20 000 and Friesian crossbreds (Friesian × local) 5 300. Countries of origin were the USA, UK, Zimbabwe and the Netherlands. The average milk yield (in 305 days) on commercial farms was 3 500 litres per cow. The breeding strategy was to produce an F_1 Friesian × Sahiwal local cross. At present semen is being imported from the USA, UK, Australia and the Netherlands. Earlier Hultnäs (1973) reported that 6 150 doses of semen of beef and dairy breeds were imported from UK, USA and Kenya and 6 408 doses were sold in the same year. Of the 4 620 doses of semen from dairy breeds, 3 768 (80 percent) were Friesian.

Diseases that caused the greatest losses to the cattle industry and reduced the country's milk and beef production were trypanosomiasis, tick-borne diseases, East Coast fever, foot-and-mouth disease and contagious bovine pleuropneumonia (Teferra, 1983).

At the experimental station of Mazabuka, crossbreeding results of Angoni-Barotse and Boran dams with Friesian, Hereford, Boran, Angoni and Barotse sires were studied under ranching conditions (Thorpe *et al.*, 1980). In this study, based on large numbers of animals, the mean calf weaning weights for progeny (of all types) from Angoni dams were 147 kg, from Barotse dams 167 kg and from Boran dams 170 kg. Hereford and Friesian bulls were similar in siring progeny which were approximately 25 kg heavier at weaning than those sired by Angoni bulls and between 10 and 18 kg heavier at weaning than those with Boran sires.

5.3.17 Zimbabwe

In recent years, Zimbabwe has made much economic progress mainly due to its very effective agricultural policy. The native cattle breeds (Sanga type) are the Mashona, Nkone and Tuli (Ward, 1982). These cattle in the subsistence sector are used for draught purposes, milk production, for manure and occasionally for meat. Only a very small proportion (less than 2 percent) of the substantial cattle population are regarded as dairy cows with, for the region, a very high average milk production of nearly 1 700 kg per cow allowing 22 kg per person of the human population. Imports added only negligibly in 1985 (Table 5.42). The milk yield of indigenous cattle reaches 650 kg of milk in 250 days' lactation. Commercial beef and milk production are concentrated on big farms which cover 50 percent of agricultural land. The number of dairy cattle is 115 000 head originating from UK and South Africa. Commercial farms utilize mostly purebred cattle, the majority being Black and White Friesians. Average yield is 4 600 kg of milk with a calving interval of 403 days.

With a view to introducing small-scale commercial milk production into the smallholder section, a programme was recently initiated to grade up indigenous stock with Holstein-Friesian. At present the smallholder makes no impact on dairy production in Zimbabwe.

In 1985, some 26 000 straws of local Black and White cattle semen and 17 000 straws of imported semen were sold to the dairy industry. The source of the semen was predominantly USA Holstein-Friesian, but some were from British Friesian and Canadian and Holstein sires. The demand for imported semen far exceeds the foreign currency available for its purchase.

The commercial sector of the industry is modern and highly sophisticated and is well-supported by services and organized marketing. The industry is based largely on homegrown, high-quality forages, improved pastures (dryland and irrigated) and usually homegrown grain (maize) with some purchased concentrate. In the drier areas, more use is made of exclusively purchased concentrate rations for milk production. The growing season is five to six months of the year (summer rainfall), this being naturally the period of maximum production. Heavy reliance is placed on the use of silage (mostly maize) for winter (dry season) feeding.

Important aspects of herd health are tick control (compulsory dipping), vaccination against contagious abortion, tuberculosis, blackleg, in some cases

Rift Valley fever and anthrax. Dairy hygiene is controlled by government inspectors (Department of Research and Specialist Services-Dairy Services) and so is milk quality (J.L. Grant, 1986, personal communication).

5.4 LATIN AMERICA AND THE CARIBBEAN

In Latin America and the Caribbean region, livestock production, particularly that devoted to ruminants, has traditionally been one of the most important components of the agricultural sector supported by an estimated 550 million ha of permanent pasture of which 15 percent is improved and cultivated. Climate and vegetation (Figure 5.5) vary greatly, rainfall, elevation and latitude being the major influencing factors. Rainy seasons differ in length, duration, and distribution over the seasons of the year.

Cattle are by far the most important component of the region's livestock industry since the vast grazing areas are more conducive to the production of cattle than of other ruminants and preference has traditionally been given to cattle over small ruminants. The distribution of cattle is shown on the world map (Figure 1.2).

General information on the cattle population and its productivity in the countries of Latin America is given in Table 5.65. As may be seen from Table 5.66 there have been changes in the region, both in respect of numbers and total output from cattle, all having increased over the period 1975 to 1985. Table 5.67 shows, however, that these total increases have not been matched by increases in milk or beef production in relation to the size of the human population — production from cattle per person has declined and production per animal has declined for beef and veal and has been no more than maintained for milk. As a result (as shown in Table 5.68) the region has become a net importer of beef and veal where formerly it was a net exporter, and net imports of milk and dairy products have increased substantially since 1975.

The low level of cattle productivity in Latin America is due mainly to low reproductive rates, high mortality rates and prolonged fattening periods. First calving normally occurs at 3 to 4 years of age and the calving interval exceeds 400 days. This low reproductive efficiency combined with high calf mortality makes it almost impossible to raise productivity because it results in: (1) an insufficient number of animals for sale to justify investments for the better management of the herd, and (2) too small a number of replacement females and a long generation interval which limit genetic progress.

There are approximately 36 million dairy cows in the region yielding an average of less than 1 000 litres of milk per cow. Most of the dairy farms in the lowland tropical zone do not specialize in milk and most of the cows used for milk production are dual purpose, milked seasonally and in many instances only once a day.

Fig. 5.5 Latin America and the Caribbean — main agro–vegetation zones

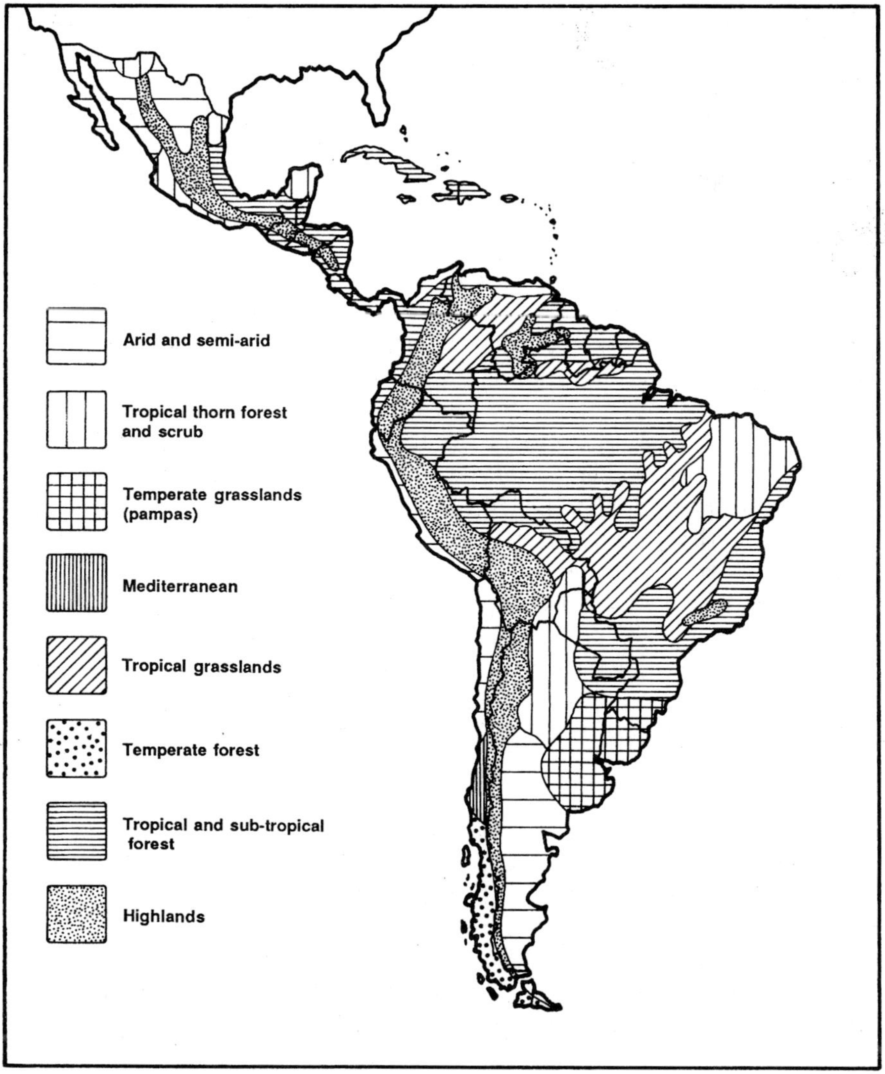

Table 5.65 **Cattle production in the countries of Latin America and the Caribbean** (from FAO, 1985)

Country	No. of cattle	No. of animals slaughtered	Beef and Veal Meat production	Beef and Veal Carcass wt. per animal	No. of dairy cows	Milk yield per cow	Milk produced available per person	Net imports (milk equivalent)
	('000)	*('000)*	*('000 MT)*	*(kg)*	*('000)*	*(kg)*	*(kg)*	*(US$ millions)*
Antigua and Barbuda	17	3.3	0.45	152	6	1 000	75.0	2.4
Argentina	54 800	13 800	2 700	196	2 950	1 898	183.2	11.0[1]
Bolivia	5 851	600	94	156	56	1 418	12.6	11.4
Brazil	34 500	10 234	2 136	208	14 700	729	76.7	19.2
Chile	3 400	800	180	225	660	1 576	84.1	5.4
Colombia	21 935	3 470	643	185	2 850	983	97.5	6.5
Costa Rica	2 553	311	62	200	290	1 279	142.7	1.8
Cuba	6 400	880	145	165	680	1 470	99.6	86.9
Ecuador	3 378	637	115	180	720	1 372	105.3	11.1
Guatemala	2 587	300	57	190	404	825	41.8	10.0
Honduras	2 508	267	39	146	430	651	64.0	15.8
Mexico	37 450	3 500	665	190	8 500	814	87.6	144.2
Nicaragua	1 890	300	59	197	200	625	38.2	9.4
Paraguay	6 400	526	100	190	90	1 889	51.6	0.5
Peru	3 900	680	101	148	690	1 172	41.1	33.9
Uruguay	9 948	1 950	350	180	530	1 604	296.5	21.6[1]
Venezuela	12 486	1 719	320	186	1 410	1 087	88.5	115.2
Caribbean								
Barbados	18	1.5	0.28	187	7	1 296	36.9	4.4
Dominican Republic	2 420	329	58	177	249	2 000	79.8	11.4
Jamaica	321	78	16	205	50	1 000	21.2	14.2
Puerto Rico	580	192	24	161	202	1 797	105.2	—
St. Christopher and Nevis	6.4	1.7	0.30	181	—	—	—	1.7
Trinidad	77	8	1.4	175	6	1 750	8.8	50.7

[1] Net exports

Table 5.66 **Cattle numbers and beef and milk production in Latin America and the Caribbean, 1975-85**

	No. of cattle	No. of cows	Total meat	Beef & veal	Total milk	Cows milk
	(Million head)		*(Million metric tons)*			
1975	259.4	32.1	11.7	7.3	31.1	30.6
1980	297.4	37.7	14.2	8.0	35.5	35.1
1985	313.4	36.3	14.9	8.0	35.3	34.7

Table 5.67 **Milk and beef production per head of human population and cattle productivity in Latin America and the Caribbean, 1975-85**

	Production per caput *(kg)*				Productivity per animal	
	Total meat	Beef & veal	Total milk	Cows milk	Beef & veal	Cows milk
1975	36.5	22.6	96.7	95.3	28.1	953
1980	39.3	22.0	98.4	97.0	26.9	931
1985	36.9	19.8	87.1	85.8	25.5	956

Table 5.68 **Net trade in meat and milk in Latin America and the Caribbean, 1975-85**

	Total meat		Total milk		Beef and Veal	
	('000 MT)	*(million US$)*	*('000 MT)*	*(million US$)*	*('000 MT)*	*(million US$)*
1975	+ 139.2	+ 111.4	-1 891	-339.6	+ 71.2	+ 49.9
1980	+ 204.0	+ 340.3	-5 293	-935.8	+ 85.5	+ 160.9
1985	+ 209.0	+ 156.3	-4 120	-656.3	-33.5	-54.8

\+ Exports - Imports

Table 5.69 **Milk and fat yield (kg) of Holstein-Friesians of different origin in Latin America** (from de Vaccaro, 1982)

Country		Cows imported from				Home-bred cows (descendants from imported animals)			
		USA	Canada	Netherlands	Origin Sire Semen:	USA	Canada	Local	Unspecified
Mexico									
[1]		5 451 (742)	4 924 (1 124)			5 194 (2 124)	4 884 (5 053)	4 460 (2 955)	
[2]		6 138 (199)	6 035 (992)						5 909 (12 863)
Peru[3]		5 125 (617)	5 263 (529)						4 926 (3 036)
Venezuela[4]		4 090[a]				3 849		3 812	
		4 311[b] (139)				3 779 (30)		3 665 (101)	
Brazil[5]	Milk yield	3 958		3 785					
	Fat yield	138 (537)		144 (2 509)					

Sources:
[1] McDowell et al. (1976)
[2] Powell and Dickinson (1977)
[3] Sarmiento (1970)
[4] McDowell (1974)
[5] Alves Netto *et al.* (1967)
No. of animals in parentheses

[a] First lactation
[b] Later lactations

The dairy cow population in the temperate zones is much more homogeneous and Holstein-Friesians predominate. Producers tend to specialize in dairy production with cows milked twice a day and calves reared apart from the cows.

The dairy cows in Latin America comprise the distinct native criollo breeds such as the Central American Criollo Lechero and Colombian Costeno con Cuernos; the well-established specialized zebu breeds such as the Brazilian Gir; the nondescript criollo and zebu types which are common in Peru and Ecuador, and the specialized North American and European dairy breeds (de Vaccaro, 1982).

Information on Holstein-Friesian cattle is based mainly on experiments dealing with comparison of different imported (North American and European) dairy breeds and crosses among imported breeds and local cattle reviewed by de Vaccaro (1982). The information presented is far from complete, but illustrates the production capacity of imported dairy breeds and crossbreds with local criollo or zebu breeds as well as problems of cattle management in tropical conditions.

Holstein-Friesians of different origin have been compared in some countries (Table 5.69) and though the cows of USA origin showed the best yields, strain differences were quite small. The breeding values of the Black and White sires used in the Latin American environment show high positive correlation with the corresponding progeny test results in their countries of origin (de Vaccaro, 1982).

Some comparative data on breeds and crosses are presented in Table 5.70 for milk production and in Table 5.71 for aspects of fertility and for survival.

The success of the introduction of European breeds into the region has depended greatly on management and nutrition and studies reviewed by de Vaccaro (1982) show that additional inputs of feed and management, including housing, led to extra productivity. However, breed differences are also apparent. Both Holstein-Friesian and Brown Swiss breeds have been used to produce large numbers of crosses and the weight of evidence now suggests that the Holstein-Friesian crosses give the better performance in most respects, although differences in specific comparisons are often not statistically significant. Some results are shown in Table 5.72.

North American and European cattle, although numerically in a minority, produce most of the milk sold in such countries as Mexico, Peru and Chile especially under the more temperate conditions. These breeds are thus recognized as having a vital economic role in the region (de Vaccaro, 1982).

Table 5.70 **Performance of dairy cattle of different genotypes in Latin America** (from de Vaccaro, 1982)

Country		Breeds[1]	Proportion of Holstein-Friesian genes						Source
			0	1/2	5/8	3/4	7/8	Pure	
Brazil	Milk recorded	Holstein		2 970		3 235	3 906	3 579	Alves Netto
	herds, Sao Paulo	Grades		(79)		(374)	(916)	(7 720)	*et al* (1967)
	”	Holstein				3 140	3 290	3 437	”
		Grades				(108)	(104)	(1 560)	”
		Holstein						3 605	
		Holstein						(21 144)	
								3 454	
								(3 211)	
Colombia	Turipana	Costeno con	896	2 000		1 832		1 950	Rubio (1976)
		cuernos	(652)	(187)		(35)		(162)	
		Holstein							
	El Nus	Blanco		1 039					Botero (1976)
	Milk with	Orejinegro		(31)					
	calves	Holstein							
Cuba	Grazing +	Holstein	500	1 811	3 1				
					00	3 339		3 554	Prada (1978)
	concentrates	Zebu		(1 300)	(200)	(65)		(high)	

Table 5.70 *Continued*

Country		Breeds[1]	Proportion of Holstein-Friesian genes						Source
			0	1/2	5/8	3/4	7/8	Pure	
Venezuela	Farm 1	Holstein				2 600		2 708	Verde
		Native				(1 376)		(90)	*et al.* (1970)
	Farm 2	Holstein		3 087		3 560	3 643	4 018	,,
		Native		(1 132)		(877)	(183)	(205)	
	Farm 3	Holstein		2 792		2 959		3 766	Verde
	Least squares means, 574 heifers	Native							*et al.* (1971)
	6 herds, 1 200	Zebu	1 670	Holstein		2 022			Cerrada
	mm rain, breeds	Holstein		type cross:					*et al.* (1977)
	class yield	B. Swiss		Brown Swiss		1 970			
	appearance (n = 2 351)			type cross:					
	part-stabled	Native	2 632[a]	undefined cross:		2 597[a]		3 961[a]	Martinez
	Highland Region (n = 382)	Holstein	3 089[b]	,, ,,		3 013[b]		4 050[b]	*et al.* (1977a)

[a] 1st lactation

[b] later lactations

[1] pure Holstein-Friesian or, more generally, crosses of specified or unspecified local cattle with Holstein-Friesian

No. of animals in parentheses

Table 5.71 **Aspects of fertility and survival of dairy cattle of different genotypes in Latin America** (from de Vaccaro, 1982)

Country		Breeds[1]	Proportion of Holstein-Friesian genes						Source
			0	1/2	5/8	3/4	7/8	Pure	
Colombia	Turipana	Costeno con cuernos	C.I.	453	426			505	Rubio (1976)
				(3 554)	(47)			(14)	
		Holstein	S.C.	1.7	1.9			3.0	Huertas (1972)
				(475)	(90)			(216)	
Colombia	El Nus, Grazing	Blanco Orejinegro Holstein	C.I.		471				Botero (1976)
Cuba	Grazing + concentrates	Holstein	C.I.	480	382	365	375	433	Prada (1978)
		Zebu	C.I.		(1300)	(200)	(65)	(high)	

Table 5.71 (Continued)

Country		Breeds[1]	Proportion of Holstein-Friesian genes						Source
			0	1/2	5/8	3/4	7/8	Pure	
Venezuela (1977)	Lake Maracaibo region (n = 1 349)	Criollo B. Swiss	C.I	400 (criollo) ±0.16			(B. Swiss)	397 ±0.37	Contreras *et al.* (1977)
		Holstein Zebu		415 (zebu) 0.33				419 ±4.42	
	6 herds, 1 200 mm rain breeds classified by appearance (n = 2 351)	Zebu Holstein B. Swiss	C.I.	411	Holstein type cross:		422		Cerrada *et al.*. (1977)
					Brown Swiss type cross:		445		
	Part-stabled Highland Region (n = 576)	Native Holstein	Days Open	108	Undefined cross:		97	113	Martinez *et al.* (1977b)
		Native Holstein	S.C.	1.74	Undefined cross:		1.82	183	"

No. of animals in parentheses C.I.: Calving interval S.C: Services per conception

[1] Pure Holstein-Friesian or, more generally, crosses of specified or unspecified local cattle with Holstein-Friesian

Table 5.72 **Comparison of Holstein-Friesian and Brown Swiss crosses with zebu and criollo in various Latin American studies** (from Vaccaro, 1984; Capriles *et al.*, 1981 and Reaves *et al.*, 1985)

	Country	Crossbreds			
		Numbers	Holstein-Friesian	Brown Swiss	Level of significance of difference
Stillbirths (%)					
	Venezuela	1 350	2.3	2.8	NS
Death and involuntary culling (%)					
0-9 months	Venezuela	1 656	13.4	18.4	.01
9 months-1st calving	,,	665	9.2	10.9	NS
Service-1st calving	,,	1 958	22.5	30.8	.01
Herd life					
Calvings/ lifetime	,,	577	2.6	2.7	NS
Body weight (kg)					
10 months old	Colombia	255	134	128	NS
adult cows	,,	687	382	366	.01
Age at 1st calving (months)					
	Colombia	62	31.0	35.0	—
	Venezuela	996	32.7	35.6	.01
	El Salvador	548	34.6	34.4	—
Reproduction					
Birth rate (%)	Colombia	201	71.9	67.7	—
Calving interval (days)	Venezuela	625	432	439	NS
	,,	297	441	445	NS
	,,	1 349	418	397	NS
	,,	2 351	422	445	NS
	,,	217	386	399	NS
	,,	608	424	443	NS
	Colombia	507	421	474	.01
Service period (days)	Venezuela	65	66	104	.01
	El Salvador	1 640	156	151	—
Milk production (kg/lactation)					
	Colombia	154	720	730	NS
	,,	240	2 000	1 318	—
	El Salvador	1 640	2 710	2 797	—
	Venezuela	94	1 816	1 705	NS
	,,	3 002	3 409	3 148	.01
	,,	464	1 457	1 664	.05
	,,	1 063	1 486	1 689	.05
	,,	2 351	2 022	1 970	NS

5.4.1 Argentina

As shown in Table 5.65, the total number of dairy cows, which includes dual-purpose animals, was close to 3 million out of a total cattle population of nearly 55 million in 1985. Milk production is distributed over about 38 000 dairy farms (FAO, 1984). Average production per cow in 1985 was 1 898 kg and there was a useful net export of milk or milk products (Table 5.65).

At the end of the nineteenth century milk was produced using the criollo, which still has an established role on marginal land, and the Durham breed (a Shorthorn type of dairy animal). The beginning of the twentieth century saw the start of importations of Dutch Black and White cattle and the establishment of pedigree herds in the Buenos Aires province. Herds of Black and White cattle built up through grading-up with Dutch cattle which were started in Santa Fé province, and the first Holstein-Friesians began to appear around the same time. Between the two world wars importations were primarily from Europe and thereafter from Canada and the USA (A.J. Nader, personal communication, 1987).

In 1919 the Sociedad Rural Argentina opened a register of Friesians, establishing the Holando Argentino breed, 50 000 of which are now registered as pedigree animals with the Society. There are about 417 000 cows of that breed on 3 500 dairy farms which are officially recorded in 70 regional groups. It is estimated that about 35 percent of these cows are served by AI. The development of a dual-purpose type of animal which is useful for both milk and meat production, based on the Black and White breed, is now also attracting attention (A.J. Nader, personal communication, 1987).

Black and White cattle constitute a major part of the country's dairy cattle population. Most of the Black and White population is in the provinces of Buenos Aires, Santa Fé and Córdoba and much of the milk is processed industrially for urban use.

A recent analysis of 15 891 first lactation records showed a mean milk yield of 3 742 $\pm$1 246 kg and a fat yield of 123.6 $\pm$40.2 kg (Musi and Bidart, 1983). Average age at first calving was 33.3 $\pm$4.7 months). These results were obtained from cattle kept principally on grass in the temperate "pampas" region, with only seasonal grain supplementation. The same study showed the mean predicted difference in genetic merit of 110 local AI bulls to be +35 kg milk, compared with +100 kg for 20 sires used via imported semen, but the maximum predicted difference among 64 bulls proven locally with more than 50 daughters was +783 kg milk which shows excellent prospects for continued genetic improvement.

5.4.2 Bolivia

As seen from Table 5.65 less than one percent of the cattle population of nearly six million was classified as dairy cows in 1985 with an average yield of more than 1 400 kg per cow per year. The milk available per head of the human population from that produced in Bolivia is almost the lowest in the South American continent (at about 13 kg milk per person). Imports not quite doubled that amount in 1985.

Holstein-Friesian cattle were first imported to Bolivia some fifty years ago for use in the Cochabamba valley. The valley, which has an altitude of 2 700 m and is at latitude 17°S, is semi-arid and as a result has a mild temperate climate. The cattle population, which contains much "Carnation" blood, numbers about 10 000 head. Cattle are fed on lucerne grown under irrigation and numbers are not likely to increase as land suitable for irrigation is being lost to urban development (J.V. Wilkins, 1985, personal communication). After the 1952 land reform most of the herds were split up into smaller units with a milk producers' association supporting an AI and milk recording programme which is still operating. In recent years almost all semen has been purchased from the USA (Atlantic Breeders Corp.).

The lactation yield amounts to 3 000-4 000 litres (J.V. Wilkins, 1985, personal communication). Whilst tropical parasites and diseases are virtually unknown in the valley, brucellosis (and tuberculosis) are considered to be serious.

One Holstein-Friesian herd in the Cochabamba valley achieved the following results over a ten year period (540 lactation records).

Age at first calving	38.3 months
First lactation yield	3393 kg
Yield — all lactations	3385 kg
Calving interval	14.5 months

Other herds in this valley have similar performance (Agreda Claros, 1985 cited by J.V. Wilkins, 1985, personal communication).

The success in establishing this Black and White population has been followed, on a much smaller scale, in a similar environment in the valleys of Tarija with Black and White cattle bought from Uruguay and Argentina.

The Black and White cattle of Santa Cruz are descended from American Holstein-Friesian cows of Cochabamba and Dutch Friesians from Argentina. Semen has been exclusively imported from the USA (J.V. Wilkins, 1985, personal communication).

At altitudes above 3 000 m Black and White cattle are numerically unimportant.

Wilkins *et al.* (1979) compared the production of purebred Holstein-Friesian with crossbred Holstein-Friesian × Mestizo under three different management systems in Santa Cruz, Boliva. Management systems were as follows:

1. Cows maintained permanently in shaded yards; cut fodder from improved pastures and forages; calves weaned in first week; milked twice per day.
2. Cows on improved pastures, fed concentrates; milked twice per day; calves weaned in first week.
3. Cows on improved pastures and forages; without concentrates; calves not weaned; milked once per day when milk surplus.

Although the differences among genotypes within the system were statistically not significant they show a tendency towards superiority of F_1 Holstein-Friesian × Mestizo crossbreds over both pure Holstein-Friesian and Mestizo on management systems (2) and (3) (Table 5.73).

5.4.3 Brazil

Milk is produced mainly in the tropical south-east and in the southern regions of Brazil which contribute about 57 percent and 25 percent of total production, respectively.

Virtually all the milk and milk products available for the human population were produced in the country and amounted to 77 kg per person in 1985. The net imports of milk and milk products (US$19 million in 1985) were small relative to the size of the population. Currently the milk yield per cow is among the lowest in this region. The total cattle population (more than 134 million) is the largest on the continent and similar in magnitude to the human population, but dairy cows represent little more than 10 percent of all cattle (Table 5.65).

In the south, cattle are predominantly crosses of high grade Holstein-Friesian with mixed breeds. A high proportion of milk is produced on small farms. Production is mainly based on natural pastures, with some utilization of small winter pasture paddocks — inputs of concentrates are low. As a result, average yield per cow is low. However, there are some farms where more concentrates are used and where higher productivity is obtained. One example of these farms is in the Castro area of the State of Parana (Dutch immigrants, cooperative system). In 29 herds with a total of 2 152 cows the yield in 305 days' lactation was 5 086 kg of milk, 180 kg of fat and 3.5 percent fat content.

Table 5.73 **Milk production and fertility and mortality of purebred Holstein-Friesian, Holstein-Friesian × Mestizo and Mestizo under different management systems in Bolivia** (from Wilkins *et al.*, 1979)

System	Breed or cross	Cows (No.)	Milk yield/cow/year (kg)	Estimated calving interval (days)	Calves born (No.)	Calf mortality (%)	Adults (No.)	Adult mortality (%)
1	Holstein-Friesian	84	3 091 ± 123	417	152	5	114	4
2	Holstein-Friesian	20	2 380 ± 90	379	30	20	26	15
	Holstein cross Mestizo	11	2 744 ± 246	355	21	10	17	0
3	Holstein-Friesian	17	543 ± 98	514	12	42	23	13
	Holstein cross Mestizo[1]	21	840 ± 65	480	16	19	19	5
	Mestizo	15	525 ± 76	545	10	0	15	0

[1] Mestizo = zebu + criollo

The age at first calving was 33.2 months and the calving interval 416 days.

In dairy herds Holstein-Friesians and their crosses predominate but censuses do not give the breed or type of cattle. As a result only numbers of registered Holstein-Friesian cattle can be given. In the last 20 years this number has grown ten-fold and amounted to 410 974 head in 1977-78 (Breeders' Association, quoted by F.E. Madalena, 1985, personal communication). For example, in the Zona da Mata (Minas Gerais) area, Holstein-Friesians and their crosses were kept on 42 percent of small farms, on 86 percent medium sized farms and on 92 percent large farms.

Officially recorded national average milk yield figures for Holstein-Friesian cattle (considered a biassed sample as in general only the best animals are recorded) (Source: Breeders' Association) are shown in Table 5.74.

The popularity of the Holstein-Friesian in Brazil is indicated by the extent of importations both of live animals and frozen semen. Data from the Breeders' Association (F.E. Madalena, 1985, personal communication) show the importation of close to 7 000 purebred animals in the period 1978-81 of which more than 2 000 came from Canada, between 1 and 2 000 from each of Argentina, USA and Uruguay and smaller numbers (totalling about 300) from 4 European countries. In addition, 51 000 and nearly 17 000 grade Holstein-Friesians were imported from Uruguay and Argentina respectively and 83 000 units of frozen semen mostly from North America and the rest from Europe.

The extensive use of AI and the popularity of the Holstein-Friesian (HF) (close to a third of the total) is shown in Table 5.75 in terms of sales to farmers of doses of semen from AI centres.

Costa *et al.* (1974) have published information on the performance of Holstein-Friesian cattle in Brazil in comparison with other breeds. For 157 HF cows, milk yield of cows calving in winter, spring, autumn and summer averaged 3 849, 3 767, 3 720 and 3 523 kg and fat yield 124, 123, 129 and 124 kg respectively. The corresponding values for 61 Jerseys were 3 227, 2 915, 2 972 and 2 830 kg milk and 122, 112, 116 and 114 kg fat.

Sarmiento (1979) studied the fertility of pure Guzerat, and both 50 percent HF and 75 percent HF crosses with Guzerat and found that age at first calving averaged 47.0, 33.2 and 33.0 months respectively, length of service period 8.0, 4.7 and 4.8 months and length of calving interval 17.2, 14.1 and 14.2 months.

Data on 2 000 lactations of cows in Brazil with varying amounts of HF blood were analysed by Neiva *et al.* (1979). Milk yield averaged 3 000 $\pm$1 010 kg, and lactation length 301 $\pm$61 days. Lactation length accounted for 41.5 percent of the total variation in milk yield. Year had a significant effect on milk yield, as did month of calving. Cows calving in May-October yielded 158 kg more milk, on average, than those calving in November-April.

Costa *et al.* (1982) found the number of inseminations per conception of 493 HF cows in Brazil averaged 1.59 ±0.04, and was significantly affected by year of calving. Gestation length averaged 276.1 ±0.73 days, and was significantly affected by season and sex of calf; it was longer for calves born in winter and spring than for those born in summer and autumn, and longer for male than for female calves. Length of calving interval averaged 406.5 ±4.2 days, and was significantly affected by year ($P < 0.05$).

Freitas *et al.* (1983) studied 2 987 lactation records, collected between 1962 and 1978, from 864 cows. Average lactation length was 318 days. Least squares means for milk yield, fat yield and fat percentage were, respectively 4 334 ±230

Table 5.74 **Holstein-Friesian milk yields [1] in Brazil (305 days ME 2x)** (F.E. Madalena, 1985, personal communication)

	No. of lactations	Milk yield *(kg)*	Fat yield *(kg)*
Black and White	5 644	5 442	191
Red and White	1 533	4 957	176

[1] Selected sample

Table 5.75 **Numbers of semen doses sold to farmers in Brazil** (from Brazilian Ministry of Agriculture; F.E. Madalena, 1985, personal communication)

Type of cattle	Sales of semen *(No. per annum)*
Black and White Holstein-Friesians	415 000
Red and White Holstein-Friesians	54 000
All breeds	1 540 000

All breeds, 5 year average for frozen semen

(1979-83): 63% beef; 32% dairy; 5% dual purpose

kg, 160 ±7 kg and 3.7 ±0.1. Month of calving had a significant effect on lactation yield, with maximum yields related to calving in May or June.

In many crossbreeding experiments Holstein-Friesians were used for improving local zebu breeds. The results of some of these experiments are presented below.

Production results for Black and White Holstein-Friesian and crossbred cows on a private farm at Lavras in Minas Gerais State are given in Table 5.76 (Madalena *et al.*, 1983). They show that the difference in milk yield of pure Holsteins and of the cattle highly upgraded to Holstein-Friesians was small.

Data for the south-east region of Brazil on milk yield of Holstein-Friesians (and other exotics) and their crosses are given in Table 5.77 and suggest that the pure exotic breeds and grades above 50 percent were inferior to the first crosses in yield.

A smaller experimental group of Black and White Holstein-Friesian and HF × Gir cows maintained in a hilly region of Rio de Janeiro State, and covering 634 lactations, indicated that crossbreds had a higher initial production and higher persistency than purebred cows and F_1 had a higher initial production than the first backcross — though groups had a similar persistency of lactation (Madalena *et al.*, 1979a).

In yet another report by Madalena *et al.* (1979b) the results indicated that crossbreds were the most useful in tropical areas where feed, management and health impose constraints (Table 5.78).

Table 5.76 **Least squares means (± s.e.) for some production traits of crossbreds with different levels of Holstein-Friesian (HF) blood on Lavras farm, Brazil** (from Madalena *et al.*, 1983)

Proportion of HF blood	Lactation length *(days)*	Lactation milk yield *(kg)*	Calving interval *(days)*
3/4 HF	305 ± 14	4 034 ± 252	406 ± 19
7/8 HF	301 ± 15	3 849 ± 263	431 ± 19
pure HF	318 ± 18	4 149 ± 318	433 ± 23

Table 5.77 **Milk yield of different European breeds × zebu crossbreds in Brazil** (from Madalena, 1981)

Proportion of European blood	Holstein[1] × Guzera (S. Jose do Rio Pardo Sao Paulo)	Holstein[2] × Gir (Valenca Rio de Janeiro)	Brown[3] Swiss Holstein Jersey × zebu crossbreds (Florestal, Minas Gerais)
	Yield per cow per year *(kg)*	300 day yield *(kg)*	Lactation yield *(kg)*
Zebu	1 582 (34)	—	—
1/4	1 992 (46)	—	—
3/8	2 238 (18)	—	—
1/2	2 527 (69)	2 471 (350)	2 887[4] (163)
5/8	2 567 (73)	—	—
3/4	2 425 (144)	2 347 (164)	—
7/8	2 336 (32)	—	2 819[5] (553)
European	2 332	1 898	2 760[6]

[1] Vencovsky, Dias and Ricardo (1970) — [2] Madalena, Freitas and Martinez (1978) [3] Reis (1977) [4] Groups 1/2 to 5/8 — [5] Groups 3/4 to 15/16 — [6] 15/16.

Numbers in parentheses

Table 5.78 **Least squares means (±s.e.) for milk yield and calving interval of groups with differing proportions of Holstein-Friesian (HF) blood (at Valenca, Rio de Janeiro, Brazil** (from Madalena *et al.* 1979b)

Breed	Lactation yield *(kg)*	Lactation length *(days)*
F_1 (HF × Gir)	[a]2 471 ± 138	[a]262 ± 10
B_1 (3/4 HF:1/4 Gir)	[a]2 347 ± 145	[b]246 ± 10
HF (Black & White pure)	[b]1 898 ± 162	[c]218 ± 11

[a,b,c,] means with different superscript differ significantly ($P < 0.05$)

5.4.4 Chile

Out of a total of 3.4 million cattle, only 660 000 are milking cows mainly of a dual-purpose (beef and milk) Friesian type. They originated both from Europe and the USA. About 20 percent of the milk cows are artificially inseminated, the semen being part of local origin with the rest imported mainly from the USA, New Zealand and the UK. Milk production on commercial dairy farms averages around 2 800 kg per cow per year with management and nutrition presenting the major problems, whilst better farms yield from 4 000 to 6 000 kg per cow per year. Machine milking is practised in the central area of the country. The average calving interval is reported as 410 days (E. Madrid, 1986, personal communication). Feeding comprises mainly pasture in the spring and summer whilst in winter hay, silage and concentrates are used. Corn silage and sugarbeet by-products are used in some areas. The country is reported to be free from foot-and-mouth disease. A few farmers accept purebred Black and White cattle whilst acceptance of crossbred cattle depends on the price (E. Madrid, 1986, personal communication).

Ureclay *et al.* (1981) found no differences in the fertility between Holstein Friesian cows of North American origin and Black and White cows of European origin. Jahn *et al.* (1976) in studying the influence of the level of nutrition during the dry period on milk yield obtained yields of over 4 000 kg per lactation with Holstein-Friesians under Chilean conditions.

5.4.5 Colombia

The dairy cattle population of Colombia represents 13 percent of the total cattle population of 22 million. Imports of milk and dairy products in 1985 represented only a negligible proportion of the quantity available from home production which amounted to a little below 100 kg per head of the human population (Table 5.65).

The most important Holstein-Friesian population is found in the highland plains around Bogotá (2600 m). Cattle are kept chiefly on grazing with supplementary concentrate feeding. A recent analysis of 24 134 records showed the following average values: milk yield per lactation (kg): 4 281 $\pm$1 981; days dry: 88 $\pm$68; calving interval (days): 421 $\pm$100 and age at first calving (months): 36.6 $\pm$10.8 (Abubakar *et al.*, 1986). Colombian daughters of North American bulls with high predicted differences in genetic merit for milk had 31 days longer calving intervals and 4.3 percent fewer survivors completing a third lactation than daughters of bulls with low predicted differences but they produced 852 kg more milk per lactation (Abubakar, 1985).

A study on the reproductive behaviour of 693 Holstein-Friesian cows (with 1 144 records) in Colombia was made by Ramos (1978) on three experimental farms sited in different areas (Bogotá, Cauca Valley and near the city of Pasto in the south). The number of "days open" was reported to be higher than that in temperate zones but that good management reduced postpartum problems and thus the number of days open. The services per conception were found to be significantly affected by rainfall ($P < 0.05$), number of parturitions ($P < 0.01$) and by the season when the parturition took place ($P < 0.01$). The low conception rate at first and second service was considered due to a series of factors including the level of production in the first 100 days, age, number of parturitions and season of the year ($P < 0.01$).

5.4.6 Costa Rica

Most Black and White cattle are of USA origin. However, even after the third generation, Black and White cattle kept above 1 500 m on volcanic slopes are reported as not doing well (J. de Alba, 1985, personal communication). Although systematic observations have not been made, it is presumed that the cattle may be suffering from chronic anaplasmosis. It has been noted that, in the same conditions, Jerseys are not found to be anaemic. At lower altitudes it is found that all European breeds suffer from climatic stress as well as blood diseases which do not seem to affect tropical cattle breeds such as criollo or zebu (J. de Alba, 1985, personal communication).

Just over eleven percent of the total cattle population of 2.6 million was classed as dairy cows in 1985 with an average annual production of 1 279 kg per cow (Table 5.65). The total production from cows amounted to 143 kg per head of the human population and imports added very little.

In a study on dairy ranching, with combined milk and meat production of crossbreds, where the cows were milked once a day and stripped by the calves, good results have been reported for Holstein-Friesian crosses (Kropf *et al.*, 1983). In this system the cows remained on permanent pastures and occasionally received small quantities of concentrates. The system is considered to represent a method for diversifying activities, improving returns and intensifying labourer requirements.

5.4.7 Cuba

The dairy cattle population represented, in 1985, just over 10 percent of the total cattle population of over 6 million. Average production per cow was 1 576 kg which provided 100 kg per head of the human population — about two thirds of the estimated consumption of milk and milk products in 1985 (Table 5.65).

Large-scale importations of Holstein-Friesians from Canada were initiated in 1965 aimed mainly at crossbreeding with the local zebu (Ribas *et al.*, 1978). Milk and butterfat production, butterfat content and length of lactation for three lactations of 141 imported Holstein-Friesian cows in 1968-69 were: 3 610 kg, 108 kg, 3.01 percent, and 307 days respectively. No significant differences in economically important traits between Holstein-Friesian cows born in Cuba and those imported from Canada were found (Ribas *et al.*, 1980). At present there are 64 000 purebred Holstein-Friesian cows and 300 000 F_1 and F_2 (FAO, 1986c).

Results presented by Prada (1978) also show satisfactory production levels of Holstein-Friesians in Cuba but indicate the existence of some fertility problems as shown in Table 5.79.

A study on the adaptability of Holstein-Friesian cattle to the hot humid climate of Cuba indicated that cows underwent an acclimatization process (Garcia *et al.*, 1979a).

Tick resistance of Holstein-Friesian cattle was studied using *Boophilus microplus* larvae applications on four 8-9 month old Holstein-Friesian calves (de la Vega *et al.*, 1980). The results indicated that, after three infestations, considerable resistance was obtained which is directly proportional to the quality of the diet.

Holstein-Friesian × zebu crossbreds in the conditions of Cuba are superior to purebred Holstein-Friesians if the management and feeding limit the milk production level to 3 000-3 500 kg/lactation. Table 5.80 shows that fertility of Holstein-Friesians was poorer than that of crossbreds with zebu (de Vaccaro, 1982).

Table 5.79 **Performance traits of Holstein-Friesian cattle in Cuba** (from Prada, 1978)

	1976	1977	1978
Gestation length (days)	283	284	283
Calving interval (days)	433	436	430
AI index	2.4	2.2	2.3
Age at first calving (months)	32	31	31
Milk production (kg)	3 638	3 259	3 766
Lactation length (days)	322	299	320
Butterfat %	—	—	3.25
Protein %	—	—	3.45

Great importance has been given by Cuba to the development of the "Siboney", a synthetic breed with 5/8 Holstein-Friesian and 3/8 zebu inheritance. Recent publications show mean production levels of 2 960 kg milk, with a calving interval of 380 days and 1.8 services per conception (Menéndez, 1985). Combined with a mean fat percentage of 4.2 (Ponce and Bell, 1981), this means that the Siboney exceeds the Holstein-Friesian in fat production per day of calving interval, even though the pure Holstein-Friesian is favoured in such comparisons by more intensive standards of feeding and management. Considerable numbers of Siboney cattle have been exported by Cuba to other countries in the region.

Similar results were obtained by Leon *et al.* (1982) who showed that Holstein-Friesian cows had longer first lactations but gave less milk in 100, 200 and 244 days than Holstein-Friesian × zebu cows with 3/4 and 5/8 Holstein-Friesian genes. However, body weight at first calving increased with the degree of zebu inheritance. Hernandez Gonzalez (1982) studied meat production of the zebu and Holstein-Friesian × zebu crosses. For 8 zebu and 8 Holstein-Friesian × zebu animals, body weight at the end of fattening averaged 423 and 488 kg respectively ($P < 0.01$), and dressing percentage 56.3 and 58.3 ($P < 0.05$). Carcass weight was significatly

Table 5.80 **Performance of Holstein-Friesians and Holstein-Friesian × zebu crossbreds in Cuba** (from Prada, 1978; quoted by de Vaccaro, 1982)

	Proportion of Holstein-Friesian genes			
	1/2	**5/8**	**3/4**	**Pure**
(n)	**(1 300)**	**(200)**	**(65)**	**(large)**
Age at 1st calving (months)	31	31	31	31
Interval (days):				
Calving — 1st subsequent heat	77	66	71	92
Between calvings	382	365	375	433
Services per conception	1.6	1.7	2.0	2.3
Milk yield (kg):				
Per lactation	1 811	3 100	3 339	3 554
Per day of calving interval	4.7	8.5	8.9	8.2

higher for the crossbreds than for the zebus (128 vs. 107 kg), as was thc weight of lean in the carcass (97 vs. 81 kg), but there were no significant differences between purebreds and crossbreds in the percentages of lean, fat and bone or the weight of carcass fat.

A further study using F_1 (Holstein-Friesian × zebu) and R_1 (3/4 Holstein-Friesian × 1/4 zebu) using a limited number of crossbreds concluded that the crossbreds in general show a persistence of heat tolerance derived from the zebu (Garcia *et al.*, 1979b).

Preston and Ugarte (1972) studied the effect of restricted suckling of the Holstein-Friesian breed and its crosses compared with removal of the calf from the cow at birth. The results showed that the mortality in the restricted suckling system was zero, the incidence of mastitis was lower and none of the crossbred cows went dry before weaning, whilst 17 percent of the control crossbred cows (only milked) went dry before reaching 70 days of lactation. The results imply that cows that are on a milking and suckling regime are more efficient milk producers than those managed conventionally, i.e. milked only.

5.4.8 Guatemala

The dairy cattle population of around 400 000 out of a total of 2.5 million cattle provided for more than 80 percent of consumption of milk and milk products in 1985 with an average production per cow of 825 kg (Table 5.65).

Holstein-Friesians have so far only made a limited impact on the dairy cattle population. The production of Holstein-Friesians in the upland tropical zone of Guatemala (Agua Tibia) was studied using data over the period 1960 to 1970 (749 records for 319 cows). The overall production averages for adult cows were 5 277 kg milk and 174 kg fat (Perozo *et al.*, 1974).

5.4.9 Honduras

Dairy cows in general numbered approximately 430 000 in 1985 with an average yield of 651 kg per cow (Table 5.65). Home production was supplemented by imports of about US$16 million worth of milk and dairy products — 17 percent of the total of these products available in 1985, thus providing an incentive for increased productivity from cattle and to make more available per person.

In a review of the position of Holstein-Friesians in Honduras, F.A.F. Castro (1986, personal communication) indicated that from 1967 to 1985, 8 677 animals had been registered officially, including about 1 000 purebred bulls, others were upgraded females. In 1956 an AI service was started under the aegis of an FAO project at San Pedro Sula. The use of liquid semen has been replaced by frozen semen imported from the USA. Frozen semen is available both commercially and through the government AI service. The source of animals has also been from the USA (F.A.F. Castro, 1986, personal communication). The policy on crossbreeding is to produce an F_1 and then to proceed to a backcross with 75 percent exotic blood as a maximum. Only exceptionally would higher backcrosses be used when the conditions for maintaining the cattle were sufficiently good. In the future, the AI service will be used to improve the national herd genetically.

Table 5.81 **Milk production and aspects of fertility (mean and s.d.) of imported (from USA) and homebred Holstein-Friesians and Brown Swiss in Honduras** (from Bravo, 1982).

	Holstein-Friesian					
	Imported			Home-bred		
	n	mean	s.d.	n	mean	s.d.
Full lactation milk yield (kg)	183	3 827	1 652	277	3 426	1 161
305-day milk yield (kg)	183	3 256	1 083	277	3 150	995
Age at first calving (days)	66	1 021	269	109	1 041	227
Calving interval (days)	146	520	193	214	464	127
	Brown Swiss					
Full lactation milk yield (kg)	344	3 697	1 522	461	3 266	1 171
305-day milk yield (kg)	335	3 041	995	461	2 912	862
Age at first calving (days)	90	1 052	210	163	1 148	246
Calving interval (days)	279	535	212	356	500	186

Brown Swiss and Holstein-Friesians are the two most popular breeds. The latter is popular with the farmers specializing in milk production. In a study on the Holstein-Friesian dairy herd at the Centro Nacional de Agricultura y Ganaderia, Comayagua, of the Ministry of Natural Resources, the data analysed were for the period 1964 to 1980 (Bravo, 1982) (Table 5.81). Some 460 Holstein-Friesians and 805 Brown Swiss cattle were covered by this study. The site was 600 m above sea level (912 mm rainfall, 23.3°C average temperature and 73 percent humidity). The major causes of mortality were diarrhoea, pneumonia, infections, accidents and toxicities. The average calf mortality for Holstein-Friesians was 11.2 percent. Bravo (1982) concluded that with purebreds in hot tropical zones in Honduras it was possible to obtain high levels of milk yield using high levels of inputs (nutrition, health control) although economically the system was questionable. The reproductive efficiency was considered to be low but the Holstein-Friesian breed had, according to Bravo (1982), somewhat better milk yield (on average) and reproductive characteristics and viability than the Brown Swiss.

5.4.10 Mexico

The total cattle population with over 37 million is the third largest in this region and approaching a quarter of them are classed as dairy cows — among the higher proportions for the countries of this region. Average yield per cow was 814 kg in 1985 (Table 5.65) and imports had the highest value (US$ equivalent) of all the countries in the region in 1985, though like for some of its neighbours, this accounted, in 1985, for only 17 percent of the total milk and dairy products available.

Holstein-Friesians and Holstein-Friesian grades make up 60 percent of the total dairy cattle population in Mexico. In 1980 there were 57 350 cows registered with the Holstein-Friesian Association of Mexico and their average yield was 5 783 kg of milk and 3.40 percent butterfat content. McDowell *et al.* (1976a) analysed 17 255 lactation records from 48 herds. They have found that the effects of herd, year, age and season on the performance of Holstein-Friesians in subtropical regions of Mexico did not differ markedly from temperate areas. Comparisons of sires (with five or more daughters) used in artificial insemination showed that 74 percent of those imported from the USA, but only 34 percent of those imported from Canada, and 40 percent of locally bred sires, had positive values in the progeny tests. Mexican sire comparisons carried out in Mexico were highly correlated with USA or Canadian sire proofs. Holstein-Friesian cows bred in Mexico from imported dams and sires produced similar quantities of milk to their imported dams, but yield decreased with the number of generations beyond the importation (McDowell *et al.*, 1976b).

Table 5.82 **Milk yield and fertility characteristics (means and s.d.) of Holstein-Friesian and Brown Swiss cattle in Mexico**

		Holstein-Friesian						Brown Swiss					
Author	**Number and type of record**	**Milk yield** *(kg)*	**Lact. length** *(days)*	**Calving interval** *(days)*	**Semen doses per concep-tion**	**Days open**	**Days betw. calving and next oestrus**	**Milk yield** *(kg)*	**Lact. length** *(days)*	**Calving interval** *(days)*	**Semen doses per concep-tion**	**Days open**	**Days betw. calving and next oestrus**
Garza *et al.* 1973	118 lact	3 661 ±883						2 722 ±667					
Ayala *et al.* 1976	414 cows 1 067 lact	5 492 1 342	293 ±28	395 ±70	2.03 ±1.44								
Frias and Diaz 1980	59 herds			412	2.11	134							
	71 herds	5 456 ±374											
Flores Trujillo 1980	15 years' obser-vation	3 212		451	3.4	166	65	2 737		422	2.3	126	42

In Table 5.82 information is given on Holstein-Friesian and Brown Swiss milk yield and fertility in Mexico. Garza *et al.* (1973) found significant differences in milk yield between these breeds, in favour of the Holstein-Friesian in the semi-humid tropical climate of Paso del Toro, Ver. This agrees with the findings of Flores Trujillo (1980) which also refer to a tropical climate. In dry tropical zones (Hermosillo, Sonora) good production and fertility of Holstein-Friesians were achieved (Ayala *et al.*, 1976). The same applies to the results of Frias and Diaz (1980).

Rodríguez López (1979) evaluated the reproductive performance in a Holstein-Friesian dairy herd of 626 lactating cows in the city of Hermosillo, Sonora, Mexico. Over a 9-month period, the first postpartum oestrus occurred at an average of 51.8 days after parturition, but only 15.4 percent of mated cows conceived at this oestrus with 43 percent of cows mated at the second oestrus, which occurred, on average, 119.2 days after calving. Cows mated in January or February, at average temperatures of 15.9-16.9°C, had significantly better conception rates than those mated in July-August at average temperatures of 31.5-32.3°C.

Saucedo *et al.* (1980) studied 46 crossbred (Holstein-Friesian × zebu and Brown Swiss × zebu) cows in early lactation in a trial to evaluate leucaena forage grown as a pure stand and concluded that leucaena offered considerable possibilities for increasing milk production on tropical pasture.

Studies by Alvarez *et al.* (1980) on the effect of calf suckling on milk production of Holstein-Friesian × zebu and Brown Swiss × zebu indicated the beneficial effects of suckling similar to those reported by Preston and Ugarte (1972) and showed that suckling also reduced calf mortality significantly.

Román Ponce *et al.* (1978) in a tropical climate in Mexico, compared a total of 296 lactations from 58 Holstein-Friesians (HF), 56 American Brown Swiss (BS) and 18 Jersey (J) cows over the period 1964 to 1971. The cows were permanently housed throughout the period. Milk yield differed significantly between breeds, averaging 3 545 kg for HF, 2 752 for BS and 2511 for J (325, 315 and 318 days of lactation, respectively). The increase in milk yield from first lactation to the lactation with maximum yield was 19, 24 and 34 percent for HF, BS and J, respectively. There were significant positive correlations between milk yield and lactation length, length of previous dry period, days open, body weight, and age at calving. There was no evidence of breed × year, breed × season or breed × year × season interactions. Milk yield was higher in October-March than in April-September (3 287 vs. 2 604 kg), but the difference was not significant. The repeatability of milk yield for HF and BS was 0.38. The authors concluded that the performance of these breeds, properly managed under tropical conditions, was highly acceptable (similar to their performance in temperate climates).

5.4.11 Peru

Dairy cows, as classified in Table 5.65 numbered 690 000 in 1985 and their production at about 1 200 kg per cow provided 41 kg of milk per person, one of the lower figures for countries in Latin America. Imports in 1985 added a further significant quantity.

The main populations of Holstein-Friesians in Peru are found in the temperate areas around Arequipa, Lima and Cajamarca, as well as in the hot, dry region of Trujillo in the north. In 1982, 16 199 cows were enrolled in the national milk recording scheme, of which the majority were Holstein-Friesians. The mean yields were 12.0 kg milk/day, 3.4 percent fat, 91 days dry period and 14.1 months calving interval (Pallete, 1982). The highest milk yields were obtained from the Arequipa region (15.8 kg milk/day) where feeding is based on irrigated alfalfa, followed by Cajamarca (12.6 kg/day) where cattle graze ryegrass/clover pastures. Information on the Lima area was given by Vaccaro *et al.* (1979a) and shows mean first lactation yields of 4 170 $\pm$831 kg milk, 136 $\pm$27 kg fat and an average first calving age of 32 months. Following a study which showed that productive criteria had not been of chief importance in the importation of semen since 1957 (Vaccaro *et al.*, 1979b), effective steps have been taken to ensure that young sires bred locally for AI are sons of outstanding dams and imported semen from bulls with the highest possible predicted breeding values for milk.

The productive capacity of Holstein-Friesians, Brown Swiss and crossbred Brown Swiss × zebu cattle in the semi-humid tropics of Peru has been examined by La Hoz and Rosemberg (1978) and is shown in Table 5.83. The herds were sited in Tarapoto, San Martin Department (23°C annual temperature, 1 200 mm rainfall at 400 m). Females grazed pasture from 6 months of age and received supplements of chopped grass and concentrates and were housed after calving. Appropriate health measures (vaccination, dipping, and oral treatment) were applied. Only the age at first calving of Brown Swiss females was found to be significantly later from that of the others. In the view of the authors both reproduction and milk yield from the breeds and grades in these semi-humid tropical conditions were low.

5.4.12 Venezuela

Venezuela has one of the larger total cattle populations (12.5 million) in relation to its human population (17.3 million) but only 11 percent of the cattle were classified as dairy animals. Production per cow in 1985 was of the order of 1 100 kg and this provided 88 kg per person in the country but in addition there were substantial imports of milk and dairy products in 1985 (Table 5.65).

Table 5.83 **Reproduction and production results of Holstein-Friesian and Brown Swiss and Brown Swiss × zebu cattle in Peru** (from La Hoz and Rosemberg, 1978)

	Age at first calving *(days)*	**Calving interval** *(days)*	**Lactation yield** *(litres)*	**Lactation period** *(days)*
Holstein-Friesian	1 259 (51)	560 (89)	2 274 (150)	305
Brown Swiss	1 509 (80)	518 (138)	2 547 (215)	345
3/4 Brown Swiss 1/4 zebu	1 150 (29)	500 (60)	2 643 (89)	349
7/8 Brown Swiss 1/8 zebu	1 100 (9)	416 (7)	2 522 (16)	319

No. of animals in parentheses

Table 5.84 **Pre-and postnatal losses in purebred European cattle and their crosses in Venezuela** (from de Vaccaro, 1986 and Hahn, 1986)

	Number		**Weighted**	
	Farms	**Animals**	**mean** *(%)*	**Range** *(%)*[1]
Losses	a) *imported females*			
From arrival to -1st calving	7	1 217	11.6	4.2-22.9
Abortions	5	1 605	7.5	3.7-11.3
Stillbirths	3	730	8.6	7.7-12.2
	b) *Locally born females*			
Birth to 9 or 12 months	8	4 278	17.7	13.4-81.3
9 or 12 months to 1st calving	8	3 577	21.1	8.1-30.8

[1] Range of mean values per farm

Table 5.85 **Reproduction and production data on Holstein-Friesian and Brown Swiss in Venezuela** (pooled results) (from Bodisco *et al.*, 1977)

Generation	Age at calving *(months)*	Weight at calving *(kg)*	Lactation length *(days)*	Total milk yield *(kg)*	305 days' milk yield *(kg)*	Days open	Services per conception
1st	33.1	423.8	375	3 525	3 003	270	4.4
2nd	30.2	437.5	317	3 130	2 778	224	2.9
3rd	36.2	441.4	337	3 277	2 803	219	1.6

Peru. Women are responsible for milking these Black and White cows

Animals of the Holstein-Friesian and Brown Swiss breeds have been continually imported by Venezuela since the 1940s to improve milk yield by crossing with native cattle (Bodisco *et al.*, 1977).

In 1986, the official recorded 244-day yield of 45 666 Holstein-Friesians was 4 048 kg milk, exceeding the Brown Swiss average by 924 kg (Hahn, 1986). In contrast to other countries in the region, purebred European cattle in Venezuela are kept in large numbers in the lowland tropics in intensive systems with very high levels of concentrate feeding. This has been possible only because the cost of the imported grains has been subsidized.

Despite their high yields of milk per lactation, serious problems have been reported with regard to survival and reproduction. A summary of pre and postnatal losses in imported and local females is given in Table 5.84. The data referring to imported cattle are all from pure Holstein-Friesian, while the local data refer mainly to Holstein-Friesian, a few Holstein crosses and a very few Brown Swiss crosses.

Data from 385 pure Holstein-Friesian cows (1 058 lactations between 1961 and 1969) in a herd in the north-central region of Venezuela analysed by Garroni and Verde (1974) showed a milk yield of 4 224 kg in a 305-day lactation.

Results for the first three generations of Holstein-Friesian and Brown Swiss cattle from a study by Bodisco *et al.* (1977) are shown in Table 5.85. The first generation was from calves imported from the USA in 1964 and the subsequent two generations were born in Venezuela. Results span the years 1966 to 1976. Although there appeared to be a decline in milk production from generation 1 to 3 the difference was not significant nor was there a significant breed difference. The only significant effect ($P < 0.01$) was the reduction in the number of services per conception with advancing generations, attributed by Bodisco *et al.* (1977) to adaptation.

De Villalobos *et al.* (1975) studied heat and humidity tolerance on a limited number of cattle of imported and native Holstein-Friesian cows and Brahman × Brown Swiss crossbreds in Jusepin, Venezuela by measuring body temperature and respiration rate as the factors. The crossbreds were superior to the imported breeds but there was little difference between the adaptability of imported and native Holstein-Friesians.

The production of milk at high altitudes, using various breeds including Holstein-Friesians and their crossbreds, has been reported for the Andean region of Venezuela (Martínez *et al.*, 1978). Milk production based on 382, 305-day records is shown in Table 5.86. The average age at calving was 30 months, lactation length 329 days and dry period 133 days.

In recent papers (Martínez *et al.*, 1982a; Combellas and Martínez, 1982; Martínez *et al.*, 1982b) the records of the Holstein-Friesian herd maintained at the Institute of Animal Production, Maracay were analysed. Average body weight was 606 kg and total milk production 4 731 kg for the cows imported from the USA and

Table 5.86 **Average milk production of cattle at high altitudes in Venezuela** (from Martinez *et al.*, 1978)

Breed	**Lactation**	**Milk production** *(kg)*
Holstein-Friesian	First	3 961
	Later	4 050
Jersey	First	2 459
	Later	2 959
Holstein-Friesian	First	2 597
× Mestizo	Later	3 013
Native	First	2 632
	Later	3 089

Years covered: 1971-1976; number of calvings 1 to 15

Table 5.87 **Least squares means (±s.e.) for productive and reproductive characteristics of imported Holstein-Friesian animals and those born in Maracay, Venezuela** (from Martinez *et al.*, 1982a)

	Imported cows	**Cows born in Maracay**
Production		
No. of records	181	412
Lactation milk yield (kg)	4 731 ± 96	4 213 ± 120**
305-day milk yield (kg)	4 564 ± 73	4 041 ± 91**
Lactation length (days)	323 ± 5	326 ± 7NS
Liveweight (kg)	606 ± 6	517 ± 6**
Reproduction		
No. of records	72	473
First postpartum oestrus (days)	69 ± 5	63 ± 3NS
Days open	151 ± 10	157 ± 6NS
Services per conception	2.32 ± 0.28	3.10 ± 0.17**

** Significance of differences $P < 0.01$

517 kg and 4 213 kg respectively for those born in Maracay. The Holstein-Friesian herd was not grazed, but fed freshly cut forages and concentrates. The reproductive efficiency was lowest in those animals which calved at the beginning of the year and were served during the rainy season. Calving problems decreased reproductive efficiency without affecting milk production (Martínez *et al.*, 1982a). In a study of 425 records from the Holstein-Friesian herd at Maracay it was found that cows lost weight during the first weeks of lactation. Their fertility depended considerably both on weight at calving and on the extent of liveweight loss during this period (Martínez *et al.*, 1982b). Table 5.87 gives a comparison of imported Holstein-Friesian cows with those born in Maracay. The authors concluded that although in some respects the cows born in Maracay performed less well than the imported cows, the levels of production and reproduction were acceptable and higher than previously reported for tropical countries. However, the application of these production methods in practice might be limited by economic, social and other factors.

Bodisco *et al.* (1973) compared heat and humidity tolerance of different exotic dairy breeds in Venezuela. Body temperature, respiration rate and daily milk yield of Holstein-Friesian (HF), Brown Swiss (BS) and Guernsey (G) cows — numbering 21 in total — were measured for 15 consecutive days during the dry and rainy seasons; mean air temperatures were 17.5 and 27.3°C respectively, and mean air humidity was 61.2 and 69.7 percent. During the dry season, body weight and daily milk yield averaged, respectively, 498 and 14.4 kg for HF, 483 and 11.6 kg for BS and 423 and 12.1 kg for G; the corresponding figures during the rainy season were 483 and 14.2 kg, 474 and 11.6 kg and 410 and 11.1 kg respectively.

Rodríguez Voigt *et al.* (1974) gave production data for 31 farms in the area south of Lake Maracaibo with 1 138 milk records and 517 calving intervals. Grade BS, HF and zebu crossbred cows and indeterminate-type cows were used to study the effects of genotype, lactation number and season of calving on milk yield and fertility. Lactation length averaged 286 days and milk yield 1 665 kg for the BS type, 280 days and 1 433 kg respectively for HF, 269 days and 1 434 kg for zebus, and 288 days and 1 378 kg for the cows of indeterminate type. Differences between lactations in yield and length were not significant. Lactation length was shortest (268 days) for cows calving in December-February and longest (294 days) for cows calving in June-August. Cows calving in March-May had the lowest average lactation yield (1 439 kg) and cows calving in September-November the highest (1 587 kg). There were significant differences between farms in milk yield.

Rodríguez *et al.* (1978) studied the production of different crossbreds of dairy breeds in Venezuela with 2 510 lactation records of predominantly BS, HF and zebu cows, and of cows of indeterminate type. Overall, lactation length averaged 302 days and lactation milk yield 1 534 kg. Lactation milk yield was 155 kg higher than average in the BS cows, and 48, 90 and 16 kg lower in the HF, zebu and indeterminate cows respectively. Calving interval did not differ significantly between breed type but significantly affected milk yield of BS cows, but not of cows of the other types.

Results on HF crosses with criollo cattle for the dry tropics of Venezuela are provided by Cerrada *et al.* (1978) for 2 351 lactations of zebu × criollo × HF, zebu × criollo × BS, criollo × HF and criollo × BS cows. Lactation length averaged 311 days, lactation milk yield 1 858 kg and calving interval 424 days. For cows with BS blood the averages were 333 days, 1 970 kg and 445 days respectively and for those with HF blood 321, 2 022 and 422. There was a significant decrease in calving interval from the first to the third parturition (from 480 to 399 days).

Capriles *et al.* (1981) compared the production and reproductive characteristics of HF and BS crosses with native breeds in Venezuela. For HF × Perijano criollo (PC) and BS × PC cows, milked twice daily and under restricted suckling management, the production of saleable milk averaged 1 618 and 1 511 kg respectively, the difference between the two breed groups being non-significant. In the two groups, the conception rate was 72.2 and 72.4 percent respectively, and the service period averaged 66 and 104 days (P (((0.01). All pregnant HF crossbreds conceived in less than 100 days after parturition as against only 53.8 percent of the BS crossbreds.

Vaccaro and de Vaccaro (1982a) presented information on 1 441 services of 400 HF crossbreds in an intensive tropical milk production system. In 1977-80, age at first calving averaged 30.8 ($\pm$4) months for HF crossbreds; heifers born in the rainy seasons were younger at first calving than those born in the dry seasons (P (((0.05). The number of services per conception averaged 1.65, 1.65, 2.36 and 2.76 respectively for HF crossbred heifers in the two dry and two rainy seasons each year. Overall, there were significant differences between years in the percentage of abortions (3.5-11.1 percent) and stillbirths (1.2-11.1 percent). For 190 HF crossbred heifers, the intervals between first and second calvings averaged 432 ($\pm$97) days and those between second and third calvings 441 ($\pm$109) days.

Vaccaro and de Vaccaro (1982b) analysed data on 146 HF cows, 120 of which completed at least 2 lactations. First lactation milk yield was 3 722 ($\pm$728) kg. A lower proportion (23 percent of the total) of heifers calving at the beginning of the rains in April-July was culled than of those calving in December-March or August-November. Lactation milk yield of cows culled due to infertility after the first lactation exceeded the first lactation yield of cows calving a second time by 285 kg.

5.4.13 Caribbean

The Caribbean islands have had dairy cattle breeding as a comparatively recent undertaking and though cattle were introduced into Jamaica as early as 1494 they were mainly used for the production of leather, beef and manure, or for draught purposes. It was not until the twentieth century that dairy herds were developed (Wellington, 1982) mainly through importation of dairy cattle of European origin. The Holstein-Friesian ranks first, though Jerseys, Brown Swiss and Guernseys have also been imported.

Many Caribbean islands now have a dairy industry. The most notable are Puerto Rico, Jamaica, Trinidad and Barbados.

The Holstein-Friesian breed is the most numerous and in some countries this breed has been crossed with *Bos indicus* cattle to impart hardiness and adaptability to the conditions in the humid tropics (Wellington, 1982). Taking the Caribbean islands as a group the performance of Holstein-Friesians in these islands is shown in Table 5.88.

Table 5.88 **Performance of purebred and grade Holstein-Friesian cattle in some Caribbean countries** (from Wellington, 1982)

Country	Breed type	Age at first calving *(months)*	305-day milk yield *(kg)*	Lactation length *(days)*	Calving interval *(months)*
Barbados	grade	34.4	3 032	323	15.3
Barbados	purebred HF	31.6	3 297	334	14.7
Barbados	purebred HF	32.8	4 051	357	15.8
Cuba	62.5% HF	30.7	3 100	270	12.0
Cuba	75% HF	31.0	3 339	301	12.5
Cuba	purebred HF	32.0	3 638	322	14.4
Jamaica	purebred & grade	33.0	1 965	370	17.5
Jamaica	purebred & grade	—	3 178	313	14.5
Puerto Rico	local	29.0	3 679	295	15.0
Puerto Rico	imported HF	28.9	4 050	289	15.1
St. Kitts	grade	42.0	1 700	327	15.5
Trinidad	grade	33.3	2 317	231	14.4
Trinidad	purebred HF	39.7	2 819	273	16.4

a) Antigua

The introduction of Black and White Holstein-Friesian cattle on the island of Antigua started in 1936 within a programme of livestock breeding based on the importation of purebred and crossbred Holstein-Friesian bulls from Trinidad. This was continued with the importation of purebred Holstein-Friesian bulls from Canada. As a result of the Government's breeding programme several farmers have herds of 5/8 Holstein-Friesians and many animals of local breeds have some Holstein-Friesian blood. Farmers at present interested in milk production choose Black and White strains. The recently introduced AI programme includes semen from Black and White bulls (J.L. Robinson, 1985, personal communication).

b) Dominican Republic

Information from the Livestock Division of the Ministry of Agriculture (S. Fernandez-Baca, 1986, personal communication) shows that of a total cattle population of 2 million in 1984, 19 percent were pure Holstein-Friesian and a further 29 percent included Holstein-Friesian blood. The other principal exotic breed was Brown Swiss. The average milk production per dairy type cow is about 2 000 kg per year. After the boom of the Holstein-Friesian in the 1970s and the beginning of the 1980s the economic condition forced breeders to change their policy.

The present trend on most dairy farms is to mate Holstein-Friesian or Brown Swiss cows to zebu bulls in order to get crossbred progeny of lower production but

Table 5.89 **Performance of Holstein-Friesian cattle on three commercial farms in Jamaica during 1975-76** (from Wellington, 1982)

	Farm a	Farm b	Farm c
No. of cows	245	98	304
Milk yield (kg)	2 542	1 861	1 534
Age at first calving (months)	30	35	34
Services per conception	1.6	1.7	1.3
Calving interval (months)	17	17	18
Percentage cows in milk	69	69	55
Days dry	147	169	146
Days open	260	252	283
Butterfat (%)	3.7	3.7	3.7

capable of producing milk without the use of concentrates since the latter has gone up considerably in price whilst milk prices have remained controlled by the government. Notwithstanding, substantial importations of semen of dairy breeds continue. These numbered approximately 75 000 in the five years to 1984 at a cost of US$ 220 000.

One experiment with Holstein-Friesian was reported in which crossbred Holstein × zebu (50:50) cows were subjected to feeding trials involving comparisons of pangola pasture grazing, drylot feeding with chopped sugarcane and molasses/urea and concentrate mineral supplements, with and without limited grazing. It was concluded that there was no difference between unsupplemented grazing and restricted grazing plus supplements (MacLeod *et al.*, 1976). The production of milk and weaned calves under the same system led to the same conclusion (Morales *et al.*, 1976).

c) Guadeloupe

Since 1971, INRA (National Institute for Agricultural Research) has been active at the Gardel estate, with an average rainfall of 1100-1500 mm and an average temperature of 25.4°C.

Data from 70 French Friesian cows (imported from France or their descendants) show age at first calving, 30-40 months; average liveweight, 345 kg; an average of 2.26 services per conception, a calving rate for heifers of 92.3 percent and for cows of 75 percent. The average milk yield was 3 291 kg for 166 lactations recorded between 1971-75 with a lactation length of 290 days.

Table 5.90 **Reproductive performance and body weight of Jamaica Hope and Holstein-Friesian cows on a non-concentrate diet at Bodles, Jamaica 1976-78** (from Wellington, 1982)

Trait	**Jamaica Hope**		**Holstein-Friesian**	
	Lact. 1	Lact. 2	Lact. 1	Lact. 2
No. of records	28	28	10	8
No. of days open	132	122	160	172
No. services/conception	2.1	2.2	2.2	3.5
Calving interval (days)	416	406	420	—
Calf birthweight (kg)	27.0	27.8	28.9	—
Cow weight to start of lactation (kg)	367	388	442	453
Cow weight at the end of lactation (kg)	390	419	453	470

The most serious health problems in these humid tropics were pathological and those linked with breeding and rearing of heifers. Forage crop species in Guadeloupe (pangola grass, Merket guinea grass, forage sorghum) gave good yields if properly managed (France, 1978).

d) Jamaica

There are about 320 000 cattle in the island with an average production of 1 000 kg of milk per cow. Importations of dairy products are substantial (see Table 5.65).

The native Jamaica Hope breed has been created as a tropically adapted dairy breed derived from Jersey, Sahiwal and Holstein-Friesian by crossbreeding. Table 5.89 gives the performance of Holstein-Friesian cattle for three commercial farms in Jamaica, showing long calving intervals, a low percentage of cows in milk, long average dry period and excessive average days open (Wellington, 1982). How much of this is due to poor management and how much to other environmental conditions is difficult to determine.

Despite the apparent success in other warm climates in the countries in which high levels of concentrates are fed (e.g. Israel, Puerto Rico and Cuba), Holstein-Friesian cattle have never been considered to perform satisfactorily in Jamaica although several relatively large-scale importations were made between 1917 and 1970. Dairy production in Jamaica is based on grass feeding regimes without concentrates and under these circumstances Holstein-Friesians have lower production, and fertility problems occur. Table 5.90 provides some comparative information on body weights and aspects of reproduction of Jamaica Hope and Holstein-Friesians. In general Jamaica Hope does better in Jamaica on non-concentrate diets than imported Holstein-Friesians (Table 5.91).

Table 5.91 **Performance traits of Holstein-Friesian cattle on Alcan Jamaica farms 1969-73** (from Wellington, 1982)

Year	Milk yield *(kg)*	Days in milk	Calving interval *(days)*	Milk yield/day of C.I. *(kg)*
1969	3 453	315	400	8.6
1970	3 973	353	448	8.9
1971	3 525	311	395	8.8
1972	2 874	297	374	7.7
1973	2 440	305	390	6.3

Health problems which affected the Alcan Jamaica herds as a whole in the early years included failure of imported animals to acclimatize easily to local conditions with the result that mortality exceeded 20 percent. Lameness was frequent.

Causes of death were anaplasmosis, piroplasmosis, parasitism and misadventure. Numerous calf rearing problems were encountered over the same period — enteritis, pneumonia and parasitism were the most significant with many purchased calves dying (Wellington, 1982).

e) Puerto Rico

This island has more than half a million cattle with about 200 000 dairy cows. Average milk production per cow and year is close to 1 800 kg (Table 5.65). Milk production per head of human population is high, nearly 105 kg per year, which allows Puerto Rico to be independent of dairy product importations.

Table 5.92 gives results from reports on 25 Puerto Rican herds of Holstein-Friesian cattle. Milk yield was satisfactory but the lactation lengths averaged 265 days only. It was concluded that factors affecting Holstein performance in Puerto Rico did not differ from those in temperate countries and more importantly that the total variation appeared large enough to permit selection for higher milk and butterfat yields (Wellington, 1982).

f) Trinidad and Tobago

The island has only 70 000 cattle and depends greatly on importations of dairy products (Table 5.65).

Table 5.92 **Holstein-Friesian cattle production in 25 herds in Puerto Rico** (from Camoens *et al.*, 1976; quoted by Wellington, 1982)

Lactation milk yield (kg)	3 844
Butterfat yield (kg)	121
Butterfat content (%)	3.2
Lactation length (days)	265
Days dry	102
Days open	137

Table 5.93 gives milk production data on Holstein-Friesian × zebu cattle in Trinidad and Table 5.94 the results on reproductive performance for these crossbreds. They indicate a short lactation of less than 300 days and milk yields of less than 3 000 kg. Though the number of services per conception were within a satisfactory range, the length of the dry period was high and the calving intervals exceeded 400 days in most cases. The net result was a milk yield per day of calving interval being in the low range of 5.3 to 8.6 kg (Wellington, 1982), which is most likely to reflect inadequate feeding and management of cattle.

Table 5.93 **Lactation data and milk yield of Holstein-Friesian × zebu cattle in Trinidad** (from Howard, 1976; quoted by Wellington, 1982)

Lactation No.	**No. of records**	**Lactation length** *(days)*	**No. of records**	**305-day milk yield** *(kg)*
1	260	283	221	2 403
2	185	256	145	2 794
3	122	280	104	3 421
4	84	279	77	2 671
5	58	298	53	2 976

Table 5.94 **Number of services per conception, dry period (days) and calving intervals in Holstein-Friesian × zebu cattle in Trinidad** (from Howard, 1976; quoted by Wellington, 1982)

	Services per conception		**Days dry**		**Calving interval**		**Yield per day of calving interval** *(kg)*
Lactation Sequence	**No. of Records**	**Mean**	**No. of Records**	**Mean**	**No. of Records**	**Mean** *(days)*	
1-2	277	1.8	195	133	200	444	5.3
2-3	195	2.2	131	166	138	434	6.4
3-4	135	2.4	104	163	103	395	8.6
4-5	88	2.3	72	132	73	396	6.8
5-6	69	1.8	53	146	54	439	5.7

References

ABDALLA, Z.M. *Livestock production, Zambia.* Report of assignment. DP/ZAM/74/006. FAO, Rome.
1978

ABDEL AZIZ, A.S., OBEIDAH, A., ABDEL AZIZ, M. and MOSTAGEER, A. A statistical study on the
1979 heat tolerance in Friesian cattle and buffaloes. *Egypt J. Anim. Prod.*, 19: 11-64.

ABDEL-GHANI, W. and FAHMY, S.K. Productivity of Friesian and its crosses in UAR. *The Agric. Res.*
1966 *Rev.*, 44: 37-42.

ABLAHAD, B.S. *Study on the productive and physiological efficiency of dairy cattle imported from different sources under some environmental conditions in Iraq.* MSc Thesis (Animal Production), Mosul University, pp 1-11.
1980

ABUBAKAR, B.Y. *Evaluation of the performance of Holsteins in Mexico and Colombia.* PhD thesis,
1985 Cornell University, Ithaca, N.Y. pp 106.

ABUBAKAR, B.Y., MCDOWELL, R.E. and VAN VLECK, L.D. Genetic evaluation of Holsteins in Col-
1986 ombia. *J. Dairy Sci.*, 69: 1081-86.

ACHARYA, R.M. *Cattle crossbreeding to increase milk production: an instrument to rural development*
1986 *— a success story in India.* FAO Regional Office, Bangkok.

ACHARYA, R.M. and BHAT, P.N. *Livestock and poultry genetic resources in India.* IVRI Research
1984 Bull. No. 1. Indian Veterinary Research Institute, Izatnagar, UP.

AGYEMANG, K. and NKHONJERA, L.P. *Evaluation of the productivity of crossbred dairy cattle on*
1986 *smallholder and government farms in the Republic of Malawi.* International Livestock Centre for Africa Research report No. 12. Addis Ababa, Ethiopia.

AJAYI, S.A., FABIYI, J.P. and UMO, I. Clinical bovine anaplasmosis and babesiosis in Friesian cattle.
1982 *Wld Anim. Rev.*, 43: 41.

AL-ANI, L.M. *Estimation of some genetic parameters in dairy cattle in Central Iraq.* MSc Thesis (Animal
1980 Production), University of Baghdad, pp 111-116.

ALBERRO, M. Comparative performance of Holstein-Friesian, Dutch Friesian and Friesian × Africander
1980 heifers in the coastal belt of Mozambique. *Anim. Prod.*, 31: 43-49.

ALBERRO, M. Friesian dairy herd in the coastal belt of Mozambique. *Wld Anim. Rev.*, 37: 20-24.
1981

AL-CHALABI, F.A., MEHDI, A.W.R., AL-KAFAWI, A.A., INJIDI, M.H., LAFTA, H.D. and
1978 DESSOUKY, F. Preliminary study on some semen characteristics of Friesian bulls raised in three climatic regions in Iraq. *Iraqi J. Vet. Med.*, 2.

ALI, M.H. *On the incidence of mastitis in purebred and crossbred Friesian cattle.* MSc Thesis, University of Baghdad, pp 4-9.
1977

ALLEN, C.J. and KRISTENSEN, J. *Third annual report of the Malawi milk recording scheme.* FAO/UNDP Project MLW/70/020. Working paper No. 2. Lilongwe.
1977

AL-RAWI, A.A. and SAID, S.I. Some genetic and non-genetic factors associated with milk yield of Friesian, indigenous and their crosses in Iraq. *Mesopotamia J. Agric.*, 15: 21-29.
1980a

AL-RAWI, A.A. and SAID, S.I. Significance of the age at first calving and lactation length on the first lactation yield. *Mesopotamia J. Agric.*, 15: 33-41.
1980b

AL-RAWI, A.A. and SAID, S.I. Some sources of variation associated with pre-weaning growth in purebred and crossbred Friesian. *Mesopotamia J. Agric.*, 15: 5-23.
1980c

AL-RAWI, A.A. and AL-ANI, L.M. Evaluation of imported vs native-born Friesian bulls in Iraq. *Ind. J. Anim. Sci.*, 51: 395-397.
1981

AL TASSIM, A.F. *Effect of clipping of coat and sprinkling on milk production from dairy cows.* MSc Thesis (Animal Resources/Animal Production), University of Baghdad, pp a-c.
1982

ALVAREZ, F.J., SAUCEDO, G., ARRIAGA, A. and PRESTON, T.R. Effect on milk production and calf performance on milking crossbred European/zebu cattle in the absence or presence of the calf, and of rearing their calves artificially. *Trop. Anim. Prod.*, 5: 25-37.
1980

ANSELL, R.H. *Observations on the reactions of British Friesian cattle to the high ambient temperatures of the United Arab Emirates.* University of Berne (Thesis).
1974

ANSELL, R.H. Maintaining European dairy cattle in the Near East. *Wld Anim. Rev.*, 20: 1-7.
1976

ANSELL, R.H. Cattle breeding in the tropics. *Wld Anim. Rev.*, 54: 30-38.
1985

ARIA, G. and CRISTOFORI, F. Reproductive traits of zebu cattle in Somalia. *Atti della Società Italiana delle Scienze Veterinarie,* 34: 205.
1980

ARMOUR, J., LEE, R.P. and ROSS, J.G. Observations on a cross breeding experiment with cattle in Nigeria. *Trop. Agric., Trin.*, 38 (4): 319-323.
1961

ARORA, D.N. and SHARMA, J.S. Factors affecting some of the economic traits in Holstein-Friesian cattle. *Ind. Vet. J.*, 60: 820-823.
1983

ASKER, A.A., JUMA, K.H. and KASSIR, S.A. Factors affecting milk production in crossbred cattle in Iraq. Faculty of Agric., Ain Shams University, Cairo. *Annals of Agric. Sci.*, 10: 47-62.
1965a

ASKER, A.A., JUMA, K.H. and KASSIR, S.A. Dairy characters of Friesian, Ayrshire, native and crossbred cattle in Iraq. Faculty of Agric., Ain Shams University, Cairo. *Annals of Agric. Sci.*, 10: 29-45.
1965b

AYALA, F.J., BERRUECOS, J.M., VAZQUEZ, C.G., CORELLA, E. and ZEPEDA, J.U. Estimate of
1976 reproductive behaviour and milk production of a Holstein herd in the dry tropics. *Técnica Pecuaria,* 30: 95.

BABONA, B.P. *Genetic parameters of some economic traits in imported Friesian cattle in Central Iraq.*
1981 MSc Thesis (Animal Resources/Animal Breeding), University of Baghdad, pp 1-4.

BAGHDASAR, G.A. *Genetic analysis of some physiological characters, milk production and coefficient*
1980 *of heat tolerance in dairy cattle.* MSc Thesis (Animal Resources/Animal Breeding), University of Baghdad, pp 7.

BAHMAN, A.R.M. *Improvement of milk cows in the State of Kuwait.* A study presented to the Seminar
c. 1983 for the Assessment of the Local and Foreign Species of Cows in the Arab World. Agriculture Affairs and Fisheries Authority, State of Kuwait, pp 17.

BHAT, P.N., SINGH, G., BISWAS, J.C., SINGH, R.A., GARG, R.C. and KUMAR, S. Performance of
1978 Holstein-Friesian cattle in India. *Indian J. Anim. Sci.*, 48: 820-823.

BENLEKHAL, A. *Rapport du Royaume du Maroc.* FAO Expert Consultation on Artificial Insemination
1985 and Embryo Transfer, Rome, 9-12 December, pp 23.

BODISCO, V., MANRIQUE, U., VALLE, A. and CAVALLOS, E. Tolerance of heat and atmospheric
1973 humidity in Holstein Brown Swiss and Guernsey cows. *Agron. Trop.*, 23: 241-261.

BODISCO, V., RODRIGUEZ-VOIGT, A., ALFARO, E.C. and MENDOZA, S. First lactation of three
1977 generations of Holstein and Brown Swiss in Maracay, Venezuela. *Agron. Trop.*, 27: 591-600.

BONG KUG OHH. *National report of Korea.* SABRAO Workshop on Animal Genetic Resources.
1981

BOTSWANA. *Ten years of animal production and range research in Botswana.* Ministry of Agriculture,
1980 Animal Production Research Unit, Gaborone.

BOUJENANE, I. and MATY BA . The performance of Moroccan Holstein-Friesian cows on breeding
1986 and milk production. *Rev. Elev. Vét. Pays trop.*, 39 (1): 145-149.

BRAHMIA, A. and BEN MEHDI, M. *Performances de troupeaux inscrits au contrôle laitier. Lactations*
1982 *terminées en 1982.* Ministère de l'Agriculture. Office de l'Elevage et de Pâturages, Contrôle de Performances, pp 55.

BRÄNNÄNG, E., MESKEL, L.B., SCHAAR, J. and SWENSSON, C. Breeding activities of the Ethio-
1980 Swedish integrated rural development project. Part I. *Wld Anim. Rev.*, 36: 34-36.

BRAVO, O.M. *Production, reproduction and mortality of the Holstein and Brown Swiss breeds in Com-*
1982 *ayagua, Honduras.* MSc Thesis, University of Costa Rica, pp 74.

BREINHOLT, L.A., GOWEN, F.A. and NWOSU, C.C. Influencia de los factores ambientales y animales
1981 sobre el pastoreo diurno y nocturno de vacas importadós Holstein-Friesian en las tierras bajas del trópico húmedo de Nigeria. *Prod. Anim. Trop.*, 6: 328-336.

BUVANENDRAN, V. and MAHADEVAN, P. Crossbreeding for milk production in Sri Lanka. *Wld Anim.*
1975 *Rev.*, 15: 7-13.

BUVANENDRAN, V., OLAYIWOLE, M.B., POTROWSKA, K.I. and OYEJOLA, B.A. A comparison of
1981 milk production traits in Friesian × White Fulani crossbred cattle. *Anim. Prod.*, 32: 165-170.

CAPRILES, M., PAZ, M., ROJAS, D. and RIVERO, A. Preliminary report on production and reproduc-
1981 tion in Holstein-Friesian × Perijano Criollo and Brown Swiss × Perijano Criollo cows in humid tropical conditions. *Memoria Asoc. Latinoam. Prod. Anim.*, 16: 155-156.

CASINABE, G.R. and BOLORON, L.B. Ability and breeding performance of imported American Brahmin
1981 and Holstein-Friesian cattle of CMU, Philippines. *CMU J. Agric. Fd. Nutr.*, 3: 113-127.

CASTILLO, L.S., RAMIN, B.B., CUEVAS, R.L., OCAMPO, P.O., MARTIN, O.C. and PALAO, O.A.
1979 *Imported Holsteins in semi-commercial dairy farms — a strategy for dairy development.* Philippines Society of Animal Science.

CERRADA, G., BODISCO, V., BRUN, A. and ABREU, O. Milk yield of commercial crossbreds in the
1978 dry tropics. *Memoria Asoc. Latinoam. Prod. Anim.*, 13: 142.

CHEAH, P.F. and KUMAR, R.A. Preliminary observations on the performance of Sahiwal × *Bos taurus*
1984 dairy cattle. *Kajian Vet.*, 16: 1-7.

CHOUEIRI, E., BAR, A.M. and CHAABAN, R., The adaptability of Holstein-Friesian cattle in Lebanon.
1966 *Magon Série Technique,* 5: 1-19.

COMBELLAS, J. and MARTINEZ, N. Intake and milk production in cows fed chopped elephant grass
1982 (*Pennisetum purpureum*) and concentrate. *Trop. Anim. Prod.*, 7: 57-60.

COSTA, L.Z., LOPEZ, J., OLIVEIRA, W.M. and DE MACHADO, A.C.P. *Anuario Tecnico do Instituto*
1974 *de Pesquisas Zootecnicas, Porto Alegre,* Estacao Experimental Zootecnica de Montenegro, Rio Grande do Sul, Brazil. pp 201-227.

COSTA, C.N., MILAGRES, J.C., GARCIA, J.A. and CASTRO, A.D.G. Genetic and environmental fac-
1982 tors affecting reproductive performance in a herd of Holstein-Friesians. *Rev. Soc. Bras. Zootec.*, 11 (1).

CÔTE D'IVOIRE. *Stratégie et plan d'action pour le développement de l'élevage en Côte d'Ivoire*
1986 *1986-2000.* FAO, Rome.

COWICONSULT. *Prefeasibility study for milk collection centres, transport and processing of milk at*
1980 *dairies, Malaysia.* COWIconsult Consulting Engineers and Planners AS, pp 18-24.

DABDOUB, S.A.M. *Studies on some physiological and haematological aspects in purebred and crossbred*
1975 *Friesian cattle.* MSc Thesis (Animal Production/Animal Breeding), University of Baghdad, pp 80-88.

DAHAL, I.A.M. *Comparison of feedlot performance, carcass traits and financial returns among Iraqi,*
1976 *Friesian and water buffalo yearling bull calves.* MSc Thesis (Meat Production), Mosul University, pp 53-57.

DAPLYN, M.G., BESSELL, J.E. and RABEN, B.A. *Estimate of the commercial and state dairy herd,*
1975 1975-80. UNGZAMI.

DE HAEN, H. and LANGHOLZ, H-J. Experiences with Friesian cattle imported into Syria. *Anim. Res.*
1981 *and Dev. (Tübingen),* 13: 75-87.

DE LA VEGA, R., GUERRERO, A. and DIAZ, G. Resistance of Holstein cattle to *Boophilus microplus*
1980 infestation. *Cuban J. Agric. Sci.* 14: 165.

DESSOUKY, F. and JUMA, K.H. Seasonal variation in semen characteristics of Friesian bulls in Iraq.
1968 *J. Agric. Sci., Camb.*, 71: 37-40.

DE VACCARO, L.P. The performance of dairy cattle breeds in tropical Latin America and programmes
1982 for their improvement. In *Dairy: cattle breeding in the humid tropics.* FAO/GOI Expert Consultation, Hissar, India, 12-17 February 1979 (Haryana Agricultural University), pp 159-182.

DE VACCARO, L.P. The comparative performance of Holstein Friesian and Brown Swiss breeds in
1984 crosses with tropical cattle: a review of the literature. *Trop. Anim. Prod.*, 9: 86-94.

DE VACCARO, L.P. *Sistemas de producción bovina predominantes en el trópico latinoamericano.*
1986 Seminario Internacional de Sistemas de Producción Doble Propósito en el Trópico, Bogota. Mimeo, pp. 17.

DEVKOTA, B.M. *Performance and economic evaluation of Holstein-Friesian cows in semi-commercial*
1982 *farms of Batanagar province, Philippines.* MS Thesis (Animal Science), University of the Philippines at Los Baños College, Laguna, pp 116.

DE VILLALOBOS, M., SUNIAGA, J. and PULIDO, F. Tolerancia al calor y humedad atmosférica de vacas
1975 Holstein y mestizas Pardo Suizo-Cebu, en Jusepin, Estado Monagas, Venezuela. *Agron. Trop,* 25: 243-256.

DOUMANDJI, N. Stock breeding centre planning, stable building and management systems. *In:* 1st In-
1982 ternational DLG Symposium on Animal Production — *Milk Production in Northern Africa and the Middle East,* 16-21 May. Inf. No. 4, pp 14.

DUC, N.U. and TANEJA, V.K. Performance of purebred and crossbred grades for some reproductive
1984a traits in India. *Indian J. Anim. Sci.*, 54: 141-144.

DUC, N.U. and TANEJA, V.K. Comparative performance of purebred and crossbred grades in India.
1984b *Indian J. Anim. Sci.*, 54: 1023-28.

ECONOMIDES, S. *The effect of dried citrus pulp and grape marc on milk yield and milk composition*
1974 *of dairy cows.* Agricultural Research Institute, Ministry of Agriculture and Natural Resources, Nicosia, Cyprus. Technical Paper No. 7, pp 9.

ECONOMIDES, S., PAPACHRISTOFOROU, C. and GEORGHIADES, E. *The use of cereal grains in ruminant diets.* Agricultural Research Institute, Ministry of Agriculture and Natural Resources, Nicosia, Cyprus. Miscellaneous Report No. 2, pp 7.
1982

EL-DESSOUKY, F.I., YOUSSEF, N.B., ALI KHAN, A., AL-JAFF, F.H., AL-HAKIM, M.K., JUMA, K.H. and KASSIR, S.M. Studies on some infertility problems in cows. *Iraqi J. Agric. Sci.*, 4: 66.
1969

EL-DESSOUKY, F.I., HUSSEIN, M.M., ISHAQ, M., KHAWAJA, A. and ENEMARK, L. *UNDP/FAO Project "Development of Livestock Production in Northern Iraq" (IRQ/71/542), Mosul.* Technical Report No. 5, pp 5.
1973

EL-DESSOUKY, F.I., LAFTAH, H.D., AL JANABI, S.I., RAJEB, B.A. and ALI, A.H. Some reproductive aspects of imported Friesian bulls in Iraq. *Iraqi J. Vet. Med.*, 1: 63-76.
1977a

EL-DESSOUKY, F.I., ENEMARK, L., AL-SALMAN, M.H. and AL-DOORI, S.A. Alleviating heat stress on a group of imported Friesian cows in their first lactation. *Iraqi J. Agric. Sci.*, 12: 143-156.
1977b

ELTAWIL, E.A., JUMA, K.H. and ALI, M.H. Evaluation of the factors affecting milk production from purebred and crossbred Friesian cattle in Iraq. *Iraqi J. Agric. Sci.*, 12: 175-185.
1977

EUROPEAN ECONOMIC COMMUNITY (EEC). *Livestock development study in Liberia.* BDPA publication No. 86 001.
1986

FAHMY, S.K., SHAHEEN, M.A., YOUSSEF, F.M. and SOLIMAN, A.M. A comparative study of herd milk production in Friesian and buffaloes. *Agric. Res. Rev., Cairo,* 53: 19-32.
1975

FAHMY, S.K., HABIB, R. and ATTIA, M.M. Life span of an imported Dutch Friesian herd in Egypt. *Agric. Res. Rev., Cairo,* 56: 17-22.
1978

FAO. *East African Livestock Survey,* Vol. II. UNDP/FAO, Rome.
1967

FAO. *Dairy development in Uganda.* UNDP/FAO mission report.
1975

FAO. *International Scheme for the Coordination of Dairy Development.* Final Report to the Government of Korea. AGA: ISCDD/Korea, Rome.
1976a

FAO. *International Scheme for the Coordination of Dairy Development.* Report to the Government of the Syrian Arab Republic. AGA: ISCDD/SYR/1976, Rome.
1976b

FAO. *International Meat Development Scheme and International Scheme for the Coordination of Dairy Development.* Report to the Government of the Republic of Turkey. IMDS/76/2, Rome.
1976c

FAO. *Programmme international de coordination du développement laitier.* Rapport final au gouvernement de la République démocratique de Madagascar. AGA: ISCDD/MAG, Rome.
1976d

FAO. *Facts of AI industry in Korea.* Paper presented at FAO/SIDA Breeding and AI Seminar, Sri Lanka.
1977a

FAO. *Artificial insemination in Malawi.* Paper presented at FAO/SIDA Seminar on AI and Breeding,
1977b Kenya. Department of Animal Health and Industry, Malawi.

FAO. *International Scheme for the Coordination of Dairy Development and International Meat Develop-*
1979a *ment Scheme.* Final Report to the Government of Liberia. AGA: ISCDD/IMDS/Liberia, Rome.

FAO. *International Scheme for the Coordination of Dairy Development and International Meat Develop-*
1979b *ment Scheme.* Report to the Government of Tanzania. AGA: ISCDD/IMDS/Tanzania, Rome.

FAO. *Syria 2352: Assistance for the development of the milk industry.* Report of interim evaluation of
1980a WFP-assisted project, Rome.

FAO. *Trypanotolerant livestock in West and Central Africa, Vol. 2. Country studies.* FAO Animal Pro-
1980b duction and Health Paper No. 20/2, Rome.

FAO. *International Scheme for the Coordination of Dairy Development.* Draft final report to the Govern-
1981 ment of Kenya. AGA: ISCDD/Kenya, Rome.

FAO. *Improvement of milk and milk yield of Shami cattle and Shami goats.* AG: DP/SYR/77/007. Ter-
1982a minal report, Rome.

FAO. *International Scheme for the Coordination of Dairy Development.* Draft final report to the Govern-
1982b ment of Botswana. AGA: ISCDD/BOT, Rome.

FAO. *Production yearbook,* Vol. 38, Rome
1984

FAO. *Production yearbook,* Vol. 39, Rome.
1985

FAO. *Asian livestock,* XI: 8. August.
1986a

FAO. *African agriculture: the next 25 years,* Rome.
1986b

FAO. Six-monthly report by FAO Representative, Cuba. July.
1986c

FLORES TRUJILLO, L. Reproductive performance of Holstein-Friesian and Brown Swiss cows in a
1980 tropical climate of Mexico. *Veterinaria, Mexico,* 11: 117.

FORDE, B. *Organizing artificial insemination in cattle,* Part II. UTF/TUR/028, Rome.
1984

FRANCE. French cattle breeds in tropical countries. *Bulletin de l'élevage français,* 12: 2-24.
1978

FREITAS, M.A.R. DE, LOBO, R.B., NAUFEL, F. and DUARTE, F.A.M. Effects of various non-genetic
1983 factors on milk production of Holstein. *Arquivo Brasileiro de Medicina Veterinaria e Zootecnia,* 35: 575-590.

FRIAS, E.C. and DIAZ, P.R. Productivity characteristics of Holstein-Friesian cattle in the control of
1980 milk production. *Tec. Pecuar.*, 39: 38.

GALAL, E.S.E., BEYENE, K. and TEGEGN, A. A study on the reproduction of local zebu and F_1
1981 crossbred (European × zebu) cows. II. Age at first calving and calf production. *Eth. J. Agric. Sci.*, 3: 97-104.

GARCIA, L., RODRIGUEZ, N., CORZO, J. and CAMA, M. Adaptability of Holstein cattle to the hot
1979a humid Cuban climate. *Rev. Cub. Cienc. Vet.*, 10: 59-65.

GARCIA, L., RODRIGUEZ, O.N. and PLANAS, M.T. Comparative study of the heat tolerance of Hols-
1979b tein × zebu crosses (F_1 and F_2) in a hot humid climate. *Rev. Cub. Cienc. Vet.*, 10: 143-149.

GARRONI, J. and VERDE, O. Partial and total production of pure Holstein cattle. *Memoria Asoc.*
1974 *Latinoam. Prod. Anim.*, 9: 70-71.

GARZA, F.A., PONCE, H.R. and MERINO Z., M.S.H. Efficiency and nutrition index of milk cattle
1973 of Holstein and Swiss breeds in a tropical climate. *Tec. Pecuar.*, 24: 17.

GEORGIADES, E., ECONOMIDES, S. and HADJIPANAYIOTOU, M. *The use of soyabean meal and urea*
1979 *in diets of growing heifers.* Agricultural Research Institute, Ministry of Agriculture and Natural Resources, Nicosia. Technical paper 18.

GIORGETTI, A., POLI, B.M. and SABRIE, M.E. Productive and reproductive characteristics of Dauara
1983 (Garre) cattle. *Riv. Agric. Subtrop. e Trop.*, 77: 4.

HABERS, L.H., SCHALLES, R.R., OKOYE, B.I.C. and ZEMMELINK, G. Comparative study of White
1972 Fulani, Friesian and White Fulani-Friesian bulls for meat production. *Trop. Agric. Trin.*, 49: 1-8.

HAHN, M. *El ROPL en la producción de leche. Registro Oficial de Producción de Leche,* MAC/UCV,
1986 Maracay, Venezuela. Mimeo, pp.6

HAIDER, I., ZAMAN, M., FAROOQ, M.A. and SHAH, S.K. Productive and reproductive performance
1984 of crossbred cows under the sub-tropical environmental conditions of the Punjab. *Pak. J. Agric. Res.*, 5.

HALL, J.M. Beef production in Tunisia. An example of a development project. *Wld Anim. Rev.*, 23:
1977 32-36.

HALPIN, J.M. *Report on livestock and dairy production in Somalia with suggested course of action.*
1978 USAID.

HAN, Y. and HONG, M. The present status and prospects of Black-and-White cattle in Jilin province.
1981 *J. Jilin Univ. Agriculture,* 4: 36-41.

HERMOSURA, S.R. and MORDENO, R.A. Comparative performance of Philippine-born Holstein and
1982 Jersey cows at Bagnio Dairy Farm with reference to milk production, reproductive efficiency and heat tolerance. *Philippine J. Vet Med.*, 21: (1-2).

HERNANDEZ GONZALEZ, M.F. A comparison of meat production in zebu and Holstein-Friesian ×
1982 zebu F_1 cattle on pasture. *In:* Proceedings of 2nd world congress on genetics applied to animal production, Madrid, 4-8 Oct. 8. Symposia (2).

HOSTE, C. *L'amélioration génétique des ruminants en Côte d'Ivoire.* Projet sur l'amélioration, la
1986 multiplication et la conservation du bétail trypanotolérant en Afrique de l'ouest. GCP/RAF/190/ITA.

HULTNÄS, C.A. *International bull semen donation scheme (IBSDS).* Report on investigations of pro-
1973 ject potentials under the IBSDS in Zambia. FAO, Rome.

INSTITUTE OF AGRICULTURAL RESEARCH (IAR). *Second annual progress report, April 1978-March*
1980 *1979.* Animal Science, Forage and Range Management Department, Addis Ababa.

JAGUN, A.G. Calf mortality on Shika farm during 1975-79. *Bull. Anim. Hlth Prod. Afr.*, 30 (2).
1982

JAHN, E.B., KLEE, G.G., BUSTAMANTE, J.M. and RUIZ, I.N. Effect of weight gains during the dry
1976 period on milk production of Holstein Friesian cows. *Agric. Tec.*, 36: 37-42.

JAJO, S.H. *Some productive and reproductive features and sweating rate in imported and locally born*
1984 *Friesian cows.* MSc Thesis (Animal Resources/ Animal Breeding), University of Baghdad, pp A-F.

JALATHGE, E.F.A. *Breeding for improved milk production.* Mimeo.
1969

JUMA, K.H. and DESSOUKY, F. Deterioration in some semen characteristics of Friesian bulls raised
1969 in Iraq. *Trop. Agric. Trin.*, 46: 63-68.

KAMARA, J.K. *Report of a study for the establishment of a Mano River Institute for the training of middle-*
1985 *level manpower in the livestock sub-sector.* FAO Regional Office for Africa, Accra.

KAMAU, C.A. Crossbreeding schemes for beef production in Kenya. *Proceedings of the Anim. Prod.*
1975 *Society of Kenya,* 8: 15-22.

KASSIR, S.A. and JUMA, K.H. A preliminary report on the performance of the Friesian and its crosses
1968 in Iraq. *Ind. J. Vet. Sci.*, 38: 541-545.

KATPATAL, B.G. *Cattle breeding research and application in India. Present and future in dairy cattle*
1982 *breeding in the humid tropics.* FAO/GOI Expert Consultation, 12-17 February 1979 (Haryana Agricultural University), pp 31-91.

KAZIMI, S.E. and SHAH, S.K. Effect of foot-and-mouth disease on production performance in crossbred
1981 cattle. *Pak. J. Anim. Sci.*, 3: 41-48.

KEBEDE, B. and GALAL, E.S.E. A study of body weight from birth to 1 year of age in European-zebu
1982 crossbred cattle in Ethiopia. *Anim. Prod.*, 34: 85-94.

KENYA. *Annual report 1980.* MOLD, Animal Production Division.
1980a

KENYA. Milk records annual report 1979. *In* Oudia, G. *Seminar on performance recording systems*
1980b *of livestock development,* Gaborone, Botswana, 14-24 October.

KIWUWA, G.H. Production characteristics of Friesian and Jersey dairy cattle on privately owned farms
1974 in Kenya. *E. Afr. Agric. For. J.*, 39: 289-297.

KIWUWA, G.H., TRAIL, J.C.M., KURTU, M.Y., WORKU, G., ANDERSON, F.M. and DURKIN, J.
1983 *Crossbred dairy cattle productivity in Arsi region, Ethiopia.* ILCA Research Report No. 11. Addis Ababa.

KNUDSEN, P.B. and SOAHEL, A.S. The Vom herd: a study of the performance of a mixed Friesian/zebu
1970 herd in a tropical environment. *Trop. Agric. Trin.*, 47: 189-203.

KOTBY, E.A. SALMAN, A.A., EL-SARAFY, A.M. and KHISHIN, S.S. Physiological responses of Frie-
1978 sian cattle under arid environment. 3. Influence of source of animals on body temperature and cardio-respiratory activities in response to diurnal climatic variations during the hot summer. *Agric. Res. Rev., Cairo,* 56: 19-24.

KROPF, W., ARAGON, A., KÜNZI, N. and HAGNAUER, W. Dairy ranching in Costa Rica. *Wld Anim.*
1983 *Rev.*, 45: 23-27.

KROSTITZ, W. *International Scheme for the Coordination of Dairy Development (ISCDD).* Draft desk
1974 study on Madagascar. FAO, Rome.

KUMAR, S. and SINGH, H. *Holstein-Friesian at IVRI, Izatnagar.* Livestock Production Research (Cat-
1979 tle and Buffalo), Indian Veterinary Research Institute, Izatnagar, UP.

KUMAR, R.A. and CHEAH, P.F. *Performance of Sahiwal* × Bos taurus *crossbreds in a milk collection*
1984 *centre.* Malaysian 9th Annual Conference of the Malaysian Society of Animal Production. pp 1-5.

LA HOZ, E. and ROSEMBERG, M. Capacidad productiva de las razas Holstein y Brown Swiss en el
1978 trópico peruano. *Memoria Asoc. Latinoam. Prod. Anim.*, 13: 152.

LEON, R.P. DE, BIEN, R. DE and CARAM, N. Milk yield during 1st lactation of Holstein-Friesian cows
1982 and Holstein-Friesian × zebu cows with 3/4 and 5/8 Holstein-Friesian inheritance. *In:* Proceedings 2nd world congress on genetics applied to animal production. Madrid, 4-8 Oct. Vol. VIII.

LOPEZ, R.R., ARAMBULO, B.R.S., BAUTISTA, J.A.N., FLORES, S.M. and BELTRAN, E.D. *The UPLB-*
1985 *Laguna dairy demonstration farm.* DTRI, University of the Philippines.

LUBBADEH, W. Internal report, Faculty of Agriculture, University of Jordan.
1985

MACLEOD, N.A., MORALES, G. and PRESTON, T.R. Milk production by dual-purpose cows grazing
1976 unsupplemented pangola or fed in drylot on sugarcane and molasses/urea based diets. Trop. *Anim. Prod.*, 1: 128-137.

MADALENA, F.E. Crossbreeding strategies for dairy cattle in Brazil. *Wld Anim. Rev.*, 38: 23-30.
1981

MADALENA, F.E., MARTINEZ, M.L. and FREITAS, A.F. Lactation curves of Holstein-Friesian and
1979a Holstein-Friesian × Gir cows. *Anim. Prod.*, 29: 101-107.

MADALENA, F.E., FREITAS, A.F. and MARTINEZ, M.L. Comparative evaluation of milk production
1979b of Holstein-Friesian and Holstein-Friesian × Gir cows. *In:* Verde, L.S. and Fernandez, A., eds. Proceedings 4th world conference on animal production. Vol. 2: 650-657.

MADALENA, F.E., VALENTE, J., TEODOR, R.L. and MONTEIRO, J.B.N. Milk yield and calving in-
1983 tervals of Holstein-Friesian and crossbred Holstein-Friesian × Gir cows. *Pesq. Agropec. Bras.*, 18: 195-200.

MANGRURKAR, B.R., GOKHALE, S.B., SHINDEY, D.N., PANDE, A.B. and PHADNIS, Y.P. Reproduc-
1985 tion performance of Holstein-Friesian and Jersey purebred cows in a herd in India. *Indian J. Anim. Sci.*, 55: 893-97.

MARPLES, H.J.S. and TRAIL, J.C.M. An analysis of a commercial herd of dairy cattle in Uganda. *Trop.*
1967 *Agric. (Trin.)* 44: 69-75.

MARTINEZ, N.D., VERDE, O.L. and FENTON, F.R. G18 production of milk cattle in the high zone
1978 of Venezuela. *Memoria Asoc. Latinoam. Prod. Anim.*, 13: 144.

MARTINEZ, N.D., DE RAMIREZ, S.G. and COMBELLAS, J. Reproductive and productive performance
1982a of a Holstein herd in Maracay, Venezuela. *Trop. Anim. Prod.*, 7: 31-39.

MARTINEZ, N.D., VERDE, O., FENTON, F.R. and LOPEZ, S. Liveweight change during the first 90
1982b days of lactation and its effect on reproduction. *Trop. Anim. Prod.*, 7: 105-112.

MASON, I.L. *Conservation and management of indigenous cattle breeds in Iran.* Report to Government
1985 of Iran. FAO, Rome.

MCDOWELL, R.E., CAMOENS, J.K., VAN VLECK, L.D., CHRISTENSEN, E. and CABELLO FRIAS,
1976a E. Factors affecting performance of Holsteins in subtropical regions of Mexico. *J. Dairy Sci.*, 59: 722-729.

MCDOWELL, R.E., WIGGANS, G.R., CAMOENS, J.K., VAN VLECK, L.D. and ST. LOUIS, D.G. Sire
1976b comparisons for Holsteins in Mexico versus the United States and Canada. *J. Dairy Sci.*, 59: 298-304.

MCKONNEN, G. Preliminary evaluation of Holstein breed and their half-breeds for milk production.
1983 *Eth. J. Agric. Sci.*, 5: 43-49.

MENÉNDEZ, A. *Breve resumen sobre la productividad del vacuno en Cuba.* II Jornadas sobre Ganadería
1985 de Doble Propósito, Maracaibo, Venezuela. Mimeo, pp 22.

MEYN, K. and WILKINS, J.V. Breeding for milk in Kenya, with particular reference to the Sahiwal
1974 stud. *Wld Anim. Rev.*, 11: 24-30.

MOHAMED, A.A., ABDEL AZIZ, A.S. and BADRAN, M.H. Some reproductive aspects of Friesian cat-
1985 tle under semi-arid environment. Faculty of Agriculture, Al Azhar Univ., Egypt, *Al Azhar Agric. Res. J.*

MONGODIN, C. *Le développement de la production laitière en Côte d'Ivoire.*
1985

MORALES, S., MACLEOD, N.A. and PRESTON, T.R. Production of milk and weaned calves
1976 in dual-purpose systems based on unsupplemented pangola pasture or chopped sugar-cane and molasses/urea in drylot with restricted grazing. *Trop. Anim. Prod*, 1, Paper 17.

MOTASSEM, M. Livestock cooperatives in the Syrian Arab Republic. *Wld Anim. Rev.* 1: 37-41.
1972

MUSI, D.O. and BIDART, J.B. *Evaluación de toros lecheros en Argentina.* IX Meeting, Asocia-
1983 ción Latinoamericana de Producción Animal, Santiago, Chile. GM-41 (Abst.).

MUTIGA, E.R., OGAA, J.S. and AGUMBAH, G.J.O. Dystocia in Kenyan dairy cattle. *Modern*
1981 *Vet. Practice*, 62: 111-113.

MWANDOTTO, B.A.J. Comparison in milk yield of Sahiwal, Friesian and Friesian × Sahiwal
1980 crosses under extensive management in Naivasha, Kenya. *Egerton College Agric. Bull.*, 4: 58-61.

MWANDOTTO, B.A.J. Some factors affecting birth weights of cattle in Kiboko, Kenya. *Bull.*
1981 *Anim. Hlth Prod. Afr.*, 29: 225-235.

NATIONAL DAIRY RESEARCH INSTITUTE (NDRI). *Animal genetic resources in India.*
1981

NDUMBE, R.D. *FAO/DANIDA Expert Consultation on the Role of Dairy Education in Na-*
1981 *tional Dairy Development.* Institute of Animal Research (IRZ), Yaoundé, Cameroun.

NEIVA, R.S., SILVA, H.M. and SAMPAIO, I.B.M. Some environmental factors affecting milk
1979 yield in a herd of Holstein-Friesian cattle in southern Minas Gerais. *Arquivos da Escola de Veterinaria da Universidade Federal de Minas Gerais*, 31: 263-273.

OCAMPO, P.O., QUILLOY, M.B., BELTRAN, E.D., TORRETA, N.K., SIAPNO, M.E., FLORES,
1979 S.M. and CASTILLO, L.S. *Study series: introducing dairy animals to backyard farms.* DTRI, University of the Philippines.

OVERSEAS DEVELOPMENT ADMINISTRATION (ODA). *The dairy industry in Peninsular Malaysia*
1982 *(with particular reference to smallholder programmes).* UK/ODA — Colombo Plan Technical Cooperation Programme.

PALLETE, A. *Servicio Oficial de Productividad Lechera:* Informativo Mensual Año XIV: No. 1. Universidad Nacional Agraria, Lima, Perú. Mimeo, pp 15.
1982

PANDA, P.B. and SADHU, D.P. Some genetic and non-genetic factors affecting milk production in crossbreds of Holstein and Jersey with Hariana and Deshi Bengal cows. *Ind. Vet. J.*, 60: 99-106.
1983

PEROZO, O.T., DEATON, O. and MUNOZ, H. Production of Holsteins in an upland zone of Guatemala. *Memoria Asoc. Latinoam. Prod. Anim.*, 9: 70.
1974

POIRIER, P. and SHAH, V.H. *Preliminary appraisal of project request. China 2647. Expansion of dairy development in major cities.* FAO, Rome. pp 12-19.
1985

PONCE, P. and BELL, L. *Composición de la leche y estudio de la lactancia en el Siboney de Cuba y sus parentales.* VIII Meeting Asociación Latinoamericana de Producción Animal, Santo Domingo. G-35 (Astr.).
1981

PRADA, N. *Selection programme for dairy and beef production through artificial insemination in Cuba.* Paper presented at FAO/SIDA Seminar on Breeding and Artificial Insemination, 30 Oct.-4 Nov.
1978

PRAJAKSAKUL, T. *Effect of ambient temperature on physiological responses of lactating cows (Holstein crossbred and Brown Swiss crossbred).* MS Thesis (Agriculture), Kasetsart University, Bangkok, pp 108.
1981

PRESTON, T.R. and UGARTE, J. Rearing dairy calves by restricted suckling. *Wld Anim. Rev.*, 3: 28-30.
1972

PROVOST, A. *Report to the Government of Liberia on an FAO consultancy on contagious bovine pleuropneumonia.* FAO Regional Office for Africa, Accra, Ghana.
1977

RAGAB, M.T., ABDEL-AZIZ, A.S. and MORAD, H.M. Phenotypic and genetic parameters of some productive traits in a herd of Friesian cattle in the Tahreer Province. *Egypt J. Anim. Prod.*, 13: 1-8.
1973

RAMOS D., J. I. Reproductive behaviour of Holstein cattle in three zones of Colombia. *Rev. ICA,* 13: 99-105.
1978

REAVES, C.W., WILCOX, C.J., SALAZAR, J.M. and ADKINSON, R.W. Factors affecting productive and reproductive performance of dairy cows in El Salvador. *J. Dairy Sci.*, 68: 3104-3109.
1985

RIBAS, M., PONCE DE LEONE, R. and PORTAL, N. Preliminary study on non-genetic factors and correlations in milk and fat production in Holstein in Cuba. *Cub. J. Agric. Sci.*, 12: 217.
1978

RIBAS, M., SANKHARE, B. and FERNANDEZ, M. A comparison of the performance of Holstein cows born in Cuba or imported from Canada. *Cub. J. Agric. Sci.*, 14: 113.
1980

ROBERT, J.P. *Développement laitier á Madagascar.* Bureau Central Laitier, Antananarive. Direction de la coopération au développement, Berne.
1979

RODRIGUEZ LOPEZ, M.A. Evaluation of reproductive performance in a Holstein-Friesian dairy herd
1979 (lactating cows) in the town of Hermosillo, Sonora, Mexico. *Veterinaria,* 10: 149.

RODRIGUEZ VOIGT, A., BODISCO, V., RAMIREZ, M. and GARCIA, E. Production of crossbred dairy
1974 cattle south of Lake Maracaibo in 1974. *Agron. Trop.*, 24: 201-217.

RODRIGUEZ, VOIGT A., BODISCO, V., RAMIREZ, M. and GARCIA, E. Productivity of different types
1978 of crossbred dairy cows. *Memoria Asoc. Latinoam. Prod. Anim.*, 13: 187-194.

ROMAN PONCE, H., CABELLO FRIAS, E. and WILCOX, C.J. Milk yield of Holstein-Friesian, American
1978 Brown Swiss and Jersey cows in a tropical climate of Mexico. *Investigaciones Pecuarias. Tec. Pecuar. Mex.*, 34: 21-33

ROMDHANE, B. Support of small and medium scale milk production through extension. In *Milk pro-*
1982 *duction in Northern Africa and the Middle East.* 1st International DLG Symposium on Animal Production, 16-21 May, pp 1-7.

SADANA, D.K. and BASU, S.B. *Performance of exotic breeds of cattle in Northern India.* NDRI An-
1980 nual report. pp. 148.

SADANA, D.K. and BASU, S.B. Reproductive performance of exotic breeds in India. *Indian J. Dairy*
1983 *Sci.*, 36: 121-124.

SÁRMIENTO, G.G. A study on the reproductive efficiency of crossbred Friesian × Guzerat cows at
1979 Sete Lagoas, Minas Gerais. *Arquivos da Escola de Veterinaria da Universidade Federal de Minas Gerais,* 31: 493-494.

SAUCEDO, G., ALVAREZ, F.J., JIMENEZ, N. and ARRIAGA, A. *Leucaena leucocephala* as a supple-
1980 ment for milk production on tropical pastures for dual-purpose animals. *Trop. Anim Prod.*, 5: 38-42.

SEKASIDDHI, P., SINGHAJAN, S., VADHANAKUL, N., CHANTARAPRATEEP, P., WATTCHAI, K. and
1981 TONGLUA, W. *Infertile investigation project of crossbred dairy cattle. 1. Calving rate of crossbred dairy cattle in Ratchaburi Province. 2. Relationship between percentage of exotic blood and calving rate in crossbred dairy cows in Ratchaburi Province.*

SEKONI, V.O. Terminal sterility in a Friesian bull naturally infected with chronic scrotal cutaneous strep-
1983 tothricosis (Kirchi). *Theriogen.*, 20: 27-36.

SEKSIT, P., SONGSASEN. P. and CHANTARAPRATEEP, P. *Preliminary results of milk records of*
1977 *crossbred dairy cattle in Ratchaburi Province.*

SEKSIT, P., SARIKAPUTI, P., JITJUMROONCHOKCHAI, J., SONGSASE, P., TODSODSAENG, S. and
1980 CHANTARAPRATEEP, P. *Some observations on reproductive performances of crossbred dairy cows during 1973-79 in Chiang Mai Province.*

SERRES, H. *Attempt at beef production from young bulls in Madagascar. L'embouche intensive des bovins*
1974 *en pays tropicaux.* Actes du colloque, Dakar, 4-8 Dec. 1973, pp 237-238.

SHAH, S.K. and CHAUNDHRY, M.Z. *Progeny testing of Holstein-Friesian bulls used for cross-breeeding.*
1979 LPRI, Bahadurnagar. First annual report 1978-79.

SHAH, S.K., ZAMAN, M., SHAH, I.H. and FAROOQ, M. *Adaptability.* Mimeo.
1980

SIVARAJASINGAM, S., SUDIN, Y., NGAH, M., SIVASUPRAMANIAM, G., OTHNAN, A.M. and JAMAK,
1982 E. Dairy cattle production in Peninsular Malaysia. *Mardi Res. Bull.* 973-987.

SLOANE, P.H. *Population supporting capacity study for Kenya.* FAO, Rome.
1985

SONI, B.K. *Cattle breeding in India — the conservation and development of indigenous breeds of the*
1982 *Indian sub-continent.* FAO/GOI Expert Consultation, Hissar, India, 12-17 February 1979 (Haryana Agricultural University), pp 12-19.

STOTZ, D. *Smallholder dairy development in the past, present and future in Kenya.* Hohenheim.
1979

STOTZ, D. *Production techniques and economics of smallholder livestock production systems in Kenya.*
1983 Farm management handbook of Kenya, Vol. IV. Ministry of Agriculture.

TALO, W.J.H. *Studies on some factors affecting economic traits of Friesian heifers.* MSc Thesis, Univer-
1978 sity of Baghdad, pp 1-9.

TANEJA, V.K. and BHAT, P.N. Additive and non-additive genetic effects for various economic traits
1978 among Sahiwal and Sahiwal × Holstein crossbred grades. *Ind. J. Anim. Sci.*, 48: 845-852.

TANEJA, V.K. and BHAT, P.N. *Milk and beef production in a tropical environment.* Proc. 4th World
1986 Congress on Genetics Applied to Livestock Production. 17-20 July 1986. Nebraska, USA.

TEFERRA, Y. *Project findings and recommendations, Zambia.* AG: DP/ZAM/74/006 and AG:
1983 DP/ZAM/82/009. FAO, Rome.

TEGEGN, A., GALAL, E.S.E. and KEBEDE, B. A study on the reproduction of local zebu and F_1
1981 crossbred (European × zebu) cows. *Eth. J. Agric. Sci.*, 3: 1-14.

THORPE, W., CRUICKSHANK, D.K.R. and THOMPSON, R. Genetic and environmental influences on
1980 beef cattle production in Zambia. 1. Factors affecting weaner production from Angoni, Barotse and Boran dams. *Anim. Prod.*, 80: 217-234.

TILAKARATNE, N. Indigenous breeds of livestock: a perspective. *Anim. Prod. Hlth Bull.*, 13: 14-17.
1980

TONN, R. *The suitability of the Boran and its crosses with European breeds for range beef production*
1974 *in the tropics, investigated on ranches in the Rift Valley in Kenya.* Thesis, Georg-August-Universität, Göttingen.

TOURÉ, M. *Santé animale et médicine vétérinaire préventive en Côte d'Ivoire, bilan actuel (1985) et*
1986 *prévisions de politique sanitaire.* Unité sous-régionale d'appui au développement dans

le cadre du programme FAO de lutte contre la trypanosomiase animale africaine et de mise en valeur des zones concernées (GCP/RAF/191/ITA), Ouagadougou.

TUNISIA. MINISTÈRE DE L'AGRICULTURE. OFFICE DE L'ÉLEVAGE ET DES PÂTURAGES. *Contrôle*
1983 *des performances. Performances des troupeaux inscrits au contrôle laitier. Lactations terminées en 1983.*

UMOH, J.U. and ADEWUYI, A.A. Some epizootiological features of clinical infectious keratoconjunc-
1984 tivitis in a university livestock farm, Zaria, Nigeria. *Bull. Anim. Hlth. Prod. Afr.*

URECLAY, V.S., MORALES, M. and NUNES, G. Fertility in Holstein-Friesian and Black and White
1981 (Black Pied) breeds. Population aspects. *Agric. Tec.*, 41: 261-263.

VACCARO, R., PALLETE, A. and CORDERO, A. Parámetros fenotípicos de vacunos Holstein de la cuenca
1979a lechera de Lima. *Memoria Asoc. Latinoam. Prod. Anim.*, 14: G-21 (Abstr.).

VACCARO, R., PALLETE, A. and FLORES, A. Producción y tipo en toros de semen importado. *Memoria*
1979b *Asoc. Latinoam. Prod. Anim.*, 14: g-14 (Abstr.).

VACCARO, R. and VACCARO, L. DE. Age at first calving, reproduction and prenatal survival in Holstein-
1982a Friesian and Brown Swiss crossbreds in an intensive tropical milk production system. *Trop. Anim. Prod.*, 7: 191-197.

VACCARO, R. and VACCARO, L. DE. *Production level in the first lactation and culling due to reproduc-*
1982b *tive performance in Holstein-Friesian cows.* Informe anual 1982, Instituto de Producción Animal, Universidad Central de Venezuela, Maracay, pp 67-68.

WARD, H.K. *Livestock production, breeds and breed evaluation in Zimbabwe.*
1982

WASFY, A. Support of small and medium scale milk production in Egypt through extension. In *Milk*
1982 *production in Northern Africa and the Middle East.* 1st International DLG Symposium on Animal Production, 19-21 May, pp 7.

WELLINGTON, K.E. A review of breeding programmes and genetic resources of dairy cattle and buf-
1982 faloes in the Caribbean. In *Dairy cattle breeding in the humid tropics.* FAO/GOI Expert Consultation, Hissar, India, 12-17 February 1979 (Haryana Agricultural University), pp 133-158.

WIENER, G. *An investigational framework to assist livestock development in Ethiopia with special*
1972 *reference to breeding.* Institute of Agricultural Research. AGP: SF/ETH 533. FAO, Rome.

WILKINS, J.V., PEREYRA, G., ALI, A. and AYALA, S. Milk production in the tropical lowlands of
1979 Bolivia. *Wld Anim. Rev.*, 32: 25-32.

YALÇIN, B.C. *Local cattle and sheep breeds in Turkey and their performance.* First Expert Consulta-
1977 tion on Breed Evaluation and Crossbreeding, 30 March-1 April. FAO, Rome.

YOUNG KUM SHIM. Livestock development in Korea: issues and policies. In Fine, J.C. and Lattimore,
1982 R.G., eds. *Livestock in Asia: issues and policies.* IDRC, Ottawa, Canada.

YOUNIS, A.A., ABOUL-NAGA, A., KOTBY, S. and SALMAN, A.A. A study of some reproductive traits
1976 in a Friesian herd kept under Kuwait conditions. *Ind. J. Anim. Sci.*, 46: 576-579.

YOUSIF, N.B. *Theileria annulata* infection in Friesian cattle imported to Iraq from Europe. *Vet. Rec.*,
1969 5 April, pp 360-363.

ZAMAN, M., HAIDER, I., FAROOQ, M.A. and SHAH, S.K. Production performance and adaptability
1983 of crossbred cows. *Pak. J. Agric. Res.*, 4: 180-189.

ZAMBIA. *Dairy development in Eastern and Lusaka Provinces.*
1977

ZEINSTRA, J.S. *Development and management of a large scale dairy operation. Case study from Saudi*
1984 *Arabia in dairy cattle production in intensive agriculture.* 2nd International DLG Symposium on Animal Production, 27-29 May, pp 22.

ZEMMELINK, G., DE LEEUW, P.N. and HABERS, L.H. Effect of plane of nutrition on comparative
1973 weight gains of White Fulani and White Fulani Friesian crossbred bulls. *Trop. Anim. Hlth* Prod., 5: 57-64.

6. Concluding Considerations

In this book, we have tried to place the results of the international comparison of strains of Friesian cattle into the context of worldwide perspectives. In organizing this major experiment, FAO always had in mind the potential usefulness of the results for a large number of countries, including those in the developing world. However, these concluding remarks can do little more than point to a number of possible directions or situations in which the results may find application. Different people, from what it is hoped will be a very diverse readership, may pick out their own, possibly different, conclusions depending on their interests and experience.

As far as the experiment itself is concerned, the first point to be stressed is that considerable variation has been found among the ten strains of Friesian cattle tested. While not all this variation is important in practice — even where statistically "significant" — much of it has practical and biological significance as it concerns major production traits such as body size and related growth rate, carcass composition, milk yield and milk composition, and aspects of health and milking characteristics.

A second important conclusion from the results is that there is no overall best or worst strain in respect of all characteristics. Every strain had strong and weak features and it is precisely this aspect of the results which will provide objective criteria for choice to breeders with different aims.

As far as growth and beef production traits are concerned, the Holstein-Friesian strains, as the largest animals both in weight and body frame, had the highest rates of gain and lean gain, with the USA paternal strain also having the best conversion of feed energy to lean gain. However, even here, generalization is difficult, because the Polish strain, for example, one of the smallest and slowest-growing of the strains, was well above average in lean gain and in the efficiency of lean gain. The proportion of valuable cuts, and the proportions of lean, fat and bone varied among some of the strains as did the ratios of tissues relative to each other. Although, as might be expected, lean varied inversely to fat to a large extent (e.g. the New Zealand strain with most fat was the least lean), the relationship was not absolute. Moreover, the proportion of bone also played a role, so that animals of the Polish strain, for example, which had a relatively low proportion of bone in the carcass, had the best lean bone ratio. Since profit to farmers from producing beef is not necessarily based on such slaughter criteria, but often on a combination of body weight and yield of carcass and on the costs of production, it is of interest that strains varied in dressing percentage (with British, Dutch and Swedish Friesians particularly good), and also in the energy intake for overall gain, which provided a different ranking again.

As far as milk performance was concerned, the Holstein-Friesian strains produced, as expected, the highest quantity of liquid milk both in field and intensive conditions. But for milk composition the picture was different, with the New Zealand strain in particular, but also several of the European strains, having appreciably higher fat, protein and solid-not-fat content in their milk. Although not formally calculated, there is little evidence from the results that there were important interactions between paternal strain and treatment (field or intensive). Thus it is of considerable interest that progeny of the New Zealand strain, a strain developed almost exclusively under pastoral conditions, had a total fat and protein yield close to that of the Holstein-Friesian strains under the field conditions but was, on balance, the best yielder of fat and protein (combined) under intensive feeding conditions (even though not significantly better than the Holstein-Friesian strains) that is to say, in the environment in which it was *not* selected.

Under most circumstances it is a combination of the merits of different traits which will be required, and not excellence in a single characteristic alone. Thus a combination of small body size relative to milk yield may be considered advantageous if feed costs for maintenance are considered important. The relative importance to be placed on milk, milk composition, growth rate and carcass composition and quality, when combining traits, will depend on the marketing and pricing structure for the products. Unfortunately for farmers, changes in political and economic dictates affecting animal production can and do occur more rapidly than the biological time-scale required by livestock breeders to respond; but at least the process is helped by the availablility of breeds and strains differing in merit for different characteristics.

It would seem unwise, to us, therefore, if the genetic diversity, which is the hallmark of the present world population of Friesian cattle, were lost by the domination of that population by perhaps a single strain. For example, in recent years many European breeding programmes have seemed set to be dominated by the now far-advanced introduction of the Holstein-Friesian. The rapid rate at which this can happen was quoted in an earlier chapter. Yet the oversupply of milk and dairy products in EEC countries and the introduction of milk production quotas may necessitate changes from the direction that the improvement of Friesians has taken in the relatively recent past. Perhaps there may emerge in the future a re-emphasis of the dual-purpose nature of several of the Friesian strains. Moreover, as profitability from milk production alone, or even from a combination of milk and meat, becomes more marginal, other factors may assume greater economic importance than before and might have to be given greater weight in selection indices. Such factors might include a refinement of "beefing qualities"; health and fertility including a reduction in calf mortality; improvements in longevity (the average number of lactations achieved by cows in dairy herds is still remarkably low with consequently high replacement costs); and factors which could ease management, for example udder shape, milk let-down, mastitis susceptibility, or the quality of the feet. It is interesting that the experiment has been able to show strain variation for a number of such traits, often regarded in the past as of minor importance, but perhaps more critical in future. It will not escape attention that a negative result, the absence of strain variation, should also be of assistance to breeders in making decisions on future breeding policy.

One of the difficulties faced by breeders using imported bulls, or their semen, has been the absence of good information on how to evaluate the merit of a bull for their own conditions when all that is usually known are its merits in the country of origin. This problem is likely to be greatly eased in future by guidelines which have been developed, in part using results from the present experiment, from an initiative of the International Dairy Federation and the European Association for Animal Production. The resulting publications "Interbull" (Philipsson, 1986a and b) should facilitate greatly the comparison of breeding values across national boundaries — although, at least initially, this is likely to benefit mainly the developed countries, and European in particular, where sufficient progeny of different bulls from different origins co-exist.

To turn now to considerations more particularly related to the developing world, which arise from the surveys and reviews of results involving Friesian cattle in these countries.

The first point to emerge is the remarkable spread, in the developing countries, of the Black and White strains of cattle-Friesian and, more recently, Holstein-Friesian. The popularity of the Friesian arises at least partly from the fact that in comparison with other breeds used for crossing such as the Jersey, Ayrshire and Brown Swiss, in tropical and subtropical countries, the Friesian crosses have had the highest milk production. Friesians are now found, or at least being tried out, as a pure breed or as a crossing sire in all corners of the world — corners for which the originators of Friesians can have had little thought. The fact that highly improved yields relative to native stocks are usually obtained from crossbreeding with exotics — and in the present context the Friesians — even under relatively poor conditions, says a great deal about the capacity of different genotypes to adapt.

It also reaffirms the point that it was the power of genetic selection which created different types. Selection, however, is a relatively slow process so that progress is helped by starting from a good baseline. There are very few breeds in tropical and subtropical countries which would appear, on their own, to provide a good potential for increasing milk production by selection in a reasonable span of time. The Sahiwal and Red Sindhi may be two such breeds now, but in other cases the baseline of productivity of indigenous breeds may be too low for the process of selection alone to give sufficiently rapid improvement to an acceptable level. Thus for dairy production in particular, crossbreeding, perhaps followed by selection and the development of a new breed type, is likely to be the practical solution in most cases. Although, as always, the interests of national schemes have to be reconciled with the interests, abilities and resources of individual farmers. For example, continuous grading-up, even if quite inappropriate as a national programme, might be easier for a small farmer to perceive as a personal challenge to improve his cattle management and nutrition, than an otherwise more sensible selection programme.

On a national basis, selection of dairy cattle in the developed countries of Europe and North America has for several decades relied on the widespread use of progeny testing in large cattle populations and depended on high technical inputs with artificial insemination, recording, and computerized evaluation systems. It should be possible to overcome some of the problems inherent in such large-scale operations by the adoption of schemes of dam selection in ''nucleus'' herds and, as advocated by Nicholas and Smith (1983), involving multiple ovulation and embryo transfer (MOET). Their scheme, though requiring high technical sophistication, is at least circumscribed in scale — and, in theory, can produce greater rates of progress than traditional progeny testing. Much more limited progress, but at least progress, could also come from the adoption of other variants on closed or open nucleus herd schemes. Such schemes seem set, therefore, to be particularly attractive for use in developing countries, since technical inputs can be concentrated in the nucleus herd.

However, both in a crossbreeding programme and in the development of new breeds, one of the questions to be answered concerns the optimum proportion of exotic blood (Friesian in the present context). As the results from developing countries show, there is no single answer, because of different climatic, nutritional and managerial circumstances. However, many studies with milking cattle have shown that around 50 percent exotic blood is optimal in most situations in the developing world and that this optimum is higher only where conditions of feeding and management are exceptionally good. In our view, it would not be right to advocate that the feeding or management, perhaps apart from health, should be improved to those highest levels in order to exploit an attainable genetic potential, unless perhaps in some exceptional situations. To do so would be to lose sight of the one overwhelming advantage of ruminants — that they can utilize feed and other resources which do not compete, or compete only minimally, with land and resources more directly usable for human needs.

The part of the book dealing with the developing world has also highlighted the dearth of good information on two important points. The first of these relates to the importance or otherwise of heterosis in contributing to the level of F_1 performance. Answers to this are important for decisions on breeding strategy and it is worth making the point in some detail. As found from the strain comparisons in Poland and the review of crossbreeding in developed countries, heterosis from crossing of strains makes, at best, only a small contribution to F_1 performance under such relatively favourable farming conditions. Indications from several relatively small studies in developing countries suggest, however, that heterosis is positive and significant in crosses between breeds of indigenous and exotic cattle (usually *Bos indicus* with *Bos taurus*) and particularly in climatically and nutritionally unfavourable conditions and even more especially when disease risks are also high. Hence in generations subsequent to F_1 (e.g. F_2) there is some loss in performance due to a loss of some favourable gene combinations (recombinant loss). But it is important to know much more precisely than at present the actual extent of such loss in performance in particular circumstances. Such knowledge should be available before making a choice between, for example, the continuous re-creation of first crosses on the one hand or, on the other hand, the development of a new breed by interbreeding the crosses and selecting within that population. The former might be preferred if the contribution of heterosis is very high so that first-cross performance exceeds that of later crosses to such an extent that this superiority more than makes up for the disadvantage of having also to maintain a large, lower-producing indigenous female population (which is needed to produce the crossbreds). Of course if the indigenous breed fills a different environmental niche from the crossbreds, the presence of two populations can be advantageous.

If heterosis has been found to be important for particular circumstances, creating a new breed in that situation will of course lead to an initial loss of performance relative to the F_1. Nonetheless this procedure allows a single population to be maintained at an improved level of production. When combined with selection, there will be further steady improvement and the selection can eliminate any unwanted variability, if indeed this occurs in the first place. Again this is, of necessity, considering the matter in the context of a cattle population. As referred to earlier, in a related context, the problem may present itself differently at individual farmer level. There may be greater practical difficulties in discarding unwanted animals (selection) if it involves one of perhaps only two or three animals belonging to a small farmer than would appear from considering the entire population of cattle.

Other alternative breeding strategies, such as continuous grading-up (only advisable if management, health and nutrition can be improved in step, and if this is justified on economic grounds) and rotational crossing, also involve consideration of the level of heterosis and hence the extent of recombinant loss. The review of the use of Friesian cattle in the developing world points to the need for further deliberate studies on the role of non-additive genetic variation, particularly for the less favoured environments, so that the consequences of the various breeding strategies can be more confidently predicted.

A second point in relation to cattle breeding on which the review shows that there is a dearth of information, again particularly in the developing world, is whether the performance of crossbred progeny of different sires from a particular breed, such as the Friesian, ranks in the developing country in the same order as progeny tests suggest in the country of origin. That, by and large, they did so in the Polish trials provides little guide to whether this would be so when the larger differences of climate, nutrition, disease risks and cattle management between receiving and contributing country of the bulls (or semen) are taken into account. Up to the present there is little information on this aspect from developing countries. What few results there are suggest that interactions between breeding values of different bulls and environment are unimportant in practice, but that differences are also less important because nutrition and environment are inadequate to show up or exploit potential differences in most cases. However, until answers are provided on a convincing scale and over a wider range of conditions than at present, developing countries and individual breeders will continue to pay high prices for ''proven'' sires in the hope that they are purchasing ''the best'' for their conditions. However, taking into consideration the point made above, that in developing countries only a proportion of the genetic production potential of imported breeds can generally be reached, the use of semen from ''average-plus'' bulls, or from young bulls in the testing programme (the as yet untested progeny of the best of the previous generation) will have more to commend it, if it incurs lower cost and allows a broadening of the genetic base for future selection.

As a final point one can express the hope that the international comparisons of strains in Poland and the results reviewed from both developed countries and the developing regions will help breeders across the world to have better information on which to base decisions on using Friesian cattle either pure or in crossbreeding. The experiments in Poland, done with two feeding regimes and covering a wide range of traits, have increased enormously the available information on which rational decisions can be based concerning choice among strains. The experiments have also stimulated similar comparisons elsewhere. Thus it will become easier to answer questions such as whether the large expansion of the Holstein-Friesian strains, at the expense of the ''older'' strains of Friesian, is appropriate to all conditions. It will also provide breeders with better information on the relative contributions from different strains for making improvements in different components of production — milk, milk composition, beef characteristics, and so on. It is likely that the performance of some strains may have proceeded further in the direction of the Holstein-Friesian and that other changes may have taken place in national herds since the time when the semen for the trials in Poland was donated by the participating countries. However the principles shown by the comparisons are likely to remain and it would be surprising if the direction of the differences recorded had drastically changed.

If, therefore, the international comparisons in Poland, and this book, help to answer questions and assist rational decision-making, by replacing opinion with fact, the objectives of the undertaking will have been achieved.

References

NICHOLAS, F.W. and SMITH, C. Increased rates of genetic change in dairy cattle by embryo transfer
1983 and splitting. *Anim. Prod.*, 36: 341-353.

PHILIPSSON, J. *Procedures for international comparisons of dairy sires — current practice and evaluation of methods.* Interbull, Bulletin No. 2. Dept. of Anim. Brdg and Genetics, Uppsala, Sweden.
1986a

PHILIPSSON, J. *Sire evaluation procedures for dairy production traits practised in various countries — 1985.* Interbull, Bulletin No. 1. Dept. of Anim. Brdg and Genetics, Uppsala, Sweden.
1986b